GEOLOGICAL SOCIETY SPECIAL PUBLICATION NO. 52

Phosphorite Research and Development

EDITED BY

A. J. G. NOTHOLT
Mineral Resource Consultant,
Ickenham, UK

&

I. JARVIS
School of Geological Sciences, Kingston Polytechnic, UK

1990

Published by

The Geological Society

London

Geological Society Special Publications
Series Editor K. COE

THE GEOLOGICAL SOCIETY

The Geological Society of London was founded in 1807 for the purposes of 'investigating the mineral structures of the earth'. It received its Royal Charter in 1825. The Society promotes all aspects of geological science by means of meetings, special lectures and courses, discussions, specialist groups, publications and library services.

It is expected that candidates for Fellowship will be graduates in geology or another earth science, or have equivalent qualifications or experience. All Fellows are entitled to receive for their subscription one of the Society's three journals: *The Quarterly Journal of Engineering Geology*, the *Journal of the Geological Society* or *Marine and Petroleum Geology*. On payment of an additional sum on the annual subscription, members may obtain copies of another journal.

Membership of the specialist groups is open to all Fellows without additional charge. Enquiries concerning Fellowship of the Society and membership of the specialist groups should be directed to the Executive Secretary, The Geological Society, Burlington House, Piccadilly, London W1V 0JU.

Published by the Geological Society from:
The Geological Society Publishing House
Unit 7
Brassmill Enterprise Centre
Brassmill Lane
Bath
Avon BA1 3JN
UK
(*Orders*: Tel. 0225 445046)

Distributor USA:
AAPG Bookstore
PO Box 979
Tulsa
Oklahoma 74101−0979
USA
(*Orders*: Tel. 918−584−2555)

First published 1990

© The Geological Society 1990. All rights reserved. No reproduction, copy or transmission of this publication may be made without written permission. No paragraph of this publication may be reproduced, copied or transmitted save with the written permission or in accordance with the provisions of the Copyright Act 1956 (as Amended) or under the terms of any licence permitting limited copying issued by the Copyright Licensing Agency, 33−34 Alfred Place, London WC1E 7DP. Users registered with Copyright Clearance Center: this publication is registered with CCC, 27 Congress St., Salem, MA 01970, USA.) 0305−8719/90 $03.00.

British Library Cataloguing in Publication Data
Phosphorite research and development.
1. Phosphate deposits
I. Notholt, A. J. G. (Arthur John George)
II. Jarvis, I. (Ian)   III. Geological Society of London   IV. Series
553.64

ISBN 0−903317−53−2

Printed in Great Britain at the Alden Press, Oxford

Contents

Preface   v

COOK, P. J., SHERGOLD, J. H., BURNETT, W. C. & RIGGS, S. R. Phosphorite research: a historical overview   1

MCCLELLAN, G. H. & VAN KAUWENBERGH, S. J. Mineralogy of sedimentary apatites   23

LUCAS, J., EL FALEH, E. M. & PREVOT, L. Experimental study of the substitution of Ca by Sr and Ba in synthetic apatites   33

NATHAN, Y. Humic substances in phosphorites: occurrence, characterization and significance   49

O'BRIEN, G. W., MILNES, A. R., VEEH, H. H., HEGGIE, D. T., RIGGS, S. R., CULLEN, D. J., MARSHALL, J. F. & COOK, P. J. Sedimentation dynamics and redox iron-cycling: controlling factors for the apatite-glauconite association on the East Australian continental margin   61

HEGGIE, D. T., SKYRING, G. W., O'BRIEN, G. W., REIMERS, C., HERCZEG, A., MORIARTY, D. J. W., BURNETT, W. C., & MILNES, A. R. Organic carbon cycling and modern phosphorite formation on the East Australian continental margin: an overview   87

MCARTHUR, J. M., & HERCZEG, A. Diagenetic stability of the isotopic composition of phosphate-oxygen: palaeoenvironmental implications   119

BALSON, P. C. Episodes of phosphogenesis and phosphorite concretion formation in the North Sea Tertiary   125

VAN KAUWENBERGH, S. J. & MCCLELLAN, G. H. Comparative geology and mineralogy of the southeastern United States and Togo phosphorites   139

LAMBOY, M. Microbial mediation in phosphatogenesis: new data from the Cretaceous phosphatic chalks of northern France   157

LEWY, Z. Pebbly phosphate and granular phosphorite (Late Cretaceous, southern Israel) and their bearing on phosphatization processes   169

NATHAN, Y., SOUDRY, D. & AVIGOUR, A. Geological significance of carbonate substitution in apatites: Israeli phosphorites as an example   179

ABED, A. M. & FAKHOURI, K. Role of microbial processes in the genesis of Jordanian Upper Cretaceous phosphorites   193

GLENN, C. R. Depositional sequences of the Duwi, Sibaiya and Phosphate Formations, Egypt: phosphogenesis and glauconitization in a Late Cretaceous epeiric sea   205

ZANIN, YU. N., ZVEREV, K. V. & SOLOTCHINA, E. P. Clay minerals and phosphorite genesis in the Upper Cretaceous of the northern Siberian Platform   223

FÖLLMI, K. B. Condensation and phosphogenesis: example of the Helvetic mid-Cretaceous (northern Tethyan margin)   237

ILYIN, A. V. & HEINSALU, H. N. Early Ordovician shelly phosphorites of the Baltic Basin   253

HOWARD, P. F. The distribution of phosphatic facies in the Georgia, Wiso and Daly River Basins, Northern Australia   261

DONNELLY, T. H., SHERGOLD, J. H., SOUTHGATE, P. N. & BARNES, C. J. Events leading to global phosphogenesis around the Proterozoic/Cambrian boundary   273

BRASIER, M. D. Phosphogenic events and skeletal preservation across the Precambrian–Cambrian boundary interval   289

CHOUDHURI, R. Two decades of phosphorite investigations in India 305

SISODIA, M. S. & CHAUHAN, D. S. The influence of magnesium ions during the formation of stromatolitic phosphorites of Udaipur, Rajasthan, India 313

Index 321

Preface

This book contains a selection of the papers contributed to the 11th and Final International Field Workshop and Symposium of International Geological Correlation Programme (IGCP) Project 156: Phosphorites, which was held at Hertford College, Oxford, England, 5–8 September, 1988. The main aim of the meeting, which attracted over 70 delegates and speakers representing 14 countries (Notholt & Jarvis 1989), and the aim of this volume, is to present the scientific results and ideas that have arisen from research carried out in recent years by geological surveys, universities and other establishments in many parts of the world. Phosphorites provide the basis for the world's fertilizer industry, the backbone of modern intensive agriculture; their occurrence in a sedimentary succession is almost certainly indicative of dramatic changes in palaeoenvironment and sedimentation. Few naturally-occurring raw materials offer such a combination of great socio-economic importance and fundamental scientific significance.

During its ten-year existence, Project 156: Phosphorites became known as one of the largest and most successful of the long-term interdisciplinary geological research projects supported by the IGCP. It owes its creation in 1977 to Peter J. Cook, of the Bureau of Mineral Resources, Geology and Geophysics, Canberra, who hypothesized that a major Proterozoic–Cambrian phosphogenic province extends throughout the Asian and Australian region, one that was perhaps comparable in size (and therefore economic importance) with the Late Cretaceous–Eocene Tethyan province of northern Africa and the Middle East. Project 156 was established, therefore, as a research programme on Proterozoic–Cambrian phosphorites of Asia and Australia, but the international interest that the Project generated led rapidly to the formation of three other Working Groups, devoted respectively to an International Phosphate Resource Data Base; Young Phosphogenic Systems; and Cretaceous–Eocene Phosphorites. Much of the research carried out under the aegis of these working groups has to date formed the basis of three major publications on phosphorites (Cook & Shergold 1986; Notholt, Sheldon & Davidson 1989; Burnett & Riggs 1990).

The present volume starts with a comprehensive review of phosphorite research by **Cook** *et al.*, the former leaders of Project 156, emphasizing the considerable value to phosphate exploration of Kazakov's 'Upwelling Hypothesis' which came into prominence in the 1930s, followed by the 'Phosphoria Model', developed as a result of extensive studies by the US Geological Survey, which involves upwelling on the flanks of open ocean basins at water depths of 200–300 m. These models remain fundamental to much modern thinking on phosphogenesis. The importance of global factors such as sea-level changes and plate tectonics, and their interaction with the processes of phosphate formation has become better understood in recent years, and links between periods of phosphogenesis and major evolutionary changes have also been established.

The following contribution by **McClellan & van Kauwenberg** on the mineralogy of sedimentary carbonate fluorapatites (francolite) emphasizes the importance of CO_3^{2-} substitution for PO_4^{3-}, variations in francolite composition with stratigraphical position and particle size as possibly indicators of the extent and type of post-depositional alteration. Geochemical composition may provide additional evidence of depositional, early diagenetic and post-burial history. It is well known that numerous elements may be incorporated into the francolite structure, including the substitution of Sr for Ca, and the experiments described by **Lucas** *et al.* support the conclusion that Sr may be a valuable indicator of the genetic environment of apatite, although variations in the Sr/Ca ratio also may reflect differences in post-depositional history. Studies of carbonate substitution in apatite (francolite) from phosphorites of Santonian (Late Cretaceous) to Eocene age in Israel, reveal a significant decrease in CO_2 concentrations in samples that have been subjected to epigenetic alteration and/or weathering (**Nathan** *et al.*). Carbon, oxygen and sulphur isotopes have also been used to study the diagenetic history of phosphorites, and have made possible differentiation between those formed in oxic/suboxic and anoxic conditions. However, again, the effects of diagenetic alteration need to be taken into account if sound palaeoenvironmental interpretations are to be made based on analysis of the isotopic composition of the phosphate-oxygen bond (**McArthur & Herczeg**). The organic geochemistry of phosphorites has also attracted considerable interest. Notably significant is the recognition that humic substances form a significant component of organic matter in many phosphorites of various ages. The humic material may be primary, as in the Tunisian phosphorites, or appear to be secondary, as in phosphorites from the western United States, where it has probably formed by degradation of kerogen (**Nathan**). It seems generally agreed that humic substances and phosphorites are closely associated genetically.

The East Australian continental margin is one of only four areas where modern (Neogene to Holocene) marine phosphorite nodules are known to be forming at present. These phosphorites were the subject of a detailed, multidisciplinary study mounted in 1987 by the Australian Bureau of Mineral Resources, which included an examination of the processes controlling the cycling of such elements as organic carbon, nitrogen, phosphorus, sulphur and iron, and also of the role of these processes in the formation of modern phosphorites. Sediment data from samples collected between 29° and 32° S indicate that organic carbon oxidation by oxygen and sulphate reduction, iron and phosphorus recycling between oxic and anoxic sediments, and sediment mixing by bioturbation are three key processes in their formation (**Heggie** *et al*.). Ultimately, the compositions of the phosphorites reflect this interaciton between redox-driven diagenetic processes and multiple periods of reworking (**O'Brien** *et al*.). Ferruginous Neogene phosphorites occur within glauconitic, foraminiferal sands and are concentrated within depths of 200–300 m, where they form discrete nodules and massive hardground units; Quaternary phosphorite nodules occur in deeper water (250–460 m), in sands of similar composition, but notably they exhibit lower goethite contents.

The development of phosphorite concretions in the marine, clastic sequence of the southern North Sea Tertiary Basin displays a clear cyclicity which corresponds to sea-level 'highstands' (**Balson**). In particular, the concretions distributed sparsely in Early Eocene mudstones occur just above levels which indicate initiation of more open connection between the Atlantic Ocean and the relatively enclosed North Sea Basin. Thus, the development of phosphorite concretions appears to be linked to large-scale, possibly global, cycles in addition to changes in the local diagenetic environment.

Several major phosphogenic episodes of global extent are now known to occur, together containing deposits with resources estimated to be at least 133 000 Mt of all grades and varieties of phosphate rock. Among the largest and economically the most important deposits in the world are those contained in the Neogene succession of the southeastern USA. There the Miocene Pungo River and Miocene–Pliocene Hawthorn Formations together represent a major phosphogenic episode which also coincided with an extensive global rise in sea level associated with the Miocene transgression. Cores penetrating phosphatic sediments of the Hawthorn Formation in south-central Florida record multiple phosphogenic episodes separated by quartz-rich sands and indurated carbonates. Comparison of these deposits with those of Eocene age in southern Togo, confirm that both have been produced by a combination of primary depositional and secondary diagenetic processes (**van Kauwenbergh & McClellan**). The latter account for the removal primarily of carbonates and the decarbonatization of francolite (carbonate–fluorapatite), as well as the development of Fe-Al phosphate minerals and clay mineral alteration profiles. The North Carolina deposits appear to be relatively unaltered or merely in the early stages of alteration.

The Late Cretaceous–Eocene Tethyan Phosphogenic Province is also one of the most extensive in the world, with major deposits situated throughout north Africa and the Middle East. Those of Late Cretaceous age in Egypt are associated with porcellanites, cherts, dark-coloured shales, glauconitic sandstones and bioclastic fine-grained carbonate of the Duwi Group. Although oceanic upwelling is usually advocated as the source of the phosphorus in most phosphorite deposits, at least some Tethyan occurrences may have been strongly influenced by an input of local fluvially-derived phosphorus (**Glenn**). These Egyptian occurrences are believed to be the result of current winnowing and reconcentration of authigenic grains initially precipitated in reducing shales and biosiliceous sediments. Studies of Upper Cretaceous (Turonian–Maastrichtian) nodular phosphorites in the Yenisei Mouth Depression of northern Siberia (**Zanin** *et al*.) demonstrate that these phosphate-bearing strata contain a consistent and distinctive clay mineral assemblage of kaolinite and chlorite, with a small but constant admixture of gibbsite, and in some cases montmorillonite. The mineralogical assemblage indicates that, in these deposits at least, phosphogenesis accompanied periods of intense weathering of the adjacent continental landmass under warm and humid climatic conditions, a process which provided a local, possibly fluvially-derived, source of phosphorus. The development of many massive phosphorite deposits appear to be caused by condensation. Very low accumulation rates ($2-10$ m Ma^{-1}) are indicated, for example, by the occurrence, on the Alpine Helvetic Shelf, of condensed phosphatic beds of Aptian to early Cenomanian age, formed within a stable westward-flowing current system along the northern Tethyan margin (**Föllmi**). These beds are thin, generally less than 50 cm, and consist of closely packed phosphatic particles and crudely laminated crusts embedded in glauconitic sands, marls, and pelagic micrites.

In spite of a considerable increase in data, divergent views continue to be expressed on the precise mechanism of phosphate precipitation. Some petrographic and SEM studies have emphasized the significant role played by microbial communities in phosphogenesis. For example, the Upper Cretaceous (Santonian–Campanian) phosphatic chalks of northern France are now shown to consist essentially of mineralized microbial remains (**Lamboy**). In the Negev Desert of Israel, the phosphate nodules found at specific levels in the chalky, locally bituminous, Ghareb Formation of Campanian–Maastrichtian age are also believed to have been microbially generated, the globular microstructures commonly found in the phosphate cements being regarded as mineralized cells of endobenthic microorganisms (**Lewy**). Similarly, cynobacterial sheaths have been identified within the laminae of phosphorites of Maastrichtian age in Jordan. These phosphorites are interpreted as having formed as algal mats in shallow-water, subtidal to intertidal environments (**Abed & Fakhouri**).

Palaeozoic phosphorite deposits and occurrences have been identified in recent years from several parts of the world. Major resources have been delineated within the Baltic Phosphorite Basin where, for example, the predominantly lower Tremadoc (basal Ordovician) Rakvere deposits in Estonia contain the equivalent of 400 Mt P (**Ilyin & Heinsalu**). The Cambrian phosphorites of the Georgina Basin in western Queensland and eastern Northern Territory, Australia, are among the most intensely studied deposits of their type in the world. These have been assigned collectively to the Beetle Creek Formation, which is of Ordian and/or Templetonian (earliest middle Cambrian) age. The carbonate–siltstone–chert phosphatic facies bordering the Georgina, as well as the Wiso and Daly River basins is similarly mainly of Ordian age (**Howard**). In addition to the 18 or so deposits delineated within the Georgina Basin over a distance of approximately 1000 km, borehole logging together with modelling of aeromagnetic data has led to the discovery of two new phosphorite areas: the Lady Judith in the Wiso Basin, which rests on volcanics and interdigitates with members of the Montejinni Limestone; and the Ammaroo deposit, in the southwestern part of the Georgina Basin, which occurs within a depression bounded by limestones of the Arthur Creek Formation.

Among the most notable achievements of Project 156 is the recognition of the Proterozoic–Cambrian as a major phosphogenic episode of global extent, particularly in central and southern Asia. It is noteworthy also that the Precambrian-Cambrian boundary provides the strongest evidence for a close link between major phosphogenic events and important evolutionary changes (**Cook *et al.***). In this context, it has been suggested (**Brasier**) that the phosphogenic events were themselves related to the first widespread appearance of small shell spaces and faecal/dead organic substrates which provided suitable environments for the formation of authigenic phosphate (and glauconite) minerals. Interactions between early Cambrian ecosystems and phosphogenesis are well preserved in South China, where preservation reaches a climax in the Zhongyicun Member (lower Meishucunian Stage, basal Cambrian). In India, Proterozoic basins offer the most important phosphate exploration potential and several outstanding phosphate discoveries have been made since the mid-1960s (**Choudhuri**). The Proterozoic phosphorite-bearing sequences of the three main areas around Udaipur, Jhabua, and Hirapur–Lalitpur all lie on Archaean basement and comprise essentially a basal quartzite member succeeded by phosphorite–carbonate sequences. The Jhamarkotra deposit near Udaipur, in Rajasthan, is now a major source of phosphate rock in India.

Major palaeoceanographic changes occurred during the Late Proterozoic, with the temporary development of a stagnant oceanic regime (**Donnelly *et al.***). Enhanced burial of organic matter must have caused significant increases in atmospheric oxygen, which may have stimulated the evolution of new metazoan groups. Increased rates of oceanic turnover during the earliest Cambrian would have caused the influx of P-rich, deep-ocean waters into epicontinental seas, thereby promoting a phosphogenic event. Oxidation associated with the development of better mixed oceans would have caused an increase in atmospheric CO_2, which may have been an additional factor in promoting the appearance of mineralized skeletons in many fossil groups.

Important insights into prevailing depositional conditions during the formation of Proterozoic phosphorites are provided by their almost ubiquitous association with stromatolites. Probably the best known examples of stromatolitic phosphorites in the world are those found near Udaipur, where they exhibit both carbonate (mainly dolomite) and phosphate (francolite) mineralogies. It has been proposed (**Sisodia & Chauhan**) that the differences in mineralogy were caused by differences in the Mg contents of the primary sediments, which reflected varying depositional environments. The high Mg content of intertidal stromatolites may have prevented extensive phosphatization in this facies.

It may be concluded that our understanding of the origin, nature, and distribution of phosphorites has progressed considerably in recent years. Nevertheless, it is evident also that many problems remain unresolved and that phosphorites will continue to provide a fascinating and fruitful field of scientific research for many years to come.

References

BURNETT, W. C. & RIGGS, S. R. (eds). 1990. *Phosphate deposits of the world, Vol. 3, Neogene to Recent phosphorites*. Cambridge University Press, Cambridge.

COOK, P. J. & SHERGOLD, J. H. (eds). 1986. *Phosphate deposits of the world, Vol. 1, Proterozoic and Cambrian phosphorites*. Cambridge University Press, Cambridge.

NOTHOLT, A. J. G. & JARVIS, I. 1989. A decade of phosphorite research and development. *Journal of the Geological Society, London*, **146**, 873–876.

NOTHOLT, A. J. G., SHELDON, R. P. & DAVIDSON, D. F. (eds). 1989. *Phosphate deposits of the world, Vol. 2, Phosphate rock resources*. Cambridge University Press, Cambridge.

Acknowledgements

We would like to acknowledge the generous support of the Geological Society of London and the Royal Society, and also of Albright & Wilson Ltd, the British Sulphur Corporation Ltd, and the International Fertilizer Industry Association, Paris. Following four days of plenary sessions at Oxford, there were two three-day field trips: to the Cretaceous phosphorites of southern England, led by Dr Ian Jarvis, Kingston Polytechnic, and to the Cretaceous phosphorites of northern France and Belgium, led by Drs Francis Robaszynski, Faculté Polytechnique de Mons, and Gérard Sustrac, Bureau de Recherches Géologiques et Minières, Orleans. In this context, the logistical support of these three research establishments is also gratefully acknowledged.

We are greatly indebted to the 43 contributors for their collaboration, and to the following for their assitance in producing this volume: Shimshon Axelrod, James B. Cathcart, Robert E. Garrison, Christopher V. Jeans, Hugh C. Jenkyns, William J. Kennedy, James R. Lehr, James F. Marshall, H. Martyn Pedley, Terence G. Powell, Liliane Prévôt, Adrian W. A. Rushton, Mark W. Sandstrom, Richard P. Sheldon, Graham B. Shimmield, Maurice Slansky, John Thomson, and Jean Trichet. Finally, the considerable efforts of Angharad Hills and other staff of the Geological Society Publishing House, Bath, are much appreciated.

London
January 1990

A. J. G. Notholt
I. Jarvis

List of Contributors

A.M. Abed, Department of Geology & Mineralogy, University of Jordan. Amman, Jordan

A. Avigour, Negev Phosphate Company, Tel Aviv 6100, Israel

P.S. Balson, British Geological Survey, Keyworth, Nottingham, NG12 5GG, UK

C.J. Barnes, CSIRO Division of Water Resources, GPO Box 1666, Canberra, ACT 2601, Australia

M.D. Brasier, Department of Earth Sciences, The University, Parks Road, Oxford OX1 3PR, UK

W.C. Burnett, Department of Oceanography, Florida State University, Tallahassee, Florida, USA

D.S. Chauhan, Department of Geology, University of Jodhpur, Jodhpur 342001, India

R. Choudhuri, Rajasthan State Mines & Minerals Ltd, Udaipur, Rajasthan, India

P.J. Cook, Division of Continental Geology, Bureau of Mineral Resources, P.O. Box 378, Canberra, Australia

D.J. Cullen, New Zealand Oceanographic Institute, DSIR, Private Bag, Kilbirnie, Wellington, New Zealand

T.H. Donnelly, CSIRO Division of Water Resources, GPO Box 1666, Canberra, ACT 2601, Australia

E.M. El Faleh, Département des Sciences de la Terre, Université Louis Pasteur, Centre de Geochimie de la Surface, CNRS, 1 Rue Blessig, F-67084, Strasbourg Cedex, France

K. Fakhouri, Department of Geology & Mineralogy, University of Jordan, Amman, Jordan

K.B. Föllmi, Geological Institute, ETH-Zentrum, CH 8092 Zürich, Switzerland

C.R. Glenn, Department of Geology & Geophysics, Hawaii Institute of Geophysics, University of Hawaii, Honolulu, HI 96822, USA

D.T. Heggie, Bureau of Mineral Resources, GPO Box 378, Canberra City, ACT 2601, Australia

H.N. Heinsalu, Geological Institute, Estonian Academy of Sciences, Tallinn 200101, Estonia, USSR

A. Herczeg, CSIRO, Division of Water Resources, Private Bag #2, Glen Osmond, SA 5064, Australia

P.F. Howard, 9 Leeds Place, Turramurra, Sydney, NSW 2074, Australia

A.V. Ilyin, Institute of the Lithosphere, USSR Academy of Sciences, Moscow, 109180, USSR

M. Lamboy, Department of Geology, University of Rouen, BP 118, 76134 Mont Saint Aignan, France

Z. Lewy, Geological Survey of Israel, 30 Malkhe Yisrael Street, Jerusalem, 95501, Israel

J. Lucas, Département des Sciences de la Terre, Université Louis Pasteur, Centre de Geochimie de la Surface, CNRS, 1 Rue Blessig, F-67084, Strasbourg Cedex, France

J.F. Marshall, Bureau of Mineral Resources, GPO Box 378, Canberra City, ACT 2601, Australia

J.M. McArthur, Department of Geological Sciences, University College London, Gower Street, London, WC1E 6BT, UK

G.H. McClellan, Department of Geology, University of Florida, Gainesville, FL 32611, USA

A.R. Milnes, CSIRO Division of Soils, Private Bag No. 2, Glen Osmond, South Australia 5064, Australia

D.J.W. Moriarty, CSIRO, Division of Fisheries, P.O. Box 120, Cleveland, Queensland 4163, Australia

Y. Nathan, Geological Survey of Israel, 30 Malkhei Israel Street, 95 501, Jerusalem, Israel

G.W. O'Brien, Bureau of Mineral Resources, GPO Box 378, Canberra City, ACT 2601, Australia

L. Prévôt, Département des Sciences de la Terre, Université Louis Pasteur, Centre de Geochimie de la Surface, CNRS, 1 Rue Blessig, F-67084, Strasbourg Cedex, France

C. Reimers, Geological Research Division, Scripps Institution of Oceanography, La Jolla, CA 92093, USA

S.R. Riggs, Department of Geology, East Carolina State University, Greenville, North Carolina, USA

J.H. Shergold, Division of Continental Geology, Bureau of Mineral Resources, P.O. Box 378, Canberra, Australia

M.S. Sisodia, Department of Geology, University of Jodhpur, Jodhpur 342001, India

G.W. Skyring, CSIRO, Division of Water Resources, GPO Box 1666, Canberra, ACT, Australia

E.P. Solotchina, Institute of Geology & Geophysics, Siberian Branch of the Academy of Sciences of the USSR, 630090, Novosibirsk, USSR

D. Soudry, Geological Survey of Israel, 30 Malkhei Israel Street, Jerusalem, 95501, Israel

P.N. Southgate, BMR Division of Continental Geology, GPO Box 378, Canberra, ACT 2601, Australia

S.J. Van Kauwenbergh, Fertilizer Technology Division, International Fertilizer Development Center (IFDC), Muscle Shoals, AL 35662, USA

H.H. Veeh, School of Earth Sciences, Flinders University of South Australia, Bedford Park, SA 5042, Australia

Yu.N. Zanin, Institute of Geology & Geophysics, Siberian Branch of the Academy of Sciences of the USSR, 630090, Novosibirsk, USSR

K.V. Zverev, Institute of Geology & Geophysics, Siberian Branch of the Academy of Sciences of the USSR, 630090, Novosibirsk, USSR

Phosphorite research: a historical overview

P. J. COOK[1], J. H. SHERGOLD[1], W. C. BURNETT[2] & S. R. RIGGS[3]

[1] *Division of Continental Geology, Bureau of Mineral Resources, PO Box 378, Canberra, Australia*
[2] *Department of Oceanography, Florida State University, Tallahassee, Florida, USA*
[3] *Department of Geology, East Carolina State University, Greenville, North Carolina, USA*

Abstract: The nature and origin of phosphorites have been a matter of much speculation since were they first discovered more than 150 years ago. They are also of wide scientific interest in that they provide evidence of past changes in the biology and chemistry of the world's oceans.

In the past ten years or so, due to a considerable extent to the efforts of the many participants in IGCP Project 156, our knowledge and understanding of phosphorites has increased enormously: it is now evident that there were times in earth history when phosphorites were preferentially deposited in many parts of the world. This probably was the consequence of the interaction of a number of factors including oceanography, changes of sea-level, and plate tectonics. Perhaps one of the most intriguing features to have become apparent in recent years is the link between times of phosphogenesis and periods of major evolutionary change. Not only has our knowledge of ancient deposits increased but much information now exists on present-day deposits, such as those off Chile−Peru and the more enigmatic deposits off southern Australia. In addition, there are better data and a greater understanding of some of the details of phosphorite formation and phosphate hardgrounds, on the role of bacteria in phosphogenesis, and on the petrology, mineralogy, geochemistry and biology of phosphorites.

Whilst a great deal has been achieved in the past 10 years, there still remains much to be done: there is scope for grass-root phosphate exploration in countries presently lacking their own sources of phosphate; the nature of the relationship between phosphate and some diagenetic minerals such as zeolites has yet to be elucidated; similarly the link between phosphogenesis and the formation of petroleum source rocks needs to be better understood and much more work is required on Precambrian phosphorites.

This Special Publication of the Geological Society of London marks the final meeting of the International Geological Correlation Program (IGCP) Project 156: Phosphorites, and provides the opportunity to review a decade of phosphate research (1978−1988). It is also an appropriate time to recognize the work on phosphorites undertaken in the previous decades, particularly in view of the Oxford venue for the final meeting of Project 156.

Early phosphate research

The earliest descriptions of phosphate rock appear to have been of the occurrences in the Estremadura region of eastern Portugal. They were described in a book by Bowles published in 1782 (and cited by Daubeny & Widdrington 1845) on *The natural history and physical geography of Spain* as 'a vein of phosphoric stone'. The first detailed study of phosphate-rich sedimentary rocks was published in 1829, by William Buckland, Professor of Mineralogy at Oxford.

As pointed out by Folk (1965), this paper is also a benchmark for another reason: it is the earliest recognition of coprolites. Even by the standards of eloquence for that time, Buckland was outstanding, describing coprolites as 'Records of warfare waged by successive generations of inhabitants of our planet on one another, the imperishable phosphate of lime derived from their digested skeletons having become embalmed in the substance and foundations of the everlasting hills'!

In 1837 or 1838 Buckland showed the famous German chemist Liebig the Liassic phosphorites of the Bristol area. Shortly after, Liebig suggested that phosphorites could be used as a fertilizer in much the same way as bone and guano were used at that time. This proposal was enthusiastically taken up by agriculturalists, and according to Buckland (1843), phosphorite was sold under the name of 'coprolite manure'. However, the major breakthrough in phosphate usage for agriculture came from the work of John Lawes, a gentleman farmer from Hertfordshire, England, who first dissolved phosphate rock in sulphuric acid to produce a more soluble phosphatic fertilizer. In 1842 he took out a patent for 'chemically decomposing for purposes of manure, by means of sulphuric acid, bones, bone-ash or bone dust or apatite or phosphorite or any other substance containing phosphoric acid'. The discovery by Lawes of a process to produce a more soluble phosphatic fertilizer marked the beginning of the phosphate industry, and of a tremendous upsurge in the exploration for and investigation of phosphorites.

Following on Buckland's original work, major emphasis was placed on the nodular phosphorites. Many publications were to appear on the 'Greensand' deposits of southern England, including those by Austen (1848) who considered the phosphate to be of organic origin. However, he argued against a direct coprolitic origin for most of the nodules, suggesting that many of them had formed diagenetically, drawing an analogy with chert nodules, a hypothesis supported by de la Beche (1849). Other publications which discussed various new phosphate discoveries in Great Britain include those by Brodie (1866), Keeping (1868), Sollas (1872), and Fisher (1873).

It was soon realized that phosphatic sequences were present in France, Belgium and elsewhere, and their study, and exploitation, soon overtook that in Britain. Publications by Meugy (1855), Aoust (1864), de Luna (1865), Davies (1868), and Gruner (1871), testify to this quickening of interest in phosphorites in Western Europe. This interest extended east into Russia where phosphorites were first recognized in 1845 by Chodneff. By 1873 Yermalow, had determined that the Cretaceous phosphorites of Central Russia formed the most extensive phosphate deposits in Europe. Over the next 50 years vigorous exploration and research programmes were undertaken by Russian geologists, culminating in the work of Kazakov and his colleagues in the 1930s (e.g. Kazakov 1937).

North American phosphorites were first reported by Smith (1845), from the Tertiary of South Carolina, where phosphate mining commenced in 1867 (Shepard 1869; Holmes 1870). Subsequent discoveries of major phosphate deposits in North Carolina in 1883 (Day 1885), and in Florida by Lawrence Johnson in 1884 (Day 1887), were to confirm the United States' position as the world's major phosphate producer. The most exhaustive publication on phosphorites during this early period was that by Penrose (1888). Whilst much of the emphasis in the United States and elsewhere was on Tertiary or Mesozoic phosphorites, this was not so in Canada where phosphorites were discovered in the Lower Palaeozoic of eastern Canada by Sir William Logan (Hunt 1852; Logan & Hunt 1854). Detailed work by Davies (1875) showed that phosphorites also occur in the Cambrian of Wales and he ascribed an organic origin to them. The relationship between phosphorites and organic matter was also pursued by Hicks (1875) and Hudleston (1875), and they too concluded that Cambrian rocks were no less phosphatic than younger rocks despite the apparent lack of fossil remains in many of them. In their opinion, the phosphate was derived directly from shells particularly trilobite carapaces. Work in Canada by Dawson (1876) finally led to the realization that phosphorites, were relatively abundant not only in the Cambrian but also in the Precambrian (Laurentian).

At much the same time that Dawson was discovering Precambrian phosphorites another Canadian scientist, the biologist Sir James Murray was busy collecting the first systematic information on marine sediments, including phosphorites, during the course of the Challenger Expedition of 1872–1876. The occurrence of phosphorites was documented by Murray & Renard (1891). Murray & Renard made the perceptive observation that 'phosphatic nodules are apparently more abundant in the deposits along coasts where there are great and rapid changes of temperature, arising from the meeting of cold and warm currents'. Therefore by 1876 phosphorites had been found in sediments of all ages ranging from Precambrian to the Holocene.

The discovery of the Tunisian and Algerian phosphate deposits in 1873 (Thomas 1885; 1887) and the Moroccan deposits in 1908 (Brives 1908) attracted considerable attention amongst European and especially French geologists such as Cayeux (1896, 1932, 1934). With the commencement of mining operations in Algeria in 1889, Tunisia in 1899, and Morocco in 1921, considerable economic interest was also generated.

However, it was discoveries in the United States which were to have perhaps the greatest influence on ideas at this time. The deposits of the southeastern USA were becoming increasingly important (Wyatt 1891; Millar 1892; Darton 1891; Sellards 1915). New and different deposits associated with Ordovician limestones and their weathering products were found in Tennessee, (Hayes 1896; Safford 1902; van Horn 1909). However, the Tennessee discoveries were eclipsed by the finding of the Permian Western Phosphate Field in the 1880s. It was left to the US Geological Survey to undertake a thorough geological assessment of the area. A large number of publications resulted from these investigations, most notably those by Weeks & Ferrier (1907), Weeks (1908), Gale & Richards (1909), Blackwelder (1910, 1911, 1915, 1916), Breger (1911), Richards & Mansfield (1914), Mansfield (1916, 1918, 1920, 1927) and Pardee (1917). The publications of Blackwelder and Mansfield were perhaps the most significant.

By the 1930s a number of hypotheses on the origin of phosphorites had been put forward, but the only generally accepted feature was that there was a relationship between the biota and phosphorites. It was realized that the coprolithic theory first put forward by Buckland was only a small part of the total story. Mass mortality, particularly of fish, was regarded as one of the important mechanisms involved in the formation of phosphorites. There was, however, a wide divergence of opinion on the cause of such mass mortalities. Also, as a result of investigations on the Western Phosphate Field, the idea of direct precipitation of apatite from marine bottom waters was receiving some attention.

Kazakov and after

The work of A. V. Kazakov, one of the founders of the Scientific Institute for Fertilizers and Insectofungicides in the USSR, came into prominence in the 1930s. Kazakov investigated the $P_2O_5-CaO-H_2O$ system in an attempt to better understand the genesis of phosphorites, but he did not neglect the 'classical' approach to deposits, studying their lithology, facies relationships, and palaeogeography. Kazakov (1937) proposed that phosphorites formed as a result of direct inorganic precipitation of apatite, from ascending phosphorus-rich deep ocean waters. A decrease in the partial pressure of CO_2, accompanied by an increase in pH, and a decrease in phosphate solubility, were suggested as the mechanisms of precipitation. Kazakov further proposed that 'The chemical sedimentation of phosphates takes place in the upper and middle portions of the shelf, i.e. at a depth of 50–200 m approximately.' Kazakov's 'Upwelling Hypothesis' provided an elegant integrated theory which explained many of the features of phosphorites, notably the regular facies distributions associated with many phosphorites, and particularly the distribution of carbonates in relation to phosphate deposits.

At first, Kazakov's ideas received little attention, until the US Geological Survey commenced a new phase of research on the Phosphoria Formation. In the late 1940s, as a result of increased interest in uraniferous sediments, the US Geological Survey embarked upon a major study of the Western Phosphate Field under the leadership of Vincent McKelvey. This work was to spawn a flood of publications. It was also to exert very great influence on phosphate geology in general, becoming a yardstick against which many other deposits were judged i.e. whether or not they conformed to the 'Phosphoria Model'. The popularity of this model can be traced to the seminal paper of McKelvey et al. (1953) in which they proposed that 'The facies changes and composition of these (Phosphoria) deposits reflect marine deposition of phosphorites controlled primarily by depth, temperature, pH and CO_2 content of the water, much as previously indicated by Kazakov', thus placing the stamp of approval on a modified version of the 'Upwelling Hypothesis' of Kazakov.

Early experimental work was undertaken on phosphorites by Warington (1866), Cameron & Hurst (1904) and Cameron & Siedel (1904) but it was only in the 1950s that meaningful experimental and theoretical studies got underway, in part because new and better analytical techniques had by then become available. The additional impetus for undertaking experimental and theoretical work was a consequence of Kazakov's upwelling hypothesis, which could be tested. Important work followed by Smirnov et al. (1958, 1962), Smirnov (1964), Ames (1959), Kramer (1964), Roberson (1966), Simpson (1967), McConnell et al. (1961), McConnell (1965). Much of this early experimental work is summarized by Gulbrandsen (1969).

A thorough study of phosphate equilibria was undertaken by Atlas (1975) who examined the chemistry of phosphate in sea water in terms of solution and solubility equilibria, using experimental and theoretical approaches. Atlas & Pytkowicz (1977) showed that the solubility behaviour of apatite in sea water is best described in terms of complex reaction on the apatite surface rather than by simple solubility theory. Martens & Harriss (1970) showed that apatite precipitation is strongly inhibited by magnesium ions. This question of Mg inhibition was further pursued by Nathan & Lucas (1972, 1976) who showed that the constraints on apatite precipitation are more complicated than originally envisaged by Martens & Harriss with, for example, the critical Ca/Mg ratio varying with pH. They also considered that abundant organic matter must be present.

For the past 30 years, the 'Upwelling Hypothesis' has undoubtedly been the favoured theory of phosphogenesis, but there have been, and still are, many who question its applicability and indeed have questioned the whole idea of the ocean being the primary source of the phosphorus. Pevear (1966) has advocated a fluvial source for the phosphorites of the southeastern United States. Bushinskii has also proposed that the continents, via the rivers, are the main source of phosphate in the ocean, suggesting (Bushinskii 1966) a direct fluvial/estuarine origin for the phosphate without passing through what could be regarded as a marine intermediary. He also provided abundant evidence that many phosphorites form under very shallow conditions (Bushinskii 1935; 1964). Interestingly, Bushinskii used the facies distribution in the Phosphoria Formation as evidence of a fluvial source for the phosphate, considering the Shedhorn Sandstone as a deltaic sand. Unfortunately for Bushinskii's fluvial hypothesis, detailed work by Sheldon (1963) has shown fairly conclusively that the Shedhorn Sandstone was deposited as a barrier sand, flanked on its eastern (landward) margin by a lagoonal complex.

Hite (1978) attempted to genetically link evaporites, phosphorites and iron-rich sediments through a dynamic sea-water system with landward movement of shallow seawater into an evaporative basin and the phosphorous derived from the ensuing brine. Henderson *et al.* (1979) proposed a diagenetic reflux type of mechanism for the Georgina Basin of Queensland with phosphatization of calcium carbonate by the phosphate-rich brine.

A link between phosphogenesis and volcanism has been proposed by many investigators. Mansfield (1940) was one of the early advocates of the importance of volcanism to phosphogenesis. He proposed 'a close temporal relationship between periods of volcanism and the formation of major phosphate deposits'. Bidaut (1953) proposed a fumarolic source for the phosphorus (and silica) in the Devonian phosphorites of southeastern France. A number of Soviet geologists, notably Shatskii (1955), Strakhov (1960) and Brodskaya & Ilyinskaya (1968, 1970), proposed a link between phosphorites and volcanism. Rooney & Kerr (1967) proposed a similar link for the Miocene phosphorites of North Carolina, taking the presence of abundant quantities of clinoptilolite to indicate the relationship to volcanism. However, recent stratigraphic and petrographic work by Snyder *et al.* (1984) and Riggs & Mallette (1990) has demonstrated that the clinoptilolite is formed in foraminiferal tests through diagenesis associated with the development of stratigraphic condensed sections.

Throughout the 1950s and 1960s there can be no doubt that, in the English language literature at least, the ideas generated from work on the Phosphoria Formation were especially important. Publications such as those of McKelvey *et al.* (1959) and Sheldon (1963) were to have great influence. Their model for phosphate deposition involved upwelling on the flanks of an open ocean basin at water depths of 200–300 m. This model and the presence of a black shale-chert-phosphorite assemblage, coupled with an understanding of palaeogeography (including an appreciation of the importance of palaeolatitude as a consequence of the pioneering work by Sheldon (1964) on palaeolatitudes), was to lead to important phosphate discoveries in the 1950s and 1960s. The Georgina Basin deposit of northern Australia was one such discovery, although it has since been shown that much of the chert has been diagenetically remobilized silica rather than a primary chert.

One feature that could not always be clearly extrapolated from the Phosphoria model was the water depth. Evidence for shallow water environments comes from the Cambrian phosphate deposits of the Georgina Basin, for example (de Keyser & Cook 1973; Howard & Cooney 1976; Howard & Perrino 1976; Rogers & Keevers 1976; Soudry & South-gate 1989). There, sedimentary structures and facies relationships provide abundant evidence of shallow nearshore environments of deposition. Some of the best evidence for shallow water phosphogenesis comes from stromatolitic phosphorites, such as the stromatolitic phosphorites of Rajasthan, India (Nath & Sant 1967; Banerjee 1971; Choudhouri 1990). They provide quite compelling evidence that, by analogy with the modern stromatolites of Shark Bay (Logan *et al.* 1964), these stromatolitic phosphorites formed in water depths of no more than a few metres.

Evidence of structural and bathymetric/topographic controls on phosphate deposition is apparent from a number of deposits particularly those of the Middle East and North Africa. Bentor (1953) and later workers such as Reiss (1962) and Würzburger (1968) showed that the Upper Campanian phosphorites (the principal deposits in Israel) are found almost exclusively in the synclinal axes of folds, which started to develop in the Turonian but which continued to grow during Campanian sedimentation, exerting a profound influence on the distribution of phosphorites. Elsewhere, Wilcox (1953) related the Santonian-Lower Campanian phosphatic chalk occurrences of England to synclinal troughs. Jarvis (1980*a*) reinterpreted these structures as elongate erosional troughs ('cuvettes') which he considered exerted a major influence on the genesis and distribution of phosphatic chalks both in southern England and throughout northern France. Similarly, for the Duchess phosphate deposits of northern Australia, Russell & Trueman (1971) have shown that 'the bathymetric, lithofacies and isochemical data of the Duchess deposit reveal a close control of sedimentation by basinal topography. . . . Bioclastic phosphatic carbonates were developed on the most elevated basement highs, the chert and phosphorite beds developed on the flanks and over weaker positive basement features and the fetid micritic phosphatic carbonates were formed in the deeper troughs. Sedimentation was therefore similar to that of the Mishash of Israel where active fold structures controlled the type of sediment deposited.'

On both the regional and local scales associated with individual deposits throughout the southeastern United States there is a strong structural control by the structural framework and palaeotopography of the depositional system during the Upper Cenozoic. Cathcart & McGreevy (1959) and Altschuler et al. (1964) suggested a structural relationship for the phosphate deposits in central Florida. They believe that the 'reworked phosphorites' of the Bone Valley Formation were derived from the underlying 'weathered residuum and limestone' of the Miocene Hawthorn Formation and were redeposited in shallow-water, marine and estuarine environments by a major marine transgression. Within central Florida, Riggs (1967, 1979a) clearly demonstrated the primary sedimentological and stratigraphic relationships of the Miocene and Pliocene phosphorites, and suggested that their marine formation and deposition was intimately controlled by changing conditions within highly variable depositional environments determined by the palaeotopography. The case for structural control of phosphate formation on a regional basis was clearly established for Florida by Freas & Riggs (1965, 1968), Freas (1968), and Riggs (1979b) and for southeastern US by Riggs (1979b, 1984). The Ocala and Sanford Highs in Florida and the Carolina Platform High in the Carolinas determined (1) the locations of associated coastal environments, shelf platforms and deeper depositional basins; (2) loci, type, and abundance of phosphate formation and accumulation; and (3) types and abundance of associated sediments.

The genetic meaning of the structural relationships in the southeastern US continued to evolve. By the mid-1960s most workers agreed that phosphorite formation represented primary marine sedimentation and was in direct response to upwelling (Freas & Riggs 1965, 1968; Riggs 1967; Freas 1968; Cathcart 1968). It was clear that primary phosphate formed in both shallow-water environments on and around the palaeotopographic highs and within deeper depositional basins, and that the type of phosphate and associated sediments are controlled by specific environments of formation. Initial interpretations involved interaction between the warm-water Gulf Stream and cold-water Labrador currents with upwelling over and around the palaeotopographic features. However, mechanisms of upwelling associated with western boundary currents were then poorly understood.

Whilst the majority of phosphate research and exploration during the 1950s and mid-1970s concentrated on onshore deposits, there was also a significant level of offshore activity. This was triggered by the realization that a better understanding of modern onshore deposits would provide a better understanding of ancient deposits, and also by the need to assess the extent of offshore phosphate resources. Important studies included those by Dietz et al. (1942) and Pasho (1972) on the offshore deposits of southern California. Previously unknown deposits were discovered off southeastern Australia (von der Borch 1970), a location notable for its lack of large-scale oceanic upwelling. Investigations off the southeastern United States (Manheim et al. 1975) showed the extent of major deposits on the Blake Plateau. These and other deposits were shown by Kolodny (1969) to be residual, with pre-Holocene ages. Phosphorites from the Tertiary of the southern North Sea Basin have been described by Balson (1990).

Subsequent work, however, showed that at least two deposits (those off Chile/Peru and southwest Africa) were modern (Baturin et al. 1972). By the mid-1970s a vast amount of phosphate research had been under-taken, many valuable ideas had been generated and numerous deposits found as a result. However, there was no mechanism for bringing together the many phosphate workers and consequently no easy way of developing a common 'language' for phosphate geology.

IGCP Project 156

The idea of a major international phosphate project was developed in the mid-1970s. An initial impetus came from the suggestion in de Keyser & Cook (1973) of a 'Cambrian Austral-Asiatic phosphogenic province' which was compared to the Cretaceous–Eocene Tethyan phosphogenic province. There was obviously a need to test this model but at that time there was little opportunity to do so until following extensive discussions at the International Geological Congress in 1976. In 1977, P. J. Cook and J. H. Shergold decided to put forward a proposal to IGCP for a study of the late Proterozoic–Cambrian phosphorites of Australia and Asia. That proposal was accepted by IGCP and the first meeting of the project was held in Australia in 1978. It soon became apparent that there was a need to expand the scope of the project to encompass phosphorites of all ages. This was done through the establishment of Working Groups concerned with various phosphogenic episodes, together with one working group concerned with phosphate resources in general (Table 1).

For the first six years (1978–1984) the Project was led by P. J. Cook and J. H. Shergold. W. C. Burnett & S. R. Riggs then took over for the remaining four years (1984–1988). The work of the various Working Groups is drawn together in series of major volumes by Cook & Shergold (1986a), Notholt et al. (1989) and Burnett & Riggs (1990). However, during the course of the Project, many individual publications were produced, and also a number of volumes related to specific project meetings (Table 2) including those by Cook & Shergold (1979a, b) Burnett & Sheldon (1979), Notholt (1980a, b), Ilyin & Bjamba (1980), Sheldon & Burnett (1980), Indian National Committee for IGCP (1981a, b), China National Committee for IGCP (1982, 1984), Lucas & Prévôt (1983); Snyder et al. (1985); Hine & Riggs (1986); Riggs et al. (1986); Rodriguez et al. (1986); Shergold & Southgate (1986); Belayouni & Beji-Sassi (1987); Abed (1988); Dunbar & Baker (1988); Jarvis (1988); Notholt et al. (1988); and Robaszynski & Sustrac (1988).

Additionally, a number of major phosphate volumes have been produced by various members of IGCP 156 over the past decade and include the following: Notholt (1978, 1980b); Howard (1979); Bentor (1980); Banerjee & Khan (1981); Notholt & Hartley (1983); Bridges et al. (1983); Nriagu & Moore (1984); Lucas & Prévôt (1985a); Sheldon et al. (1985); Slansky (1986); Zanin (1987); Eganov (1988); and Burnett & Froelich (1988).

These and many other volumes, together with hundreds of individual papers published elsewhere have had a profound influence on our understanding of the nature and origin of phosphorites. The series of international phosphate seminars and field workshops held over the last ten years (Table 2) have also provided outstanding opportunities to exchange knowledge.

In general terms, the achievements of IGCP Project 156 have included:
  a major increase in our knowledge and understanding of phosphorites;
  a major improvement in communications through the holding of meetings, through publications, the project newsletter and equally importantly, the development of a 'network' between institutions and scientists;

Table 1. *Organizational structure of IGCP Project 156: Phosphorites*

Project Leader: 1978–1984	P. J. Cook & J. H. Shergold
1984–1988	W. C. Burnett & S. R. Riggs
Working Group 1: Proterozoic & Cambrian Phosphorites	P. J. Cook & J. H. Shergold
Working Group 2: International Phosphate Resource Data Base	R. P. Sheldon & A. J. G. Notholt
Working Group 3: Young Phosphogenic Systems	W. C. Burnett & S. R. Riggs
Working Group 4: Cretaceous–Eocene Phosphorites	K. Al-Bassam & J. Lucas
Committee on Rock Phosphate Standards (CORPS)	Z. S. Altschuler, L. Prévôt & Y. Zanin
Working Group on Petrography of Phosphorites	L. Prévôt, P. N. Southgate, G. Ratnikova & D. Giot
Committee on Igneous Phosphates	A. J. G. Notholt, S. M. Punukollu, O. Harmala/F. G. Theuri, R. Choudhuri & P. van Stratten
Newsletter Editors	A. J. G. Notholt & W. C. Burnett

Table 2. *IGCP 156 international and regional field workshops and symposia, 1978–1988*

#	Year/Location	Topic
1.	1978 Australia	Cambrian Phosphorites of the Georgina Basin, Western Queensland.
2.	1979 Hawaii	Marine Phosphatic Sediments Workshop, Honolulu.
3.	1979 USA	Permian Phosphorites of the Rocky Mountains, western USA.
4.	1979 Hawaii	Fertilizer Raw Materials Resources Workshop, Honolulu.
5.	1980 England	Workshop on Phosphatic and Glauconitic Sediments, London.
6.	1980 Mongolia P.R.	Precambrian–Cambrian Phosphorites of the Lake Khubsugul District.
7.	1981 Mexico	Upper Cenozoic Phosphorites of Baja California.
8.	1981 India	Precambrian Stromatolitic Phosphorites of Central Rajasthan.
9.	1981 India	Palaeozoic Phosphorites of the Musoorie District, Lesser Himalaya.
10.	1982 China	Cambrian Phosphorites of Guizhou Province.
11.	1982 China	Cambrian Phosphorites of Yunnan Province.
12.	1983 Morocco	Cretaceous and Paleogene Phosphorites of Western Tethys.
13.	1983 Senegal	Weathered Tertiary Phosphorites of the West African Margin.
14.	1984 USSR	Symposium at 27th Session of IGC, Moscow: Phosphate Ore Deposits.
15.	1984 USSR	Cambrian Phosphorites of Lesser Karatau, Kazakhstan.
16.	1985 USA	Neogene Phosphorites of the Southeastern US Continental Margin.
17.	1986 Venezuela	Cretaceous and Neogene Phosphorites of Northern South America.
18.	1986 Australia	Phosphate Research Program: Cambrian Phosphorites of Georgina Basin, Upper Cenozoic Phosphorites and Glauconites of the East Australian Margin and Southern Victoria.
19.	1986 Australia	Symposia at 12th ISC, Canberra: (1) Cyclicity and Phosphogenesis; (2) Phosphogenesis and Global Processes.
20.	1986 Australia	Cambrian Lithofacies and Depositional Environments of Eastern Georgina Basin.
21.	1986 USA	Workshops at SEPM: (1) Phosphorites on the N.C. Continental Shelf; (2) Cyclic Sedimentation of Neogene Phosphorites in N.C.
22.	1987 Tunisia	Tethyan Phosphorites and Associated Petroleum Source Rocks.
23.	1987 Iraq	Cretaceous and Tertiary Phosphorites of Iraq.
24.	1988 Jordan	Cretaceous and Tertiary Phosphorites of Jordan.
25.	1988 Peru	Cenozoic Phosphorites and Associated Organic-rich Sediments of the Peruvian Continental Margin.
26.	1988 Israel	Tethyan Phosphorites of the Negev Desert.
27.	1988 England	Symposium: IGCP 156, A Decade of Phosphorite Research and Development, Oxford.
28.	1988 England	Cretaceous Phosphorites of Southern England.
29.	1988 France/Belgium	Cretaceous Phosphorites of Northern France and Belgium.

international co-operation has been facilitated through the project, with more than 60 countries involved in a range of activities;

training and education has been of particular importance, with training courses being given in Australia, Thailand, Senegal, USA (North Carolina), Venezuela, Brazil and Burma (Table 3);

the Project has made a significant contribution to phosphate exploration through the development of new models and concepts;

discoveries in Nepal, Spain, Argentina, and Chile have been attributed directly to ideas developed through the Project;

over the past decade, there can be no doubt that the 'profile' of phosphate geology has been raised; there is now an acceptance that an understanding of phosphate geology and the distribution of phosphate through time is essential to developing a full understanding of the evolution of the hydrosphere and biosphere;

the Project has directly facilitated research through programs such as the Phosphate Research Project (PHOSREP) in Australia.

Returning to some of the recent achievements in phosphate research, it is not possible to adequately summarize all the information and ideas generated over those ten years. However,

Table 3. *IGCP 156 training programme*

1. The Geology of Phosphorites for geologists from Asia; convener Prof. P. F. Howard, Sydney, Australia, May 1982.
2. Phosphate Resource Assessment of Southeast Asia; convener Dr R. P. Sheldon, Bangkok, Thailand, January 1983.
3. The Geology of Sedimentary Phosphorites of West Africa; convener Dr M. Slansky, Dakar, Senegal, November 1983.
4. Phosphate Potential of the Caribbean Basin and Central America; conveners Dr R. P. Sheldon and Dr S. R. Riggs, US Geological Survey and East Carolina University, Greenville, N.C., July 1984.
5. Geology of Phosphate Deposits; conveners Dr R. P. Sheldon and Dr S. R. Riggs, US Geological Survey and East Carolina University, Greenville, N.C., May 1985.
6. Indigenous Phosphate Deposits; convenor Dr G. H. McClellan, International Fertilizer Development Center, Muscle Shoals, Alabama, May 1985.
7. The Geology and Exploration for Phosphate Deposits; conveners Prof. P. F. Howard and Dr P. N. Southgate, The Australian Development Assistance Bureau — Burma Aid Program in Rangoon, Burma, June–July 1985.
8. Exploration and Evaluation of Phosphorite Deposits in Latin America; convener Dr Simon Rodriguez; Ministry of Energy and Mines of Venezuela, in Merida, Venezuela, March 1986.
9. Phosphorite and Offshore Minerals Workshop; conveners Drs W. C. Burnett and S. R. Riggs, Co-Directors of IGCP 156; IGCP and IOC of UNESCO, Porto Alegre, Brazil, November 1988.

Geologists from 49 countries participated in the training courses:

Asia (12)	Africa (16)	Middle East (4)	Latin America (17)
Bangladesh	Algeria	Iran	Argentina
Burma	Benin	Iraq	Bolivia
China	Cameroons	Jordan	Brazil
India	Egypt	Syria	Chile
Indonesia	Ethiopia		Colombia
Malaysia	Guinea Bissau		Costa Rica
Mongolia	Mali		Dominican Republic
Nepal	Morocco		Ecuador
Pakistan	Nigeria		Guatemala
Philippines	Senegal		Haiti W.I.
Thailand	Tanzania		Honduras
Vietnam	Togo		Jamaica
	Tunisia		Mexico
	Uganda		Paraguay
	Zaire		Peru
	Zambia		Uruguay
			Venezuela

some of the major features can be outlined. Certainly a common language in phosphate geology has developed. This has not necessarily lead to a uniformity of ideas and interpretations but it has meant that terms such as 'phoscrete' or 'phosphatic hardground' now convey the same picture to most people.

Distribution of phosphorites in space and time

Prior to the mid-1970s our knowledge of the distribution of phosphorites was sketchy at best. Geijer (1962) argued that there was an absence of Precambrian phosphorites for example. However, the recent compilations of Cook & McElhinny (1979), Notholt (1980a, b), Lee (1980), Cook & Shergold (1986a), Notholt et al. (1989) and Burnett & Riggs (1990) have provided us with a definitive picture of the spatial/temporal distribution of phosphorites and enabled us to test the model that represented the starting point for Project 156; namely that there were preferred times in earth history when phosphorites were deposited in many parts of the world. Information collected over the past 10 years has validated the model.

Palaeoceanography

The understanding of palaeoceanography and palaeoclimatology that has grown in recent years has to a considerable degree developed independently of an understanding of phosphorites.

However phosphorites and upwelling systems have been used by Parrish (1981, 1982), Parrish *et al.* (1986) and others to constrain the models. The link between oceanic anoxic events and phosphogenesis has been a matter of considerable speculation: Fischer & Arthur (1977), Cook & McElhinny (1979) and Arthur & Jenkyns (1981) have proposed a genetic correlation. Data collected as a result of Project 156 activities lend support to the hypothesis. It was not until the late 1970s and early 1980s that physical oceanographers studying the Gulf Stream developed an understanding of modern upwelling dynamics associated with western boundary currents (Brooks & Bane 1978; Pietrafesa *et al.* 1978; Bane & Brooks 1979; Janowitz & Pietrafesa 1980, 1982; Pietrafesa 1983). This understanding within the modern continental margin system has led to new interpretations of the inter-relationships between continental margin palaeotopography, upwelling, and phosphorite formation and sedimentation during the Upper Cenozoic (Riggs 1984; Riggs & Mallette 1990; Popenoe 1990; Snyder *et al.* 1990).

Sea-level change

The concept that the formation of phosphorites is linked to changes of sea level is not new. The occurrence of phosphorites on many disconformity and unconformity surfaces had been noted by some of the early workers on phosphorites and Mansfield (1927) in particular, proposed that phosphatic horizons in the Phosphoria Formation may correlate with Permian glacial events. Based on patterns of marine inundation of the continent Arthur & Jenkyns (1981), Sheldon (1980) and Cook (1983) were also able to show a sea-level-phosphogenesis correlation. However, it was undoubtedly the development of a global sea-level curve by Vail and his colleagues (Vail *et al.* 1977a, b; Vail 1987) and the subsequent work by Riggs and his colleagues on the Atlantic continental margin of the USA (Riggs 1984; Riggs *et al.* 1982a, b; Riggs *et al.* 1985), that demonstrated that a major Miocene phosphogenic episode occurred during the Miocene transgression, and that phosphogenesis reached a maximum on the mid portion of the rising sea level. They also showed that the duration of high stands is an important factor in providing adequate time for phosphogenesis to accumulate significant concentrations of phosphate within the optimum environment. Latest sea-level modelling (Vail 1987) indicates that phosphorites correlate with the maximum flooding surface, i.e. the downlap surface, and coincides with condensed sequences (see e.g. Glenn 1990, Föllmi 1990).

Plate tectonics and phosphogenesis

At an early stage in the project Cook & McElhinny (1979) proposed a series of models to explain the timing of phosphogeneis in relation to the various phases of sea-floor spreading. That model has since been extended and refined (Sheldon 1980; Parrish 1982) and can indeed be applied to many deposits. At the same time, like all models, subsequent work has shown that it is too simplistic and that there are many exceptions. This is inevitable, given the new knowledge on the nature and distribution of phosphorites that has been gathered over the past decade, leading in turn to a need to further refine the earlier models. The models developed by Project 156 have led to important new concepts for phosphate deposition.

Evolution

Perhaps one of the most spectacular developments in recent years has been our new understanding of the links between phosphogenesis and major evolutionary events. The most significant of these is perhaps the appearance of shelly faunas around the Precambrian–Cambrian boundary (Brasier 1979, 1982, 1985, 1990; Donnelly *et al.* 1990). Cook & Shergold (1984a, b) were able to demonstrate that the major phosphogenic event coincides closely with the major evolutionary step involved in the development of hard exoskeletons which initially were commonly phosphatic (possibly when the oceans were high in phosphate) before giving way to overwhelmingly calcareous hard parts. Whilst the strongest evidence for this relationship is found near the Precambrian–Cambrian boundary, the possibility exists, as pointed out by Cook & Cook (1985), that a similar association may also occur around the Cretaceous–Tertiary and other boundaries where major biotic changes have also occurred.

Modern phosphorites

During the life of Project 156 there has been new interest in offshore phosphorites (Bentor 1980; Burnett *et al.* 1980; Burnett & Froelich 1988; Cullen 1980; Cook & Marshall 1981; Kolodny 1981; Lucas *et al.* 1978; Marshall & Cook 1980; McArthur *et al.* 1988; O'Brien & Veeh 1980; O'Brien *et al.* 1981; Riggs 1984; Veeh & Burnett 1982). The deposits of offshore Chile–Peru and southwest Africa had been well documented and they provided excellent models for many ancient deposits. However, even these well-known deposits are still yielding much new and important information (Burnett & Froelich 1988). Particularly significant is the radiometric work of Burnett et al. (1982) and Kim & Burnett (1986) in which use is made of thorium isotopes to measure the growth of phosphate nodules and crusts in the Chile−Peru System today.

The offshore deposits of southeastern Australia and southeastern United States' on the other hand, were poorly known in the mid-1970s and even more poorly understood. The work of Riggs and co-workers (Riggs 1984; Riggs et al. 1982a, b, 1985; Riggs & Snyder 1986) on the North Carolina continental margin has been spectacularly successful in providing a complete picture of the sedimentary suite into which the phosphorites fit, and has provided insights on the controls (such as sea-level change) on that suite (Riggs 1984; Riggs et al. 1982a, b, 1985). However, modern phosphorites have yet to be found amongst the deposits of the offshore United States.

The Australian deposits, on the other hand, have yielded modern phosphorites (O'Brien & Veeh 1980). They have also produced many questions. These include the nature of the iron−phosphorus relationship, and the mechanism by which phosphorus can be concentrated in sediments notable for their lack of organic matter. These and other questions on the Australian offshore deposits are addressed in some of the papers in this volume (Heggie et al. 1990; O'Brien et al. 1990). The role played by Project 156, through PHOSREP, in stimulating the eastern Australian work is certainly one of the important achievements of the Project.

Facies models and palaeogeography

Palaeogeography has always been one of the most important tools for phosphate exploration and, consequently, geologists have expended a considerable amount of effort in reconstructing the palaeogeography of phosphogenic (or potentially phosphogenic) provinces. In the English literature it was the Phosphoria model of McKelvey et al. (1953) which was dominant whereas in the French literature, for example, the models were predominantly developed from the North African deposits. In the past decade, the most significant advances to our understanding of the influence of palaeogeography on phosphogenesis has come from the previously-mentioned studies of the Miocene deposits of the USA (Riggs 1979a, b, 1984) and the Cambrian deposits of Asia and Australia (see Howard 1990). For example the work of Chinese geologists, documented by Li (1986) and Yeh et al. (1986), has shown convincingly the importance of palaeogeography to the localization of phosphate deposits. This, and work elsewhere, has demonstrated that many (perhaps most) economic phosphate deposits form (or the phosphate grains accumulate) in water depths of no more than a few tens of metres, and commonly no deeper than a few metres.

Phosphate hardgrounds

Phosphatic hardgrounds were for many years the 'cinderellas' of the phosphate world! The work by Kennedy & Garrison (1975) was a notable exception. Riggs (1967, 1979a) demonstrated, however, that in central Florida, hardgrounds (which he termed microsphorite: in situ microcrystalline carbonate−fluorapatite mud) formed extensively throughout the shallow-water, marine environments around the Ocala High. Discrete beds of mud were first deposited, burrowed, pelletized and subsequently indurated into hardgrounds that were bored, abraded, and polished. Repetition of the process produced interbedded hardground units that ranged from a few millimetres up to 1 m in thickness. Subsequent biological and storm activity periodically broke up the hardgrounds to produce the most dominant shallow-water phosphate grain type, intraclastic phosphate that includes the extensive phosphate gravels that led to the old name of 'land pebble phosphate district'.

During the past decade there has been a major increase in our interest in and understanding of phosphatic hardgrounds, particularly through the work of Jarvis (1980a, b) and Southgate (1986). These researchers have documented the range of textures associated with hardgrounds and have shown not only how they fit into the mosaic of facies in a phosphogenic system, but also how hardgrounds give critical insights into prevailing depositional conditions during phosphogenesis and how they can be related to sea-level change. Phosphatic stromatolites are also associated with some hardground surfaces (Schmitt & Southgate 1982). Proterozoic phosphatic stromatolites have been particularly well documented (Banerjee 1971; Sisodia & Chauhan 1990), but it is only recently that the Phanerozoic occurrences have been described in detail. As a consequence of detailed petrographic work, it is now realized that mudstone phosphorites (microphosphorites), coated surfaces, hardgrounds and phoscretes contain many of the same features, may in part be products of similar processes, and appear to be important (and in many cases previously unrecognized) features of many phosphate deposits.

Mineral associations

The phosphorite-black shale-chert assemblage has been well documented from the Phosphoria Formation by McKelvey et al. (1953) and elsewhere by others. However, we now know that there are many different types of phosphate deposits; each type has specific mineral associations with other authigenic/diagenetic sediments. The major types of phosphate assemblages have been summarized by Riggs (1986) and others and include the following:
  phosphate−black shale−chert
  phosphate−dolomite−magnesium clays−chert
  phosphate−glauconite
  phosphate−banded iron oxides−chert
  phosphate−manganese and ferromanganese oxides.

Such associations occur in deposits ranging in age from Precambrian to the present. Some are a consequence of a low rate of marine clastic sedimentation, which allows chemical or biochemical sedimentation to become dominant. Others may be, in part, the result of the prevailing biota. Subaerial exposure is a critical element for some. However, many such assemblages are not fully understood, and the relationships are little more than empirical for the present, but we now have much better information on their nature and extent.

One association on which there has been much speculation in recent years has been that proposed between phosphorites and tillites. It has been shown that many of late Proterozoic−Cambrian deposits are underlain by tillites (Cook & Shergold 1979, 1986b). It has been suggested by Cook & Shergold (1984a, b) that there is a genetic association with oceanic cooling causing increased rates of oceanic overturn, and hence increased rates of phosphate cycling. A major difficulty in this hypothesis is that, despite the apparent stratigraphic juxtaposition of phosphate and tillite, there is commonly an age gap of millions to tens of million of years between the two events. This suggests that either a direct genetic association between phosphorites and tillites is unlikely, or that some assumed ages for the phosphorites or tillites are gravely in error. Riggs (1986) suggested that this disparity should be expected since it reflects major changes in global, first-order, palaeoclimatic and palaeoceanographic events. The phosphorite assemblages occurs throughout the geologic column within the transition zone between thicker and more uniform sequences of carbonate and siliciclastic or volcaniclastic sediments. This stratigraphic succession, including the transition zone sediments, occur in concert with major changes in sea level leading into or coming out of glacial episodes.

Petrography and mineralogy

Over the past decade, detailed petrographic studies of phosphate have lead to many additional classification schemes being proposed including those of Slansky (1980), Prévôt (1982), Prévôt & Lucas (1986), Elgueta (1981), and Southgate (1983). Riggs (1979a) developed a detailed classification scheme for phosphate-rich sediments, macroscopic phosphate grain types, and the microscopic components occurring within these complex sediments. This classification was modelled after the Folk classification of carbonate sediments (Folk 1959). Cook & Shergold (1986b, c) employed a classification system based on Dunham's classification of carbonates (Dunham 1962) using terms such as grainstone phosphorite, packstone phosphorite, and mudstone phosphorite. There are, therefore, many successful and workable classifications but, in spite of much effort to date, there is no universal support for any one set of terminology.

Because of its economic significance there has been a considerable amount of work in recent years on the mineralogy of phosphorites and of the effects of weathering on mineralogy (Nathan et al. 1990; Van Kauwenbergh & McClellan 1990). Work by the International Fertilizer Development Center has been particularly notable (Lehr et al. 1967; McClellan 1980; McClellan & Gremillion 1980; McClellan & Lehr 1969; Smith & Lehr 1966; McClellan & Van Kauwenbergh 1990; Van Kauwenbergh & McClellan 1990). Work on the weathering of phosphorites has tended to be concentrated on the deeply weathered deposits of Florida (Altschuler 1973) and Africa (Flicoteaux et al. 1977) and particularly with the development of aluminium phosphates as a result of intense weathering.

Role of bacteria in phosphogenesis

The role of bacterial mats in the fixation of phosphate in the contemporary sedimentary system is an exciting recently developed field of phosphate research. Without a doubt some of the most innovative investigations undertaken on phosphate systems in the past decade has been through the geobiological experimental work of Lucas and his colleagues at Strasbourg (Nathan & Lucas 1972, 1976; Lucas & Prévôt 1981a, b, 1984, 1985b; Lucas et al. 1990). What they have shown is that much of the earlier experimental and theoretical work was too simplistic. Furthermore, they have been able to demonstrate that processes of direct phosphate precipitation and phosphatization of carbonate are facilitated by the presence of organic material, whether in the form of micro-organisms or as DNA. They have also shown that the precipitation can occur in a range of aqueous environments, but that well-crystallized apatite only forms in sea-water.

Taken together, these and other experiments clearly show that the essentially non-biological experiments of earlier workers were of little relevance to the conditions prevailing in most natural systems, and also that the inhibiting effects of Mg, whilst significant, can be readily overcome by the action of the microbiota. This research is important in the light of recent discoveries in the Middle Cambrian of the Georgina Basin, northern Australia, where bacterial mats have been recently reported by Soudry & Southgate (1989). Comparison can also be made to the microbially generated phosphate nodules of late Cretaceous age described by Soudry & Lewy (1988) from southern Israel (see also Lamboy 1990; Lewy 1990; Abed & Fakhouri 1990). Therefore, it is evident that benthic microbial populations were as important in the past as they are today in the formation of mudstone phosphorite, phosphate nodules and phosphatic crusts.

Geochemistry of phosphorites

Our knowledge of isotope geochemistry has greatly improved over the past decade. Phosphorites are of course almost unique in the range of isotopes that they contain, including O (carbonate and phosphate), C (carbonate and organic carbon), S (sulphate and sulphide) and Sr. There are few other chemical or biochemical sediments with such a suite of environmentally significant isotopes. The work of Kolodny et al. (1983) and Shemesh et al. (1983) has shown that the oxygen in phosphorites is less exchangable than the oxygen in carbonates or cherts. From this it can be shown that the temperatures at which phosphorites have formed in the past have varied quite markedly. McArthur and co-workers (McArthur et al. 1980, 1986, 1987; McArthur & Herczeg 1990) have used carbon,

oxygen and sulphur isotopes to study the diagenetic history of a wide range of phosphorites, and have differentiated between those formed in oxic/suboxic and anoxic conditions. Piper & Kolodny (1987) have been able to successfully use isotopes to support the hypothesis that phosphorites in the Phosphoria Formation were deposited on a continental shelf in an area of intense oceanic upwelling during several episodes of sea level change.

Much more data are now available on the major and trace-element geochemistry of phosphorites. Indeed, so many papers have been published on this topic that it is difficult to single out any papers for particular mention. However, recent papers of particular note include those by Banerjee et al. (1980); Howard & Hough (1979); Jarvis (1980b, 1984); Lucas et al. (1980a, b, c); McArthur (1978, 1985); McClellan & Saavedra (1986); Nathan et al. (1979); Prévôt & Lucas (1980); Donnelly et al. (1988a, b).

In the same way there has been an increase in our knowledge and understanding of the organic geochemistry of phosphorites through the work of Amit & Bein (1982), Belayouni & Trichet (1983, 1984), Sandstrom (1986) and others. One of the notable features has been the recognition of the abundance of humic acids in phosphorites (Nathan 1990) and the occurrence of an aliphatic hydrocarbon distribution indicative of extensive microbial activity.

Phosphorite research in the future

After 10 years of concentrated research into many facets of phosphate geology, a number of important questions have been answered but, as is inevitable, many remain to be answered. So what are some of the directions for the future?

Without a doubt there is a need for broader application of the IGCP 156 results, particularly to phosphate exploration. Many countries in the world still fall into the phosphate 'have-not' category. Application of Project 156 results may help to change this.

Despite a decade of work there is a need to collect yet more scientific data. Nowhere is this more evident than in the Ocean Drilling Project where there is routine analysis of parameters such as organic carbon and carbonate carbon, but not of phosphate, which is so commonly the biolimiting element. A large database of C/P analyses would be of great value in determining variations in the P flux through time.

The link between phosphorus and other diagenetic minerals, some of them (e.g. zeolites) of potential economic significance, is yet to be fully understood. There is a need to study both modern and ancient P-diagenetic mineral assemblages.

As pointed out earlier, there have been great advances in the understanding of the geochemistry of phosphorites in recent years, but there is need for more work in areas such as:
- cadmium and other 'biologically active' elements;
- phosphorites as providers of palaeothermometry and palaeoenvironmental data;
- experimental systems including a range of micro-organisms.

Further study of the link between phosphogenesis and petroleum source rocks is likely to lead to new insights into petroleum generation in high-productivity systems.

A number of phosphogenic provinces and episodes have been extensively studied in the past decade. However, there is still scope for further work:
- the Tethyan province requires more research, particularly on biochronology and regional palaeogeography;
- the Precambrian phosphorites are still not fully understood; for example, their ages are imperfectly known and their association with banded iron formations poorly understood;
- the Neogene is one of the major phosphogenic episodes, yet in some parts of the world, such as Australia and New Zealand, sequences of the right type and age exist but have yet to yield major phosphorites;
- similarly, the Devonian may yet prove to contain a phosphogenic episode, but again more work is required.

In conclusion then, much has been achieved in phosphate research but much remains to be done. Phosphorites have revealed many of their secrets in the past decade, but they are likely to continue to fascinate and frustrate geologists for many decades to come. As the world's population increases, so will the importance of phosphorites for there is no other source of phosphorus in fertilizer manufacture. There is also no substitute for good science and the efforts of dedicated scientists. We are confident that in this volume there will be ample evidence of this.

We thank the Australian National University, the Bureau of Mineral Resources, East Carolina University, and Florida State University for their support over the past 10 years. The National Science Foundation and the Australian Bilateral Science and Technology agreements have provided essential funding. In addition, a large number of companies and organizations have assisted us over the years. We particularly thank the International Geological Correlation Programme, UNESCO, and IUGS for their support and encouragement.

Finally we extend our special thanks to all our friends and colleagues in Project 156 who have so generously contributed their time, knowledge, and friendship over an exciting and rewarding ten years.

P. J. Cook and J. H. Shergold publish with the permission of the Director of the Bureau of Mineral Resources, Geology and Geophysics, Canberra. P. N. Southgate (Bureau of Mineral Resources, Geology and Geophysics, Canberra) kindly provided editorial comment and the assistance of M. Bowden (Bureau of Mineral Resources, Geology and Geophysics, Canberra) is gratefully acknowledged.

References

ABED, A. M. 1988. *Regional IGCP 156 field workshop and symposium: Cretaceous and Tertiary phosphorites of Jordan.* Geology Department, East Carolina University, Greenville, N.C.

—— & FAKHOURI, K. 1990. Role of microbial processes in the genesis of Jordanian Upper Cretaceous phosphorites. *In*: NOTHOLT, A. J. G. & JARVIS, I. (eds) *Phosphorite Research and Development.* Geological Society, London, Special Publication, **52**, 193–204.

ALTSCHULER, S., CATHCART, J. B. & YOUNG, E. 1964. *The geology and geochemistry of the Bone Valley Formation and its phosphate deposits, west-central Florida.* Geological Society America Annual Meeting, Miami Beach, Guidebook 6.

AMES, L. L. 1959. The genesis of carbonate apatites. *Economic Geology*, **54**, 829–841.

AMIT, O. & BEIN, A. 1982. Organic matter in Senonian phosphorites from Israel — origin and diagenesis. *Chemical Geology*, **37**, 277–287.

AOUST, V. 1864. Formation des couches de grès par transports moleculaires de la matière du ciment. *Bulletin de la Société géologique de France*, 2nd Series, **22**, 130–138.

ARTHUR, M. A. & JENKYNS, H. C. 1981. Phosphorites and palaeoceanography. *Oceanologia Acta*, Special Publication, 83–96.

ATLAS, E. 1975. *Phosphate equilibria in seawater and interstitial waters.* PhD thesis, Oregon State University.

—— & PYTKOWICZ, R. M. 1977. Solubility behaviour of apatites in seawater. *Limnology & Oceanography*, **22**, 290–300.

AUSTEN, R. A. C. 1848. On the position in the Cretaceous series of beds containing phosphate of lime. *Quarterly Journal of the Geological Society of London*, **16**, 256–262.

BALSON, P. C. 1990. Episodes of phosphogenesis and phosphorite concretion formation in the North Sea Tertiary. *In*: NOTHOLT, A. J. G. & JARVIS, I. (eds) *Phosphorite Research and Development.* Geological Society, London, Special Publication, **52**, 125–137.

BANE, J. M. & BROOKS, D. A. 1979. Gulf Stream meanders along the continental margin from the Florida Straits to Cape Hatteras. *Geophysical Research Letters*, **6**, 280–282.

BANERJEE, D. M. 1971. Precambrian stromatolitic phosphorite of Udaipur, Rajasthan, India. *Geological Society of America Bulletin*, **82**, 2319–2330.

—— & KHAN, N. W. Y. 1981. *Annotated bibliography of phosphate geology in India*. Department of Geology University of Delhi, Publication 3.

——, BASU, P. C. & SRIVASTAVA, N. 1980. Petrology, mineralogy and origin of the Precambrian Aravallian phosphorite deposits of Udaipur and Jhabua, India. *Economic Geology*, **75**, 8, 1181–1199.

BATURIN, G. N., MERKULOVA, K. I. & CHALOV, P. I. 1972. Radiometric evidence for recent formation of phosphatic nodules in marine shelf sediments. *Marine Geology*, **13**, M37–M41.

BELAYOUNI, H. & BEJI-SASSI, A. 1987. *Tenth international IGCP 156 field workshop and symposium: genesis of Tethyan phosphorites and associated petroleum source-rocks*. Faculty of Sciences, University of Tunis, Tunisia.

—— & TRICHET, J. 1983. Preliminary data on the origin and diagenesis of the organic matter in the phosphate basin of Gafsa (Tunisia). *In*: BJOROY, M. (ed.) *Advances in Organic Geochemistry 1981*. Pergamon Press, Oxford, 328–335.

—— & —— 1984. Hydrocarbons in phosphatised and non-phosphatised sediments from the phosphate basin of Gafsa. *Organic Geochemistry*, **6**, 741–754.

BENTOR, Y. K. 1953. Relations entre la tectonique et les dépôts de phosphates dans le Neguev israélien. *19th Session of the International Geological Congress, Algiers, 1952, Section 11*, 93–101.

—— 1980. Phosphorites – the unsolved problems. *In*: Y. K. Bentor (ed.) *Marine Phosphorites. Geochemistry, Occurrence, Genesis*. Society of Economic Paleontologists and Mineralogists, Special Publication, **29**, 3–18.

BIDAUT, H. 1953. Note préliminaire sur un mode de formation possible des phosphates dinantiens des Pyrénées. *19th Session of the International Geological Congress, Algiers, 1952, Section 11*, 185–190.

BLACKWELDER, E. 1910. Phosphate deposits east of Ogden, Utah. *Bulletin of the US Geological Survey*, **430**, 536–551.

—— 1911. A reconnaissance of the phosphate deposits in western Wyoming. *Bulletin of the US Geological Survey*, **470**, 452–481.

—— 1915. Origin of the Rocky Mountain phosphate deposits (Abstracts). *Geological Society of America Bulletin*, **26**, 100–101.

—— 1916. The geological role of phosphorus. *American Journal of Science*, series 4, **42**, 282–298.

BRASIER, M. D. 1979. The Cambrian radiation event. *In*: HOUSE, M. R. (ed.) *The origin of major invertebrate groups*. Systematics Association Special Volume, **12**, 103–159.

—— 1982. Sea level changes, facies changes, and the Late Precambrian–Early Cambrian evolutionary explosion. *Precambrian Research*, **17**, 105–123.

—— 1985. Evolutionary and geological events across the Precambrian–Cambrian boundary. *Geology Today*, Sept–Oct 1985, 141–146.

—— 1990. Phosphogenic events and skeletal preservation across the Precambrian/Cambrian boundary interval. *In*: NOTHOLT, A. J. G. & JARVIS, I. (eds) *Phosphorite Research and Development*. Geological Society, London, Special Publication, **52**, 289–303.

BREGER, C. L. 1911. Origin of Lander oil and Western Phosphate. *Mining and Engineering World*, **35**, 631–633.

BRIDGES, N. J., JONES, M. Y., LEE, A. I. N., BOWEN, R. W. & SHELDON, R. P. 1983. *Bibliography of the geology of sedimentary phosphorite and igneous apatite*. US Geological Survey, Open File Report, 83–841.

BRIVES, A. 1908. Sur le Sénonien et l'Eocène de la bordure nord de l'atlas marocain. *Comptes Rendus de l'Académie des Sciences, Paris*, **146**, 873–875.

BRODIE, P. B. 1866. On a deposit of phosphatic nodules in the Lower Greensand, at Sandy, Bedfordshire. *Geological Magazine*, **3**, 153–155.

BRODSKAYA, N. G. & ILYINSKAYA, M. N. 1968. Phosphate accumulation in volcanic regions. *In*: *Sedimentology and useful fossils of past volcanic regions, 2*. Proceedings of the Geological Institute, USSR Academy of Sciences, **196**, 193–292.

—— & —— 1970. Principal genetic types of phosphate deposits associated with an endogenic source of phosphorus. *In*: *State and problems of Soviet lithology*. All Union Lithological Conference, 8th Report, **3**, 257–262.

BROOKS, D. A. & BANE, J. M. 1978. Gulf Stream deflection by a bottom feature off Charleston, South Carolina. *Science*, **201**, 1225–1226.

BUCKLAND, W. 1829. On the discovery of coprolites, or fossil faeces, in the Lias at Lyme Regis, and in other formations. *Transactions of the Geological Society of London*, **3**, 223–236.

—— 1843. On the causes of the general presence of phosphates in the strata of the earth and in all fertile soils. *Journal of the Royal Agricultural Society*, **10**, 520–525.

BURNETT, W. C., BEERS, M. J. & ROE, K. K. 1982. Growth rates of phosphate nodules from the continental margin off Peru. *Science*, **215**, 1616–1618.

—— & FROELICH, P. N. 1988. Preface. *In*: BURNETT, W. C. & FROELICH, P. N. (eds) *The Origin of Marine Phosphorite. The Results of the R. V. Robert D. Conrad Cruise 23–06 to the Peru Shelf*. Marine Geology, **80**, iii–vi.

—— & RIGGS, S. R. (eds) 1990. *Neogene to Modern Phosphorites. Phosphate Deposits of the World* Vol. 3, Cambridge University Press, Cambridge. (in press).

—— & SHELDON, R. P. 1979. *Report on the Marine Phosphatic Sediments Workshop*. Resource Sys-

tems Institute, East-West Center, Honolulu.
——, VEEH, H. H. & SOUTAR, A. 1980. U-series, oceanographic, and sedimentary evidence in support of contemporary formation of phosphate nodules off Peru. *Society of Economic Paleontologists and Mineralogists, Special Publication*, **29**, 61–71.
BUSHINSKII, G. I. 1935. Structure and origin of the phosphorites of the USSR. *Journal of Sedimentary Petrology*, **5**, 2, 81–92.
—— 1964. On shallow water origin of phosphorite sediments. *In*: VAN STRAATEN, L. M. J. (ed.) *Deltaic and shallow marine deposits*. Elsevier, Amsterdam, 62–70.
—— 1966. *Drevniye fosfority azii i ikh genezis*. (Old phosphorites of Asia and their genesis). USSR Academy of Sciences, Geological Institute, Transactions, 149.
CAMERON, F. K. & HURST, L. A. 1904. Action of water and saline solutions upon certain slightly soluble phosphates. *Journal of the American Chemical Society*, **26**, 855.
—— & SIEDEL, A. 1904. Action of water on phosphates of calcium. *Journal of the American Chemical Society*, **26**, 14–54.
CATHCART, J. B. 1968. Phosphate in the Atlantic and Gulf coastal plains. 4th Forum on geology of industrial minerals. *Bureau of Economic Geology University of Texas, Proceedings*, 23–34.
—— & McGREEVY, K. J. 1959. Results of geologic exploration by core drilling, 1953 land-pebble phosphate district, Florida. *Bulletin of the US Geological Survey*, **1046-K**, K221–K298.
CAYEUX, L. 1896. Note préliminaire sur la constitution des phosphates de chaux Senonien du Sud de la Tunisie. *Comptes Rendus de l'Académie des Sciences, Paris*, **123**, 273–276.
—— 1932. Les manières d'être de la glauconie en milieu calcaire, *Comptes Rendus de l'Académie des Sciences, Paris*, **195**, 1050–1052.
—— 1934. The phosphate nodules of Agulhas Bank. *South African Museum Annual*, **31**, 105–136.
CHINA NATIONAL COMMITTEE FOR IGCP. 1982. *Guidebook to Field Excursions*. Fifth International Field Workshop and Seminar on Phosphorite, Kunming, China, November 17–24, 1982.
—— 1984. *International Geological Correlation Programme Project 156-Phosphorite. Symposium of 5th International Field Workshop and Seminar on Phosphorite*. Vols 1 and 2. Geological Publishing House, Beijing, China.
CHOUDHURI, R. 1990. Two decades of phosphorite investigations in India. *In*: NOTHOLT, A. J. G. & JARVIS, I. (eds) *Phosphorite Research and Development*. Geological Society, London, Special Publication, **52**, 305–311.
COOK, P. J. 1982. World availability of phosphorus: an Australian perspective. *In*: COSTIN, A. B. & WILLIAMS, C. H. (eds), *Phosphorus in Australia*. Centre for Resource and Environmental Studies, Australian National University, Monograph, **8**, 6–41.
—— & COOK, J. R. 1985. Marine biological changes and phosphogenesis around the Cretaceous–Tertiary boundary. *Sciences Géologiques, Mémoire*, **77**, 105–108.
—— & MARSHALL, J. F. 1981. Geochemistry of iron and phosphorus-rich nodules from the east Australian continental shelf. *Marine Geology*, **41**, 205–221.
—— & McELHINNY, M. W. 1979. A re-evaluation of the spatial and temporal distribution of sedimentary phosphate deposits in the light of plate tectonics. *Economic Geology*, **74**, 315–330.
—— & SHERGOLD, J. H. 1979a. The field workshop. *In*: COOK, P. J. & SHERGOLD, J. H. (eds) *Proterozoic–Cambrian phosphorites*. Australian National University Press, Canberra, 1–17.
—— & —— 1979b. Proterozoic and Cambrian phosphorites of Australia and Asia – a progress report. *In*: SHELDON, R. P. & BURNETT, W. C. (eds). *Fertilizer Mineral Resource Potential in Asia and the Pacific*. Proceedings of the Fertilizer Raw Materials Resource Workshop, August 24–29, Honolulu, Resource Systems Institute, East–West Center Honolulu, 207–233.
—— & —— 1984a. Phosphorus, phosphorites, and skeletal evolution at the Precambrian–Cambrian boundary, *Nature*, **308**, 231–236.
—— & —— 1984b. *Late Proterozoic–Cambrian phosphorites and phosphogenesis*. Proceedings of the 27th International Geological Congress, VNU Press, Utrecht, 397–444.
—— & —— (eds) 1986a. *Phosphate Deposits of the World. Volume 1, Proterozoic and Cambrian Phosphorites*. Cambridge University Press, Cambridge.
—— & —— 1986b. Proterozoic and Cambrian phosphorites – nature and origin. *In*: COOK, P. J. & SHERGOLD, J. H. (eds) *Phosphate Deposits of the World, 1: Proterozoic and Cambrian Phosphorites*. Cambridge University Press, Cambridge, 369–386.
—— & —— 1986c. Introduction (Chapter 1). *In*: COOK, P. J. & SHERGOLD, J. H. (eds), *Phosphate Deposits of the World, 1: Proterozoic and Cambrian Phosphorites*. Cambridge University Press, Cambridge, 1–8.
DARTON, N. H. 1891. Notes on the geology of the Florida phosphate deposits. *American Journal of Science*, **141**, 102–105.
DAUBENY, C. & WIDDRINGTON, C. 1845. On the occurrence of phosphorite in Estremadura. *Quarterly Journal of the Geological Society of London*, **1**, 52–55.
DAVIES, D. C. 1868. On the deposits of lime recently discovered in Nassau, North Germany. *Geological Magazine*, **5**, 262–266.
—— 1875. The phosphorite deposits of North Wales. *Quarterly Journal of the Geological Society of London*, **31**, 357–364.
DAWSON, J. W. 1876. Note on the phosphates of the Laurentian and Cambrian rocks of Canada. *Quarterly Journal of the Geological Society of London*, **32**, 285–291.
DAY, D. T. 1885. *Mineral resources of the United States, 1883 and 1884*. Government Printing Office, Washington, 783–808.

—— 1887. *Mineral resources of the United States. Calendar year 1886.* Government Printing Office, Washington, 606–620.

DE KEYSER, F. & COOK, P. J. 1973. The geology of the Middle Cambrian phosphorites and associated sediments of northwest Queensland. *Bureau of Mineral Resources of Australia, Bulletin*, **138**.

DE LA BECHE, H. 1849. Phosphate of lime in greensand and marl. *American Journal of Science*, **58**, 422–424.

DE LUNA, M. R. 1865. On considerable deposits of phosphate of lime at Caceres, Estremadura. *Geological Magazine*, **2**, 446.

DIETZ, R. S., EMERY, K. O. & SHEPARD, F. P. 1942. Phosphorite deposits on the sea floor off California. *Geological Society of America Bulletin*, **53**, 815–848.

DONNELLY, T. H., SHERGOLD, J. H. & SOUTHGATE, P. N. 1988a. Pyrite and organic matter in normal marine sediments of Middle Cambrian age, southern Georgina Basin, Australia. *Geochimica et Cosmochimica Acta*, **52**, 259–263.

——, —— & —— 1988b. Anomalous geochemical signals from phosphatic Middle Cambrian rocks in the southern Georgina Basin, Australia. *Sedimentology*, **35**, 549–570.

——, ——, —— & BARNES, C. J. 1990. Events leading to global phosphogenesis around the Precambrian/Cambrian boundary. *In*: NOTHOLT, A. J. G. & JARVIS, I. (eds) *Phosphorite Research and Development.* Geological Society, London, Special Publication, **52**, 273–287.

DUNBAR, R. B. & BARKER, P. A. 1988. *Regional IGCP 156 field workshop: genesis of Cenozoic phosphorites and associated organic-rich sediments: Peruvian continental margin.* Geology Department, Rice University, Houston, Texas.

DUNHAM, R. J. 1962. Classification of carbonate rocks according to depositional texture. In Classification of carbonate rocks – a symposium. *American Association of Petroleum Geologists, Memoir* **1**, 108–121.

EGANOV, E. A. 1988. *Phosphate deposition and stromatolites.* Academy of Sciences of the USSR, Siberian Division, Institute of Geology and Geophysics. (In Russian).

ELGUETA, S. 1981. *Sedimentological study of the western zone of the Lady Annie phosphate deposit, Queensland, Australia.* MSc thesis, Australian National University.

FISCHER, A. G. & ARTHUR, M. A. 1977. Secular variations in the pelagic realm. *Society of Economic Paleontologists and Mineralogists, Special Publication*, **25**, 19–50.

FISHER, O. 1873. On the phosphatic nodules of the Cretaceous rock of Cambridgeshire. *Quarterly Journal of the Geological Society of London*, **29**, 52–63.

FLICOTEAUX, R. 1982. *Génèse des phosphates alumineux du Sénégal occidental – Etapes et guides de l'altération.* Sciences Geologiques, Mémoires, 67.

——, NAHON, D. & PAQUET, H. 1977. Génèse des phosphates alumineux à partir des sédiments argilo-phosphatés du Tertiaire de Lam Lam (Sénégal). *Sciences Géologiques, Bulletin*, **30**, 153–174.

FOLK, R. L. 1959. Practical petrographic classification of limestones. *American Association of Petroleum Geologists, Bulletin*, **43**(1), 1–38.

—— 1965. On the earliest recognition of coprolites. *Journal of Sedimentary Petrology*, **35**, 272–273.

FÖLLMI, K. B. 1990. Condensation and phosphogenesis: example of the Helvetic mid-Cretaceous (northern Tethyan margin). *In*: NOTHOLT, A. J. G. & JARVIS, I. (eds) *Phosphorite Research and Development.* Geological Society, London, Special Publication, **52**, 237–252.

FREAS, D. H. 1968. Exploration for Florida phosphate deposits. *In: Proceedings of the seminar on sources of mineral raw material for the fertilizer industry in Asia and the Far East.* United Nations Mineral Resources Development Series, **32**, 187–200.

FREAS, D. L. & RIGGS, S. R. 1965. Stratigraphy and sedimentation of phosphorite in the Central Florida phosphate district. *American Institute of Mining Engineers*, **65H84**, 1–13.

FREAS, D. L. & RIGGS, S. R. 1968. Environments of phosphorite deposition in Central Florida phosphate district. *In*: BROWN, L. F. (ed.) *Fourth forum on geology of industrial minerals.* University of Texas at Austin, and Bureau of Economic Geology, 117–128.

GALE, H. S. & RICHARDS, R. W. 1909. Preliminary report on the phosphate deposits of southeastern Idaho and adjacent parts of Wyoming and Utah. *Bulletin of the US Geological Survey*, **430**, 457–535.

GEIJER, P. 1962. Some aspects of phosphorus in Precambrian sedimentation. *Arkiv for Mineralogie och Geologi*, **3**, 165–86.

GLENN, C. R. 1990. Depositional sequences of the Duwi, Sibaiya and Phosphate Formations, Egypt: phosphogenesis and glauconitization in a Late Cretaceous epeiric sea. *In*: NOTHOLT, A. J. G. & JARVIS, I. (eds) *Phosphorite Research and Development.* Geological Society, London, Special Publication, **52**, 205–222.

GRUNER, M. L. 1871. Note sur les nodules phosphates de la Porte du Rhône. *Bulletin de la Sociétié géologique de France*, 2nd Series, **28**, 62–72.

GULBRANDSEN, R. A. 1969. Physical and chemical factors in the formation of marine apatite. *Economic Geology*, **64**, 365–382.

HAYES, C. W. 1896. *The Tennessee phosphates.* 17th Annual Report of the US Geological Survey.

HEGGIE, D. T., SKYRING, G. W., O'BRIEN, G. W., REIMERS, C., HERCZEG, A., MORIARTY, D. J. W., BURNETT, W., & MILNES, A. R. 1990. Organic carbon cycling and modern phosphorite formation on the East Australian continental margin: an overview. *In*: NOTHOLT, A. J. G. & JARVIS, I. (eds) *Phosphorite Research and Development.* Geological Society, London, Special Publication, **52**, 87–117.

HENDERSON, R. A., CUFF, C. & SOUTHGATE, P. N. 1979. A brine-leaching mechanism of phosphorite genesis. *In*: COOK, P. J. & SHERGOLD, J. H. (eds)

Proterozoic and Cambrian Phosphorites. Australian National University Press, Canberra, 22.

HICKS, H. 1875. On the occurrence of phosphates in the Cambrian rocks. *Quarterly Journal of the Geological Society of London*, **31**, 368–383.

HINE, A. C. & RIGGS, S. R. 1986. Geologic framework, Cenozoic history and modern processes of sedimentation on the North Carolina continental margin. *In*: TEXTORIS, D. A. (ed.) *SEPM Field Guidebooks, Southeastern United States Third Annual Meeting*. Society of Economic Paleontologists & Mineralogists, Tulsa, 129–194.

HITE, R. J. 1978. Possible genetic relationships between evaporites, phosphorites, and iron-rich sediments. *Mountain Geologist*, **14**, 97–107.

HOLMES, F. S. 1870. *Phosphate rocks of south Carolina and the great Carolina marl bed*. Holmes Book House, Charleston, S. C.

HOWARD, P. F. 1979. Phosphate. *Economic Geology*, **74**, 192–194.

—— 1990. The distribution of phosphatic facies in the Georgia, Wiso and Daly River Basins, Northern Australia. *In*: NOTHOLT, A. J. G. & JARVIS, I. (eds) *Phosphorite Research and Development*. Geological Society, London, Special Publication, **52**, 261–272.

—— & COONEY, A. M. 1976. D Tree phosphate deposit, Georgina Basin, Queensland. *In*: KNIGHT, C. L. (ed.) *Economic Geology of Australia and Papua New Guinea, 4, Industrial Minerals and Rocks*. Australasian Institute of Mining and Metallurgy, Monograph, **8**, 265–273.

—— & HOUGH, M. J. 1979. On the geochemistry and origin of the D Tree, Wonarah, and Sherrin Creek phosphorite deposits of the Georgina Basin, northern Australia. *Economic Geology*, **74**, 260–284.

—— & PERRINO, F. A. 1976. Wonarah phosphate deposits, Georgina Basin, Northern Territory. *In*: KNIGHT, C. L. (ed.) *Economic Geology of Australia and Papua New Guinea, 4, Industrial Minerals and Rocks*. Australasian Institute of Mining and Metallurgy, Monograph, **8**, 273–277.

HUDLESTON, W. H. 1875. On the chemical analyses of the rocks (appendix to paper by Henry Hicks). *Quarterly Journal of the Geological Society of London*, **31**, 376–385.

HUNT, T. S. 1852. Examinations of phosphatic matters, supposed bones and coprolites, occurring in Lower Silurian rocks of Canada. *Quarterly Journal of the Geological Society of London*, **8**, 209–210.

ILYIN, A. V. & BJAMBA, Z. 1980. *Guidebook for Excursion, Phosphorites of the Khubsugul Basin, Mongolian People's Republic*. Field conference of IGCP Project 156. Geological Institute, USSR, Academy of Sciences, Moscow (In Russian and English).

INDIAN NATIONAL COMMITTEE FOR IGCP 1981*a*. *Guidebook for Excursion, Aravalli Phosphorites around Udaipur, Rajasthan, India*. Fourth International Field Workshop and Seminar of IGCP Project 156 — Phosphorite, Geological Survey of India, Jaipur.

—— 1981*b*. *Excursion Guidebook on Dehradun–Mussoorie area*. Fourth International Field Workshop and Seminar on Phosphorite. International Geological Correlation Programme Project 156 — Phosphorites, Geological Survey of India, Northern Region.

JANOWITZ, G. S. PIETRAFESA, L. J. 1980. A model and observations of time dependent upwelling over the mid-shelf and slope. *Journal of Physical Oceanography*, **10**, 1574–1583.

—— & —— 1982. The effects of a longshore variation in bottom topography on a boundary current or topographically induced upwelling. *Continental Shelf Research*, **1**, 123–141.

JARVIS, I. 1980*a*. Geochemistry of phosphatic chalks and hardgrounds from the Santonian to early Campanian (Cretaceous) of northern France. *Journal of the Geological Society of London*, **137**(6), 705–722.

—— 1980*b*. The initiation of phosphatic chalk sedimentation — the Senonian (Cretaceous) of the Anglo-Paris Basin. *In*: BENTOR, Y. K. (ed.) *Marine Phosphorites*. Society of Economic Paleontologists and Mineralogists Special Publication, **29**, 167–192.

—— 1984. Rare-earth element geochemistry of late Cretaceous chalks and phosphorites from northern France. *Geological Survey of India, Special Publication*, **17**, 179–190.

—— 1988. *Cretaceous phosphorites of southern England*. Project 156 Field excursion guidebook. 11th Field Workshop/Seminar, International Geological Correlation Program.

KAZAKOV, A. V. 1937. The phosphorite facies and the genesis of phosphorites. In Geological Investigations of Agricultural Ores. *Transactions of the Scientific Institute for Fertilizers and Insectofungicides*, **142**, 93–113 (in Russian).

KEEPING, H. 1868. Discovery of Gault with phosphatic stratum at Upware. *Geological Magazine*, **5**, 272–273.

KENNEDY, W. J. & GARRISON, R. E. 1975. Morphology and genesis of nodular chalks and hardgrounds in the Upper Cretaceous of southern England. *Sedimentology*, **22**, 311–386.

KIM, K. H. & BURNETT, W. C. 1986. Uranium-series growth history of a Quaternary phosphatic crust from the Peruvian continental margin. *Chemical Geology*, **58**, 227–244.

KOLODNY, Y. 1969. Are marine phosphorites forming today? *Nature*, **224**, 1017–1019.

—— 1981. Phosphorites. *In*: EMILIANI, C. (ed.) *The Sea, ideas and observations on progress in the study of the seas*. Wiley Interscience, New York. **7**, 981–1023.

——, LUZ, B. & NAVON, O. 1983. Oxygen isotope variations in phosphate of biogenic apatites I. Fish bone apatite-rechecking the rules of the game. *Earth & Planetary Science Letters*, **64**, 398–404.

KRAMER, J. R. 1964. Sea water: saturation with apatites and carbonates. *Science*, **146**, 637–638.

LAMBOY, M. 1990. Microbial mediation in phosphatogenesis: new data from the Cretaceous phosphatic chalks, northern France. *In*: NOTHOLT, A. J. G. & JARVIS, I. (eds) *Phosphorite Research and Development*. Geological Society, London, Special Publication, **52**, 157–167.

LEE, A. I. N. (ed) 1980. *Fertilizer Mineral Occurrences in the Asia-Pacific Region*. East-West Resources Systems Institute, East-West Center, Honolulu.

LEHR, J. R., MCCLELLAN, G. H., SMITH, J. P. & FRAZIER, A. W. 1967. Characterization of apatites in commercial phosphate rocke. *In*: *Proceedings of International Colloquium of Solid Inorganic Phosphates, Toulouse, May 16–20*, Société Chimique de France, Paris. **12**, 29–44.

LEWY, Z. 1990. Pebbly phosphate and granular phosphorite (Late Cretaceous, southern Israel) and their bearing on phosphatization processes. *In*: NOTHOLT, A. J. G. & JARVIS, I. (eds) *Phosphorite Research and Development*. Geological Society, London, Special Publication, **52**, 169–178.

LI, Y., 1986. Proterozoic and Cambrian phosphorites — regional review: China. *In*: COOK, P. J. & SHERGOLD, J. H. (eds), *Phosphate Deposits of the World, 1: Proterozoic and Cambrian Phosphorites*. Cambridge University Press, Cambridge, 42–62.

LOGAN, B. W., REZAK, R. & GINSBURG, R. N. 1964. Classification and environmental significance of algal stromatolites. *Journal of Geology*, **72**, 68–83.

LOGAN, W. E. & HUNT, T. S. 1854. On the chemical composition of Recent and fossil Lingulae and some other shells. *American Journal of Science*, **67**, 235–239.

LUCAS, J. & PRÉVÔT, L. 1981*a*. Caractères pétrographiques et microchimiques de phosphorites pre-Cambriennes du Bassin des Volta (Afrique de l'Ouest). Considérations génétiques. *Bulletin de la Société géologiques de France*, (7), XXIII(5), 5–14.

—— & —— 1981*b*. Synthèse de apatite à partir de matière phosphorée (ARN) et de calcite par voie bactérienne. *Comptes Rendus de l'Académie des Sciences, Paris*, **292**, Series II, 1203–1208.

—— & —— 1984. Synthèse de l'apatite par voie bactérienne à partir de matière organique phosphatée et de divers-carbonates de calcium dans des eaux douce et marine naturelles. *Chemical Geology*, **42**, 101–118.

—— & —— 1985*a*. Phosphorites. 6th International Field Workshop and Seminar on Phosphorites, IGCP 156. Sciences Géologiques, Mémoire, 77.

—— & —— 1985*b*. The synthesis of apatite by bacterial activity: mechanism. *In*: LUCAS, J. & PRÉVÔT, L. (eds) *Phosphorites*. Sciences Géologiques, Mémoire, **77**, 83–92.

LUCAS, J., EL FALEH, E. M. & PRÉVÔT, L. 1990. Experimental study of the substitution of Ca by Sr and Ba in synthetic apatites. *In*: NOTHOLT, A. J. G. & JARVIS, I. (eds) *Phosphorite Research and Development*. Geological Society, London, Special Publication, **52**, 33–47.

——, FLICOTEAUX, R., NATHAN, Y., PRÉVÔT, L. & SHAHAR, Y. 1980*a*. Different aspects of phosphorites weathering. *Society of Economic Paleontologists and Mineralogists, Special Publication*, **29**, 41–51.

——, PRÉVÔT, L., ATAMAN, G. & GUNDOGDU, N. 1980*b*. Mineralogical and geochemical studies of the phosphatic formations in southeastern Turkey (Mazidagi-Mardin). *In*: BENTOR, T. K. (ed.) *Marine Phosphorites*. Society of Economic Paleontologists and Mineralogists, Special Publication, **29**, 149–152.

——, —— & LAMBOY, M. 1978. Les phosphorites de la marge nord de l'Espagne. Chimie, minéralogie, génèse. *Oceanologica Acta*, **1**, 55–72.

——, —— & TROMPETTE, R. 1980*c*. Petrology, mineralogy and geochemistry of the late Precambrian phosphate deposits of Upper Volta (W Africa). *Journal of the Geological Society of London*, **137**, 787–792.

MANHEIM, F., ROWE, G. T. & JIPA, D. 1975. Marine phosphorite formation off Peru. *Journal of Sedimentary Petrology*, **45**, 743–751.

MANSFIELD, G. R. 1916. A reconnaissance for phosphate in the Salt River Range, Wyoming. *Bulletin of the US Geological Survey*, **620**, 331–349.

—— 1918. Origin of the Western phosphates of the United States. *American Journal of Science*, **46**, 592–598.

—— 1920. Geography, geology and mineral resources of the Fort Hall Indian Reservation, Idaho. *Bulletin of the US Geological Survey*, **713**.

—— 1927. Geography, geology and mineral resources of part of southeastern Idaho. *Bulletin of the US Geological Survey*, **152**.

—— 1940. The role of fluorine in phosphate deposition. *American Journal of Science*, **238**, 863–879.

MARSHALL, J. F. & COOK P. J. 1980. Petrology of iron and phosphorus-rich nodules from the East Australian continental shelf. *Journal of the Geological Society, London*, **137**, 765–771.

MARTENS, C. S. & HARRISS, R. C. 1970. Inhibition of apatite precipitation in the marine environment by magnesium ions. *Geochimica et Cosmochimica Acta*, **34**, 621–625.

MCARTHUR, J. M. 1978. Systematic variations in the contents of Na, Si, CO_3 and SO_4 in marine carbonate fluorapatite and their relation to weathering. *Chemical Geology*, **21**, 89–112.

—— 1985. Francolite, geochemistry: compositional controls during formation, diagenesis, metamorphism and weathering. *Geochimica et Cosmochimica Acta*, **49**, 23–35.

——, BENMORE, R. A., COLEMAN, M. L., SOLDI, C., YEH, H. W. & O'BRIEN, G. W. 1986. Stable isotope characterization of francolite formation. *Earth and Planetary Science Letters*, **77**, 20–34.

——, COLEMAN, M. L. & BREMNER, J. M. 1980. Carbon and oxygen isotope composition of structural CO_2 in sedimentary francolite. *Journal of the Geological Society, London*, **137**, 669–673.

——, HAMILTON, P. J., GREENSMITH, J. T, BOYCE, A. J., FALLICK, A. E., BIRCH, G., WALSH, J. N., BENMORE, R. A. & COLEMAN, M. L. 1987. Phosphorite geochemistry; isotopic evidence for meteoric alteration of francolite on a local scale. *Chemical Geology*, **65**, 415–425.

——, THOMSON, J., JARVIS, I., FALLICK, A. E. & BIRCH, G. F. 1988. Eocene to Pleistocene phosphogenesis off western South Africa. *Marine Geology*, **85**, 41–63.

MCCLELLAN, G. H. 1980. Mineralogy of carbonate fluorapatite. *Journal of the Geological Society, London*, **137**, 675–682.

—— & GREMILLION, L. R. 1980. Evaluation of phosphatic raw materials. *In*: KHASAWNEH, F. E., SAMPLE, E. C. & KAMPRATH, E. J. (eds) *The Role of Phosphorus in Agriculture*. Madison, Wisconsin, 43–80.

—— & SAAVEDRA, F. N. 1986. Proterozoic and Cambrian phosphorites — specialist studies: chemical and mineral characteristics of some Cambrian and Precambrian phosphorites. *In*: COOK, P. J. & SHERGOLD, J. H. (eds) *Phosphate Deposits of the World, 1: Proterozoic and Cambrian Phosphorites*. Cambridge University Press, Cambridge, 244–267.

—— & VAN KAUWENBERGH, S. J. 1990. Mineralogy of sedimentary apatites. *In*: NOTHOLT, A. J. G. & JARVIS, I. (eds) *Phosphorite Research and Development*. Geological Society, London, Special Publication, **52**, 23–31.

MCCONNELL, D. 1965. Precipitation of phosphates in sea-water. *Economic Geology*, **60**, 1059–1062.

—— 1973. *Apatite: its crystal chemistry, mineralogy, utilization, and geologic and biologic occurrences*. Springer-Verlag, New York.

——, FRAJOLA, W. J. & DEAMER, D. W. 1961. Relation between inorganic chemistry and biochemistry of bone mineralization. *Science*, **133**, 28–282.

MCKELVEY, V. E., SWANSON, R. W. & SHELDON, R. P. 1953. The Permian phosphorite deposits of western United States. *19th Session of the International Geological Congress, Algiers, 1952, Section 11*, 45–64.

MCKELVEY, V.E., WILLIAMS, J. S., SHELDON, R. P., CRESSMAN, E. R., CHENEY, T. M. & SWANSON, R. W. 1959. *The Phosphoria, Park City and Shedhorn Formations in the Western phosphate field*. US Geological Survey, Professional Paper, 313-A.

MEUGY, A. 1855. Phosphate de chaux en nodules dans la craie de Rethel (Ardennes). *Bulletin de la Société géologique de France Series 2*, **13**, 604–605.

MILLAR, C. C. H. 1892. *Florida, South Carolina and Canadian phosphates*. Scientific Publishing Co. New York.

MURRAY, J. & RENARD, A. F. 1891. *Scientific Results, H. M. S. Challenger, Deep Sea Deposits*. Longman, London, 391–400.

NATH MUKTI & SANT, V. N. 1967. Occurrence of algal phosphorite in the pre-Cambrian rocks of Rajasthan, *Current Science*, **36**, (23).

NATHAN, Y. 1990. Humic substances in phosphorites: occurrence characterization and significance. *In*: NOTHOLT, A. J. G. & JARVIS, I. (eds) *Phosphorite Research and Development*. Geological Society, London, Special Publication, **52**, 49–58.

—— & LUCAS, J. 1972. Synthése de l'apatite à partir du gypse; application au probléme de la formation des apatites carbonatées par précipitation directe. *Chemical Geology*, **9**, 99–112.

—— & —— 1976. Experiments on the direct precipitation of apatite in sea water; implication in the genesis of phosphorites, *Chemical Geology*, **18**, 181–186.

——, SHILONI, Y., RODED, R., GAL, I. & DEUTSCH, Y. 1979. *The geochemistry of the northern and central Negev phosphorites*. Bulletin of the Israel Geological Survey 73.

——, SOUDRY, D. & AVIGOUR, A. 1990. Geological significance of carbonate substitution in apatites: Israeli phosphorites as an example. *In*: NOTHOLT, A. J. G. & JARVIS, I. (eds) *Phosphorite Research and Development*. Geological Society, London, Special Publication, **52**, 179–191.

NOTHOLT, A. J. G. 1978. Proterozoic and Cambrian phosphorites — Inaugural Field Workshop and Seminar in Australia. *Mining Magazine*, November 1978, 497–499.

—— 1980a. Economic phosphatic sediments; mode of occurrence and stratigraphical distribution. *Journal of the Geological Society, London*, **137**, 793–805.

—— 1980b. Phosphate rock: a strategic mineral for Western Europe, *In*: JONES, M. J. (ed.) *National and international management of mineral resources*. Institution of Mining and Metallurgy, London, 281–291.

—— & HARTLEY, K. 1983. *Phosphate rock: a bibliography of world resources*. Mining Journal Books, London.

——, JARVIS, I. & BURNETT, W. C. 1988. *Eleventh international IGCP 156 field workshop and symposium: a decade of phosphorite research and development*. Hertford College, Oxford, England, (Abstracts).

——, SHELDON, R. P. & DAVIDSON, D. F. (eds) 1989. *Phosphate Deposits of the World, Volume 2, Phosphate Rock Resources*. Cambridge University Press, Cambridge, 600p.

NRIAGU, J. O. & MOORE, P. B. (eds) 1984. *Phosphate Minerals*. Springer-Verlag, Berlin.

O'BRIEN, G. W. & VEEH, H. H. 1980. Holocene phosphorite on the East Australian continental margin. *Nature*, **288**, 690–691.

——, HARRIS, J. R., MILNES, A. R. & VEEH, H. H. 1981. Bacterial origin of East Australian continental margin phosphorites. *Nature*, **294**, 442–444.

——, MILNES, A. R., VEEH, H. H., HEGGIE, D. T., RIGGS, S. R., Cullen, D. J., Marshall, J. F. & Cook, P. J. 1990. Sedimentation dynamics and redox iron-cycling: controlling factors for the apatite glauconite association on the East Australian continental margin. *In*: NOTHOLT, A. J. G. & JARVIS, I. (eds) *Phosphorite Research and Development*. Geological Society, London, Special Publication, **52**, 61–68.

PARDEE, J. T. 1917. The Garrison and Philipsburg phosphate fields, Montana. *Bulletin of the US*

Geological Survey, **640**, 195–228.
PARRISH, J. T. 1981. Global atmospheric circulation in the Mesozoic and Cenozoic. *Bulletin of the American Association of Petroleum Geology*, **65**, 969 (Abstract).
—— 1981. Upwelling and petroleum source beds, with reference to Paleozoic. *Bulletin of the American Association of Petroleum Geology*, **66**, 750–774.
——, ZIEGLER, A. M., SCOTESE, C. R., HUMPHREVILLE, R. G. & KIRSCHVINK, J. L. 1986. Proterozoic and Cambrian phosphorites – specialist studies: Early Cambrian palaeogeography, palaeoceanography and phosphorites. *In*: COOK, P. J. & SHERGOLD, J. H. (eds) *Phosphate Deposits of the World, 1: Proterozoic and Cambrian Phosphorites*. Cambridge University Press. Cambridge, 280–294.
PASHO, D. W. 1972. *Character and Origin of Marine Phosphorites* MSc thesis, University of Southern California.
PENROSE, R. A. F. 1988. Nature and origin of deposits of phosphate of lime. *Bulletin of the US Geological Survey*, **46**.
PEVEAR, D. R. 1966. The estuarine formation of United States Atlantic coastal plain phosphorites. *Economic Geology*, **61**, 251–266.
PIETRAFESA, L. J. 1983. Shelfbreak circulation, fronts, and physical oceanography: east and west coast perspectives. *Society of Economic Paleontologists and Mineralogists, Special Publication*, **33**, 233–250.
——, ATKINSON, L. P. & BLANTON, J. O. 1978. Evidence for deflection of the Gulf Stream by the Charleston Rise. *Gulf Stream*, **4**, 3–7.
PIPER, D. Z. & KOLODNY, Y. 1987. The stable isotope composition of a phosphorite deposit: delta 13 C, delta 34 S, delta 18 O. *Deep Sea Research*, **34**, 897–911.
POPENOE, P. 1990. Paleoceanography and paleogeography of the Miocene of the southeastern United States. *In*: BURNETT, W. C. & RIGGS, S. R. (eds) *Phosphate Deposits of the World, 3, Neogene to Modern Phosphorites*. Cambridge University Press, Cambridge. (in press).
PRÉVÔT, L. 1982. Proposal for a normalised easy description of the so-called palaeo-phosphorites. *Project 156, Phosphorite Newsletter*, Canberra, **10**, 24–31.
PRÉVÔT, L. & LUCAS, J. 1980 Behaviour of some trace elements in phosphatic sedimentary formations. *In*: BENTOR, Y. K. (ed.) *Marine Phosphorites*. Society of Economic Paleontologists and Mineralogists, Special Publication, **29**, 31–39.
—— & LUCAS, J. 1986. Microstructure of apatite – replacing carbonate in synthesized and natural samples. *Journal of Sedimentary Petrology*, **56**, 153–159.
REISS, Z. 1962. Stratigraphy of phosphate deposits in Israel. *Bulletin of the Israel Geological Survey*, **34**, 1–23.
RICHARDS, R. W. & MANSFIELD, G. R. 1914. Geology of the phosphate deposits, northeast of Georgetown, Idaho. *Bulletin of the US Geological Survey*, **577**.
RIGGS, S. R. 1967. *Phosphorite stratigraphy sedimentation, and petrology of the Central Florida Phosphate District*. PhD Thesis, University of Montana, Missoula.
—— 1979a. Petrology of the Tertiary phosphorite system Florida. *Economic Geology*, **74**, 195–220.
—— 1979b. Phosphorite sedimentation in Florida – a model phosphogenic system. *Economic Geology*, **74**, 285–314.
—— 1984. Paleoceanographic model of Neogene phosphorite deposition, US Atlantic continental margin. *Science*, **223**, 123–131.
—— 1986. Proterozoic and Cambrian phosphorites – specialist studies: phosphogenesis and its relationship to exploration for Proterozoic and Cambrian phosphorites. *In*: COOK, P. J. & SHERGOLD, J. H. (eds) *Phosphate Deposits of the World: 1, Proterozoic and Cambrian Phosphorites*. Cambridge University Press, Cambridge, 352–368.
—— & MALLETTE, P. M. 1990. Patterns of phosphate deposition and lithofacies relationships within the Miocene Pungo River Formation, North Carolina continental margin. *In*: BURNETT, W. C. & RIGGS, S. R. (eds) *Phosphate Deposits of the World: 3, Neogene to Modern Phosphorites*. Cambridge University Press, Cambridge. (in press).
—— & SNYDER, S. W. 1986. Patterns of cyclic sedimentation of the Upper Cenozoic section, North Carolina coastal plain. *In*: TEXTORIS, D. A. (ed.) *SEPM Field Guidebooks, Southeastern United States Third Annual Midyear Meeting*. Society of Economic Paleontologists and Mineralogists, Tulsa, 333–372.
——, LEWIS, D. W., SCARBOROUGH, A. K. & SNYDER, S. W. 1982a. Cyclic deposition of Neogene phosphorites in the Aurora area, North Carolina, and their possible relationship to global sea-level fluctuations. *Southeastern Geology*, **23**, 189–204.
——, SNYDER, S. W., ELLINGTON, M. D. & BURNETT, W. D. 1982b. Pleistocene/Holocene phosphorite formation – North Carolina continental margin. *Geological Society of America, Abstracts with programs*, **14**, 77.
——, ——, HINE, A. C., ELLINGTON, M. D. & MALLETTE, P. M. 1985. Geological framework of phosphate resources in Onslow Bay, North Carolina continental shelf. *Economic Geology*, **80**, 716–738.
ROBASZYNSKI, F. & SUSTRAC, G. 1988. *Elevenn international IGCP 156 field excursion: Cretaceous phosphorites of northern France and Belgium*. Faculté Polytechnique de Mons, Belgium.
ROBERSON, C. E. 1966. Solubility implications of apatite in seawater. *US Geological Survey, Professional Paper*, **550–D**, 178–185.
RODRIGUEZ, S., USECHE, A., CARDENAS, H., GONZALEZ, E., CESAR FIGUEROA & AROCHA, M. 1986. *Ninth international IGCP 156 field workshop, symposium and short course: phosphorite deposits of Venezuela*. Ministry of Energy and Mines, Caracas.

ROGERS, J. K. & KEEVERS, R. E. 1976. Lady Annie-Lady Jane Phosphate Deposits, Georgina Basin, Queensland. In: KNIGHT, C. L. (ed.) *Economic Geology of Australia and Papua New Guinea, 4*. Australian Institute of Mining & Metallurgy, Monograph, **8**, 251–265.

ROONEY, T. P. & KERR, P. F. 1967. Mineralogic nature and origin of phosphorite, Beaufort County, North Carolina. *Geological Society of America Bulletin*, **78**, 731–748.

RUSSELL, R. T. & TRUEMAN, N. A. 1971. The geology of the Duchess phosphate deposits, northwest Queensland, Australia. *Economic Geology*, **66**, 1186–1214.

SAFFORD, J. M. 1902. Horizons of phosphate rock in Tennessee. *Geological Society of America Bulletin*, **13**, 14–15.

SANDSTROM, M. W. 1986. Proterozoic and Cambrian phosphorites — specialist studies: geochemistry of organic matter in Middle Cambrian phosphorites from the Georgina Basin, northeastern Australia. In: COOK, P. J. & SHERGOLD, J. H. (eds) *Phosphate Deposits of the World, 1: Proterozoic and Cambrian Phosphorites*. Cambridge University Press, Cambridge, 268–279.

SCHMITT, M. & SOUTHGATE, P. N. 1982. A phosphatic stromatolite (*Ilicta composita* Sidorov) from the Middle Cambrian of northern Australia. *Alcheringa*, **6**, 175–183.

SELLARDS, E. H. 1915. The pebble phosphates of Florida. *Florida State Geological Survey 7th Annual Report*, 29–116.

SHATSKII, N. S. 1955. *Phosphorite bearing formations and classification of phosphorite deposits*. Report on Conference on Sedimentary Rocks, 2. USSR Academy of Sciences, Moscow.

SHELDON, R. P. 1963. Physical stratigraphy and mineral resources of Permian rocks in western Wyoming. *US Geological Survey, Professional Paper*, **313–B**, 49–273.

—— 1964. Paleolatitudinal and paleogeographic distribution of phosphorite. *US Geological Survey, Professional Paper*, **501-C**, 106–113.

—— 1980. Episodicity of phosphate deposition and deep ocean circulation — a hypothesis. In: BENTOR, Y. K. (ed.) *Marine Phosphorites*. Society of Economic Paleontologists and Mineralogists, Special Publication, **29**, 239–247.

—— & BURNETT, W. C. (eds) 1980. *Fertilizer Mineral Potential in Asia and the Pacific*. Proceedings of the Ferlitilizer Raw Materials Resources Workshop, Aug. 20–24, 1979. East-West Resources Systems Institute, Honolulu, Hawaii.

——, DAVIDSON, D. R., RIGGS, S. R. & BURNETT, W. C. 1985. *Undiscovered phosphate resources in the Caribbean region and their potential value for agricultural development*. US Geological Survey, Circular, 962.

SHEMESH, A., KOLODNY, Y. & LUZ, B. 1983. Oxygen isotope variation in phosphate of biogenic apatites, II. Phosphorite rocks. *Earth and Planetary Science Letters*, **64**, 405–416.

SHEPARD, C. U. 1869. Note upon the origin of the phosphatic formation. *American Journal of Science*, **97**, 338–340.

SHERGOLD, J. H. & SOUTHGATE, P. N. 1986. *Middle Cambrian phosphatic and calcareous lithofacies along the eastern margin of the Georgina Basin western Queensland*. Australian Sedimentologists Group Field Guide Series 2, Geological Society of Australia, Sydney.

SIMPSON, D. R. 1967. Effect of pH and solution concentration on the composition of carbonate apatite. *American Mineralogist*, **52**, 896–902.

SISODIA, M. S. & CHAUHAN, D. S. 1990. The influence of magnesium ions during the formation of stromatolitic phosphorites of Udaipur, Rajasthan, India. In: NOTHOLT, A. J. G. & JARVIS, I. (eds) *Phosphorite Research and Development*. Geological Society, London, Special Publication, **52**, 313–320.

SLANKSY, M. 1980. *Géologie des phosphate sédimentaires*. Bureau de Recherches Géologiques et Minières, Mémoires, 114.

—— 1986. *Geology of sedimentary phosphates*. North Oxford Academic, London.

SMIRNOV, A. I. 1964. Formation of various types of marine phosphorites. *Lithologia Polezn Iskop*, **5**, 96–104.

——, IVNITSKAYA, R. B. & ZALAVINA, T. P. 1958. Preliminary results of a study of the $CaO-P_2O_5-H_2O$ system under conditions closely approach natural conditions. *Transactions of the State Scientific Research Institute of Chemical Raw Materials*, **4**, 86–91.

——, —— & —— 1962. Experimental data on the possibility of the chemical precipitation of phosphate from sea water. *Transactions of the State Scientific Research Institute of Chemical Raw Materials*, **7**, 289–302.

SMITH, J. L. 1845. Composition of the Marl from Ashley River, S. C. *American Journal of Science*, **48**, 101–103.

SNYDER, S. W. (ed.) 1985. *Eighth international IGCP 156 field workshop and symposium: Tertiary phosphorites of southeastern United States*. Geology Department, East Carolina University, Greenville, N.C.

——, HALE, W. R., RIGGS, S. R., SPRUILL, R. K. & WATERS, V. J. 1984. Occurrence of clinoptilolite as moldic fillings of foraminiferal tests in continental margin sediments. *Geological Society of America, Abstracts with programs*, **16**, 662.

——, HINE, A. C. & RIGGS, S. R. 1990. The seismic stratigraphic record of shifting Gulf Streamflow paths in response to Miocene glacioeustacy: implications for phosphogenesis along the North Carolina continental margin. In: BURNETT, W. C. & RIGGS, S. R. (eds) *Phosphate Deposits of the World, 3: Neogene to Modern Phosphorites*. Cambridge University Press, Cambridge. (in press).

SOLLAS, W. J. 1872. Some observations on the Upper Greensand Formation of Cambridge. *Quarterly Journal of the Geological Society of London*, **28**, 397–402.

SOUDRY, D. & LEWY, Z. 1988. Microbially influenced formation of phosphate nodules and megafossil moulds (Negev, Southern Israel). *Palaeogeography, Palaeoclimatology, Palaeoecology*, **64**, 15–34.

——, & SOUTHGATE, P. N. 1989. Ultrastructure of a Middle Cambrian primary non-pelletal phosphorite, and its early transformation into phosphate vadoids: Georgina Basin, Australia. *Journal of Sedimentary Petrology*, **59**, 53–64.

SOUTHGATE, P. N. 1983. *Middle Cambrian phosphatic and calcareous depositional environments, the Undilla region of the Georgina Basin, Australia*. PhD thesis, Australian National University, Canberra.

—— 1986a. Cambrian phoscrete profiles, coated grains, and microbial processes in phosphogenesis: Georgina Basin, Australia. *Journal of Sedimentary Petrology*, **56**, 429–441.

—— 1986b. Middle Cambrian phosphatic hardgrounds, phoscrete profiles and stromatolites and their implications for phosphogenesis. *In*: COOK, P. J. & SHERGOLD, J. H. (eds) *Phosphate deposits of the World, 1: Proterozoic and Cambrian Phosphorites*. Cambridge University Press, Cambridge, 327–351.

STRAKHOV, N. M. 1960. *Fundamentals of the theory of lithogenesis*. USSR Academy of Sciences, Geological Institute, Moscow.

THOMAS, P. 1885. Sur la découverte de gisements de phosphate du chaux dans le sud de la Tunisie. *Comptes Rendus de l'Académie des Sciences, Paris*, **101**, 1184.

—— 1887. Sur la découverte de nouveaux gisements de phosphate de chaux en Tunisie. *Comptes Rendus de l'Académie des Sciences, Paris*, **104**, 1321.

VAIL, P. R. 1987. Seismic stratigraphy interpretation utilizing sequence stratigraphy. Part 1, Seismic stratigraphy interpretation procedure. *In*: BALLY, A. W. (ed.) *Atlas of Seismic Stratigraphy*. American Association of Petroleum Geologists, Studies in Geology, **27**, 1–10.

——, MITCHUM, R. M. & THOMPSON, S. 1977. Seismic stratigraphy and global changes of sea level; Part 4: global cycles of relative hydrocarbon exploration. *In*: PAYTON, C. E. (ed.) *Seismic Stratigraphy and global changes of sea level*. American Association of Petroleum Geologists, Memoir, **26**, 83–97.

——, ——, TODD, R. G., WIDMIER, J. M. THOMPSON, S., SANGREE, J. B., BUBB, J. N. & HATLELID, W. G. 1977. Seismic stratigraphy and global changes of sea level. *American Association of Petroleum Geologists, Memoir*, **26**, 49–97.

VAN HORN, F. B. 1909. The phosphate deposits of the United States. *Bulletin of the US Geological Survey*, **394**, 157–171.

VAN KAUWENBERGH, S. J. & MCCLELLAN, G. H. 1990. Comparative geology and mineralogy of the southeastern United States and Togo phosphorites. *In*: NOTHOLT, A. J. G. & JARVIS, I. (eds) *Phosphorite Research and Development*. Geological Society, London, Special Publication, **52**, 139–155.

VEEH, H. H. & BURNETT, W. C. 1982. Carbonate and phosphate sediments. *In*: IVANOVICH, M. & HARMON, R. S. (eds) *Uranium Series Disequilibrium – Applications to Environmental Problems*. Clarendon Press, Oxford, 459–480.

VON der BORCH, C. C. 1970. Phosphatic concretions and nodules from the upper continental slope northern New South Wales. *Journal of the Geological Society of Australia*, **16**, 755–759.

WARINGTON, R. 1866. Researchers on the phosphates of calcium and upon the solubility of tricalcium phosphate. *Chemical Society Journal*, **19**, 296.

WEEKS, F. B. 1908. Phosphate deposits in the Western United States. *Bulletin of the US Geological Survey*, **340**, 441–447.

—— & FERRIER, W. E. 1907. Phosphate deposits in the Western United Sates. *Bulletin of the US Geological Survey*, **315**, 449–462.

WILCOX, N. R. 1953. The origin of beds of phosphatic chalk with special reference to those at Taplow, England. *19th Session of the International Geological Congress, Algiers, Section 11*, 119–134.

WÜRZBURGER, U.S. 1968. A survey of phosphate deposits of Israel. Proceedings of a seminar on sources of mineral raw materials for the fertilizer industry in Asia and the Far East. *United Nations Mineral Resources Development Series*, **32**, 152–165.

WYATT, F. 1891. *The phosphates of America*. Scientific Publishing Co., New York.

YEH LIENTSUN, SUN SHU, CHEN QUYING, & GUO SHIZENG 1986. Proterozoic and Cambrian phosphorites – deposits: Kunyang, Yunnan, China. *In*: COOK, P. J. & SHERGOLD, J. H. (eds) *Phosphate Deposits of the World, 1: Proterozoic and Cambrian Phosphorites*. Cambridge University Press, Cambridge, 149–154.

YERMALOW, Al. S. 1873. *Recherches sur les gisements de phosphate de chaux fossile en Russie. Avec carte et tableaux analytiques*. Imprimerie Trenké & Fusnot, St Pétersbourg.

ZANIN, Y. 1987. *Ultramicrostructures of phosphorites* (atlas of pictures). Moscow, NAUKA.

Mineralogy of sedimentary apatites

G. H. McCLELLAN[1] & S. J. VAN KAUWENBERGH[2]

[1] Department of Geology, University of Florida, Gainesville, Florida 32611, USA
[2] Fertilizer Technology Division, International Fertilizer Development Center (IFDC), Muscle Schoals, Alabama 35662, USA

Abstract: Members of the apatite family are the most common phosphate minerals. Apatites in igneous and metamorphic rocks generally approach fluorapatite in composition. In sedimentary rocks, the apatite is usually the carbonate fluorapatite variety, francolite, which produces x-ray diffraction patterns similar to but distinctive from those of fluorapatite. Variations in francolite compositions and physical properties have also been identified. The substitution of CO_3^{-2} for PO_4^{-3}, from zero to about 25% of phosphate sites on a 1:1 basis, influences all other changes in composition and crystallography of francolites. Statistical models have reduced the francolite composition to a single parameter problem, the dominant factor being the variation in unit-cell a-value with increasing CO_3^{-2} substitution. Within a deposit, variations in francolite composition with stratigraphical position and particle size may indicate the extent and type of post-depositional alteration. Replacements of SO_4^{-2} for PO_4^{-3} and Ca^{+2} by Na^{+2} and Mg^{+2} are also discussed but these are of more limited significance.

The most common members of the mineral phosphates are those belonging to the apatite family. In igneous and metamorphic rocks apatites approaching fluorapatite composition, $Ca_{10}(PO_4)_6F_2$, are most common. Less abundant are apatites that contain mixtures of fluorine with hydroxyl and chlorine, which often are reported as chlorapatite and hydroxyapatite. The latter are distinctly rare as mineral species and almost never seem to occur in pure form (McConnell 1973).

In sedimentary rocks, the apatite is usually the carbonate fluorapatite variety, francolite. Previous work has shown that, although this mineral has a cryptocrystalline to microcrystalline texture, it produces x-ray diffraction patterns that are similar to, but distinctive from, those of fluorapatite (McConnell 1938; Smith & Lehr 1966; McClellan & Lehr 1969; McClellan 1980). In 1952, Altschuler et al. noted that the x-ray pattern of carbonate-apatite differed from that of fluorapatite and that the unit-cell dimensions were smaller than those reported for fluorapatite. However, their database was too small to generalize about carbonate-apatite crystal chemistry. In the 1960s, the Tennessee Valley Authority (TVA) undertook a basic research programme on phosphate rocks. Initially, the investigation focused on the francolites that occurred in about 20 samples of commercial concentrates of Florida phosphate rock. The concentrates were manually separated into various size, colour and morphological types from several geographical locations within the state. The study confirmed the observations of Altschuler et al. (1952) on the x-ray data, but no relationships could be established between the measured variation in unit-cell dimensions, chemical composition, and physical characteristics because of the narrow compositional range that existed in this small database. Before abandoning the study, TVA broadened the sample base to include materials from other geographical locations with differing geological ages and histories. Additional samples were obtained from the phosphate rock collection of the United States Department of Agriculture laboratories, Beltsville, Maryland, which expanded the sample base to include more than 100 examples. This early collection consisted of ores and concentrates from many countries including the USA (South Carolina, North Carolina, Tennessee, Utah, Wyoming), Morocco, Tunisia, Algeria, Nauru Island, and the USSR.

The study of these materials led to the discovery of the systematic variations in francolite compositions and physical properties (Smith & Lehr 1966; Lehr et al. 1968; McClellan & Lehr 1969; McClellan 1980). The experimental approach was to collect the chemical analyses, x-ray dimensions, and other physical data on a large number of samples and to use statistical methods to develop correlation models. No *a priori* assumptions were made. The statistical models that evolved from these studies are now used in the solution of many academic and practical problems in francolite mineralogy and are used as tools by the theoretical mineralogist and commercial phosphate users. During the past 20 years, some refinements have been proposed to this model. These refinements and their significance will be considered in this paper.

Francolite nomenclature

In 1939, Sandell, Hey & McConnell documented the name francolite and hoped to clarify its acceptance as a legitimate mineral species. Francolite was defined as a carbonate fluorapatite containing more than 1% fluorine and appreciable amounts of CO_2. This definition has led to some confusion because most sedimentary francolites contain more fluorine (3.77% F) than does fluorapatite, $Ca_{10}(PO_4)_6F_2$, as well as an appreciable amount of CO_2. These are the francolites containing the so-called *excess fluorine*. Terminology problems also arise with phosphates that are described as francolites but contain less than the 2 moles of F per unit-cell of fluorapatite. These 'fluorine-deficient' apatites may meet the requirements of the francolite definition, but they have properties that are quite different from the 'excess fluorine' francolites that usually occur in sedimentary rocks. The authors are not proposing a revision in the nomenclature but merely emphasize the point that most of the following discussion is directed at francolites containing 'excess fluorine'.

Francolite substitutions

The fundamental substitution in sedimentary francolites is that of CO_3^{-2} for PO_4^{-3}. This substitution requires other cation and anion substitutions to maintain electroneutrality. In 1969, McClellan & Lehr presented data for 110 francolites and some igneous apatites which showed that the CO_3^{-2} for PO_4^{-3} took place on a 1 for 1 basis. The database has been extended to 260 francolites which contain at least 2 moles of F^{-1} per unit-cell; these data do not contain any igneous, lacustrine, guano, or insular samples. The data show a linear relationship between the moles of CO_3^{-2} and moles of PO_4^{-3} per unit-cell substitution (Fig. 1) with an excellent statistical correlation ($r^2 = 0.938$). The slope of the regression line is 0.9981 (1.0 would be ideal for 1:1 substitution) with an intercept of 5.996 at zero moles CO_3^{-2} per unit-cell (ideal would be an intercept of 6 for the moles of PO_4^{-3} per unit-cell). These results reinforce the earlier findings that, in the francolite structure, CO_3^{-2} substitutes for PO_4^{-3} only on a 1:1 basis.

The fundamental substitution of one mole of

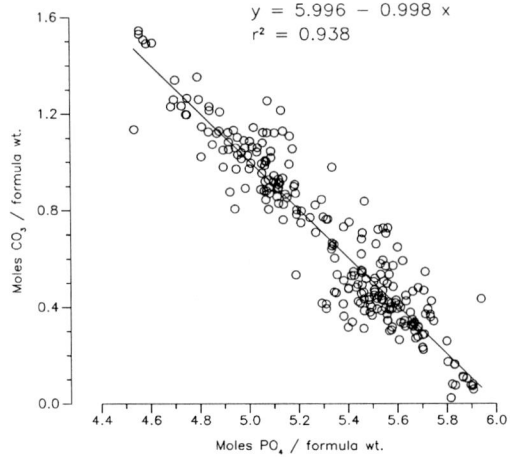

Fig. 1. Correlation of moles of CO_3^{-2} per formula weight with PO_4^{-3} per formula weight.

CO_3^{-2} for one mole of PO_4^{-3} gives rise to several questions. How does the structure maintain electroneutrality? What is the effect of substituting the planar CO_3^{-2} group for the tetrahedral PO_4^{-3} group? In 1938, Bornemann-Starinkevitch proposed the substitution of $(CO_3+F)^{-3}$ or $(CO_3+OH)^{-3}$ as distorted tetrahedral groups to replace PO_4^{-3} as a method to preserve electroneutrality. Based on the statistically significant $(CO_3+F)^{-3}$ interaction term in their models, Smith & Lehr (1966) and McClellan & Lehr (1969) supported this idea. McClellan (1980) raised some questions concerning the significance of the $(CO_3+F)^{-3}$ interaction term but concluded that the concept was correct. In 1980, Bacquet et al. presented electron paramagnetic resonance data that supported the existence of $(CO_3+F)^{-3}$ at sites in the apatite structure. This proposed association of F^{-1} with CO_3^{-2} at a substitution site would provide a logical explanation of the 'excess fluorine' known to occur in most natural francolites. As will be discussed below, this 'excess fluorine' in sedimentary francolites can be as much as 30% greater than the fluorine that occurs in fluorapatite. Thus the proposed substitution of $(CO_3+F)^{-3}$ for PO_4^{-3} can both maintain electroneutrality and explain an important chemical characteristic of sedimentary francolites. As will be discussed below, the chemical behaviour of francolites during calcination provides additional support for the proposed substitution.

The extent of CO_3^{-2} for PO_4^{-3} anion substitution is the most fundamental factor affecting francolite chemical composition and properties. In the 1969 work, McClellan & Lehr calculated that 1.395 CO_3^{-2} moles per unit-cell was the maximum carbonate substitution (23% of the sites substituted) for francolite based on 110 samples. This corresponds to about 6.3 wt% CO_2 in the sample. Similar results were obtained when the limits were taken on the models reported by McClellan (1980). The results of the study presented here indicate that the limit of CO_3^{-2} for PO_4^{-3} is about 1.5 moles per unit-cell. Synthesis work by Jahnke (1984) on carbonate fluorapatite indicated a maximum substitution of 1.4 carbonate ions per unit cell. McArthur (1978) also proposed that the limit of carbonate substitution is about 1.5 moles per unit cell. It should be noted that the range of carbonate substitution can vary from zero to about 25% of the phosphate sites in natural francolites. The extent of CO_3^{-2} substitution is a variable that is often affected by post-depositional alteration (Nathan et al. 1990).

Another systematic substitution in francolites is the replacement of Ca^{+2} by Na^{+1} and Mg^{+2}. Work by Lehr et al. (1968) and McClellan & Lehr (1969) have established that these cation substitutions are directly related to the CO_3^{-2} for PO_4^{-3} anion substitutions. When about 25% of the PO_4^{-3} is replaced by CO_3^{-2}, then about 6% of the Ca^{+2} is replaced by Na^{+1} and Mg^{+2}. McArthur (1985) did not observe this relationship in his database of selected francolites. However, the results reported here for more than 200 samples (Figs 2 and 3) clearly show increases in Na^{+1} and Mg^{+2} with increasing CO_3^{-2} for PO_4^{-3} substitution. Because the Na^{+1} may be involved in substitutions other than just CO_3^{-2} for PO_4^{-3} (for example the Na^{+1} and SO_4^{-2}), the trend in the data may be more important than the correlation. The francolite model predicts that up to 1.4 wt% of Na^{+1} can be directly substituted in the francolite structure for Ca^{+2}. Thus Na^{+1} is an important factor in the complex system of isomorphic substitutions where exact electroneutrality may not always be achieved. In the samples examined in this study, very few had a net charge imbalance of greater than 1 and the majority showed a slight excess positive charge.

Some researchers have proposed to use the Na^{+1} contents of francolites as indicators of paleosalinity or environments of deposition (Russell & Trueman 1971; Lucas et al. 1987). Studies of post-depositional alteration (Van Kauwenbergh et al. 1988) of francolites would indicate that this could be hazardous unless the geologic history of the sample is very well known.

The occurrence of the systematic substitution of magnesium is something of a problem in francolite. It is not required as part of a coupled

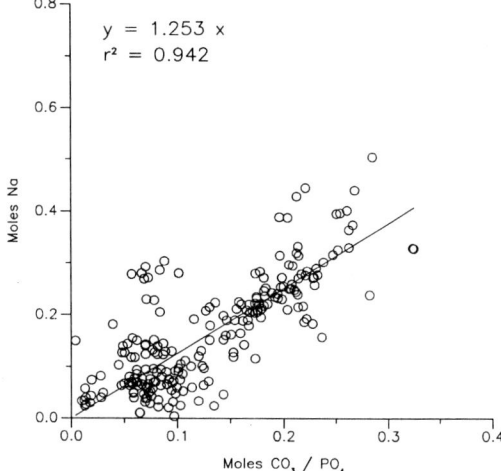

Fig. 2. Correlation of moles of Na^{+1} with increasing CO_3^{-2} for PO_4^{-3} substitution.

substitution because it has the same valence as Ca^{+2}. It seems likely that it is needed to allow the structure to physically compensate for the other substitutions. Because Mg^{+2} is much smaller than Ca^{+2} in ionic radius (about 40% less), it may be required to allow the structure to accept high levels of CO_3^{-2} for PO_4^{-3} substitution. The significance of up to 0.7 wt% MgO in the francolite structure remains unclear.

Data from natural francolites show the predicted systematic variation in CaO/P_2O_5 ratios. Fluorapatite has a CaO/P_2O_5 weight ratio of 1.318 and the most substituted francolite has a CaO/P_2O_5 weight ratio as high as 1.621. Inter-

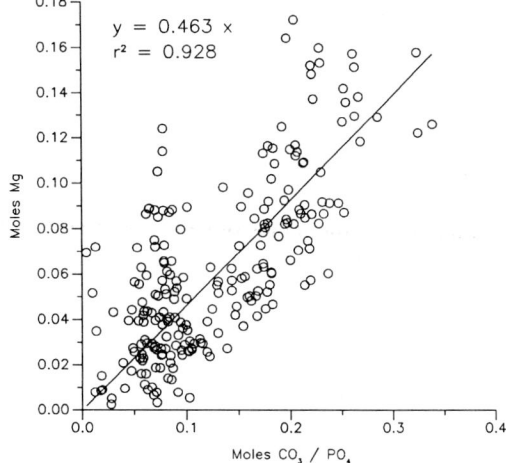

Fig. 3. Correlation of moles of Mg^{+2} with increasing CO_3^{-2} for PO_4^{-3} substitution.

mediate ratios occur in samples with lesser amounts of CO_3^{-2} for PO_4^{-3} substitution. The CO_3^{-2} content of francolites is often affected by post-depositional alteration. The CO_3^{-2} can be decreased for a variety of reasons including weathering, diagenesis, and metamorphism (McClellan 1980). In general, these alterations decrease the CO_3^{-2} content of the metastable francolite and result in a shift in composition towards fluorapatite. This alteration may go all the way to $Ca_{10}(PO_4)_6F_2$ or it may stop at some intermediate francolite composition.

The F/P_2O_5 weight ratio also has been used as an indicator of francolite composition. Because the 'excess' fluorine increases and the P_2O_5 decreases with increasing CO_3^{-2} for PO_4^{-3} substitution, the F/P_2O_5 weight ratio increases from 0.089 in fluorapatite to 0.148 in a highly substituted francolite. In some complex samples this ratio can be used as an indicator of francolite or fluorapatite.

Another anion substitution that has been proposed for francolite is SO_4^{-2} for PO_4^{-3}. Nathan (1984) reports the work of Sassi (1974) on some francolites from Gafsa, Tunisia that reportedly contain up to 3.55% SO_4^{-2} in a 28% P_2O_5 rock. Gulbrandsen (1966) also reported some high sulphate contents associated with francolites from the Phosphoria Formation. He proposed a coupled substitution of Na^{+1} or REE^{+3} and SO_4^{-2} for Ca^{+2} and PO_4^{-3} in the francolite structure to preserve electroneutrality. In 1969, McClellan & Lehr agreed that this substitution could take place but questioned the extent and importance of it. Their data showed that the average Na^{+1} to SO_4^{-2} mole ratio in a number of samples was about 4.5 and the largest quantity measured was 0.101 moles per unit-cell. An expansion of the SO_4^{-2} analytical database (Fig. 4 with 120 sample points) shows that there is a good statistical correlation ($r^2=0.944$) with the moles of Na^{+1} with a near zero intercept for these samples. The slope of the line indicates that about 44% of the moles of Na^{+1} may be coupled with the SO_4^{-2}, and that the maximum substitution of SO_4^{-2} is about 0.140 moles per formula weight (1.2% SO_4^{-2} in the original sample). Thus these data show that the SO_4^{-2} for PO_4^{-3} is a minor factor in evaluating francolites in general. This substitution may be important in some samples (such as those in Tunisia or Peru) which are associated with evaporite deposits. In the samples examined for this study, those containing more than 1.2% SO_4^{-2} contained detectable contaminants. Diamant (1970) has reported infrared spectra for a high-sulphate-containing francolite from Morocco, perhaps the only in-

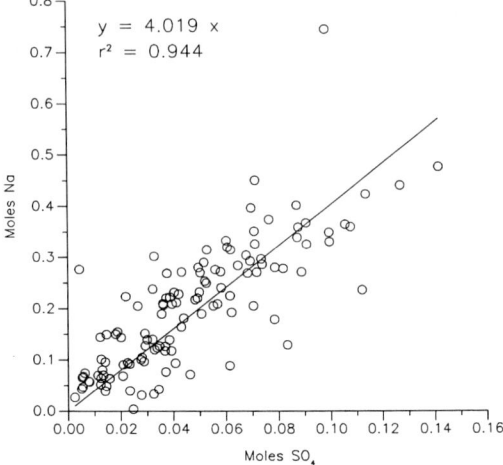

Fig. 4. Correlation of moles of Na^{+1} with SO_4^{-2} for 120 sedimentary francolites.

dependent verification of this substitution in natural francolites. Other interpretations often are based on indirect evidence, such as recalculated chemical data.

The addition of a SO_4^{-2} term to the CO_3^{-2} or $(CO_3+F)^{-3}$ models correlating francolite composition with unit-cell dimensions results in some interesting changes. The SO_4^{-2} term is statistically significant but causes only a very slight improvement in correlation (r^2 increases by 0.002%). However, the SO_4^{-2} term has a positive sign while the CO_3^{-2} and F^{-1} terms are both negative. This means that, relative to CO_3^{-2} and F^{-1}, SO_4^{-2} substitution causes an increase in the francolite unit-cell dimension. Francolites reported to contain high levels of SO_4^{-2} in some ores and concentrates (Gafsa, Tunisia for example) often have very small values for their unit-cell a-dimensions indicating that the CO_3^{-2} and F^{-1} substitutions are likely to be much more significant than the SO_4^{-2} substitution.

The idea of cation and anion vacancies at sites within the apatite structure as a result of substitutions has received some recent attention (Prévôt & Baumer 1985; Baumer et al. 1986). Vacancies associated with anion substitutions was considered in the earliest work on the occurrence of CO_3^{-2} in the structure of apatite. Gulbrandsen (1966) considered cation vacancies as an explanation of calcium deficiencies in some francolites. Lucas et al. (1987) has combined these ideas and proposed a model with both anion and cation vacancies for francolites. The EPR work of Bacquet et al. (1980) has provided some actual data confirming vacant structural sites associated with CO_3^{-2} substitution in apatite. The work on $(CO_3+F)^{-3}$ statistical models (McClellan & Lehr 1969; McClellan 1980) has shown that about 40% of the fourth tetrahedral sites are occupied by F, so some vacancies seem to be required at the oxygen sites. The cation-deficient models adapted from the studies of biological apatites may be unnecessary because the majority of sedimentary francolites have an excess positive charge. These francolite models also show an excess positive charge at maximum substitution that may be neutralized by a small, undetected quantity of OH^{-1} or other minor substitutions.

In 1980, McClellan interpreted his data to include a possible, but minor amount of $(CO_3+OH)^{-3}$ to maintain charge balance. This proposal remains problematical because no satisfactory analytical procedure has been found to unambiguously confirm the presence of small amounts of water in the francolite structure. For example, detailed Fourier transform infrared spectroscopy (Scheib et al. 1981) on a number of francolites was unable to detect any structural water of the type required by the $(CO_3+OH)^{-3}$ substitution.

There does seem to be a structural control of francolite composition. The existence of curvilinear relationships between crystallographic properties and chemical composition indicates physical limits do occur. This can be 'verified' by the lack of samples of every composition between $Ca_{10}(PO_4)_6F_2$ and $CaCO_3$. Negative evidence is not always convincing but after the many years of study of on-shore and off-shore deposits and the results of synthetic systems, it does appear that chemical and physical limits do exist for the composition of francolites.

The use of the variations in the c-value to model francolite compositions has not been very successful because of the limited range observed in these values. In general, the length of the c-axis increases with increasing CO_3^{-2} for PO_4^{-3} substitution. In spite of this increase, there is a slight decrease in unit-cell volume because the a-axis decreases more per mole of CO_3^{-2} substitution than the c-axis increases. increases.

Isomorphic substitutions in the apatite structure may also affect the crystallography. Although apatite occurs in the hexagonal crystal system, systematic substitutions may cause a decrease in the symmetry of francolite into the monoclinic system. Until sufficient data are collected to precisely determine the symmetry of francolite, it should be regarded as a hexagonal mineral.

Use of models

The x-ray diffraction reference database (patterns and d-spacings) for sedimentary francolites is surprisingly small. Because the collection of high-resolution data from randomly oriented powdered samples is the essence of the work discussed here, the x-ray patterns of some typical materials are reported for documentary purposes (Fig. 5). A comparison of the pattern of fluorapatite with that of a highly CO_3^{-2} substituted francolite from North Carolina shows the main features of interest. Because francolites are usually crypto- or microcrystalline, their patterns show less peak resolution than does the fluorapatite pattern (see specifically the separation or lack of separation of the 213 and 321 peaks). Also, there is a general decrease in peak intensity; the small crystallite size of the highly substituted francolite probably decreases the efficiency of diffraction. Finally, there is a lateral shift in the locations for nearly all the peaks; this is a clear indication that there are significant differences in the crystallographic parameters of the apatites. Following the initial report of Altschuler et al. (1952), subsequent work has clearly established that there are significant, regular variations in francolite unit-cell dimensions.

These studies have established statistical models relating crystal chemical composition with x-ray, optical, and infrared properties (Lehr et al. 1968; McClellan & Lehr 1969; McClellan 1980; Scheib et al. 1981). Of these methods, x-ray diffraction models have received the most attention and use. The statistical models that were developed have allowed the reduction of francolite composition to a single parameter problem. The dominant factor is the variation in unit-cell a-value with increasing CO_3^{-2} substitution (Fig. 6). Assuming that fluorapatite, $Ca_{10}(PO_4)_6F_2$, and francolite, $Ca_{10-x-y}Na_xMg_y(PO_4)_{6-z}(CO_3)_zF_{0.4z}F_2$, are the end-members of a limited compositional series, the values of x, y and z can be determined using the following refined formulae (McClellan 1980):

$$\frac{CO_3^{-2}}{PO_4^{-3}} = \frac{Z}{6-Z} = \frac{9.369 - a_{(obs)}}{0.185}$$

(moles Na) $x = 7.173 (9.369 - a_{(obs)})$
(moles Mg) $y = 2.784 (9.369 - a_{(obs)})$
in which
$a_{(obs)}$ = a−value, in Ångström units, determined by x-ray diffraction.
x = moles of Na^{+1} in the francolite formula.

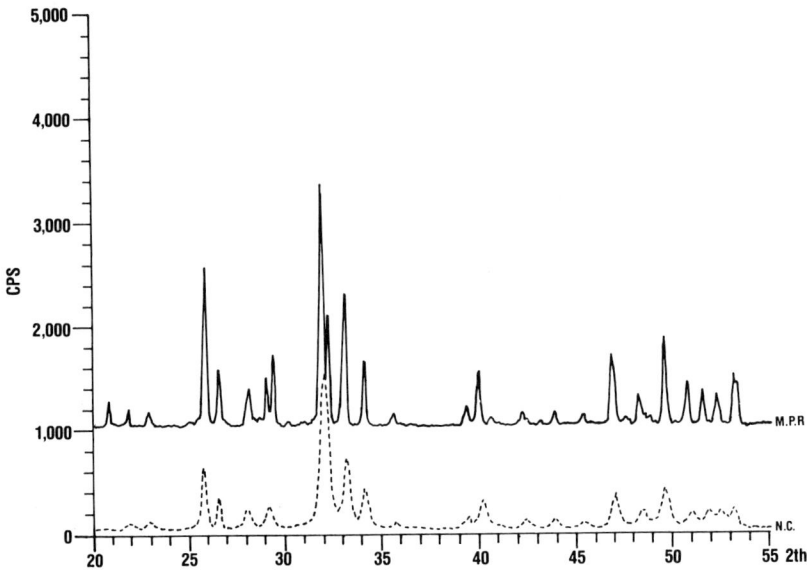

Fig. 5. X-ray diffraction patterns for fluorapatite and a highly CO_3^{-2} substituted francolite.

y = moles of Mg^{+2} in the francolite formula.
z = moles of CO_3^{-2} in the francolite formula.

Note: all constants were determined by regression analysis of the chemical and crystallographic data.

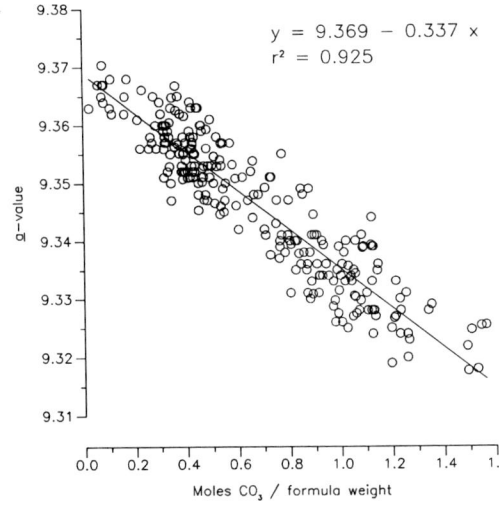

Fig. 6. Variation in unit-cell a-value with increasing CO_3^{-2} for PO_4^{-3} substitution in francolite.

Thus, the composition of any francolite can be calculated from the unit-cell a-value.

Gulbrandsen (1970) combined the data of McClellan & Lehr (1969) with data on samples from the Phosphoria Formation and developed a simple method to estimate the CO_2 of francolite based on a peak-pair separation in the x-ray diffraction pattern. The method has been widely used, but it is influenced by the presence of dolomite which interferes with the high-angle x-ray diffraction peaks of francolite. This method should *only* be used on 'excess fluorine' francolites and not on other apatites. It may underestimate the CO_2 because of the bias of the database, which has a large number of samples with low CO_2 contents, such as those from the Phosphoria Formation. Menor (1975) reports that the peak-pair method systematically overestimated the CO_3^{-2} contents of low carbonate francolites from Senegal and Brazil.

The substitution of CO_3^{-2} for PO_4^{-3} in francolites also manifests itself in the index of refraction. Increasing substitution results in a decrease in refractive index from c. 1.63 to 1.60 as the CO_3^{-2}/PO_4^{-3} mole ratio increases from 0 to 0.3 (Lehr et al. 1968; McClellan & Clayton 1980; McClellan & Gremillion 1980). This effect is of particular importance in thin sections where estimation of the index of refraction allows some knowledge of the francolite composition. Optical microscopy is particularly useful in the examination of very small samples where other methods may not be applicable. Variation in the index of refraction within grains may be an indication of an altered sample. Alteration of the exterior surfaces of grains is a common feature in sedimentary phosphorites that can be determined by microscopic study.

The crystallite size of francolites decreases with increasing CO_3^{-2} substitution. Electron microscopic studies show that the crystallites change from needle-like morphologies for fluorapatite to tabular forms in highly substituted francolites. This decrease in crystallite size is one cause for the decrease in x-ray diffraction peak intensities of francolites. The surface areas of highly substituted francolites can be as large as $20-30$ m^2 g^{-1}. This increases their potential for chemical reaction either in natural systems or during chemical processing.

Calcination of francolite results in a loss of CO_2 and some F (Lehr et al. 1968). Around 700°C, highly substituted francolites begin to lose their CO_3^{-2} and 'excess' F^{-1}. The progressive loss of carbonate and some fluorine causes the francolite to alter towards fluorapatite compositions. This results in a systematic increase in the unit-cell a-value and crystallite size. Organic matter, clay minerals, and hydrated accessories phases are also altered in this temperature range. If the sample has a very high organic matter content, the material will ignite and can raise the temperature in an uncontrolled manner. Because of the high CaO/P_2O_5 ratios in such samples, free CaO can form as well as CaF_2. In fact, CaF_2 associated with metamorphosed francolites could have such an origin. The simultaneous loss of CO_2 and the 'excess' F during calcination is interpreted as further evidence for the presence of $(CO_3+F)^{-3}$ in the francolite structure. Measured fluorine losses are usually those predicted by the francolite model.

Particle size of phosphate samples is a variable that is often overlooked. In the initial studies of Smith & Lehr (1966) and McClellan & Lehr (1969), commercial phosphate concentrates were a substantial part of the sample base. In these materials, the coarse particles (greater than 1 mm) and the fine particles (less than 0.075 mm) have been removed during beneficiation. Studies of the remaining particle sizes showed little variation in francolite composition within individual samples. In a recent study of whole ore samples from various mines and stratigraphic zones in Florida (Van Kauwenbergh et al. 1990), results show significant variations in francolite composition in vertical stratigraphic sections and with decreasing particle size. The samples studied contained francolites that varied in composition from high degrees of CO_3^{-2} for PO_4^{-3} substitution to francolites altered to fluorapatite with almost no CO_2 substitution. The intensity of alteration increased with decreasing particle size and at the tops of vertical sections. It seems likely that the fine phosphate particles in the deposits were formed by the alteration of the coarser material. Similar studies are needed in other locations to develop a better understanding of the variations in francolite compositions within each deposit.

Conclusions

The fundamental crystal chemical substitution in francolite is CO_3^{-2} for PO_4^{-3} phosphate. This single substitution influences all other systematic changes in composition and crystallography. Expansion of the database has only confirmed and refined this conclusion.

The results of chemical analysis and statistical modelling show the maximum substitution of CO_2 in francolite is between 6 and 7wt%, or about 25% of the PO_4^{-3} sites. Work by several authors has independently confirmed this limit.

Substitution of CO_3^{-2} for PO_4^{-3} causes the x-ray diffraction patterns of francolites to vary in resolution, intensity, and position. These changes in unit-cell a-values are systematic and have been related to variations in francolite crystal chemical composition. The mean refractive index of francolites also changes systematically with isomorphic substitutions.

The 'excess' fluorine of francolites is related to the amount of CO_3^{-2} for PO_4^{-3} substitution through the proposed $(CO_3+F)^{-3}$ tetrahedral group. Fluorine only occupies about 40% of the fourth tetrahedral sites when CO_3^{-2} replaces PO_4^{-3} in the apatite structure.

Calcination of francolite causes a systematic change in composition towards fluorapatite. The simultaneous loss of CO_2 and part of the F during heating is interpreted as supporting evidence for the presence of $(CO_3+F)^{-3}$ in the francolite structure. The fluorine losses produced by calcination can be predicted by the francolite model.

In francolites, Na^{+1} and Mg^{+2} increase systematically with an increase in CO_3^{-2} for PO_4^{-3} substitution. In the samples studied for this paper, Na^{+1} for Ca^{+2} substitution appears to be coupled with CO_3^{-2} and SO_4^{-2} substitution.

The extent of SO_4^{-2} substitution in francolite is much smaller than that of CO_3^{-2}. An expanded study of Na^{+1} and SO_4^{-2} substitution in more than 100 francolite samples shows a systematic increase in moles of SO_4^{-2} with moles of Na^{+1}. No simple coupled substitution is apparent because the molar Na^{+1} values are usually double the associated molar SO_4^{-2} values.

Variations in francolite composition with stratigraphic position and particle size within a deposit may be indicators of the extent and type of post-depositional alteration that the material has experienced. Detailed mineralogical studies are needed to understand the significance of this alteration.

Work on francolites requires an adequate sample base to assure that significant ranges of variables are studied. Generalizations from an inadequate or incomplete sample base can be misleading.

In the more than 20 years that have passed since the crystal chemical models of francolite were first presented, many refinements have been proposed. None of these proposals have significantly modified the original findings although minor improvements have been made. Francolite compositions are a single parameter problem that is related to the fundamental 1:1 CO_3^{-2} for PO_4^{-3} substitution. The 'survival' of this relationship through rigorous testing and fine-tuning confirms its value as a useful and reliable tool for future researchers working in this interesting area of mineralogical research.

References

ALTSCHULER, Z. S., CISNEY, E. A. & BARLOW, I. H. 1952. X-ray Evidence of the nature of carbonate apatite. *Geological Society of America Bulletin*, **63**, 1230–1231.

BACQUET, G., TRUONG, V. Q., BONEL, G. & VIGNOLES, M. 1980. Résonance paramagnétique électronique du centre F dans les fluorapatites carbonatées de type B. *Journal of Solid State Chemistry*, **33**, 189–195.

BAUMER, A., CARUBA, R. & GANTEAUME, M. 1986. Carboapatites: évolution en fonction des variables pression et température. *Onzième réunion annuelle des sciences de la terre*, March 25–27, 1986, 12.

BORNEMANN-STARINKEVITCH, I. D. 1938. On some isomorphic substitutions in apatite. *Doklady Akademii Nauk SSSR*, **22**, 89–92.

DIAMANT, R. 1970. Mise en évidence des bandes de vibration dues à l'ion SO_4^{-2} constitutif de la maille de quelques fluorapatites calciques naturelles. *Comptes Rendus de l'Academie des Sciences*, Séries B, Paris, **271**, 701–703.

GULBRANDSEN, R. A. 1966. Chemical composition of phosphorites of the Phosphoria Formation. *Geochimica et Cosmochimica Acta*, **30**, 769–778.

—— 1970. Relation of carbon dioxide content of apatite of the Phosphoria Formation to regional facies. *United States Geological Survey Professional Paper*, **700-B**, B9–B13.

JAHNKE, R. A. 1984. The synthesis and solubility of carbonate fluorapatite. *American Journal of Science*, **284**, 58–78.

LEHR, J. R., MCCLELLAN, G. H., SMITH, J. P. & FRAZIER, A. W. 1968. Characterization of apatites in commercial phosphate rocks. *In: Colloque international sur les phosphates minéraux solides*, Toulouse 1967 **2**, Société chimique de France, Paris, 29–44.

LUCAS, J., PRÉVÔT, L., BENALIOULHAJ, N. & EL FALEH, E. 1987. Relation between sodium and apatite in sedimentary phosphorites of Morocco. *Terra Cognita*, **7**, 410.

MCARTHUR, J. M. 1978. Systematic variations in the contents of Na, Sr, CO_2, and SO_4 in marine carbonate fluorapatite and their relation to weathering. *Chemical Geology*, **21**, 41–52.

—— 1985. Francolite geochemistry-compositional controls during formation, diagenesis, metamorphism and weathering. *Geochimica et Cosmochimica Acta*, **49**, 23–35.

MCCLELLAN, G. H. 1980. Mineralogy of carbonate fluorapatites. *Journal of the Geological Society, London*, **137**, 675–681.

—— & CLAYTON, W. R. 1980. Francolite: The Commercial Phosphate Mineral. *In: Proceedings of the 2nd International Congress on Phosphorus Compounds*, Boston, Mass., USA, Institut Mondial du Phosphate, Paris, 131–143.

—— & GREMILLION, L. R. 1980. Evaluation of phosphatic raw materials. *In: KHASAWNEH, F. E. SAMPLE, E. C. & KAMPRATT, E. J.* (eds) *The Role of Phosphorus in Agriculture*, ASA-CSSA-SSSA, Madison, Wisconsin, USA, 43–80.

—— & LEHR, J. R. 1969. Crystal chemical investigation of natural apatites. *American Mineralogist*, **54**, 1374–1391.

MCCONNELL, D. 1938. A structural investigation of the isomorphism of the apatite group. *American Mineralogist*, **23**, 1–19.

MCCONNELL, D. 1973. *Apatite — Its Crystal Chemistry, Mineralogy, Utilization, and Geologic and Biologic Occurrences*. Springer Verlag, Vienna-Heidelberg–New York.

MENOR, E. 1975. *La sédimentation phosphatées: pétrographie, minéralogie et géochimie des gisements de Taïba (Sénégal) et d'Olinda (Brésil)*. Thése Docteur Ingénieur, Université Louis Pasteur, Strasbourg, France.

NATHAN, Y. 1984. The mineralogy and geochemistry of phosphorites. *In: NRIAGU, J. O. & MOORE, P. B.* (eds). *Phosphate Minerals* Springer Verlag, Berlin-Heidelberg-New York-Tokyo, 275–291.

—— SOUDRY, D. & AVIGOUR, A. 1990. Geological significance of carbonate substitution in apatites: Israeli phosphorites as an example. *In: NOTHOLT, A. J. G. & JARVIS, I.* (eds) *Phosphorite Research and Development*. Geological Society, London, Special Publication, **00**, 000–000.

PRÉVÔT, L. & BAUMER, A. 1985. Structural formula and geochemistry of apatite with implications on the geology. *8th International Field Workshop and Field Symposium, IGCP Project 156 Phos-*

phorites, Greenville, North Carolina, June, 1985.
RUSSELL, R. T. & TRUEMAN, N. A. 1971. The geology of the Duchess phosphate deposits, Northwestern Queensland, Australia. *Economic Geology*, **66**, 1186–1214.
SANDELL, E. B., HEY, M. H. & MCCONNELL, D. 1939. The composition of francolite. *Mineralogical Magazine*, **25**, 395–401.
SASSI, S. 1974. *La sédimentation phosphatée au Paléocène dans le bassin phosphate de Metlaoui (Tunisie)*. Thèse Docteur Science, Université de Paris-Sud, Orsay.

SCHEIB, R. M., THRASHER, R. D. & LEHR, J. R. 1981. Chemical composition determination of francolite apatites by Fourier transform infrared spectroscopy. *Proceedings of Society of Photo-optical Instrumentation Engineers*, **289**, 289–291.
SMITH, J. P. & LEHR, J. R. 1966. An X-ray investigation of carbonate apatite. *Journal of Agricultural and Food Chemistry*, **14**, 342–349.
VAN KAUWENBERGH, S. J., MCCLELLAN, G. H. & CATHCART, J. B. 1990. Mineralogy and alteration of the phosphate deposits of Florida. *United States Geological Survey Bulletin* (in press).

Experimental study of the substitution of Ca by Sr and Ba in synthetic apatites

JACQUES LUCAS, EL MÂTI EL FALEH & LILIANE PREVOT

Département des Sciences de la Terre, Université Louis Pasteur, Centre de Géochimie de la Surface, CNRS, 1 Rue Blessig, F-67084, Strasbourg Cedex, France

Abstract: Experiments into apatite synthesis through bacterial mediation were carried out using the now well-established methods of either replacement of a carbonate precursor, or direct precipitation from solution. In one set of experiments, strontium was added in varying amounts, either to the solution, or through the carbonate precursor. In another set of experiments, barium was added. The final products were analysed by X-ray diffraction and chemical methods. It is evident that both Ba and Sr can enter the lattice of the synthetized apatite. Strontium is able to replace calcium in any proportion and there is a continuous solid-solution series between purely calcitic and purely strontic end-members. Conversely, barium can replace calcium in only small amounts (less than 10% by weight), and its introduction in the lattice soon decreases the crystallinity of the apatite. The barium phosphate which forms does not belong to the apatite group, but approximates to a phosphate having the formula $BaHPO_4$. This difference in behaviour between Sr and Ba results from their different ionic radii and may help constrain the genesis of natural phosphorites.

Geochemical studies of natural phosphorites have shown that these sediments are usually enriched in a range of elements compared to average sedimentary rocks. Phosphorites are generally composed of several mineral phases (apatite, calcite, clays, siliceous minerals, dolomite, etc.) and organic matter, and it is often difficult to associate precisely specific minor and trace elements with any one of these phases. The recent study of Prévot (1988) showed clearly that some elements are directly linked to the apatite, but that the exact type of association is often difficult to determine for most elements. This is not surprising, since it is well known that trace elements may be: (1) adsorbed onto particles surfaces, (2) coprecipitated in distinct minor mineral phases, (3) associated with organic compounds, (4) introduced in the mineral lattice as substitution for major elements. Abbas (1987) and Lucas & Abbas (1989) clearly showed, for example, that uranium, initially linked to organic matter, is subsequently precipitated and occurs predominantly as particles trapped between the tiny apatite crystallites.

Among the numerous elements which may possibly be incorporated into the apatite structure (McClellan 1980), the 'alkali earths' are very important. This is because, in contrast to the substitution of PO_4^{3-} by CO_3^{2-} or SO_4^{2-}, the replacement of Ca^{2+} by another 'alkali earth' does not modify the electroneutrality of the lattice but changes only its geometry. The content of alkali earths in marine phosphorites is very variable and seemingly not proportional to their contents in seawater. Magnesium, for example, (an element which is known to inhibit apatite formation, Martens & Harris 1970; Nathan & Lucas 1972, 1976) is much more abundant in seawater than calcium, but it constitutes only a very small component in the chemical composition of apatite. Crystallinity decreases as the magnesium content increases, which seems to point to a maximum possible magnesium content in apatite, although this has still to be determined.

Phosphorites contain one of the highest strontium contents of all rock types. The mean content for world phosphorites calculated by Swaine (in Tooms *et al.* 1969) was 1800–2000 μg g^{-1}, while Cook (1972) and McArthur (1978) gave values of between 200 and 2000 μg g^{-1} for sedimentary phosphorites. Prévot & Lucas (1979) reported mean values between 700 and 1300 μg g^{-1} for sedimentary phosphorites from various deposits.

Strontium is the only trace element which in almost all non-weathered phosphorites occurrences is closely correlated with P_2O_5 (Bliskovsky *et al.* 1967; El Mountassir 1977; Lucas *et al.* 1978; Chaabani 1978; Wadjinny 1979; Prévot & Lucas 1979; Belfkira 1980; Abbas 1987). While studying the behaviour of trace elements in phosphorites, Prévot & Lucas (1979) concluded that the distribution of trace elements between apatite and clays does not follow clear rules, except for strontium which is invariably associated with apatite, and displays little or no affinity for clay minerals.

Bliskovsky et al. (1967) have suggested that the strontium contents of calcium-bearing sedimentary minerals reflect some characteristics of sedimentary environments, but only if the element occurs structurally within the crystal lattice. Strontium incorporation into sedimentary minerals has been demonstrated experimentally for carbonates (Holland et al. 1964; Kinsman & Holland 1969) and strontium has been used as an indicator of diagenesis in limestones and dolostones (Kinsman 1969; Veizer & Demovic 1974; Veizer 1978; Renard 1972; 1975, 1979, 1984; Ambroise et al. 1977; Pomerol 1984; Puechmaille 1985).

The presence of strontium in sedimentary phosphorites is generally explained by its substitution for calcium in the apatite structure. But this explanation is not unanimously accepted. Nathan (1981), for example, in comparing analyses of phosphatic samples before and after treatment with the solution of Silvermann et al. (1952), showed that substitution of several elements (including Sr) in apatite lattice is only partial, and that the major part of them is in fact adsorbed on the surfaces of the apatite crystallites. Clearly, it is important to know whether strontium is within or outside of the lattice, particularly if this element is to be taken as an indicator of the environment of apatite formation.

In contrast to strontium, barium is difficult to link to a specific mineral phase in phosphorites. It can sometimes be related to apatite (Arrhenius et al. 1957; Khudolozhkin et al. 1973; Bashir 1977; Price & Calvet 1978) and sometimes to clays minerals (Reeves & Saadi 1971; Prévôt & Lucas 1979, 1980) or to iron oxides (Lucas et al. 1978), its adsorption capability being higher than that of strontium (Burkov in Tambiyev 1979); it also can be associated with organic matter (Puchelt 1967). Barium additionally occurs in the independant mineral phase barite (Tambiyev 1979), which may precipitate in slightly reducing environments (Brosse 1982). Barium behaves similarly in other sedimentary rocks. For example, Guilloux (1982) reported that in carbonates, barium is generally related to the abundance of clay minerals, but is also concentrated as barite and at a late stage, may be incorporated into the carbonate minerals.

Opinions are very divided over the sedimentological significance of barium. Vine (1966) and Lebedev (1967) measured barium contents in shales ranging from low values in freshwater shales to high values in those deposited in marine environments. However, Murray (1954) demonstrated the reverse situation in Indiana and Illinois sediments. Prévôt (1988) reported a wide range of barium contents in phosphorites, not only between deposits, but even from within the same deposit.

The behaviours of strontium and barium are very different, and the question is whether this difference is due only to initially different contents in the parent solutions or in solid precursors, or is controlled by relative ability of these elements to substitute in the apatite structure. The different ionic radii of Ca (1.43Å), Sr (1.48Å) and Ba (1.68Å) (cf. Whittaker & Muntus 1970), indicate that the second factor seems likely to be dominant.

We know of no experimental work that has been undertaken on Sr and Ba substitution in sedimentary phosphate, and in this paper we attempt to partly fill this gap.

Experimental procedure

In the series of experiments described here, we have tried to introduce Sr and Ba into synthetic apatites obtained by the method of Lucas & Prévôt (1981), El Faleh (1988) and Prévôt et al. (1989). These workers have demonstrated that apatite can be obtained either by replacement of a solid precursor (which acts as a calcium-donor) by dissolution-precipitation, or by direct precipitation from solution. Both methods were used to study the possible entry of Sr or Ba into the carbonate-fluorapatite structure.

The main experimental conditions, described in detail by these authors were:
(a) cations were introduced into the system as pure carbonates, $CaCO_3$, $SrCO_3$, $BaCO_3$;
(b) phosphorus was introduced via a ribonucleic acid (RNA, SIGMA type VI) containing 15.5% P_2O_5 and 0.5% CaO;
(c) freshwater for the solutions was the natural untreated tapwater of Strasbourg containing Ca: 9.6 mM l^{-1}, Mg: 2.9 mM l^{-1}, Sr: 0.014 mM l^{-1};
(d) X-ray (XRD) diffraction analyses were performed using CuKα radiation;
(e) the chemical composition of the liquid phase was obtained by inductively coupled plasma-atomic emission spectroscopy (ICP-AES) for Ca, Sr, Ba and P; chemical composition of the solid phase was obtained using ICP-AES for P and quantometry (emission spectroscopy) for Ca, Sr and Ba;
(f) all of the reactions are facilitated when the RNA is hydrolysed with a base or a salt.

Depending on the case, NaOH, CaCO$_3$, SrCO$_3$ or BaCO$_3$ were used and the solutions were respectively called RNA$_{NaOH}$, RNA$_{CaCO_3}$, etc.

Strontium experiments

Synthesis from a carbonate precursor

In these experiments, 50 g of RNA containing 15.5% P$_2$O$_5$ was hydrolysed in one litre of untreated freshwater with approximately 5.5 g of NaOH and 1.5 g NaF added. The elemental composition of this base solution was fairly constant: PO$_4$ 5000 µg ml^{-1}; Na 4000 µg ml^{-1}; F 680 µg ml^{-1}, plus 200 µg ml^{-1} of chlorine originated from the undistilled freshwater used. Another solution with the same PO$_4$ content and containing around 2650 µg ml Sr was prepared by hydrolysis of RNA using SrCO$_3$ instead of NaOH. Final solutions with varying strontium contents were obtained by mixing aliquots of the two solutions.

Chemical and mineralogical change as a function of time. Twelve identical beakers were prepared containing 100 ml of the initial solution to which 465 µg ml^{-1} Sr and 2 g of low Sr (38.5 µg g^{-1}) CaCO$_3$ were added. The experiments were stopped successively at intervals of several days and the material in each of the beakers was analysed (Table 1 and Fig. 1). The mineralization curve for phosphorus perfectly matches that described by Prévôt et al. (1989). After a period of negligible mineralization corresponding to the growth of the micro-organisms, phosphorus was then rapidly incorporated into the solid phase, and the ratio of calcite to apatite (estimated by XRD-analysis) decreased rapidly. The strontium did not follow strictly the same evolution; overall it was incorporated much more slowly than the phosphorus, but mineralization began earlier. Thus more than half of the strontium available at the beginning of the experiment was fixed before apatite formation began, which indicates that the mineralization of strontium, at least initially, was not directly associated with apatite formation. It is likely, therefore, that in the first stage a strontium phosphate is precipitated before calcite begins to dissolve, releasing soluble calcium which is then available for apatite precipitation. Only small amounts of strontium were available in the experiments, and consequently the strontium phosphate would not have been present in sufficient quantities to be detected by XRD. The final solid phase must be composed of either a mixture of apatite and strontium phosphate, or of apatite alone containing the strontium of the dissolved initial metastable strontium phosphate. The experiments described below demonstrate that the latter hypothesis is correct.

Chemical and mineralogical change as a function of strontium concentration in solution. Fourteen beakers, each containing a mixture of the initial RNA$_{NaOH}$ solution and a variable aliquot of a RNA$_{SrCO_3}$ solution were prepared, so that each beaker contained 100 ml of solution with a different strontium content (Table 2). Two grams of calcium carbonate were added to each beaker. The experiment was run for a total of 157 days, after which time the contents of each beaker were analysed. XRD analysis indicated that the initial calcium carbonate had completely disappeared and that, in all cases, the final solid

Table 1. *Replacement of calcium carbonate by apatite. Chemical evolution as a function of time, of solids obtained from the same solution (containing 465 µg ml^{-1} Sr)*

	Time (in days)	P$_2$O$_5$%	CaO%	SrO%	MgO%	Na$_2$O%	% apatite estimated by XRD
sT1	7	0.15	52.70	0.25	0.03	0.07	0
sT2	10	0.13	52.40	0.25	0.03	0.07	0
sT3	14	0.69	54.10	0.30	0.09	0.08	0
sT4	17	0.26	56.50	0.37	0.03	0.05	0
sT5	21	0.98	52.20	0.80	0.15	0.07	0
sT6	24	1.27	54.00	1.08	0.17	0.05	0
sT7	28	2.33	52.50	1.45	0.23	0.07	3
sT8	31	3.06	52.40	1.85	0.25	0.07	4
sT9	35	20.47	50.50	2.69	0.35	0.14	35
sT10	38	25.13	47.00	2.93	0.36	0.26	58
sT11	45	28.96	48.20	3.01	0.31	0.23	61
sT12	53	29.59	48.70	2.92	0.34	0.34	71

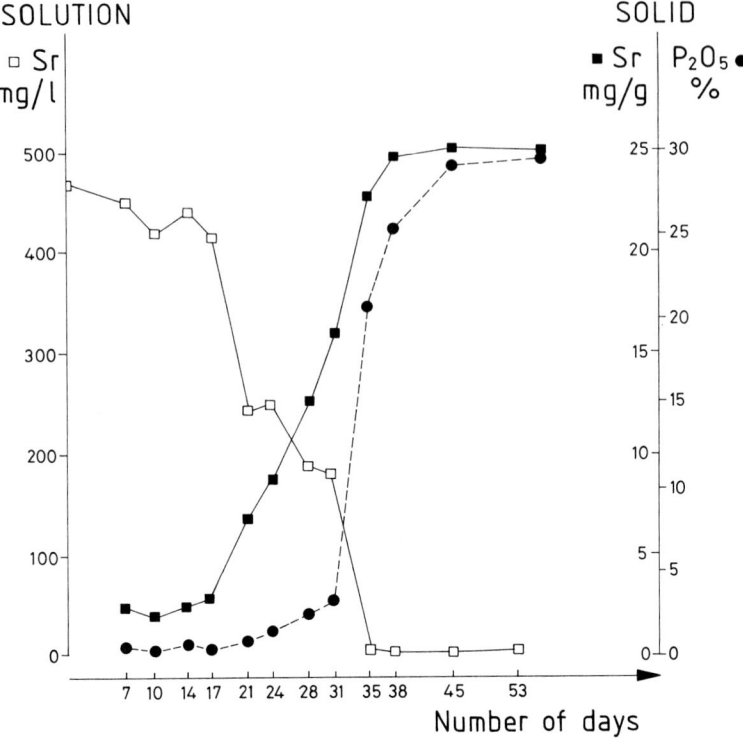

Fig. 1. Changes in the Sr contents of solid and solution and of P_2O_5 contents in the solid as a function of time.

phase contained nothing other than apatite. The compositions of the soluble and solid phases of each beaker are given in Table 2.

The following observations can be made: (1) the P_2O_5 content of the solid fraction is fairly constant, as is also the molecular ratio $(Ca + Sr)/P_2O_5$, which has a mean value of 3.57, corresponding to a CO_2-poor carbonate-fluorapatite (2.12% CO_2 after McClellan 1980); (2) all the initial strontium has been taken up in the solid phase; this is shown by the proportional relationship between available strontium and utilized strontium (Fig. 2); (3) a strong negative correlation exists between calcium and strontium (Fig. 3), indicating clearly that these elements substitute for each other in the mineral structure. These data demonstrate that strontium has substituted for calcium in the apatite formed by replacement of carbonate, and that the amount of this substitution is proportional to the amount of strontium initially available in solution.

Chemical and mineralogical change as a function of the strontium content of the initial carbonate.
A series of nine experiments were carried out using an initial RNA_{NaOH} solution without strontium. The Sr-content of the solution was varied through the addition of 2 g of natural calcium carbonates containing different amounts of Sr (Table 3) to each beaker. Again all of the initial calcite was replaced by apatite which contained all the strontium initially present in the carbonate. In spite of the low strontium concentration concerned, it is clear that the mean molecular ratio Sr/Ca is higher in the apatite than in the calcite (Table 3), which seems to indicate that strontium is taken up preferentially to calcium by apatite and therefore that replacement of carbonate by apatite occurs with conservation of strontium.

Synthesis by precipitation from solution

The two base solutions, RNA_{CaCO_3} and RNA_{SrCO_3} were mixed in varying proportions, in 13 beakers, each containing a final volume of 500 ml of solution. The Ca and Sr-contents of these solutions are given in Table 4. After 10 days, $CaCO_3$ and $SrCO_3$ were no longer detectable by XRD. Table 4 gives the contents of PO_4, Ca and Sr in the solid phase. It can be

Table 2. *Chemical composition of apatite formed by replacement of $CaCO_3$ in solutions containing different proportions of Sr*

	initial solution ($g\,l^{-1}$)	final liquid phase ($mg\,l^{-1}$)				final solid phase (%)					$\dfrac{Ca+Sr}{PO_4}(M)$
	Sr	PO_4	Ca	Sr	Mg	P_2O_5	CaO	SrO	MgO	Na_2O	
St1	0.050	290	3.80	–	0.07	34.73	48.70	0.29	0.29	0.41	3.56
St2	0.065	306	2.98	–	0.02	33.94	47.70	0.53	0.28	0.37	3.59
St3	0.092	288	2.86	–	0.01	34.47	48.40	0.74	0.27	0.41	3.51
St4	0.142	276	3.20	0.06	–	35.12	48.10	1.04	0.27	0.46	3.51
St5	0.283	266	2.66	0.10	–	34.97	49.00	1.65	0.27	0.49	3.62
St6	0.470	294	2.78	0.14	–	34.80	47.20	2.66	0.27	0.39	3.54
St7	0.672	336	3.26	0.19	–	34.12	47.00	3.57	0.26	0.25	3.64
St8	0.848	372	2.56	0.59	–	33.54	44.50	4.51	0.25	0.28	3.55
St9	1.200	332	3.48	1.48	–	32.94	43.90	5.46	0.24	0.16	3.60
St10	1.368	356	6.04	0.94	–	33.98	44.10	6.35	0.23	0.12	3.55
St11	1.580	420	6.70	1.09	–	33.01	43.50	7.24	0.23	0.10	3.65
St12	1.840	412	2.34	0.47	–	32.96	42.20	8.07	0.21	–	3.58
St13	2.160	416	4.96	7.63	0.17	33.05	41.00	8.33	0.20	0.07	3.48
St14	2.346	452	3.52	4.00	0.18	32.84	41.30	9.84	0.21	0.10	3.60
										mean =	3.57

Table 3. *Chemical data for apatite formed by replacement of $CaCO_3$ containing different proportions of Sr*

	Carbonate precursor								Synthesized solid						
	(%)		$(10^{-2}$ M kg$^{-1})$		Sr/Ca	(%)			$(10^{-2}$ M kg$^{-1})$			$\dfrac{Sr+Ca}{PO_4}$	Sr/Ca	$\dfrac{Sr/Ca_{carb.}}{Sr/Ca_{apat.}}$
	Ca	Sr	Ca	Sr	$(\times 10^3)$	Ca	Sr	PO_4	Ca	Sr	PO_4		$(\times 10^3)$	
ST1	29.17	0.02	729	0.23	0.31	29.64	0.03	21.52	741	0.34	227	3.27	0.459	0.68
ST2	37.06	0.03	926	0.34	0.37	32.50	0.04	21.42	812	0.45	225	3.61	0.554	0.66
ST3	38.37	0.05	959	0.57	0.59	32.50	0.04	20.93	812	0.45	220	3.69	0.554	0.89
ST4	37.71	0.07	943	0.80	0.85	29.00	0.05	19.56	725	0.57	206	3.52	0.79	1.07
ST5	37.29	0.07	932	0.80	0.86	34.14	0.09	22.41	853	1.02	236	3.61	1.20	0.71
ST6	34.79	0.08	870	0.91	1.05	31.29	0.10	20.74	782	1.14	218	3.59	1.458	0.72
ST7	35.71	0.08	893	0.91	1.02	30.93	0.10	21.33	773	1.14	225	3.44	1.475	0.69
ST8	37.79	0.17	945	1.93	2.04	34.21	0.19	21.38	855	2.16	225	3.81	2.526	0.95
ST9	38.21	0.88	955	10.0	10.47	35.50	0.89	21.63	887	10.11	228	3.93	11.40	0.92
												mean : 3.61		0.81

Table 4. Chemical data for apatite formed by precipitation from solutions containing different proportions of Sr

	Initial solution					Synthetized apatite										
	(g l^{-1})		(10^{-3} M l^{-1})		Sr/Ca	mass	(%)			(10^{-2} M Kg^{-1})			Ca+Sr	Sr/Ca	Sr/Ca sol.
	Ca	Sr	Ca	Sr	(× 10^3)	(g)	PO$_4$	Ca	Sr	PO$_4$	Ca	Sr	PO$_4$	(× 10^3)	Sr/Ca apat.
Sp1	2.280	0.001	57.0	0.011	0.19	3.20	21.57	30.29	—	227	797	—	3.34	—	—
Sp2	2.000	0.003	50.0	0.034	0.68	3.00	19.78	27.43	0.05	208	685	0.56	3.30	0.8	0.85
Sp3	1.980	0.005	49.0	0.057	1.16	2.93	19.85	27.43	0.09	209	686	1.02	3.29	1.49	0.78
Sp4	1.940	0.008	48.0	0.091	1.89	2.91	19.24	26.79	0.19	203	670	2.16	3.31	3.22	0.59
Sp5	1.900	0.014	47.0	0.159	3.38	2.80	20.43	28.00	0.28	215	700	3.18	3.27	4.54	0.74
Sp6	1.586	0.028	40.0	0.318	7.95	3.10	19.18	26.57	0.58	202	664	6.59	3.32	9.92	0.80
Sp7	1.580	0.046	39.5	0.523	13.24	3.00	20.04	27.64	1.00	211	691	9.09	3.32	13.15	1.00
Sp8	1.500	0.095	37.5	1.079	28.79	2.70	19.13	26.50	2.04	201	662	23.2	3.26	35.05	0.82
Sp9	1.010	0.459	35.0	5.625	160.7	2.40	21.27	25.21	9.72	224	630	110	3.30	174.6	0.92
Sp10	0.816	0.961	20.5	10.92	532.7	2.60	19.62	18.79	19.01	207	470	216	3.31	459.6	1.16
Sp11	1.358	1.410	9.0	16.02	1780	3.10	13.26	8.94	21.64	140	223	246	3.35	1103	1.61
Sp12	0.211	1.928	5.3	21.91	4134	3.00	15.52	6.21	32.07	163	155	364	3.18	2348	1.76
										(mean of Sp1 to 12)			3.296		
Sp13	0.194	2.640	4.85	30	6186	3.05	15.06	2.89	42.56	159	72	484	3.50	6722	1.09

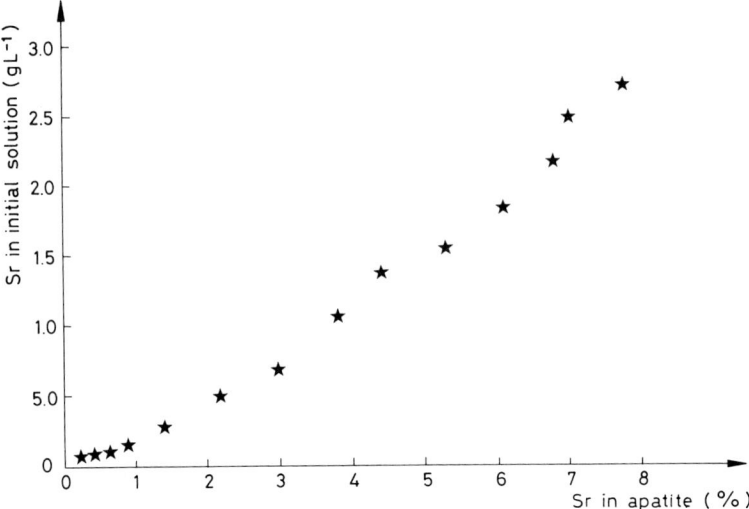

Fig. 2. Relationships between the initial Sr content in solution and the Sr content of apatites formed by replacement of carbonate.

seen from these data data that all of the calcium and the strontium available in the initial solution were incorporated into the solid phase, in which they have substituted stoichiometrically; the molecular ratio $(Ca + Sr)/PO_4$ remained fairly constant, with a mean value of 3.366.

It is worth giving special attention to experiment Sp13 which contained mainly strontium, the very low calcium content reflecting the calcium content of freshwater. The XRD-data for the solid phase from this experiment (Fig. 4 and Table 5) corresponds to a strontium phosphate of the apatite family, fairly similar to the reference mineral proposed by ASTM, with the main peak located at 2.90 Å. The differences between the synthetic mineral and that listed by ASTM are probably a consequence of some substitutions of P_2O_5 by CO_2 and of Sr by Ca. Conversely, experiment Sp1 which contained only calcium gave an XRD-pattern which is characteristic of a common carbonate–fluorapatite, with the main peak at 2.77 Å. For the experiments between these extremes, the diagrams (Fig. 5) do not correspond to a mixture of both end-members, but to intermediate minerals whose main peak (211) migrates from 2.77 to 2.90 with increasing strontium content.

It may be concluded that calcium can be replaced by strontium in apatite in all proportions, resulting in a solid-solution series which displays continuous deformation of the crystal structure between purely calcic and purely strontic end members. These minerals whose cation/anion ratios (as represented by their $(Ca + Sr)/PO_4$ contents) remain constant, can be produced by replacement of a solid precursor or by direct precipitation, with initial strontium being available either in the solid precursor or in the solution. The apatitic replacement of a strontium-containing precursor occurs with conservation of the strontium with, in some cases, the formation of strontium apatites taking place in preference to calcium apatites.

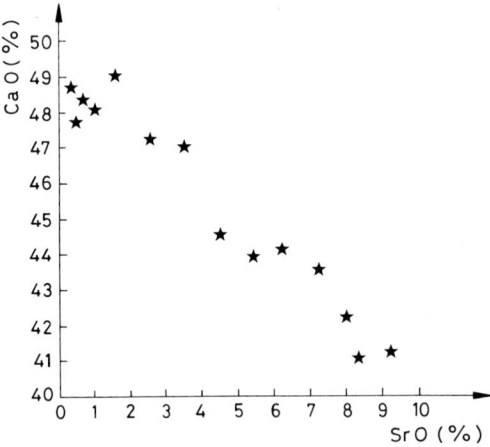

Fig. 3. Relationships between the Ca and Sr contents of apatites formed by replacement of carbonate.

Table 5. *Structural and chemical data on synthetic Sr-apatite obtained by precipitation.*

X-ray diffraction			
Synthetic Sr-apatite		$Sr_{10}(PO_4)_6 OH_2$ ASTM 14–691	
$d(Å)$	I	$d(Å)$	I
		8.45	6
		6.18	8
		4.59	8
4.21	13	4.22	16
		4.05	8
3.78	8		
3.62	24	3.63	18
3.44	8		
		3.34	18
3.30	48		
3.19	48	3.19	20
2.98	35		
2.90	100	2.91	100
		2.81	60
2.74	31	2.75	8
		2.46	6
2.38	7		
2.32	13	2.32	8
		2.29	6
2.26	3		
2.21	6	2.22	10
2.17	9	2.17	8
2.15	11		
2.10	4	2.10	6
2.01	35	2.02	20
1.955	21	1.97	14
1.92	33	1.93	20
1.86	16	1.87	14
		1.84	14
		1.83	14
1.81	31	1.82	12
1.77	6		
		1.71	4
1.685	4	1.69	4
		1.67	4
1.605	5	1.60	4
1.573	11	1.58	6
		1.53	8
1.52	14	1.52	6
1.504	12		
		1.49	8

Chemical analysis (%)	
	%
P_2O_5	20.16
SrO	50.30
CaO	4.05
MgO	0.64
Na_2O	0.16
F	1.70
LOI 1000°C	19.64
LOI 100°C	3.93
CO_2	1.80

Barium experiments

Synthesis from a carbonate precursor

It is difficult to find natural calcium carbonates with a range of barium contents, and consequently only experiments in which barium was added in solution could be used to consider the relations between apatite and barium. As in the previous strontium experiments, twelve beakers were prepared, each containing 2 g $CaCO_3$ and 100 ml of a mixture of the RNA_{NaOH} and RNA_{BaCO_3} solutions giving barium contents ranging from 11 to 2360 mg l^{-1} (Table 6). After 160 days, the experiments were stopped. XRD analyses indicated that all $CaCO_3$ had disappeared, and this was also confirmed by determination of the stability of the solid after treatment by the method of Silvermann *et al.* (1952). All the barium available in the initial solution was found in the final solid, consequently the barium content in the solid is almost proportional to the initial barium content of the solution. Chemical analyses (Table 6) indicate that the barium substitutes for calcium in the apatite lattice in the molecular ratio (Ca + Ba)/PO_4 M = 3.655, except in the final experiment with the highest barium content where the ratio deviated a little more from the mean value.

Synthesis by precipitation from solution

An RNA_{BaCO_3} solution was prepared containing negligible calcium (that found in normal freshwater). The composition (mg l^{-1}) of this solution was: 4964 PO_4; 3484 Ba; 180 Ca; 25 Mg; 2392 Na; 685 F and pH 7.6. After 110 days, the chemical composition of the precipitated solid was that of a slightly carbonated barium phosphate with some substitutions of barium by calcium. The structure of the mineral, as determined by XRD, is much closer to a barium phosphate with the formula $BaHPO_4$ (Table 7) than to a baritic apatite ($Ba_{10}(PO_4)_6F_2$), which seemed to be impossible to synthetize in our experimental conditions, whereas it can be obtained easily through a purely chemical synthesis (Akhavan-Niaki & Wallaeys 1958; Akhavan-Niaki & Montel 1959; Akhavan-Niaki 1961; Mohseni-Koutchesfehani 1961; Mohseni-Koutchesfehani & Montel 1961a, b). No trace of barium carbonate could be detected at the conclusion of the experiment.

Ten solutions with varying calcium and barium contents were obtained by mixing different proportions of the RNA_{CaCO_3} and RNA_{BaCO_3} solutions. After 260 days the final

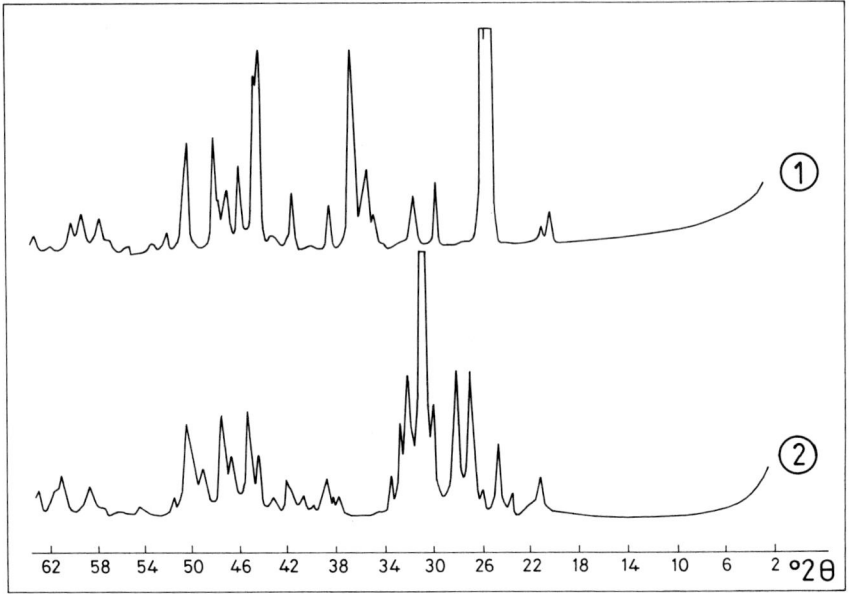

Fig. 4. X-ray diffraction (XRD) patterns of an apatite formed by precipitation (2), compared with that of strontianite ($SrCO_3$) (1).

products were analysed (Table 8 and Fig. 6). Again all the calcium and all the barium available at the beginning of the experiment were found in the solid product, which suggests that the solid phase can indeed reflect the composition of the initial environment. In experiments Bp1 to Bp8 (i.e. those with barium contents equal or less than 10% in weight of solid), apatite was the only mineral formed, and the higher the barium content of this apatite, the poorer its crystallinity (Fig. 7). In spite of the degradation of the crystallinity, the molecular ratio $(Ca + Ba)/PO_4$ remained constant, which demonstrated that barium had replaced calcium in the apatite lattice. In experiments Bp9 and Bp10, in which the barium content exceeded 10%, a barium phosphate similar to that described above was found in addition to the poorly-crystallized apatite.

It may be concluded that barium only substitutes for less than 10% of the calcium in apatite and that this substitution decreases the crystallinity of the apatite, even when the barium content is very low.

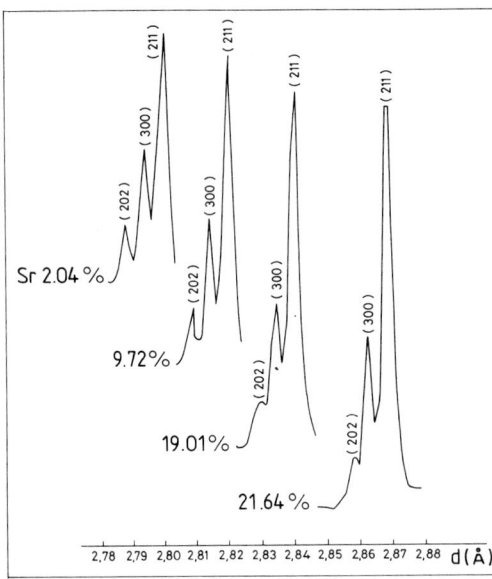

Fig. 5. Changes in the positions of (211), (300) and (202) peaks in response to changing Sr content for apatites formed by precipitation.

Discussion and conclusion

The two elements studied here, strontium and barium, are two alkali earths with chemical properties resembling those of calcium; their diadochical substitution in apatites depends essentially on the ratio of their ionic radii. Our experiments demonstrate that strontium, whose radius differs only by 3.5% from the radius of

Table 6. *Chemical compositions of phosphates formed by replacement of $CaCO_3$ in solutions containing different proportions of Ba.*

	Initial solution ($g\,l^{-1}$) Ba	Synthetized phosphate (%)					$\dfrac{Ca+Ba}{PO_4}$ (M)
		PO_4	Ca	Ba	Mg	Na	
Bt1	0.011	22.98	34.21	0.06	0.16	0.37	3.53
Bt2	0.027	23.07	33.71	0.13	0.16	0.36	3.47
Bt3	0.056	23.64	34.14	0.24	0.16	0.34	3.43
Bt4	0.096	23.62	34.50	0.41	0.17	0.43	3.48
Bt5	0.309	23.78	34.21	1.37	0.17	0.46	3.46
Bt6	0.490	21.71	34.07	2.76	0.25	–	3.80
Bt7	0.972	22.55	33.21	4.32	0.19	0.33	3.63
Bt8	1.316	21.54	32.00	5.75	0.20	0.15	3.70
Bt9	1.464	21.62	31.86	6.42	0.18	0.14	3.70
Bt10	1.752	21.56	31.00	7.57	0.18	0.04	3.66
Bt11	2.000	21.29	31.93	7.94	0.18	0.03	3.82
					(mean Bt1 to 11)		3.61
Bt12	2.360	19.59	29.93	10.02	0.21	0.03	3.98

calcium, is an easy substituant, whereas barium, because of a radius divergence of 30%, is a less suitable substituant for calcium. Strontium may substitute for calcium in apatite in all proportions while preserving the same structure, which range from the two extremes, one purely calcitic, the other purely strontic. Barium could only substitute in approximately 10% of the sites occupied by calcium in apatite. Beyond 10% substitution, the deformation of the apatite lattice becomes too great and the barium complexes with phosphorus to precipitate a specific barium phosphate mineral which does not belong to the apatite family.

Conclusions based on our experimental data are in good agreement with observations on

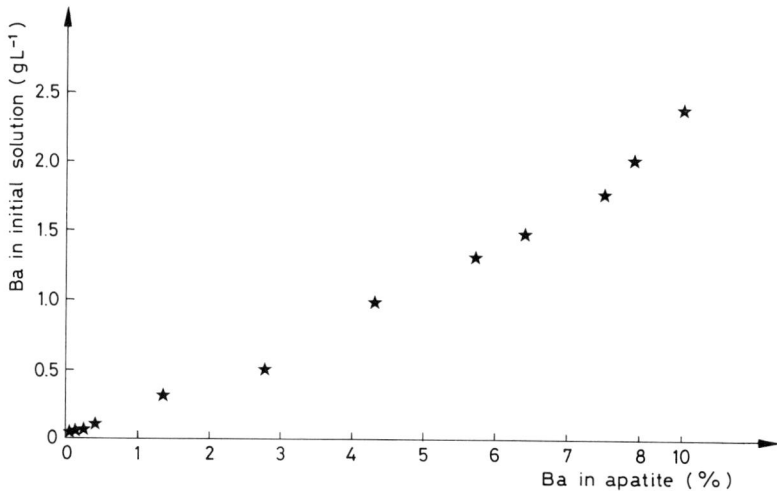

Fig. 6. Relationship between the initial Ba content in solution and the Ba content of apatites formed by replacement of carbonate.

Table 7. Structural and chemical data for synthetic Ba-phosphate obtained by precipitation.

Ba H PO$_4$ ASTM 17−929		Synthetic phosphate		Ba$_{10}$(PO$_4$)$_6$F$_2$ ASTM 16−803	
d(Å)	I	d(Å)	I	d(Å)	I
		7.047	4	5.08	2
4.45	30	4.442	44	4.40	14
		4.345	26	4.24	4
4.22	8			3.85	12
		4.027	5	3.53	35
3.89	10			3.30	30
3.85	20	3.861	29	3.07	100
3.59	100	3.577	100	3.05	95
3.54	50	3.533	78	2.932	50
		3.451	32	2.897	2
3.23	8	3.252	8	2.440	6
3.19	8	3.202	10	2.333	2
		3.109	20	2.291	12
3.06	10			2.800	4
3.00	25	3.001	28	2.120	35
2.857	35	2.851	40	2.061	40
		2.840	41	2.031	40
2.800	6			2.018	12
		2.736	13	1.952	30
		2.623	5	1.925	20
2.515	55	2.511	50	1.920	30
2.370	10			1.910	40
2.358	10			1.764	6
2.296	30	2.292	28	1.718	4
2.249	12	2.250	8	1.666	25
2.223	10	2.221	17	1.653	4
2.182	12			1.609	20
		2.172	15	1.600	18
		2.110	29	1.589	18
2.097	6			1.548	25
2.036	6			1.397	35
2.004	20	1.997	14	1.386	10
1.967	16	1.968	8	1.370	12
1.926	16	1.928	6	1.360	30
1.862	6			1.354	18
1.842	12			1.346	8
1.792	6			1.322	30
1.776	6				
1.762	16				
		1.757	9		
1.689	12	1.685	13		
1.675	10	1.671	10		
1.661	6				
1.649	10				
		1.638	5		
1.597	12				
1.575	14	1.577	7		

Chemical analysis (%)	
	%
P$_2$O$_5$	23.27
BaO	53.85
CaO	2.80
MgO	0.49
Na$_2$O	0.06
F	2.6
CO$_2$	1.8
LOI 1000°C	15.2
LOI 100°C	3.63

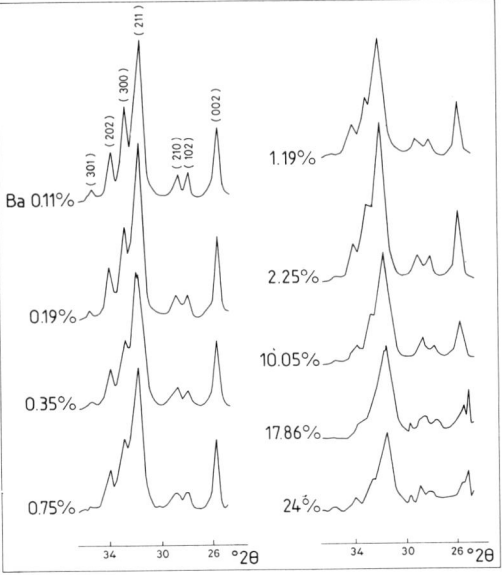

Fig. 7. Influence of Ba substitution on the crystallinity of apatites formed by precipitation. Note broadening of peaks with increasing Ba content.

natural phosphorites, in which, as has been pointed out for a long time, strontium is more abundant than barium. It can now be assumed that the good correlation between Sr and P$_2$O$_5$ always found in natural phosphorites is mainly due to substitution of calcium by strontium. Strontium seems to be taken up preferentially by calcium, even in the experiments with very low initial strontium contents. This behaviour may explain the common occurrence of natural strontic apatites such as florencite, crandallite and goyazite, which are forming during weathering of phosphorites. During the weathering process, strontium may be retained more strongly in the residual minerals or may be more easily re-utilized in the newly formed secondary minerals than calcium, which is more easily leached.

Our data contrast with the results of Nathan (1981), which suggest that at least part of the strontium in phosphorites does not occur in the apatite lattice. It is necessary to take into account this observation of Nathan (1981) and the question arises as to whether this additional strontium is adsorbed after formation of the apatite, for it is clearly important to distinguish the genesis of apatite and the deposition of phosphorite as a whole. It is suggested, therefore, that strontium could be a valuable environmental indicator since the mean Sr/Ca

Table 8. *Chemical data for Ca and Ba phosphates formed by precipitation from solutions containing different proportions of Ba.*

	Initial solutions					Precipitated phosphate						
	(g l^{-1})		(10^3 M l^{-1})			(%)			(10^{-2} M kg^{-1})			$\dfrac{Ca+Ba}{PO_4}$
	Ca	Ba	Ca	Ba	PO$_4$	Ca	Ba	PO$_4$	Ca	Ba	
Bp 1	1.800	0.003	45.00	0.022	21.39	30.93	0.06	225	773	0.43	3.44
Bp 2	1.700	0.005	42.50	0.036	21.84	30.21	0.11	230	755	0.80	3.29
Bp 3	1.650	0.008	41.25	0.058	20.26	28.07	0.19	213	702	1.37	3.30
Bp 4	1.620	0.015	40.50	0.109	20.98	29.64	0.35	221	741	2.54	3.36
Bp 5	1.606	0.029	40.15	0.21	21.83	29.79	0.75	230	745	5.43	3.41
Bp 6	1.520	0.050	38.00	0.36	22.08	29.64	1.19	232	741	8.62	3.23
Bp 7	1.500	0.097	37.50	0.70	21.27	28.50	2.25	224	712	16.30	3.25
Bp 8	1.450	0.510	36.25	3.37	20.68	25.50	10.05	218	637	72.83	3.26
Bp 9	1.220	0.999	30.50	7.24	19.15	19.71	17.86	202	477	129.4	3.00
Bp10	0.961	1.220	24.00	8.84	18.56	16.00	24.00	195	400	173.9	2.94
Bp 0	0.180	3.484	4.50	25.25	15.56	2.00	48.25	164	50	349.6	2.43

ratio of the apatite appears to reflect the original environment of the mineral genesis (which is generally fairly similar in most marine phosphorites), whereas variations of the ratio in different deposits would reflect their post-depositional history and relate to factors such as reworking or weathering.

The analyses used in this paper formed part of the doctoral thesis work of El Falch at the University of Strasbourg. We are grateful to P. Cook for his assistance in 'translating' our French English into English, and to I. Jarvis for his constructive criticism of the initial manuscript.

References

ABBAS, M. 1987. *Géochimie de l'uranium des phosphorites des Palmyrides Centrales, Syrie.* Thèse Doctorat ès Sciences, Université Louis Pasteur de Strasbourg.

AKHAVAN-NIAKI, A. N. 1961. Contribution à l'étude des substitutions dans les apatites. *Annales Chimiques*, **6**, 51–79.

—— & WALLAEYS, R. 1958. Préparation des fluorapatites strontique et barytique et des solutions solides et fluorapatite alcalino-terreuses par réaction dans l'état solide. *Comptes Rendus de l'Académie des Sciences, Paris*, **296**, 1556–1558.

—— & MONTEL, G. 1959. Sur la substitution des cations dans le réseau de la fluorapatite. *Comptes Rendus de l'Académie des Sciences, Paris*, **248**, 2486–2488.

AMBROISE, D., AZEMA, J., CHAYE D'ALBISSIN, M., FOUCAULT, A., FOURCADE, E., LEIKINE, M., MELIERES, F., MOVCHET, J. & RENARD, M. 1977. Crétacé inférieur du Montemayor d'Ayora (Province de Valence, Espagne); essai sur les conditions de sédimentation. *Bulletin de la Société Géologique de France*, **7, 19**, 1275–1284.

ARRHENIUS, G. O., BRAMLETTE, M. N. & PICCIOTTO, E. 1957. Localization of radioactive and stable heavy nuclides in oceanic sediments. *Nature*, **180**, 85–86.

BASHIR, S. 1977. *Radioactivité et état d'équilibre radioactif de quelques phosphates jordaniens. Leurs caractères pétrologique, minéralogique, cristallographique et géochimique.* Thèse Docteur-Ingénieur ENSG, Nancy.

BELFKIRA, O. 1980. *Evolutions sédimentologiques et géochimiques de la série phosphatée du Maestrichtien des Ouled Abdoun, Maroc.* Thèse de 3e cycle, Université de Grenoble.

BLISKOVSKIY, V. Z., YEFIMOVA, V. A. & ROMANOVA, L. V. 1967. On the strontium content in phosphorites. *Litologia i Poleznye Iskopaemye*, **6**.

BROSSE, E. 1982. *Contribution à la minéralogie et à la géochimie des sédiments pélagiques profonds – Comparaison des 'blacks-shales' du Crétacé dans l'Atlantique Central Nord et des dépôts du Malm et du Crétacé en Briançonnais.* Thèse Docteur-Ingénieur, Ecole Nationale Supérieure des Mines de Paris.

CHAABANI, F. 1978. *Les phosphorites de la coupe type de Foum Selja, Metlaoui, Tunisie. Une série sédimentaire séquentielle à évaporites du Paléogène.* Thèse de 3e cycle, Université Louis Pasteur de Strasbourg.

COOK, P. J. 1972. Petrology and geochemistry of the phosphate deposits of Northwest Queensland, Australia. *Economic Geology*, **67**, 1193–1213.

EL FALEH, E. M. 1988. *Les mécanismes de synthèse de l'apatite par activité bactérienne; rôle et comportement de quelques éléments minéraux. Application aux phosphates sédimentaires.* Thèse de Doctorat de l'Université de Strasbourg.

EL MOUNTASSIR, M. 1977. *La zone rubéfiée de Sidi Daoui. Altération météorique du phosphate de chaux des Ouled Abdoun, Maroc.* Thèse de 3e cycle, Université Louis Pasteur de Strasbourg.

GUILLOUX, L. 1982. Etude chimique des séries porteuses de quelques grands gisements du type Kupferschiefer. *Sciences de la Terre, Mémoire* **43**, Nancy.

GULBRANDSEN, R. A. 1966. Chemical composition of phosphorites of the Phosphoria Formation. *Geochimica et Cosmochimica Acta*, **30**, 769–778.

HOLLAND, H. D., HOLLAND, H. J. & MUNOZ, J. L. 1964. The coprecipitation of cations with $CaCO_3$ – II. The coprecipitation of Sr^{2+} with calcite between 90° C and 110° C. *Geochimica et Cosmochimica Acta*, **28**, 1287–1301.

KHUDOLOZHKIN, V. O., URUSOV, V. S. & TOBELKO, K. J. 1973. Distribution of cations between sites in the structure of Ca, Sr, Ba – apatites. *Geochemistry International*, **10**, 266–269.

KINSMAN, J. D. 1969. Interpretation of Sr^{+2} concentration in carbonate minerals and rocks. *Journal of Sedimentary Petrology*, **39**, 486–508.

—— & HOLLAND, H. D. 1969. The coprecipitation of cations with $CaCO_3$ – IV. The coprecipitation of Sr^{2+} with aragonite between 16° C and 96° C. *Geochimica et Cosmochimica Acta*, **33**, 1–17.

LEBEDEV, L. M. 1967. *Metacolloids in endogenic deposits.* Plenum Press, New York.

LUCAS, J. & ABBAS, M. 1989. Uranium in natural phosphorites: the Syrian example. *Sciences Géologiques, Bulletin* (in press).

—— & PREVOT, L. 1981. Synthèse d'apatite à partir de matière organique phosphatée (ARN) et de calcite par voie bactérienne. *Comptes Rendus de l'Académie des Sciences, Paris*, **292**, II, 1203–1208.

——, & LAMBOY, M. 1978. Les phosphorites de la marge nord de l'Espagne; chimie, minéralogie, genèse. *Oceanologica Acta*, **1**, 55–71.

MARTENS, C. S. & HARRISS, R. C. 1970. Inhibition of apatite precipitation in the marine environment by magnesium ions. *Geochimica et Cosmochimica Acta*, **34**, 621–625.

MCARTHUR, J. M. 1978. Systematic variations in the contents of Na, Sr, CO_3 and SO_4 in marine carbonate-fluorapatite and their relation to weathering. *Chemical Geology*, **21**, 89–112.

MCCLELLAN, G. H. 1980. Mineralogy of carbonate fluorapatites. *Journal of the Geological Society, London*, **137**, 675–681.

—— & CLAYTON, W. R. 1980. Francolite: The

Commercial Phosphate Mineral. *2nd International Congress on Phosphorus Compounds Proceedings, Boston*, 131–143.

MOHSENI-KOUTCHESFEHANI, S. 1961. *Contribution à l'étute des apatites barytiques*. Thèse Doctorat ès Sciences, Université de Paris, Masson.

—— & MONTEL, G. 1961*a*. Sur la préparation et quelques propriétés de l'hydroxyapatite barytique. *Comptes Rendus de l'Académie des Sciences, Paris*, 1026–1028.

—— & —— 1961*b*. Sur la synthèse de la carbonate-apatite barytique. *Comptes Rendus de l'Académie des Sciences, Paris*, 1161–1162.

MURRAY, H. H. 1954. Genesis of clay minerals in some Pennsylvanian shales of Indiana and Illinois. *Proceedings, 2nd National Conference 'Clays and clay minerals'*, 47–67.

NATHAN, Y. 1981. Structural position of trace elements in apatites. Implication for their recovery. *Israel Journal of Earth Sciences*, **30**, 31–34.

—— & LUCAS, J. 1972. Synthèse de l'apatite à partir du gypse; application au problème de la formation des apatites carbonatées par précipitation directe. *Chemical Geology*, **9**, 99–112.

—— & —— 1976. Expériences sur la précipitation directe de l'apatite dans l'eau de mer. Implication dans la genèse des phosphorites. *Chemical Geology*, **18**, 181–186.

POMEROL, B. 1984. *Géochimie des craies du bassin de Paris. Utilisation des éléments traces et des isotopes stables du carbone et de l'oxygène en sédimentologie et en paléo-océanographie*. Thèse Doctorat ès Sciences, Université Paris VI.

PREVOT, L. 1988. *Géochimie et pétrographie de la formation à phosphate des Ganntour (Maroc). Utilisation pour une explication de la genèse des phosphorites crétacé-éocènes*. Thèse Doctorat ès Sciences, Université de Strasbourg and Mémoire de la Société Géologique de France (in press).

—— & LUCAS, J. 1979. Comportement de quelques éléments traces dans les phosphorites. *Sciences Géologiques, Bulletin*, **32**, 91–105.

—— & —— 1980. Behaviour of some trace elements in phosphatic sedimentary formation. *In*: BENTOR, Y. K. (ed.) *Society of Economic Paleontologists and Mineralogists, Special Publication*, **29**, 31–39.

——, EL FALEH, E. M. & LUCAS, J. 1989. Details on synthetic apatites formed through bacterial mediation. Mineralogy and chemistry of the products. *Sciences Géologiques Bulletin* (in press).

PRICE, N. B. & CALVET, S. E. 1978. The geochemistry of phosphorites from Namibian Shelf. *Chemical Geology*, **23**, 151–170.

PUCHELT, H. 1967. Zur Geochemie des Bariums in exogenen Zyklus. *Sitzungsberichte der Heidelberger Akademie der Wissenschaften, Mathematisch-Naturwissenschaftliche Klasse*, **4**, 87–205.

PUECHMAILLE, CH. 1985. Teneurs en strontium et magnésium dans les tests de foraminifères planctoniques. Premiers indices de l'altération post-mortem. *Bulletin de l'Institut de Géologie du Bassin d'Aquitaine, Bordeaux*, **38**, 81–94.

REEVES, M. J. & SAADI, T. A. K. 1971. Factors controlling the deposition of some phosphate bearing strata from Jordan. *Economic Geology*, **66**, 451–465.

RENARD, M. 1972. Interprétation des teneurs en strontium des carbonates du Lutétien supérieur à Saint-Vaast-Les-Mello (Oise). *Bulletin d'Information de l'Association des Géologues du Bassin de Paris*, **34**, 19–29.

—— 1975. Etude géochimique de la fraction carbonatée d'un faciès de bordure de dépôt gypseux. Exemple du gypse ludien du Bassin de Paris. *Sedimentary Geology*, **13**, 191–231.

—— 1979. Aspects géochimiques de la diagenèse des carbonates. *Bulletin du Bureau de Recherches Géologiques et Minières*, section IV, 113–152.

—— 1984. *Géochimie des carbonates pélagiques. Mise en évidence des fluctuations de la composition des eaux océaniques depuis 140 M.a*. Thèse Doctorat ès Sciences, Université Paris VI.

SILVERMANN, S. R., FUYAT, R. K. & WEISER, J. D. 1952. Quantitative determination of calcite associated with carbonate-bearing apatites. *American Mineralogist*, **37**, 211–222.

TAMBIYEV, S. B. 1979. Strontium and barium in the process of oceanic phosphorite formation. *Okeanologiya*, **19** (2).

TOOMS, J. S., SUMMERHAYES, C. P. & CRONAN, D. S. 1969. Geochemistry of marine phosphate and manganese deposits. *Oceanography and Marine Biology Annual Review*, **1**, 49–100.

VEIZER, J. 1978. Simulation of limestone diagenesis-a model based on strontium depletion: Discussion. *Canadian Journal of Earth Sciences*, **15**, 1683–1685.

—— & DEMOVIC, R. 1974. Strontium as a tool in facies analysis. *Journal of Sedimentary Petrology*, **44**, 93–115.

VINE, J. D. 1966. Element distribution in some shelf and geosynclinal black shales. *United States Geological Survey Bulletin*, **1214**–E.

WADJINNY, A. 1979. *Milieu de sédimentation et mécanismes de dépôt des couches inférieures de la série phosphatée de Ben Guérir (Ganntour, Maroc); une étude séquentielle*. Thèse de 3e cycle, Université Louis Pasteur de Strasbourg.

WHITTAKER, E. J. W. & MUNTUS, R. 1970. Ionic radii for use in geochemistry. *Geochimica et Cosmochimica Acta*, **34**, 945–956.

Humic substances in phosphorites: occurrence, characterization and significance

Y. NATHAN

Geological Survey of Israel, 30 Malkhei Israel Street, 95 501, Jerusalem, Israel

Abstract: Humic substances form an appreciable part of the organic matter in many phosphorites of various ages, including some deposited as early as Cambrian times. In some phosphorites the humic substances are primary, having formed during sedimentation or early diagenesis, and seem to have been preserved as such. In others they appear to be secondary and were probably formed by oxidation of kerogen. While the presence of humic substances in phosphorites (or any sediment) does not necessarily indicate thermal immaturity, their characterization (aromaticity, size of particles, etc.) provides valuable information about their environment of deposition (oxic, anoxic) and their geological history (extent of diagenesis, burial depth, thermal history).

The results of the present work show clearly that the humic substances in the Permian phosphorites of the Western US Phosphate Field were derived from late degradation of kerogen, while those in the Eocene phosphorites from Tunisia are primary. The origin of humic substances in other fields is less clear-cut and it is possible that in some instances, for example, in the Late Cretaceous phosphorites from the Negev, Israel, they are polygenetic.

Humic substances are among the most widely distributed organic materials in the Earth. They are the principal organic component of soils and waters (Schnitzer & Kodama 1977) and form a substantial proportion of the organic matter in recent sediments (Nissenbaum & Kaplan 1972; Morris & Calvert 1977). Under special conditions they may also constitute a considerable fraction of the organic matter in older sediments. Regarding the latter, it has been stated that 'they are a valuable source of information about the original organic matter and deposit media' (Huc & Durand 1977). Nevertheless, the attitude of most organic geochemists concerning humic substances has been generally similar to that of Vandenbroucke *et al.* (1985), who wrote: 'The organic matter of ancient sediments yields only small quantities of humic substances which make their study uninteresting'.

Phosphorites appear to be an exception as most organic matter studies of them report high values of humic substances (humic substances form an appreciable part of the organic matter). Some examples are: the Cambrian Monastery Creek phosphorites of the Georgina Basin, northeastern Australia (Sandstrom 1982, 1986); the Permian Phosphoria phosphorites of the Western Mountain Field, United States (Mair 1985); the Upper Cretaceous Mishash phosphorites of the Negev, southern Israel (Bein *et al.* 1980; Amit & Bein 1982); the Eocene Metlaoui phosphorites of the Gafsa Basin, Tunisia (Belayouni & Trichet 1980; Belayouni *et al.* 1982); and the Miocene Pungo River phosphorites of North Carolina, United States (Mair 1985). The unusually high content of humic substances found in such 'ancient' phosphorites has been noted by these authors who have explained: (1) their genesis through microbial degradation (oxidation) during the early stages of diagenesis, or oxidation related to the winnowing stage of phosphorite concentration (Amit & Bein 1982; Sandstrom 1986), and (2) their preservation through: (a) mechanical protection due to inclusion of the humic substances within phosphate grains (Belayouni 1983); (b) chemical protection resulting from adsorption of the humic substances on the surface of carbonate or apatite grains (Sandstrom 1986); and (c) low thermal maturation of the phosphorite because of shallow burial (Amit & Bein 1982). All the above authors agreed that the organic matter is derived mainly from marine micro-organisms and that the origin of humic substances and phosphorite are closely associated. Belayouni *et al.* (1982) went one step further and claimed that humic substances play an active role during phosphorite genesis.

Because humic substances are mixtures of materials whose sole common characteristic is solubility in basic solutions, characterization of the humic substances from phosphorites of different ages, depositional environments and weathering profiles will further our understanding of phosphorites and their humic contents.

Material

Phosphorites from India, Israel, Jordan, Togo, Tunisia and the United States (Idaho, Utah and North Carolina), representing an age range from Palaeozoic to the Miocene, were studied. All the samples, other than those from Israel, came from the International Fertilizer Development Center (IFDC) collection. The Israeli samples were supplied by the Negev Phosphate Company. Eighteen samples, both run-of-mine ores and concentrates, which were thought to be representative of commercial phosphorites from these areas, were examined. One sample SVK-3 (Togo) represents a non-economic horizon. Four samples represent the Western US Phosphate Field. Two of these, both from Idaho (SVK-1 and USHM LLUV), represent weathered rocks; the other two, R230.60 from Utah and SVK-2 from Idaho, represent fresh material. All other samples were taken from apparently unweathered rocks. The concentrates, except for sample No. R.231.49 from North Carolina, which was treated by flotation reagents, were prepared by simple washings with water without flotation reagents or chemical washers.

Experimental procedures

Extraction and purification of the humic substances: The method used to extract the humic substances is broadly that recommended by the International Humic Substances Society (Schnitzer & Calderoni 1985), with some modifications adapting it to phosphate rock instead of soil. Fifty grams of phosphate rock (ground to less than 75 μm) were mixed with 1 litre of 2 N HCl and shaken for 16 hours at room temperature. The residue was separated by filtration (this fraction may contain a considerable portion of the fulvic acids (FA) when these are present), washed with distilled water and neutralized with a 0.1 N NaOH solution; 250 mL of 0.1 N NaOH were then added and the suspension was shaken under nitrogen for 24 hours at room temperature. The alkaline supernatant was separated from the residue by centrifugation, acidified with 6 N HCl to pH 1, and allowed to stand at room temperature for 24 hours. The supernatant (remaining part of the FA) was separated from the coagulate (humic acids, HA) by centrifugation. In two cases the FA fractions were collected and concentrated on an Amberlite XAD-2 (styrene nonionic polymeric adsorbent) column and purified by repeatedly passing them through a Rexyn 101 column. The HA fractions were purified by repeated shakings with dilute HCl-HF (5 mL 12N HCl plus 5 mL 52% HF diluted with 990 mL distilled water) solution. Finally, the HA fraction was washed with distilled water, centrifuged and freeze-dried. The procedure was repeated as many times as necessary to collect enough HA for the various analyses.

Analytical methods

Thermal analysis of samples was undertaken using a combined differential thermal analysis and thermogravimetric analysis (DTA-TGA) apparatus. The purified HA fractions were heated, in a dynamic (flowing gas) inert argon atmosphere, at a rate of 20°C per minute up to 900°C; the atmosphere was then changed to dynamic air and the samples further heated to 1100°C. Three curves were obtained as a function of the temperature: the DTA, which shows the changes of temperature in the sample relative to an inert standard (aluminium oxide); the TGA, which shows the weight changes in the sample; and the DTG (differential thermogravimetric), which is the first derivative of the TGA showing the rate of weight loss.

X-ray diffraction (XRD) analysis of both untreated samples and HA fractions were undertaken utilizing Cu Kα radiation with a graphite monochromator at 50 Kv and 20 mA. The values of francolite (carbonate-fluorapatite) a-parameters were determined by measuring the 15 most intense peaks, stripping Kα_2, manual editing of interfering peaks, and reduction by an interactive hexagonal least-squares method.

Infra-red (IR) spectra of the HAs were obtained on KBr pressed pellets. Both conventional IR and Fourier transform infra red (FTIR) spectrometers were used. A UV-visible spectrophotometer was used to measure the optical densities at 465 and 665 nm (to calculate the E4/E6 ratios). The samples for these measurements were obtained by dissolving the FAs and the HAs in 0.05 N NaHCO$_3$ solution according to the method described by Chen et al. (1977).

Carbon, hydrogen, nitrogen and ash analyses of four selected HAs were carried out by an independent commercial laboratory (Galbraith Laboratories, Knoxville, Tennessee). Inorganic chemical analyses were performed using conventional methods. Cations were determined by atomic absorption (AA) and anions by ion chromatography, apart from P$_2$O$_5$ which was determined by colorimetry.

Results

The FA content was negligible for the Idaho and Togo samples, and was low (less than 10% of the humic substances) in all the other samples except for the Israeli material, which exhibited a wide range of values from very low up to 40% of the humic substances (this work and other unpublished results). Table 1 summarizes the P$_2$O$_5$ and organic carbon contents, the carbonate-fluorapatite unit-cell a-value, the humic substances yield (humic substances C/total organic C × 100) for all samples together with information about their locations, ages and

Table 1. Summary of samples studied and their main mineralogical and geochemical characteristics

Sample No.	Location	Age	P$_2$O$_5$ (%)	value of a parameter	Organic C(%)	Humic substances yield	E4/E6
R231.62	Mussoorie, India	Early Palaeozoic?	17.0	9.359	1.4	n.d.	—
R231.74	Mussoorie, India	Early Palaeozoic?	25.0	9.355	1.8	n.d.	—
R230.60	Utah, U.S.	Permian	27.4	9.349	1.0	n.d.	—
USHM LLUV	Idaho, U.S.	Permian	33.2	9.318	1.5	48	3.7
SVK.1	Idaho, U.S.	Permian	35.4	9.356	2.4	58	3.8
SVK.2	Idaho, U.S.	Permian	21.2	9.360	3.5	n.d.	—
R231.40	El Hassa, Jordan	Late Cretaceous	32.1	9.329	2.0	8	6.5
R231.42	Ruseifa, Jordan	Late Cretaceous	30.2	9.330	2.8	n.d.	—
Z–1–12	Zin, Israel	Late Cretaceous	33.6	9.343	0.7	53	10.5
Z–1–16	Zin, Israel	Late Cretaceous	33.4	9.346	0.7	68	10.9
Z–4	Zin, Israel	Late Cretaceous	31.7	9.340	0.6	64	10.1
Z–7	Zin, Israel	Late Cretaceous	31.6	9.329	0.5	55	10.2
R231.78	Hahotoe, Togo	Eocene	35.7	9.347	2.1	26	5.0
R232.11	Hahotoe, Togo	Eocene	36.8	9.351	0.1	34	5.2
SVK.3	Bed 2/3, Togo	Eocene	4.0	9.338	0.2	28	7.8
R230.75	Gafsa, Tunisia	Eocene	29.2	9.317	0.8	62	5.1
R231.49	N. Carolina, U.S.	Miocene	29.9	9.318	3.2	55	5.1
NCUC.	N. Carolina, U.S.	Miocene	19.1	9.318	1.0	61	4.3

n.d., not detected

the E4/E6 ratios of their HA fractions. The E4/E6 ratio has been widely used by soil scientists to characterize humic substances and was considered (Kononova 1966) to vary as a function of aromaticity, a low ratio indicating high aromaticity and conversely a high ratio indicating low aromaticity. In an extensive study of the subject, Chen et al. (1977) concluded that this ratio is governed mainly by the particle size (which is related to the molecular weight), and thus is only indirectly correlated to aromaticity. Ranking the samples according to the E4/E6 ratios (particle size), the order from small to large particle size is: Israel, Togo, Tunisia, North Carolina, Idaho.

Thermal analyses

Schnitzer & Hoffman (1961) demonstrated that a highly significant negative correlation exists between the height of the DTG peak (rate of the weight loss), occurring at about 280°C when heated in static (without flow) air, and the degree of humification of the organic matter in soils. Espitalie et al. (1973) studied the total organic matter and the kerogen fractions of the Lower Jurassic (Toarcian) clays of the Paris Basin. These authors showed that the samples from greater depths (i.e., more mature) begin losing weight at higher temperatures, namely between 400 and 500°C, lose less weight and lose it more slowly (exhibiting smaller DTG peaks) when heated under nitrogen, than those from shallower depths. Therefore, in this study, the position and height of the DTG peak were used as a measure of aromaticity (for ranking purposes only). Aromaticity is a better term than humification or maturation since it is less ambiguous, is strictly descriptive and has no genetic implications. Figure 1 shows DTG curves for five selected samples and Table 2 gives the percentage weight loss (recalculated after correction for moisture and ash) for different temperature ranges in the same samples. The rank order of aromaticity established from the DTG curves, (Fig. 1), was from low to high: North Carolina Tunisia, Togo, Israel and Idaho. The Israeli sample shows a doublet (similar curves were found in three of the four Israeli samples studied), possibly indicating a mixture of HAs of different aromaticity (and origin?). According to the weight loss data (Table 2) the order is slightly different, the Tunisia sample coming before the one from North Carolina.

X-ray diffraction

XRD was used by Kodama & Schnitzer (1967) to characterize FAs while Pollack et al. (1971) used it to determine aromaticity in HAs, which supposedly is correlated with their crystallinity. The diffractograms shown in Fig. 2 are of the same samples used to illustrate the thermal analyses (Fig. 1 and Table 2) excepting the Israeli sample, which is Z-4 and not Z-7. The position shape and intensity of the peaks give a measure of the crystallinity of the samples and by implication, an indication of their aromaticity. Peaks at higher two-theta values are closer to the graphite peak value, and are therefore better crystallized, hence, more aromatic, while narrower and more intense peaks also indicate

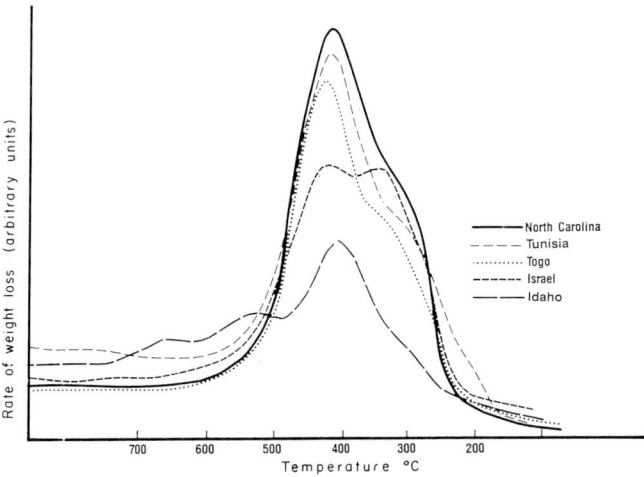

Fig. 1. DTG curves for humic acids from selected phosphorites.

Table 2. *Percentage weight loss of selected phosphorites during TGA*

Sample	Temperature (°C)		
	25–500*	500–900*	900–1100†
R230.75 (Tunisia)	60	22	18
R231.49 (N. Carolina)	55	11	34
R231.78 (Togo)	53	12	35
Z-7 (Israel)	40	15	45
SVK.1 (Idaho)	35	18	47

* inert atmosphere
† Air
Samples are in rank order (aromaticity) from lowest, Tunisia, to highest, Idaho.

better crystallinity. It is difficult to rank the North Carolina and Tunisian samples. The two-theta value of the Tunisian sample is higher but the peak of the North Carolina sample is narrower. Comparison of these results with those from the thermal analyses, however, suggests that the North Carolina and Tunisian samples probably have the same, i.e., the lowest, aromaticity, while the three other samples are ranked by X-ray diffraction in the same order of aromaticity: Togo, Israel and Idaho.

Infra-red spectroscopy

Infra-red spectroscopy is probably the most commonly used method to characterize humic substances (Kumada & Aizawa 1958; Wagner & Stevenson 1965; Theng et al. 1966; Bourboniere & Meyers 1978; Derrepe et al. 1980; Gessa et al. 1983). The IR spectra for the same five samples is shown in Fig. 3. In the range between 4000 and 2400 cm^{-1} (Fig. 3a) the spectra for the samples are in the main similar except for the Idaho sample, for which the C-H absorption bands (around 2930 cm^{-1}) are almost non-existent, and the Tunisian sample for which the OH band (around 3500 cm^{-1}) is weak (this band may be due in part to moisture). The range between 2000 and 900 cm^{-1} is more informative (Fig. 3b). The first peak, or more precisely peaks, occurring between 1720–1700 cm^{-1}, are attributed to the carbonyl C=O vibration (mainly of the carboxylic acid groups).

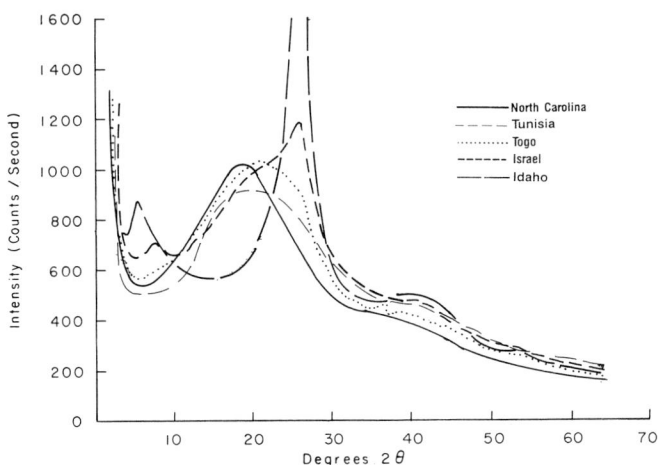

Fig. 2. X-ray diffractograms of humic acids from selected phosphorites.

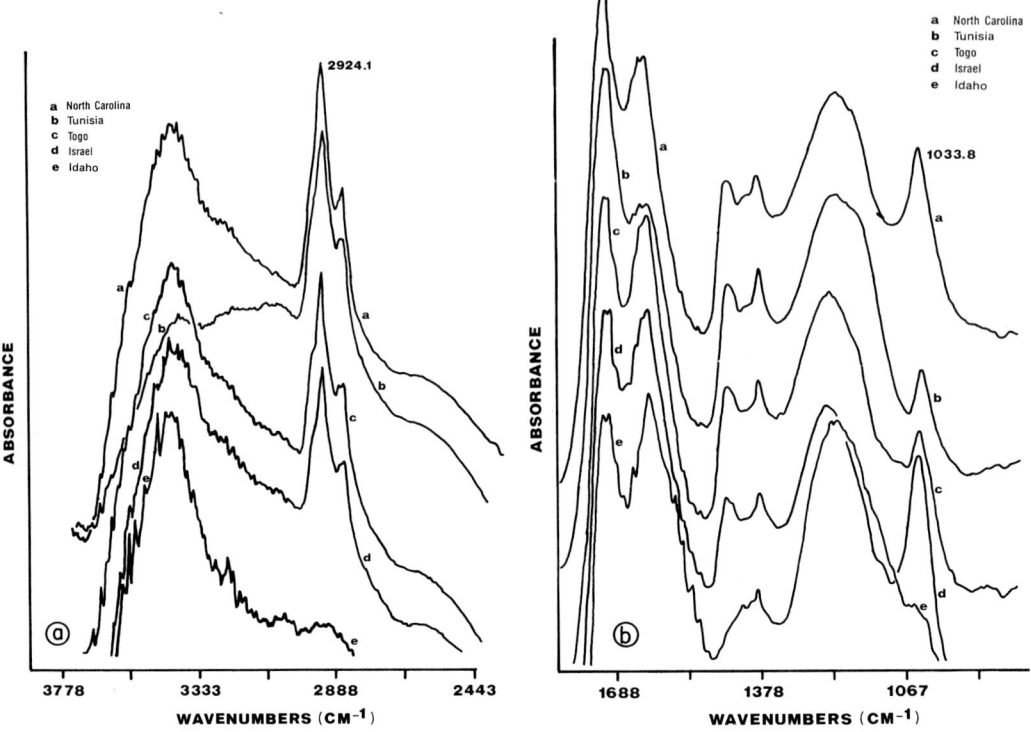

Fig. 3. Infra-red spectra of humic acids from selected phosphorites. (a) 4000 to 2400 cm^{-1} range; (b) 2000 to 900 cm^{-1} range.

These peaks occur as doublets, probably indicating two different surrounding structures for the carboxylic acid groups in the HAs (Bellamy 1975). The exact frequencies are 1712.8 and 1701.2 cm^{-1} in all cases, except for the Togo sample for which they are 1712.8 and 1705.1 cm^{-1}. The second peak is a composite of two bands, the C=O stretching which occurs between 1660–1630 cm^{-1}, and the aromatic C=C band between 1620–1600 cm^{-1} (Stevenson 1982). In the North Carolina and Tunisian spectra a doublet with similar intensities can be distinguished in this peak. In the Togo and Israeli spectra the first peak (C=O stretching) of the doublet is reduced to a shoulder for the more intense (C=C) absorption at 1620.2 cm^{-1}, while in the Idaho spectra, the doublet is resolved with two peaks at 1635.1 and 1616.3 cm^{-1} respectively, the second (C=C) being the most intense of all samples. According to the characteristics of this doublet, the samples may be divided into three groups: (a) North Carolina and Tunisia, (b) Togo and Israel and (c) Idaho; group a is the least aromatic and group c is the most aromatic. This order of aromaticity is more or less the same as that obtained by the thermal and X-ray analyses (see above).

Other band assignments (cf. Flaig et al. 1975) are as follows: 1470–1420 cm^{-1}, aliphatic C–H deformation; 1390–1330 cm^{-1}, salts of carboxylic acids; 1280–1140 cm^{-1}, C–O stretching; 1040–1020 cm^{-1}, Si–O vibration (due to silicate impurities that are almost always present). When the samples are mixed with KBr and dried for some hours at 105°C, a sharp and intense peak appears at 1390 cm^{-1}, accompanied by a corresponding decrease in the peak at 1712 cm^{-1}. These changes indicate that the potassium salt of the carboxylic acid has formed, confirming the assignment of the bands at 1390 and 1712 cm^{-1} to salts of carboxylic acids. They also indicate that the original peak at 1390 cm^{-1} may be due to the sodium salt formed during the extraction of the humic acids with NaOH.

Table 1 shows that ranking according to the particle size (E4/E6 ratio) is different from all other rankings. There is an apparent contradiction between the results e.g. the Israeli samples have the smallest particle size although they are second highest on the aromaticity scale arrived at from thermal, X-ray and infra-red analyses. It should be noted that, even in recent sediments, primary humic acids already have a high molecular weight, i.e. large particles (Cronin & Morris 1982; Poutanen & Morris 1983, 1985; Morris 1987). Moreover Cooper et al. (1986) showed that 'diagenetic alteration of humin is occurring within the first 22 cm of the sediments'. Two conclusions may be reached: (a) the present results confirm Chen et al.'s (1977) results that E4/E6 ratios do not measure aromaticity; (b) humic acids with high aromaticity and small particle size (molecular weight) are not primary (i.e., they were not formed during early diagenesis).

Elemental analyses

Elemental data for four samples (purified HA fractions) are listed in Table 3. The results when plotted on a Van Krevelen (1950) diagram (Fig. 4), can be compared with previous data for samples from Australia (Sandstrom 1986), Tunisia (Belayouni et al. 1982) Israel, (Bein et al. 1980) and the US Western Phosphate Field, Idaho (Mair 1985). This diagram shows that while the organic elemental analyses of the Tunisian samples and those from Idaho both fall in relatively narrow fields, the Israeli and Australian results are spread over a wide range.

Discussion

All of the phosphorites studied are marine and all accumulated in broadly similar depositional environments; furthermore, they probably have similar early diagenetic histories. Organic matter in phosphorites is mainly derived from marine microorganisms (Powell et al. 1975), and phosphorites are characterized by scarcity of terrestrial detritus (Slansky 1980). It is assumed, therefore, that the main elemental composition of the original humic substances in all marine phosphorites falls in the same field. A similar argument has been used by McArthur (1978) who proposed a similar primary inorganic geochemistry for marine francolites. It is further suggested that the composition of the Tunisian and North Carolina samples may approximate to this restricted original composition (Fig. 4; see also discussion below). It should be stressed that this assumption is limited to primary humic substances and is by no means meant to include all primary marine substances.

US Western Mountain (Permian) Phosphate Field

Only weathered samples from this field yielded humic substances (results from this work and Mair, pers. comm., 1986). Furthermore, the HAs contain very little hydrogen and nitrogen (Table 3, Fig. 4), and are found to be the most aromatic HAs by all methods used in this work. These properties indicate that the HAs in this field are Van Krevelen's (1981 p. 219) 'regenerated (humic) or ulmic acids', a group well known to coal scientists, which is the result of oxidation of kerogen. Samples from this area reflect a complex geological history: after deposition, the organic-rich phosphatic sediments were buried and underwent deep burial diagenesis, and in places even metamorphism, during which they liberated hydrocarbons. Later subaerial exposure, erosion and in places hydrothermal activity led to extensive weathering and alteration.

Table 3. *Elemental data for humic acids from selected phosphorites*

Sample	Element (%)					
	C	H	N	S	O	Ash
R231.49 (N. Carolina)	56.54	6.63	2.28	12.13	22.42	2.48
R231.78 (Togo)	63.60	5.49	2.56	7.00	21.35	5.95
Z−7 (Israel)	55.29	4.77	3.36	7.79	28.79	2.94
SVK.1 (Idaho)	61.90	2.63	1.37	6.18	27.90	3.18

Samples were dried at 60°C under vacuum and recalculated to a 100% after correction for the ash. Oxygen values were calculated by difference.

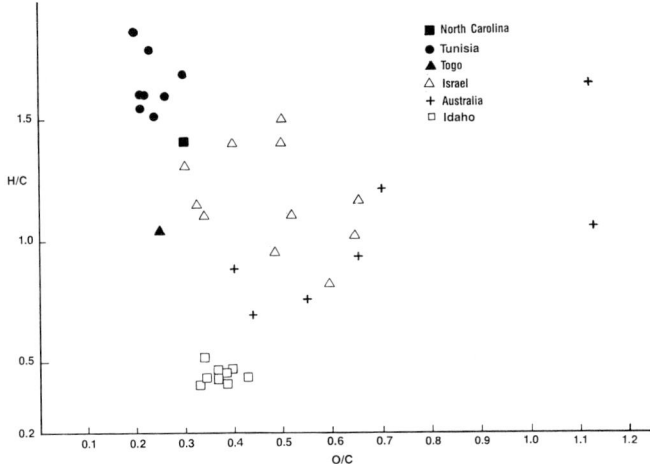

Fig. 4. Van Krevelen (1950) diagram for humic acids from selected phosphorites. Sources: Amit & Bein (1982); Bein *et al.* (1980); Belayouni *et al.* (1982); Mair (1985); Sandstrom (1986); this work.

Gafsa area, Tunisia (Eocene)

The HAs from this area are the least aromatic; furthermore, their relatively high H/C ratio and position on the Van Krevelen diagram (Fig. 4) support a primary origin. It should be stressed that although preservation of labile organic components in Eocene sediments is not common, these components are not unknown even in older sediments (Brassell *et al.* 1987). Moreover, Tunisian phosphorites have apatites with an unusually high sulphate content in the lattice (Sassi 1974; Nathan & Nielsen 1980) which indicates high concentration of sulphate in the interstitial waters and by implication high activity of sulphate-reducing bacteria. This milieu is considered favourable to the preservation of labile organic compounds (Tyson 1987 and references therein).

North Carolina, US (Miocene)

Only one sample has been studied, hence no far-reaching conclusions can be arrived at. Nevertheless, the homogeneity of the deposit (Mair 1985), its age, the fact that the sample is representative of the ore (McClellan, pers. comm., 1986) favour the possibility that the HAs in this field are primary.

Negev, Israel (Late Cretaceous)

The evidence for this area is less clear cut. On the one hand, the very small particle size of the HAs and the presence of considerable amounts of FAs in some samples supports a late epigenetic origin for the humic substances. The wide range of the HAs in the Van Kevelen diagram (Fig. 4) also supports this hypothesis for some of the samples. On the other hand, other samples appear to be in the field of 'primary' humic acids. The presence of a doublet in the DTG curve (Fig. 1) is also indicative of two different origins. Therefore, a polygenetic origin for the humic acids cannot be excluded.

Hahotoe, Togo (Eocene)

The evidence tends to indicate a 'secondary' origin for the humic acids in this field, but there are too few samples to reach a conclusion.

Georgina Basin, Australia (Cambrian)

No samples from this deposit were analysed during this study. Nevertheless, Sandstrom's (1986) data is detailed and enables discussion. The position of the humic acids in the Van Krevelen diagram points clearly to a 'secondary' origin. The good inverse correlation between CO_2 content in the francolite structure and humic acid yields (Fig. 5), strengthens the interpretation that the present HAs in the Australian phosphorites are the product of later alteration (McArthur 1978; Nathan *et al.* 1990).

The principal mechanism proposed for the formation of 'secondary' HAs in phosphorites, is oxidation of the kerogen by biodegradation, similar to that which occurs in crude oils (Philippi 1977; Sassen 1980), although the differ-

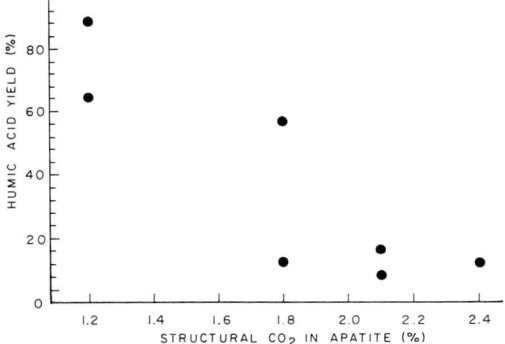

Fig. 5. Relationship between humic acid yield and francolite structural CO_2 in Australian phosphorites (plotted from data in Tables 21.2 and 21.4, Sandstrom 1986).

ences in elemental composition between phosphorite kerogens and crude oils should be taken into account. However, other mechanisms, including inorganic processes, are also possible.

The humic substance fraction of the organic matter is significantly higher in phosphorites than in other sediments (Slansky 1980; Vandenbroucke et al. 1985; this work). This is due to the abundance of both 'primary' and 'secondary' humic substances in phosphorites. The abundance of 'primary' humic substances in phosphorites may due to the following factors.

(1) The formation of phosphorites probably involves two stages (Baturin 1971). In the first stage, apatite forms in areas of high productivity and is deposited in a reducing environment together with appreciable amounts of organic matter. In the second stage, however, concentration of apatite occurs generally in oxidizing environments which decrease the amount of organic matter while favouring the formation of humic substances.

(2) The preservation of primary humic substances is probably enhanced by the inclusion of early formed humic material within phosphate grains (Belayouni 1983) and the adsorption of humic substances on the surface of apatite grains (Sandstrom 1986). The existence of calcium humate phosphorates is well known (Martinez et al. 1983). These complexes (and others) may play a further role in the conservation of humic substances.

(3) The depositional setting of phosphorites (continental margins) is conducive to shallow burial and, subsequently, lack of thermal maturation. Examples of phosphorites which contain 'primary' humic acids are those of Tunisia and North Carolina.

The abundance of 'secondary' humic substances in phosphorites may be due to the fact that many phosphorites became economic through alteration (e.g. diagenetic, hydrothermal and pedogenic processes). This natural 'beneficiation' often involves mechanical (winnowing) and chemical (dissolution of carbonates) processes. Initially insoluble organic matter (kerogen) is probably degraded during these alteration processes to form regenerated secondary humic substances. Examples of phosphorites which contain 'secondary' humic substances are those from Idaho and Australia. Both 'primary' and 'secondary' humic substances may also occur in the same phosphorite, as for example, in those from Israel.

Finally it should be emphasized that many phosphorites contain considerable amounts of organic matter that does not yield any appreciable amount of humic substances (Table 1). These are mature sediments which have not been altered.

The presence of humic substances in marine sediments has been considered to imply thermal immaturity (Nissenbaum & Kaplan 1972). The results of the present work show that more caution should be exercised in reaching this conclusion. The origin of the humic substances must first be determined, because only 'primary' humic substances imply that sediments are thermally immature.

Conclusions

(1) The humic substance fraction of the organic matter in phosphorites is significantly higher than in an average sedimentary rock.
(2) Both 'primary' and 'secondary' humic substances may occur in phosphorites.
(3) The presence of humic substances in a phosphorite (or any sedimentary rock) is not a proof of thermal immaturity.

This work was done in the laboratories of International Fertilizer Development Center (IFDC), Muscle Shoals, Alabama during a sabbatical leave from the Geological Survey of Israel. I wish to thank the management and staff of IFDC for their warm hospitality and for supplying the phosphorite samples. I also thank the German Ministry of Economic Co-operation for partial financial support. I am grateful to the Negev Phosphate Co. for supplying the Negev samples and to Rick Austin, B. Biggers, B. Hamilton, all of IFDC, and D. Thrasher of TVA for help with the analytical work, as well as to A. Bein and B. Katz, both of GSI, S. Altschuler and M. Sandstrom both of the USGS and I. Jarvis of Kingston Polytechnic for the critical reading and editing of the manuscript. Special thanks are due to G. McClellan and S. Van Kauwenbergh, both of IFDC for help in all stages of the work. The author is solely responsible for all errors.

References

AMIT, O & BEIN, A. 1982. Organic matter in Senonian phosphorites from Israel — origin and diagenesis. *Chemical Geology*, **37**, 277–287.

BEIN, A., AMIT, O. & NATHAN, Y. 1980. *Organic matter in phosphorites*. Israel Geological Survey Report MN 4/80, 16 p. (in Hebrew).

BATURIN, G. N. 1971. Stages of phosphorite formation on the ocean floor. *Nature Physical Sciences*, **232**, 61–62.

BELAYOUNI, H. 1983. *Etude de la matière organique dans la série phosphatée du bassin de Gafsa-Metlaoui (Tunisie). Application a la compréhension des mécanismes de la phosphatogenèse*. Thèse de doctorat ès-Sciences, Université d'Orléans.

—— & TRICHET, J. 1980. Contribution a la connaissance de la matière organique du bassin phosphaté de Gafsa: information fournies par l'analyse du potentiel pétrolier. *In: Geologie comparée des gisements de phosphates et de pétrole*. Documents du Bureau de Recherches Geologiques et Minieres, **24**, 37–59.

—— FAUCONIER, D., SLANSKY, M. & TRICHET, J. 1982. Etude du contenu organique des dépôts phosphatés du bassin de Gafsa (Tunisie). Documents du Bureau de Recherches Geologiques et Minieres, No. 35.

BELLAMY, L. J. 1975. *The Infra-red Spectra of Complex Molecules*. 3rd edition, Chapman and Hall, London.

BOURBONIERE, R. A. & MEYERS, P. A. 1978. Characterization of sedimentary humic matter by elemental and spectroscopic methods. *Canadian Journal of Spectrosocopy*, **23**, 35–41.

BRASSELL, S. C., EGLINTON, G. & HOWELL V. G. 1987. Palaeoenvironmental assessment of marine organic-rich sediments using molecular organic geochemistry. *In*: BROOKS, J. & FLEET, A. J. (eds) *Marine Petroleum Source Rocks*. Geological Society Special Publication, **26**, 79–98.

COOPER, W. T., HEIMAN, A. S. & YATES, R. R. 1986. Spectroscopic and chromatographic studies of organic carbon in recent marine sediments from the Peruvian upwelling zone. (International meeting on organic geochemistry, 12/1985, Julich). *Organic Geochemistry*, **10**, 725–732.

CHEN, Y., SENESI, N. & SCHNITZER, M. 1977. Information provided on humic substances by E4/E6 ratios. *Soil Science Society of American Journal*, **41**, 352–358.

CRONIN, J. R. & MORRIS, R. J. 1982. The occurrence of high molecular weight humic material in recent organic-rich sediment from the Namibian Shelf. *Estuarine and Coastal Shelf Science*, **15**, 17–27.

DEREPPE, J. M., MOREAU, C. & DEBYSER, Y. 1980. Investigation of marine and terrestrial humic substances by H and C nuclear magnetic resonance and infrared spectroscopy. *Organic Geochemistry*, **2**, 117–124.

ESPITALIE, J., DURAND, B., ROUSSEL, J. C. & SOUDRON, C. 1973. Etude de la matière organique insoluble (kerogene) des argiles du bassin de Paris. Deuxième partie. *Revue de l'Institut Francais du Pétrole*, **28** 37–66.

FLAIG, W., BEUTELSPACHER, H. & RIETZ, E. 1975. Chemical composition and physical properties of humic substances. *In*: GIESEKING, J. E. (ed.) *Soil Components, Vol. 1, Organic Components*. Springer-Verlag, Berlin, New-York, 1–211.

GESSA, C., CABRAS, M. A., MICERA, G., POLEMIO, M. & TESTINI, C. 1983. Spectroscopic characterization of extracts from humic and fulvic fractions: IR and H NMR spectra. *Plant and Soil*, **75**, 169–177.

HUC, A. Y. & DURAND, B. M. 1977. Occurrence and significance of humic acids in ancient sediments. *Fuel*, **56**, 73–80.

KODAMA, H. & SCHNITZER, M. 1967. X-ray studies of fulvic acid, a soil humic compound. *Fuel*, **46**, 87–94.

KONONOVA, M. M. 1966. *Soil Organic Matter*. 2nd English edition. Pergamon Press, Oxford, New-York.

KREVELEN, VAN D. W. 1950. Graphical statistical method for study of structure and reaction processes of coal. *Fuel*, **29**, 269–284.

—— 1981. *Coal*. 3rd. impression. Elsevier, Amsterdam, New-York.

KUMADA, K. & AIZAWA, A. 1958. The infra-red spectra of humic acids. *Soil and Plant Food*, **3**, 152–159.

MAIR, A. D. 1985. Organic matter and sulphur distribution in phosphorites: Relevance to phosphate processing. *Journal of Chemical Technology and Biotechnology*, **35A**, 135–156.

MARTINEZ, M. T., ROMER, C. & GAVILAN, J. M. 1983. Characteristics of the HA-Ca-P complexes prepared from lignite humic acids. *Fuel*, **62**, 956–958.

McARTHUR, J. M. 1978. Systematic variations in the contents of Na, Sr, CO_2, and SO_4 in marine carbonate-fluorapatite and their relation to weathering. *Chemical Geology*, **21**, 41–52.

MORRIS, R. J. 1987. The formation organic-rich deposits in two deep-water marine environments. *In*: BROOKS, J. & FLEET, A. J. (eds) *Marine Petroleum Source Rocks*. Geological Society Special Publication, **26**, 153–166.

—— & CALVERT, S. E. 1977. Geochemical studies of organic-rich sediments from the Namibian shelf, I. The organic fractions. *In*: ANGEL, M. (ed.) *A Voyage of Discovery — George Deacon 70th Anniversary Volume*. Pergamon Press, Oxford, 647–665.

NATHAN, Y. & NIELSEN, H. 1980. Sulfur isotopes in phosphorites. *In*: BENTOR, Y. K. (ed.) *Marine*

Phosphorites. Society of Economic Paleontologists and Mineralogists, Special Publication, **29**, 73–78.

——, SOUDRY, D. & AVIGUR, A. 1990. Geological significance of carbonate substitution in apatite: Israeli phosphorites as an example. In: NOTHOLT, A. J. G. & JARVIS, I (eds) *Phosphorite Research and Development* Geological Society, London, Special Publication, **52**, 179–191.

NISSENBAUM, A. & KAPLAN, I. R. 1972. Chemical and isotopic evidence for the in situ origin of marine humic substances. *Limnology and Oceanography*. **17**, 570–582.

PHILIPPI, G. T. 1977. On the depth, time and mechanism of origin of the heavy to medium-gravity naphtenic crude oils. *Geochemica et Cosmochimica Acta*, **41**, 33–52.

POLLACK, S. S., LENTZ, H. & ZIECHMAN, W. 1971. X-ray diffraction of humic acids. *Soil Science*, **112**, 318–324.

POUTANEN, E. L. 1985. Humic substances in an Arabian shelf sediment and the Si sapropel from the Eastern Mediterranean. *Chemical Geology*, **51**, 135–145.

—— & MORRIS, R. J. 1983. A study of the formation of high molecular weights compounds during the decomposition of a field diatom population. *Estuarine and Coastal Shelf Science*, **17**, 189–196.

POWELL T. G., COOK, P. G. & MCKIRDY, D. M. 1975. Organic geochemistry of phosphorites: Relevance to petroleum genesis. *American Association of Petroleum Geologists Bulletin*, **59**, 618–632.

SANDSTROM, M. W. 1982. *Organic Geochemistry of Phosphorites and Associated Sediments*. PhD Thesis, Australian National University, Canberra.

—— 1986. Proterozoic and Cambrian phosphorites — specialist studies: geochemistry of organic matter in Middle Cambrian phosphorites from the Georgina Basin, north-eastern Australia. *In*: COOK, P. J. & SHERGOLD, J. H. (eds) *Phosphate Deposits of the World. Vol. 1, Proterozoic and Cambrian Phosphorites*. Cambridge University Press, Cambridge. 268–279.

SASSEN, R. 1980. Biodegradation of crude oil and mineral deposition in a shallow Gulf Coast salt dome. *Organic Geochemistry*, **2**, 153–166.

SASSI, S. 1974. *La sédimentation phosphatée au Paléocène dans le sud et le centre ouest de la Tunisie*. Thèse de Doctorat ès-Sciences, Université de Paris-Orsay.

SCHNITZER, M. & CALDERONI, G. 1985. Some chemical characteristics of paleosol humic acids. *Chemical Geology*, **53**, 175–184.

—— & HOFFMAN, I. 1961. A thermogravimetric approach to the classification of organic soils. *Soil Science Society of America Proceedings*, **30**, 63–66.

—— & KODAMA, H. 1977. Reactions of minerals with soil humic substances. *In*: DIXON J. B. & WEED S. B. (eds) *Minerals in Soil Environments, Volume 3*, Soil Science Society of America, Madison, Wisconsin, USA, 741–770

SLANSKY, M. 1980. *Géologie des Phosphates sédimentaires*. Mémoires du Bureau de Recherches Geologiques et Minieres No. 114.

STEVENSON, F. J. 1982. *Humus Chemistry: Genesis, Composition, Reactions*. John Wiley & Sons, New-York.

THENG, B. K. G., WAKE, J. R. H. & POSNER, A. M. 1966. Infrared spectrum of humic acids. *Soil Science*, **102**, 70–72.

TYSON, R. V. 1987. The genesis and palynofacies characteristics of marine petroleum source rocks. *In*: BROOKS, J. & FLEET, A. J. (eds) *Marine Petroleum Source Rocks*. Geological Society Special Publication, **26**, 47–67.

VANDENBROUCKE, M., PELET, R. & DEBYSER, Y. 1985. Geochemistry of humic substances in marine sediments. *In*: AIKEN, G. R., McKNIGHT, D. M., WERSHAW, R. L. & MCCARTHY, P. (eds) *Humic Substances in Soil Sediment and Water*. John Wiley & Sons, New-York, 249–273.

WAGNER, G. H. & STEVENSON, F. J. 1965. Structural arrangement of functional groups in soil humic acids as revealed by infra-red analysis. *Soil Science Society of America Proceedings*, **29**, 43–48.

Sedimentation dynamics and redox iron-cycling: controlling factors for the apatite−glauconite association on the East Australian continental margin

G. W. O'BRIEN[1], A. R. MILNES[2], H. H. VEEH[3], D. T. HEGGIE[1], S. R. RIGGS[4], D. J. CULLEN[5], J. F. MARSHALL[1] & P. J. COOK[1]

[1] *Bureau of Mineral Resources, GPO Box 378, Canberra City, 2601, Australian Capital Territory, Australia*
[2] *CSIRO Division of Soils, Private Bag No. 2, Glen Osmond, South Australia, 5064, Australia*
[3] *School of Earth Sciences, Flinders University of South Australia, Bedford Park, South Australia, 5042, Australia*
[4] *Department of Geology, East Carolina University, Greenville, North Carolina 27834, USA*
[5] *New Zealand Oceanographic Institute, DSIR, Private Bag, Kilbirnie, Wellington, New Zealand*

Abstract: Detailed sedimentological and geochemical studies of phosphorites and sediments from the East Australian continental margin have shown that both apatite and glauconite are forming at a transition zone between relict, iron oxyhydroxide-rich, organic-poor (TOC<0.3%) outer shelf (200−350 m) sediments and relatively rapidly accumulating, iron oxyhydroxide-deficient, organic-rich (TOC>0.8%) deep water (460−650 m) sediments. The interaction between sediment mixing and Fe-P cycling processes (between the pore waters and the solid phase) appear critical to the formation of modern phosphorites in this area. The phosphate nodules form within the anoxic zone in the sediments at depths of approximately 10−18 cm below the sediment-seawater interface. Nodules which remain in the sediment mixed layer after they form continue to accumulate both P and Fe for up to 60 ka; during this time their apatite and iron oxyhydroxide contents more than double and the nodules become denser and more lithified. Apatite and glauconite formation are favoured by periods of high sea-level and low current velocities, as these conditions allow a relatively high organic carbon input to the sediments and thereby the maintenance of anoxia at shallow depths within the sediments. During periods of low sea-level and high current velocities, the carbon flux into the sediments decreases and the sediments become oxic. Consequently the Fe-cycling processes cease and apatite and glauconite formation stops: the glauconite is progressively transformed to goethite, and phosphorite nodules are concentrated into lag deposits and ferruginized. Alternations of high and low sea-level cycles eventually result in the formation of the massive ferruginous Neogene phosphorites that mantle much of the outer shelf. The iron enrichment processes observed in the modern to Neogene phosphorites on the East Australian continental margin provide explanations for many of the features seen in ferruginous Neogene deposits in the world's oceans.

Extensive deposits of Neogene sedimentary phosphorite occur on many continental shelves. These deposits are often relatively ferruginous, with moderate to large amounts of iron oxyhydroxide (mostly goethite) admixed with phosphate. Ferruginous phosphorite deposits occur on Agulhas Bank off South Africa (Parker & Siesser 1972; Parker 1975), on the western margin of South Africa (Birch 1979), on Danois Bank off Spain (Lucas *et al*. 1978), off Morocco (McArthur 1978) and off Eastern Australia (von der Borch 1970; Marshall & Cook 1980; Kress & Veeh 1980; O'Brien & Veeh 1980; Cook & Marshall 1981; Marshall 1983; O'Brien *et al*. 1986; 1987). In contrast, Holocene phosphate nodule formation has been documented in only three areas: off Namibia (Baturin *et al*. 1972; Veeh *et al*. 1974; Thomson 1984), off Peru−Chile (Veeh *et al*. 1973; Burnett & Veeh 1977) and off Eastern Australia (O'Brien & Veeh 1980; O'Brien *et al*. 1986; Heggie *et al*. 1990). Unlike Neogene phosphorites, Holocene phosphate nodules are invariably non-ferruginous and occur associated with either organic-rich, diatomaceous oozes (Namibia, Peru–Chile) or organic-poor, calcareous sands (Eastern Australia). The characteristic absence of significant iron oxy-hydroxide concentrations in Holocene nodules implies that either: (i) the modern phosphogenic systems do not provide readily applicable analogues for Neogene deposits or (ii) modern systems do provide analogues, but the iron in Neogene deposits is principally post-depositional.

The East Australian phosphate province is unique in that non-ferruginous Quaternary nodules co-exist with ferruginous Neogene nodules. Moreover, both nodule types appear to have formed within relatively organic-poor sediments. In this paper, we report on details of the geochemistry of the Quaternary and Neogene phosphate nodules from off Eastern Australia (Fig. 1). The geochemistry of sediments from inner shelf to mid-slope environments is also examined, with particular emphasis on sediments associated with Quaternary nodules.

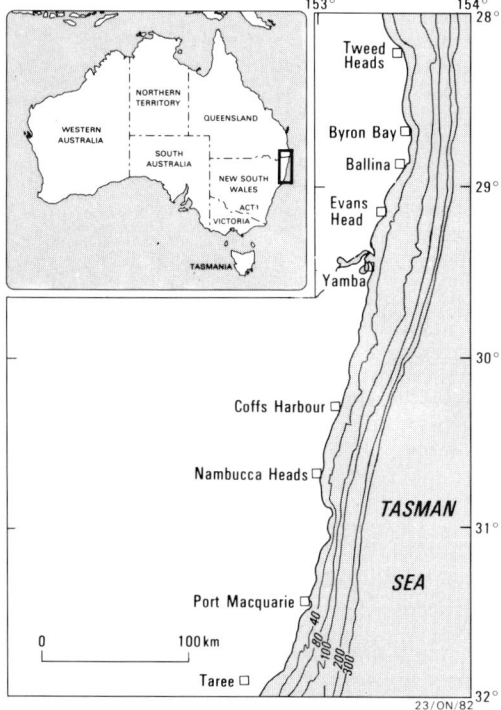

Fig. 1. Locality map showing East Australian continental margin. Modern phosphorite have been found between 29–32°S, especially at water depths between 350–460 m.

Materials and methods

The samples analysed in this study were collected during four separate cruises. *Global Marine Incorporated* samples ('G' prefix) were collected in 1966 using a rock dredge; samples collected by the *R. V. 'Sprightly'* in 1979 ('S' prefix) were obtained using a large Smith-McIntyre grab; *R. V. 'Tangaroa'* (1979) samples ('P' prefix) were sampled using a combination of dredging and piston and box coring; samples prefixed with '71' were collected during an *R. V. 'Rig Seismic'* cruise (1987). Dredge material from the BMR collection has a '15' prefix. In addition, some previously published geochemical data for shallow water (generally <300 m) sediment samples (Marshall 1980) have been integrated with our new data.

Sediments and phosphatic nodules were finely ground in a tungsten carbide ring-and-puck mill and splits of the powders were used for all analyses. Mineralogical compositions were determined by X-ray diffraction (XRD) using Philips PW1049 and PW1710 instruments with cobalt radiation and graphite monochromators. To identify clay minerals, the <2 μm fraction of samples (as-collected) was isolated by dispersion and differential settling in water columns according to Jackson (1965), followed by pipette extraction. Identification of the clay minerals essentially followed the procedures of Brindley & Brown (1984). The CO_2 content of apatite was determined from XRD data using the angular separation of the 211 and 112 peaks (Milnes, unpublished CSIRO Division of Soils Technical Memorandum 1979). An empirical relationship was established between the CO_2 content (of apatite), as measured by acid dissolution, and the separation (d) of the 211 and 112 peaks (in Å) measured from diffractometer traces for a number of apatite samples namely that:

$$\text{wt\% } CO_2 \text{ in apatite} = 17.335 - (615.524) \times d$$

Observations on apatite and goethite crystallite size were made from the broadening of diffraction peaks (Zussman 1977). Analysis of major and trace elements was carried out by X-ray fluorescence spectrometry (XRF) (Norrish & Hutton 1969) using Philips PW1540 and PW1400 instruments. Organic carbon (TOC) concentrations were determined using the Walkley & Black (1934) procedure. Rock-Eval pyrolysis (Espitalie *et al.* 1977) results were obtained using a Rock-Eval II instrument: S_2 (units = kg hydrocarbon/tonne of sediment) provides a measure of the total amount of hydrogen-rich organic material in the sediment, whereas HI (hydrogen index, units = kg hydrocarbon/gm TOC) provides a measure of the hydrogen content of the organic matter itself. Ferrous iron analyses were carried out by an in-house BMR technique (titration with potassium dichromate). Total iron was determined by XRF and ferric iron obtained by difference.

Geological setting and sample descriptions

All samples were recovered between latitudes 28° and 32°S in water depths ranging from 8 to

655 m (Fig. 1). The continental shelf in this region is fairly narrow (25–40 km), with the shelf break occurring at depths between 210–450 m (Marshall 1979). The continental slope is unusually steep, with typical gradients of 8–12°, although it is occasionally as steep as 20°.

The principal oceanographic feature of the area is the East Australian Current (EAC), a strong, southward-flowing current that greatly restricts sediment accumulation on the continental shelf and upper slope, and causes extensive mechanical reworking of shelf and slope sediments (O'Brien et al. 1986, 1987). Seasonal coastal upwelling, associated with incursions of the EAC onto the shelf, occurs within 10–15 km of the coast. However, because of the sporadic nature of the upwelling events, and the low nutrient concentrations in the EAC waters, the East Australian margin is characterized by low biological productivity, with typical organic carbon fluxes of 0.1 gm^{-2} day^{-1} (O'Brien & Veeh 1983).

Sediments and phosphorites

Sediments

The sediments show marked compositional and grainsize variations from inshore to deeper water areas, which result from complex interactions between sediment supply, the EAC, and the steepness of the upper continental slope. Very little terrigenous sediment is presently being deposited on the continental shelf, in spite of the fact that several large rivers debouch their sediment load onto the inner shelf (Marshall 1980). Sands and gravels dominate the continental shelf, with silts and muds rare to absent. It appears that the EAC, which can exceed 200 cm s^{-1} in this area, is continuously transporting the fine terrigenous fraction into deeper water. There are no significant accumulations of fine sediment on the continental shelf north of 32°S (Davies 1979), whereas south of 32°S current velocities on the shelf fall sharply (Godfrey et al. 1980) and major accumulations of fine sediment occur. The major sediment types in the study area can be loosely subdivided into five water depth-related sedimentary facies.

Facies 1 and 2. Inner shelf (Facies 1) sediments occur between 0–50 m water depth and are typically fine to medium grained, terrigenous quartz sands. These sands are predominantly relict and were probably originally deposited in fluvial environments during lower sea-level stands (Marshall 1980). The sands become progressively richer in biogenic carbonate seawards and grade laterally into Facies 2 sediments, which occur at water depths between 50–200 m. Facies 2 sediments are poorly sorted carbonate sands and gravels; foraminifera, molluscs, bryozoans, echinoderms, red algae and skeletal intraclasts dominate the skeletal carbonate fraction (Marshall 1980). The skeletal carbonate within Facies 2 contains both modern and relict components, and many of the molluscs, bryozoans and echinoid plates and spines are iron-stained. The abudance of foraminifera increases markedly towards the edge of the shelf, principally because of an increased planktonic input; shorewards the ratio of benthic to planktonic foraminifera increases (Marshall 1980).

Facies 3. Between depths of approximately 180–200 m, the iron oxide content of the Facies 2 carbonate sands increases progressively and they grade laterally into Facies 3, which extends approximately from 200–350 m. The characteristic feature of Facies 3 is that the sediments contain moderate to large amounts of goethite. The sediments are predominantly relict goethite-rich, unconsolidated glauconitic, foraminiferal sands. The goethite occurs as a coating on component grains, as discrete pellets, as infillings of body chambers within foraminifera and also apparently replacing skeletal carbonate. Glauconite, which typically comprises between 5–30% of the sediment, occurs either as an infilling of body chambers in organisms (generally foraminifera) or as discrete pellets. Where it is present as discrete pellets, the pellets clearly originally formed as infillings of body chambers. Many of the glauconite pellets are slightly to strongly altered and it is probable that a continuum exists between glauconite and goethite pellets. There is also considerable compositional variation within Facies 3: glauconite becomes progressively more abundant from shallower (200 m) to deeper (350 m) water, while the goethite content increases dramatically at water depths deeper than 200 m, peaks at 300 m and then decreases. Some of the glauconite in Facies 3 has a geochemical composition analogous to berthierine (Marshall 1983).

Ferruginous Neogene phosphorites occur almost exclusively in association with Facies 3 sediments. They are present both as discrete nodules within the sediments and also as massive hardgrounds on current-swept areas which are free of unconsolidated sediment. These hardgrounds are most common south of Coffs Harbour, but also occur off Evans Head–

Yamba (Fig. 1). At present it is uncertain whether these ferruginous units represent a lag deposit or are actually subsurface units that are outcropping on the outer shelf. The hardgrounds form an important facet of Facies 3 and are designated Facies 3'. The goethite content of Facies 3 sediments decreases at water depths >300 m and they grade into Facies 4 between 320–350 m.

Facies 4. Facies 4 sediments are glauconitic, foraminiferal sands with low goethite contents. Quaternary phosphate nodules are restricted to this facies. The sediments consist of 7–30% glauconite, 20–35% silt to fine sand-sized angular quartz and 30–60% skeletal carbonate, in which planktonic foraminifera predominate. Coccoliths are a common constituent of the clay-sized fraction. Glauconite, which occurs as pellets and infillings of foraminifera body chambers, ranges in colour from red-brown to light green, reflecting the transition from altered to unaltered; unaltered glauconite is dominant. Goethite pellets are rare, as in iron-staining of the sediments.

Facies 4 extends approximately from 350–460 m and the sediments exhibit systematic variations in composition. Both the abundance of goethite pellets and the extent of iron-staining decrease progressively from shallow (350–370 m) to deep (460 m) water, as does the ratio of altered to unaltered glauconite pellets. Similarly, sediments within 10–15 cm of the sediment–water interface have consistently higher levels of iron-staining, and a higher abundance of altered glauconite, than do deeper sediments. Sediments also show textural variations down-core: near-surface sediments are predominantly grainstones, whereas packstones predominate at depth. This textural difference may be due to sediment mixing processes (via bioturbation?) in the near-surface sediment.

Facies 5. Facies 5 sediments are laterally continuous with Facies 4 and extend from 460 m to 655 m water depth, the limit of sample coverage. They are compositionally similar to Facies 4 but finer grained. Grain-size typically ranges from silt to fine sand. Glauconite pellets are less abundant and are typically unaltered. Goethite pellets are rare to absent and iron-staining is less than in Facies 4. These sediments have considerably more matrix (mostly clay-sized) than the shallower sediments, and packstones predominate.

Phosphorites

Phosphate nodules on the East Australian continental margin range in age from Neogene to Quaternary (O'Brien *et al.* 1980; 1981; 1986). The Quaternary nodules invariably contain little geothite and are associated with Facies 4 upper slope sediments (350–460 m water depth), whereas the Neogene phosphates are very iron-rich and occur predominantly within Facies 3 on the outer shelf (200–350 m water depth). However, there is a major zone of overlap where both Neogene and Quaternary nodules coexist. This zone extends from approximately 350 to 425 m water depth, and in this zone drapes of Quaternary Facies 4 sediments overlie much older ferruginous subsurface units.

Quaternary nodules. The Quaternary phosphate nodules are usually 1–5 cm in diameter, are often flattened, and do not possess any internal structure. Holocene nodules are typically friable and porous, are white to light yellow in colour, have low P_2O_5 concentrations (5–10%) and contain negligible goethite. In contrast, nodules >60–80 ka old are often well-indurated and dense, are gray to red-brown in colour, contain 10–20% P_2O_5 and have a small but significant goethite content. Even in the most iron-rich Quaternary nodules, however, the total iron does not usually exceed 10% Fe_2O_3, and is generally between 3–5% Fe_2O_3.

The Quaternary nodules consist of fine sand-sized angular quartz, foraminifera and glauconite pellets in a very fine grained phosphatic matrix. There is no evidence of replacement of biogenic carbonate by phosphate and the nodules appear to be phosphatic equivalents of the associated non-phosphatic glauconitic, foraminiferal sands. SEM studies (O'Brien *et al.* 1981) have shown that the apatite in Quaternary nodules is present within 1.0–2.0 µm structures which resemble bacterial cells. Consequently it has been proposed that bacteria provide the principal phosphate-concentrating mechanism within these nodules. Goethite, where present, is concentrated around the margins of the nodules, but is also disseminated throughout the phosphatic matrix.

The Quaternary phosphate nodules usually comprise only a few percent of the Facies 4 sediments, though rarely they comprise up to 10–20% of the sediment (O'Brien 1983). As glauconite typically comprises 20–25% of the sediment, the East Australian margin can be considered as dominantly a 'glaucogenic' rather than 'phosphogenic' system.

Ferruginous nodules. Ferruginous nodules are dark brown to red-brown in colour and usually range from 1–15 cm in diameter. Their goethite content is highly variable, which is reflected by a variation in total iron (as Fe_2O_3) from 10–50%. They are well-indurated, commonly rounded and conglomeratic and occur most abundantly at water depths between approximately 200–300 m on relict, current-swept areas of the outer shelf (Facies 3'), particularly between Coffs Harbour and Port Macquarie (Fig. 1). Many of these highly ferruginous Facies 3' phosphorites are probably as old as middle Miocene (von der Borch 1970). Ferruginous nodules also commonly occur within Facies 3 sediments, although these nodules tend to be less ferruginous than Facies 3' phosphorites and may be younger. Similarly, ferruginous nodules are occasionally found within Facies 4 upper slope sediments. Such nodules are usually only moderately goethite-rich and are never conglomeratic.

The nodules consist of quartz, biogenic carbonate and glauconite pellets set in a matrix composed of goethite and apatite. Whereas the biogenic carbonate in many moderately ferruginous nodules is partly replaced by goethite, carbonate appears to have been completely removed from the most iron-rich examples by dissolution, indicating that these nodules experienced large fluctuations in pH during diagenesis. Pelletal glauconite exhibits progressively greater alteration to goethite in the more iron-rich nodules. In contrast to the Quaternary nodules, apatite in the ferruginous nodules is usually present as large euhedral crystals (O'Brien et al. 1981). Although goethite is principally a void-filling cement in the ferruginous nodules, it also occurs as a surficial glaze or coating which probably represents a separate phase(s) of deposition.

Mineralogy

Sediments

Facies 4 sediments associated with Quaternary phosphorite nodules have a comparatively uniform mineralogical composition. Quartz and calcite dominate; the calcite occurs in both the low- and high-Mg form, with low-Mg calcite predominating. Aragonite and plagioclase and alkali feldspars are ubiquitous but minor constituents. The acid-treated (to remove carbonate) clay-sized fraction of Facies 4 sediments contains quartz and feldspars, as well as clays which are dominated by randomly-interstratified material and kaolinite. Glauconite and a 14.5Å phase (probably chlorite) are also conspicuous. Minor goethite is present in some samples.

Quaternary nodules

The mineralogical composition of the non-ferruginous phosphate nodules is similar to that of the associated sediments except for the presence of carbonate fluorapatite:– apatite, high- and low-Mg calcite and quartz are abundant constituents, with minor feldspars, aragonite and clay minerals. The residues of acid-treated clay fractions (<2 μm) are composed dominantly of kaolinite with randomly interstratified clay minerals and glauconite. Older, well-indurated nodules usually contain detectable goethite. The apatite diffraction peaks are significantly broadened due to small crystallite size. The CO_2 content of apatite in 12 different nodules varied between 5.0–9.7% (average 7.1 ± 1.2%). These values are similar to the CO_2 content of apatite in phosphorites off Namibia (Baturin et al. 1970) and South Africa (Birch 1975), but are considerably higher than in Quaternary phosphorites off Peru–Chile (Kim & Burnett 1986) and from many on-shore deposits (McArthur 1978) including the Phosphoria Formation (Gulbrandsen 1966; 1970).

Ferruginous nodules

Highly ferruginous (Facies 3') phosphate nodules with a glazed appearance contain variable concentrations of quartz, goethite, apatite, calcite and aragonite, with minor feldspar and clay minerals, whereas nodules relatively low in goethite approach the mineralogical composition of Quaternary nodules. Carbonate minerals are scarce in ferruginous nodules. In very iron-rich samples neither calcite nor aragonite were detected. Both the goethite and apatite diffraction peaks are significantly broadened, indicating a small crystallite size. There is little difference between the CO_2 content of apatite in the ferruginous and non-ferruginous nodules, in spite of the differences in crystal morphology: CO_2 in the apatite of 5 ferruginous nodules ranged between 7.3–9.3% (average 8.3 ± 0.7%).

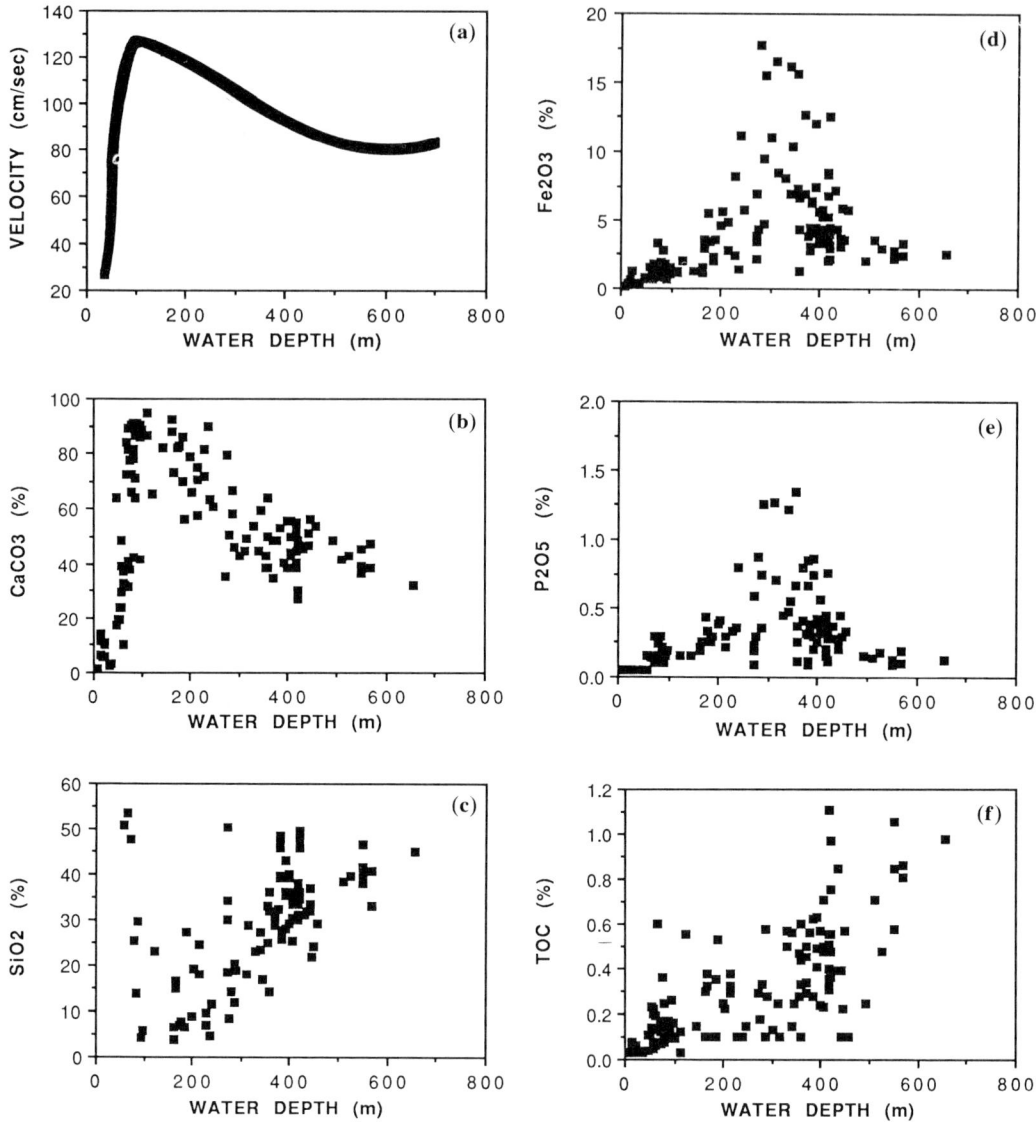

Geochemistry

Sediments

Variations across the continental margin. The variation in the surface velocity of the EAC in water depths from 0–800 m is shown in Fig. 2a, while the variations in the concentrations of $CaCO_3$, SiO_2, Fe_2O_3, P_2O_5, total organic carbon (TOC), S_2 (mgHC/g of sample), HI (hydrogen index; mgHC/g TOC) and K_2O in sediments from inner shelf to upper slope environments (0–655 m water depth) are shown in Figs 2b–2i respectively. The $CaCO_3$ concentration (Fig. 2b; Table 1) of the sediments is closely related to both water depth and current velocity. Inner shelf Facies 1 sands contain little $CaCO_3$. Carbonate increases progressively into deeper water within Facies 2 sands and gravels, peaking at >90% $CaCO_3$ in water depths of approximately 100 m. Its concentration then declines progressively into deeper water, decreasing to 32.1% at a depth of 655 m in Facies 5 (see Table 1). The distribution of SiO_2 across the margin (Fig. 2c) is the inverse of the $CaCO_3$ profile.

There is clearly a close relationship between

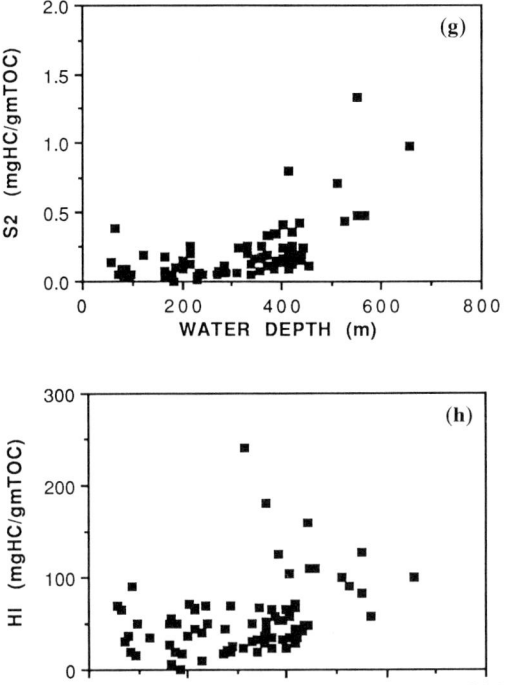

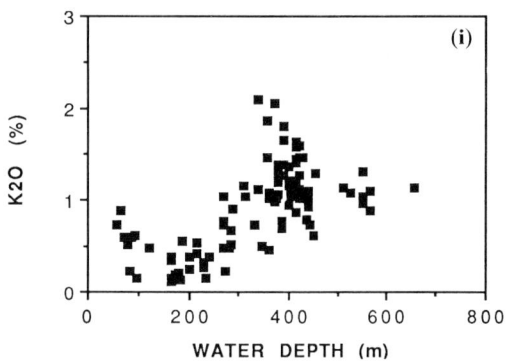

Fig. 2. Variations in the concentrations of various components within sediments across the East Australian continental margin, plotted against water depth(m) and compared with velocity of the East Australian Current. All values are in weight %, except for S_2 (kg HC/tonne) and HI (mg HC/g TOC). (a) East Australian Current velocity at 29°S. Velocities obtained from Godfrey (1973) (b) $CaCO_3$. (c) SiO_2. (d) Fe_2O_3. (e) P_2O_5. (f) TOC. (g) S_2. (h) HI. (i) K_2O.

the $CaCO_3$ and SiO_2 contents of the sediments and the velocity of the EAC. Inner shelf terrigenous sands occur in a low current regime, are relatively fine grained and have high SiO_2 but low carbonate concentrations. In deeper (>50 m) water, the current velocities increase rapidly. The current winnows the finer grained terrigenous (SiO_2) fraction and also provides an effective barrier to the seaward transport of the terrigenous sands. As a consequence, the relative abundance of skeletal carbonate begins to increase sharply. Carbonate concentration peaks at approximately 100 m water depth, where current velocities are highest. Here, the sediments are very coarse grained skeletal carbonate sands and gravels which contain little or no fine fraction. At depths >100 m, the $CaCO_3$ content decreases contemporaneously with the current velocity; the abundance of coarser grained skeletal carbonate decreases rapidly, concomitant with a sharp increase in the relative abundance of planktonic foraminifera, which dominate the skeletal fraction at water depths beyond 150–200 m. The relative abundance of planktonic foraminifera, particularly the finer grained forms, continues to increase into deeper water, but, partly as a result of the decreasing current regime, the sediments are increasingly diluted with fine grained authigenic and terrigenous components, such as goethite, glauconite, quartz and clay minerals. As a result of this dilution, the $CaCO_3$ content decreases at depths beyond about 180 m, whereas the SiO_2 concentration increases. SiO_2 and $CaCO_3$ are strongly negatively correlated in all sediments (Fig. 3a), indicating that they are major diluents for each other and also for the other components.

The current velocities on the continental shelf and upper slope are controlled by the position of the EAC with respect to the margin. This position has varied through time with sea-level variations. During periods of low sea-level, the current core moves away from the coast and the current volume is augmented, resulting in considerably higher current velocities on the shelf and slope (O'Brien et al. 1986). Current velocities on the shelf and upper slope are lowest during periods of relatively high sea-level. The relationship that presently exists between sediment composition and current velocity suggests that the various sedimentary facies migrated backwards and forwards across the shelf and slope as current velocities varied with

Table 1. *Geochemical data for various sediment facies on the East Australian continental margin*

Facies	Depth Interval (m)	CaCO₃ Average	CaCO₃ Range	TOC Average	TOC Range	Fe₂O₃ Average	Fe₂O₃ Range	P₂O₅ Average	P₂O₅ Range
Facies 1	0–50	17.3 [13]	1.4–72.4 (8) (41)	0.05 [12]	0.03–0.11 (8) (47)	0.65 [13]	0.1–2.5 (8) (41)	0.07 [13]	0.05–0.20 (8) (41)
Facies 2	50–200	66.4 [51]	10.4–94.7 (61) (113)	0.18 [47]	0.05–0.60 (59) (66)	1.52 [49]	0.60–5.54 (61) (174)	0.16 (51)	0.05–0.43 (57) (174)
Facies 3	200–350	58.6 [24]	35.2–89.9 (270) (234)	0.25 [24]	0.10–0.58 (240) (285)	7.45 [27]	1.44–17.7 (234) (280)	0.53 [25]	0.09–1.60 (270) (310)
Facies 4	350–460	46.1 [32]	26.8–64.4 (420) (360)	0.46 [42]	0.10–1.11 (360) (415)	4.97 [50]	1.34–15.6 (360) (355)	0.37 [49]	0.09–1.34 (380) (355)
Facies 5	460–655	42.0 [8]	32.1–48.6 (655) (494)	0.78 [7]	0.48–1.05 (525) (550)	2.74 [10]	2.1–3.61 (494) (510)	0.13 [9]	0.09–0.19 (550) (567)

[ ] Number of samples used in calculation.
( ) Water depth (m).
All analyses are in weight %.

eustatic sea-level changes. The observation that Quaternary Facies 4 sediments (containing Quaternary phosphates) overlie Neogene nodules at water depths between about 350–425 m supports this proposal.

The distributions of Fe_2O_3 and P_2O_5 within inner shelf to upper slope sediments are shown in Figs 2d & 2e respectively. Both are low in Facies 1 & 2, rise dramatically within goethite-rich Facies 3 sediments, and decline progressively through Facies 4 & 5. The very strong correlation between Fe_2O_3 and P_2O_5 within the sediments is evident in Fig. 3b. Since none of these sediments contain detectable apatite, the most likely explanation for the correlation is that the P_2O_5 is present as adsorbed PO_4^{3-} (Berner 1973) on the surfaces of ferric oxyhydroxides.

The distribution of the TOC within sediments across the continental margin is given in Fig. 2f. TOC is low within Facies 1 & 2, averaging 0.05 and 0.18% respectively (Table 1), but then increases progressively into deeper water, averaging 0.78% within Facies 5. The sediments which contain the Neogene (Facies 3) and Quaternary (Facies 4) phosphorites average 0.25 and 0.46% TOC respectively (Table 1). It is evident from Figs 2d–f and Table 1 that the

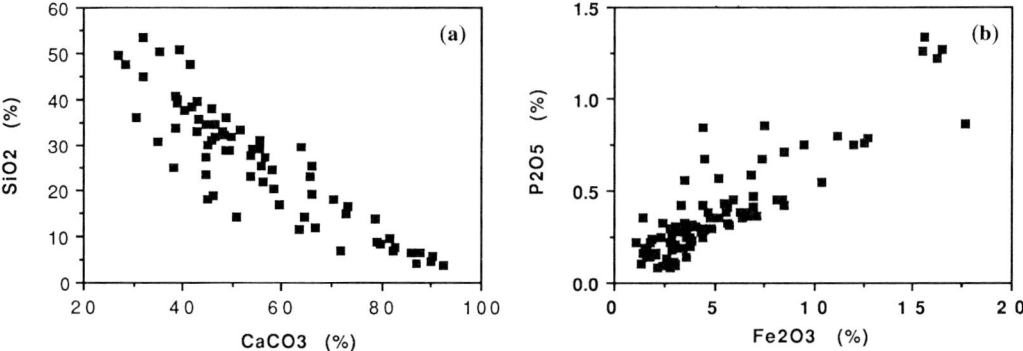

Fig. 3. (a) Total SiO_2 concentration plotted against $CaCO_3$ concentration for sediments on the East Australian continental margin. Strong negative correlation indicates that the terrigenous and biogenic components are the principal diluents for each other and other components. (b) Total P_2O_5 concentration plotted against Fe_2O_3 (total iron) concentration for sediments on the East Australian continental margin. Very strong positive correlation is the result of PO_4^{3-} adsorption by ferric oxyhydroxides.

Holocene phosphorites form at a transition between Fe- and P-rich, TOC-poor outer shelf (Facies 3) sediments and Fe- and P-poor, TOC-rich basinal (Facies 5) sediments.

Rock-Eval pyrolysis data show that the nature of the organic matter also varies across the continental margin. Both S_2 (Fig. 2g) and the hydrogen index (HI; Fig. 2h) increase markedly in water depths greater than about 350 m. The S_2 data suggest that the deeper water sediments contain greater amounts of labile organic carbon, as would be expected from their higher organic carbon content. The HI data, however, indicate that the organic carbon in the deeper water sediments is itself more labile, i.e., a greater percentage of the organic carbon in deeper water is reactive. The presence of higher TOC and more reactive organic matter in deeper water is probably the result of several interrelated factors. Firstly, the velocity of the EAC is less in deeper water and this may allow more fine grained organic matter to reach the seafloor. Secondly, ^{14}C data show that the sedimentation rate increases progressively into deeper water, from $1-2$ cm ka^{-1} in Facies 4 sediments to >5 cm ka^{-1} within Facies 5 (O'Brien unpublished data). In addition, ^{210}Pb data (O'Brien et al. 1988; Heggie et al. 1990) indicate that both the depth and rate of sediment mixing is much greater in Facies 4 sediments compared to Facies 5. A combination of lower sedimentation rates and higher sediment mixing rates in shallower water Facies 4 sediments increases the residence time of the organic matter within the mixed layer and prolongs its exposure to oxidative near-surface conditions. In contrast, the organic matter in deeper water is buried more quickly and hence near-surface oxidation is minimized. Sediment mixing is probably principally induced via bioturbation, though the effects of strong bottom currents cannot be discounted.

The distribution of K_2O in sediments across the margin is shown in Fig. 2i. K_2O concentration provides an indication of the abundance of particular clay minerals, in this case glauconite. K_2O is lowest within Facies 2 bryozoal sands, but increases into deeper water, peaking between $350-450$ m water depth. This peak incidentally corresponds to the peak in SiO_2 concentration (Fig. 2c), and represents the zone of maximum glauconite abundance. This maximum occurs within Facies 4 and corresponds precisely with the zone of modern phosphorite formation. Moreover, recent pore water and solid phase studies (O'Brien & Heggie unpublished data) have shown that glauconite is currently forming within Facies 4 sediments.

Facies 4 sediment geochemistry. Data presented in this section is limited to sediments which actually contained Quaternary phosphorite nodules. Composite vertical profiles of Fe_2O_3, P_2O_5 and F, obtained from several Facies 4 piston cores, are given in Fig. 4a. All three have a similar vertical distribution. They are depleted at depth within the sediment, then increase rapidly between about 6 cm depth and the sediment surface. This upper 6 cm corresponds to the zone in which iron-staining is evident within the cores. Recent porewater data (Heggie et al. 1988, 1990) have shown that the Facies 4 sediments become anoxic within approximately 1 cm of the sediment–seawater interface, with porewater Fe^{2+} concentrations increasing rapidly below this depth because of iron oxyhydroxide reduction. The Fe_2O_3 profile in Fig. 4a can be explained by the continuous reduction and dissolution of FeOOH within the anoxic zone and the oxidation of Fe^{2+} and reprecipitation of dissolved iron as Fe^{3+} near the oxic sediment–seawater interface. The close relationship between the solid phase Fe and P distributions is thus interpreted to be the result of PO_4^{3-} adsorption by FeOOH in the oxic zone and the loss of PO_4^{3-} from the solid phase during FeOOH reduction within the anoxic zone.

This interpretation supports Marshall's (1983) hypothesis on phosphate concentration in this region, and is in accord with the work of Heggie et al. (1990), who have shown that the geochemical behaviour of P and Fe in both the solid phase and pore waters are linked in these sediments. Their data indicate that ferric oxyhydroxides within the oxic upper few centimetres of the sediments adsorb PO_4^{3-} from the pore waters. During sediment mixing, the ferric iron is transported into the anoxic zone and reduced, leading to the dissolution of the FeOOH and liberation of both Fe^{2+} and PO_4^{3-} to the pore waters. The Fe^{2+} and some PO_4^{3-} diffuse upwards to be reprecipitated in the oxic zone, whereas much of the remaining PO_4^{3-} is taken up during apatite precipitation. Heggie et al. suggest that phosphate nodules are formed when sediment mixing rates are high and a relatively large amount of phosphate is transported into the anoxic zone (with FeOOH) for incorporation into apatite. Conversely, times of relatively slow sediment mixing correspond to periods of scattered, incipient apatite precipitation. A key part of this model is that the sediment mixed layer extends down through the zone of Fe-P cycling (i.e. the oxic–anoxic transition zone). If the mixed layer does not extend down to encompass the Fe remobilization zone, the cycling of solid phase FeOOH (with adsorbed P) into the anoxic zone is ineffective, and phosphate nodules cannot form.

The close relationship of F with Fe (Fig. 4b) and P in the solid phase suggests that F is also involved in Fe cycling between oxic and anoxic zones within the sediments. Experimental evidence has shown that F in soils (Farrah & Pickering 1986) and marine sediments (Ruttenburg 1988) is strongly scavenged by hydrous oxyhydroxides of iron. It appears, therefore that the FeOOH adsorbs both PO_4^{3-} and F in the oxic zone and releases both to the pore waters during FeOOH reduction in the anoxic zone.

Porewater silica determinations, when inte-

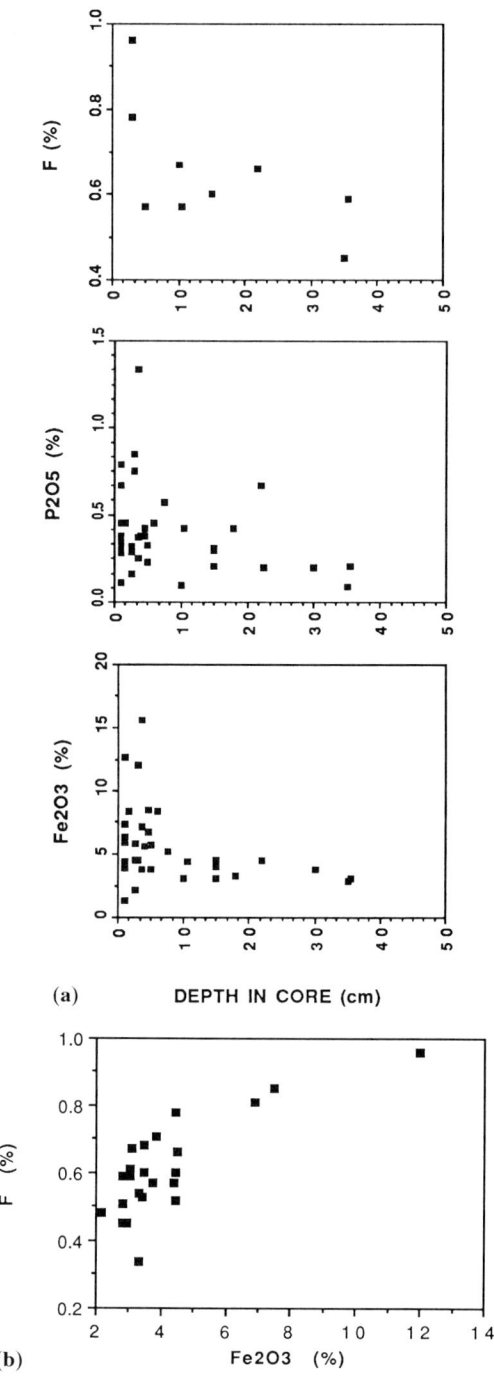

Fig. 4. (a) Down-core solid phase distribution of Fe_2O_3, P_2O_5 and F in Facies 4 sediments. Strong vertical mobilization of Fe, P and F is evident. (b) Sedimentary F plotted against Fe_2O_3 for sediments on the East Australian continental margin. A strong sympathetic relationship exists, possibly because of F adsorption by hydrous oxides of iron.

grated with scanning electron microscope (SEM) and X-ray mineralogical (XRD) studies (O'Brien & Heggie unpublished data) show that glauconite is presently forming authigenically within Facies 4 upper slope sediments. The glauconite is in fact an immature, very fine grained low-potassium, high iron-smectite which forms in the pore space within the sediments, principally at depths of 5–15 cm below the sediment–seawater interface. The upper and lower boundaries of the zone of glauconite formation appear to be defined by the availability of ferrous iron in the pore waters and ferric iron in the solid phase respectively. The principal zone of glauconite formation thus overlaps the zone of apatite precipitation. The data thus reveal that two mineral phases, apatite and glauconite, which are commonly associated in the sedimentary record, are precipitating contemporaneously on the East Australian continental margin.

Redox processes in these sediments also have a marked effect on the distribution of organic carbon. Both TOC and HI are low in Fe_2O_3-rich sediments (Figs 5a & b), presumably as a result of post-depositional oxidation. HI is highly variable in the upper 10 cm (i.e. within the mixed-layer), but decreases to <50 mgHC/gTOC at depth (Fig. 5c). This distribution is probably due to the oxidation of the organic matter within the mixed-layer, with the accompanying loss of the labile, hydrogen-rich organic compounds.

The strong vertical variation in sediment geochemistry indicate that an 'average' geochemical composition for these sediments would not be as instructive as if the sediments were divided into two groups: those from depths of ≤10 cm and those from >10 cm depth (Table 2). The average total iron (Fe_2O_3) content of the near-surface sediments (6.35%) is almost twice that of the deeper sediments, as a result of iron reduction/oxidation cycling and FeOOH precipitation in the near-surface oxic zone and its reduction at depth. While both P_2O_5 and F. are comparatively enriched in the near-surface sediments, their absolute concentrations are low. The P_2O_5 concentrations are lower than the values reported for sediments from the modern phosphogenic provinces off Peru–Chile and Namibia (Baturin 1969; Senin 1970; Romankevich & Baturin 1972; Burnett 1977; Price & Calvert 1978).

The TOC concentrations of the East Australian margin sediments are low, averaging 0.45% (Table 2), 10–40 times less than in sediments from the upwelling zones off Namibia (typically 5–20% TOC: Senin 1970;

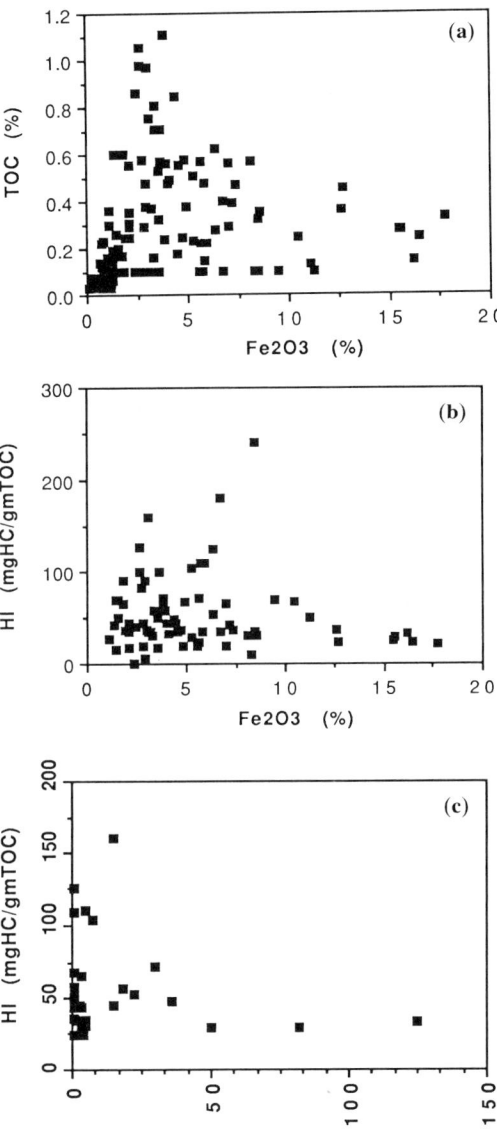

Fig. 5. Relationship between redox sensitive solid phase indicators (TOC, HI,) and the Fe_2O_3 concentration in Facies 4 sediments associated with Quaternary phosphate nodules. (a) TOC versus Fe_2O_3 (b) HI versus Fe_2O_3 (c) HI versus depth below sediment–seawater interface.

Romankevich & Baturin 1972; Veeh *et al.* 1974) and Peru–Chile (Suess 1981). Another contrast with these upwelling phosphogenic provinces is the $CaCO_3$ content of the sediments. Whereas the Peru–Chile and Namibian phosphorites are associated with organic-rich, diatomaceous

Table 2. Average geochemical data for sediments associated with Quaternary phosphorite nodules, East Australian continental margin

	All samples	<10 cm depth*	>10 cm depth†
Fe_2O_3‡	5.30 ± 2.89 (43)	6.35 ± 3.29 (25)	3.84 ± 1.14 (18)
FeO§	0.49 ± 0.08 (24)	0.47 ± 0.10 (6)	0.50 ± 0.08 (18)
MnO	0.02 ± 0.01 (43)	0.02 ± 0.01 (25)	0.02 ± 0.01 (18)
TiO_2	0.36 ± 0.09 (43)	0.31 ± 0.07 (25)	0.43 ± 0.07 (18)
CaO	25.7 ± 3.9 (43)	26.6 ± 4.2 (25)	24.4 ± 3.0 (18)
K_2O	1.23 ± 0.33 (43)	1.23 ± 0.41 (25)	1.23 ± 0.16 (18)
SO_3	0.50 ± 0.31 (43)	0.37 ± 0.24 (25)	0.67 ± 0.30 (18)
P_2O_5	0.40 ± 0.25 (43)	0.44 ± 0.27 (25)	0.34 ± 0.20 (18)
SiO_2	34.1 ± 6.8 (43)	31.2 ± 6.4 (25)	38.1 ± 5.0 (18)
Al_2O_3	4.70 ± 0.87 (43)	4.31 ± 0.79 (25)	5.27 ± 0.67 (18)
MgO	1.46 ± 0.23 (43)	1.47 ± 0.28 (25)	1.44 ± 0.13 (18)
Na_2O	3.23 ± 1.86 (43)	2.5 ± 1.5 (25)	4.32 ± 1.8 (18)
F	0.63 ± 0.14 (16)	0.79 ± 0.16 (4)	0.59 ± 0.11 (12)
$CaCO_3$	47.5 ± 7.1 (26)	47.8 ± 7.8 (20)	46.5 ± 4.0 (6)
TOC‖	0.45 ± 0.20 (33)	0.41 ± 0.17 (22)	0.53 ± 0.25 (11)
S_2	0.22 ± 0.15 (30)	0.18 ± 0.09 (22)	0.29 ± 0.21 (11)
HI	52 ± 27 (31)	54 ± 31 (22)	58 ± 41 (11)

Numbers in brackets are number of analyses.
All analyses are in weight % except S_2 (kgHC/tonne) and HI (mgHC/gTOC).
* All sediments from a depth of 10 cm or less below the sediment — seawater interface.
† All sediments from a depth greater than 10 cm below the sediment — seawater interface.
‡ Total iron expressed as Fe_2O_3.
§ Ferrous iron expressed as FeO.
‖ Total organic carbon.

oozes, the East Australian nodules form within glauconitic, foraminiferal sands which contain an average of almost 50% $CaCO_3$ (Table 2).

The major element chemistry of the sediments (Table 3) is controlled by the relative abundance of the biogenic ($CaCO_3$), terrigenous and authigenic components. High CaO concentrations reflect biogenic carbonate, whereas SiO_2 is present in detrital quartz and, to a lesser extent, glauconite, feldspar, clay, and possibly very minor biogenic silica. Fe_2O_3, Al_2O_3 and K_2O are contained principally within glauconite, with some Fe_2O_3 present as iron oxy-hydroxide. The relative abundance of these components can vary significantly, both from locality to locality and also vertically at the one location, as a result of varying terrigenous sediment supply to the respective locations on the upper slope.

Quaternary nodules

P_2O_5 concentrations in the Quaternary nodules are highly variable and range between 4.0–19.1%, though most contain 7–16% P_2O_5. No nodules studied to date contain less than 4.0% P_2O_5 and a gap appears to exist between P_2O_5 concentrations in the slightly phosphatic nodules and the essentially apatite-free sediments. The average composition of 62 Quaternary nodules is shown in Table 4 and the composition of several representative nodules is given in Table 5.

Previous work has shown that both the apatite and iron oxy-hydroxide contents of Quaternary nodules are a direct function of their residence time within the upper 20 cm of the sediments (O'Brien et al. 1987). Nodules remaining within this zone after their formation rapidly accumulate both apatite and iron oxy-hydroxide for approximately 60 ka. Those buried soon after formation gain little of either phase. Recent solid phase ^{210}Pb and ^{14}C data, in conjunction with pore water metabolite distributions, have provided an explanation for this observation (O'Brien et al. 1988; Heggie et al. 1990). ^{210}Pb data for a gravity core (which contains Quaternary phosphate nodules) is shown in Fig. 6. ^{210}Pb decreases exponentially from a near-surface (1 cm depth) maximum of 6.33 dpm g^{-1} (disintegrations per minute per gram of sediment) to a minimum of 1.16 dpm g^{-1} at a depth of 15 cm. Below about 20 cm, the ^{210}Pb activity increases, reaching a maximum of 6.06 dpm g^{-1} at a depth of 69 cm. Since the sedimentation rate in these sediments is very slow (1–2 cm ka^{-1}; O'Brien unpublished data) compared

Table 3. *Representative major element compositions for sediments (Facies IV) associated with Quaternary phosphate nodules, East Australian continental margin*

Sample No.	P895 (3 cm)−1	P895 (15 cm)−1	P895 (67 cm)−1	P895 (125 cm)−1	P884 (3 cm)−1	P884 (10 cm)−1	P884 (22 cm)−1	P884 (35 cm)−1	P884 (60 cm)−1	P884 (70 cm)−1	P884 (125 cm)−1
Fe$_2$O$_3$*	12.00	4.47	7.48	4.10	4.44	3.06	4.49	2.82	2.82	2.92	3.03
FeO†	0.48	0.60	0.66	0.54	0.54	0.63	0.54	0.57	0.43	0.54	0.56
MnO	0.02	0.02	0.02	0.02	0.02	0.03	0.02	0.02	0.02	0.02	0.02
TiO$_2$	0.29	0.45	0.48	0.49	0.45	0.55	0.42	0.50	0.37	0.48	0.50
CaO	24.4	24.6	22.6	21.4	22.9	18.0	23.4	19.0	25.2	21.1	18.5
K$_2$O	1.81	1.26	1.65	1.38	1.29	1.39	1.33	1.34	1.06	1.20	1.28
SO$_3$	0.28	0.66	0.51	0.71	0.49	1.06	0.53	1.04	0.74	0.77	0.78
P$_2$O$_5$	0.75	0.30	0.86	0.20	0.85	0.10	0.67	0.09	0.10	0.11	0.31
SiO$_2$	28.0	35.5	36.0	42.9	39.8	47.5	39.2	46.6	39.5	45.9	48.4
Al$_2$O$_3$	4.38	5.66	5.48	5.76	5.09	7.26	5.51	6.44	4.79	5.10	5.86
MgO	1.91	1.44	1.54	1.52	1.40	1.48	1.43	1.49	1.32	1.33	1.42
Na$_2$O	6.0	5.4	5.6	1.6	4.9	5.2	5.3	5.3	5.4	5.3	5.1
F	0.96	0.54	0.85	–	0.78	0.64	0.66	0.45	0.51	0.45	0.61
Loss	19.5	18.9	16.2	18.7	17.1	13.2	16.4	14.3	17.3	14.7	13.5
Total	99.3	99.2	99.3	98.6	99.5	99.4	99.4	99.4	99.1	99.4	99.3

All analyses are in weight %.
* Total iron expressed as Fe$_2$O$_3$.
† Ferrous iron expressed as FeO.
Number in brackets below sample number is depth in core (cm).
P895: 29°33.5′S, 153°49.3′E, 390 m water depth, piston core.
P884: 29°16.0′S, 153°51.2′E, 380 m water depth, piston core.

Table 4. *Average major element compositions for Quaternary phosphate nodules, 'apatite-free' component of Quaternary nodules, and associated sediments, East Australian continental margin*

	Quaternary* nodules	Quaternary[†] nodules (Apatite-free)	Associated sediments (All)	Associated sediments (< 10 cm depth)	Associated sediments (> 10 cm depth)
Fe_2O_3[‡]	4.32	6.33	5.30	6.35	3.84
FeO[§]	0.47	0.69	0.49	0.47	0.50
MnO	0.02	0.03	0.02	0.02	0.02
TiO_2	0.28	0.41	0.36	0.31	0.43
CaO	32.9	21.6	25.7	26.6	24.4
K_2O	0.90	1.32	1.23	1.23	1.23
SO_3	0.71	1.04	0.50	0.37	0.67
P_2O_5	11.2	–	0.40	0.44	0.34
SiO_2	24.4	35.8	34.1	31.2	38.1
Al_2O_3	3.58	5.25	4.70	4.31	5.24
MgO	1.36	1.64	1.46	1.47	1.44
Na_2O	1.56	2.20	3.23	2.5	4.32
F	1.5	–	0.63	0.79	0.59
$CaCO_3$	NA	–	47.5	47.8	46.5
TOC	0.55	0.81	0.45	0.41	0.53
S_2	0.27	0.40	0.22	0.18	0.29
HI	50	50	52	54	58

* Average based upon 62 analyses for major elements.
[†] Apatite-free composition calculated using ideal apatite composition of McClellan (1980).
[‡] Total iron reported as Fe_2O_3.
[§] Ferrous iron reported as FeO.
NA Not available

to the half-life of ^{210}Pb (c.22 years), this indicates that sediment mixing (probably via bioturbation) extends to depths of 15–20 cm (O'Brien *et al*. 1988; Heggie *et al*. 1990).

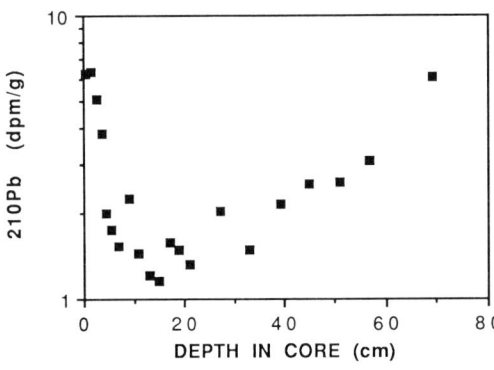

Fig. 6. Total ^{210}Pb plotted against depth in gravity core 71GC/049. ^{210}Pb decreases exponentially to minimum values at a depth of approximately 15–20 cm, suggesting that sediment mixing extends to this depth. ^{210}Pb increases below depths of 20–25 cm, probably because of ^{230}Th (and hence ^{210}Pb) ingrowth towards radioactive equilibrium with the uranium contained in apatite that has precipitated within the mixed layer.

A comparison of ^{14}C sedimentation rate data (O'Brien *et al*. 1988) and average nodule burial rates (calculated from the ^{230}Th ages of nodules (O'Brien *et al*. 1986) shows that the phosphate nodules are typically buried 10 times slower than the rate at which the surrounding sediments accumulate. Bioturbation thus appears to preferentially maintain the nodules within the mixed layer. In addition, pore water data (Heggie *et al*. 1988, 1990) indicates that apatite precipitation in these sediments takes place at depths of 10–15 cm, within the lower part of the mixed layer. Hence, nodules which are maintained within the mixed layer for long periods reside within the zone of apatite precipitation and thus continue to gain apatite. Similarly, these nodules gain iron oxy-hydroxide because of the strong remobilization of Fe^{2+} at depth within the sediments and its oxidation and precipitation near the sediment-seawater interface. The increase in ^{210}Pb activity below about 20 cm depth in Fig. 6 is probably the result of the ingrowth of ^{230}Th (and hence ^{210}Pb) from the uranium in apatite which has precipitated within the mixed layer. Since all visible phosphate nodules were removed from the sediments prior to analysis, this apatite must be very fine grained but ubiquitously disseminated throughout the sediment.

Table 5. *Chemical compositions of representative non-ferruginous phosphatic nodules from the East Australian continental margin*

	G7–8	P852–1	G7–28	P887 (0–3 cm) –1	P884 (78 cm) –1	P894–1	P895 (120 cm) –1	P897 (86–87 cm) –3	P904 (35 cm) –1	P905 (84–87 cm) –1	P906 (85 cm) –1
Lat. (°S)	29°23'	31°01.0'	29°23'	29°18.8'	29°16.0'	29°23.0'	29°23.5'	29°25.1'	29°18.0'	29°19.8'	29°20.8'
Long. (°E)	153°50'	153°18.7'	153°50'	153°50.1'	153°51.2'	153°49.7'	153°49.3'	153°49.0'	153°50.3'	153°50.4'	153°50.0'
Water depth (m)	385	450	385	370	380	415	390	400	405	420	415
Sampling Method[‡]	R.D.	B.C.	R.D.	B.C.	P.C.	B.C.	P.C.	P.C.	P.C.	P.C.	P.C.
Fe_2O_3*	4.37	2.57	8.25	4.36	1.92	2.41	2.34	4.84	2.96	3.15	2.26
MnO	0.03	0.01	0.04	0.02	0.01	0.01	0.02	0.02	0.01	0.02	0.01
TiO_2	0.37	0.31	0.26	0.32	0.26	0.15	0.25	0.27	0.30	0.26	0.23
CaO	29.6	34.4	32.7	30.4	35.8	33.9	36.0	35.9	33.4	35.4	38.5
K_2O	1.05	0.78	0.89	1.07	0.73	0.81	0.75	0.99	0.90	0.92	0.62
SO_3	0.69	0.73	0.40	0.59	0.98	0.68	0.94	0.87	0.87	0.98	1.19
P_2O_5	7.5	11.9	13.1	7.5	15.8	11.0	14.1	14.0	7.9	13.9	19.1
SiO_2	28.3	22.0	20.8	28.0	24.1	26.2	24.2	21.1	26.6	23.0	14.9
Al_2O_3	4.30	3.97	3.00	3.97	2.98	3.11	2.94	3.01	3.83	3.24	2.81
MgO	1.38	1.58	1.33	1.20	1.07	1.29	1.01	1.10	1.14	1.08	1.12
Na_2O	1.3	3.2	1.8	2.1	1.9	1.7	1.2	1.4	1.4	1.7	1.1
F	1.9	–	1.6	1.3	–	1.1	1.7	1.7	1.0	1.7	–
Loss	19.5	18.3	16.8	19.3	14.4	19.0	16.0	16.5	20.6	16.3	15.4
Total	100.29	99.75	99.37	98.83	99.95	100.26	99.75	100.00	99.91	99.95	97.24
CaO/P_2O_5	3.95	2.89	2.50	4.05	2.27	3.08	2.55	2.56	4.23	2.55	2.02
F/P_2O_5	0.25	–	0.12	0.17	–	0.10	0.12	0.12	0.13	0.12	–
Age (ka)[†]	24	2.5	66	23	>250	21	>250	177	72	209	>250

* Total Iron reported as Fe_2O_3.
[†] 230Th age, from O'Brien et al. (1986).
[‡] B. C., box core; P. C., piston core; R. D., rock dredge.
Numbers in parentheses specify the depth (below the sediment-water interface) at which the samples were recovered. All analyses are in weight %.

The differences in composition between nodules buried below the mixed layer and those maintained within it are shown in Fig. 7. These data suggest that nodules form with 5–7% P_2O_5 and 2–3% Fe_2O_3. Diagenetic alteration begins as soon as the nodules form. Nodules which remain in the mixed layer after they form continue to accumulate both P and Fe (as a result of Fe-P cycling processes within the sediments) for up to 60 ka. During this period, their P_2O_5 and Fe_2O_3 contents more than double to 16–18% and 6–8% respectively (Figs 7a & 7b) and consequently the nodules become denser and more lithified. After 60 ka, the pore space within the nodules appears to be too limited for further apatite precipitation, although colloidal iron oxyhydroxides continue to precipitate in the micro-pores, and the Fe_2O_3 content of the nodules can increase to 12–15% within 600 ka-1 Ma (O'Brien, unpublished data). In contrast, P_2O_5 and Fe_2O_3 concentrations in nodules buried below the mixed layer show no obvious relationship to nodule age.

If it is assumed that modern phosphate nodules contain about 7% P_2O_5 (equivalent to 19.4% apatite) at the time of their formation (Fig. 7a), then it appears (following recalculation of the major element chemical data on an apatite-free basis) that the nodules form within sediments containing approximately 2.5–3.7% Fe_2O_3. These Fe_2O_3 values are very similar to those obtained for sediments buried below 10 cm depth (3.84%; Tables 2 & 4), but are substantially less than those for sediments from ≤10 cm depth (6.35%). This supports the contention that nodules form in the anoxic zone, probably towards the base of the mixed layer at depths of about 10 cm or greater in the sediments, a finding in accord with pore water data (Heggie *et al.* 1988, 1990). The average apatite-free composition of the Quaternary nodules is compared to that of the associated sediments in Table 4. The compositions are very similar for all elements with the exception of SO_3 which is, on average, higher in the 'sedimentary' component of the nodules, perhaps because of SO_3 substitution in the apatite structure (Gulbrandsen 1966).

F concentrations in Quaternary phosphate nodules are between 1.0–1.9%, with F/P_2O_5 ratios ranging between 0.10–0.28 with an average of 0.15 ± 0.06 (16 analyses). Assuming all F is located in the apatite structure, calculated F/P_2O_5 ratios are higher than in pure fluorapatite. However, McClellan (1980) has shown that the F/P_2O_5 ratio of ideal francolite containing 6.3% structural CO_2 is 0.148. Apatite in the East Australian phosphatic nodules contains, on average, 7% structural CO_2. Compared with nodules off Peru–Chile (0.098; Burnett 1977) and phosphorites from the Phosphoria Formation (0.100; Gulbrandsen 1966), F/P_2O_5 ratios in the East Australian phosphatic nodules are considerably higher and more closely approach the ratios of Agulhas Bank nodules (0.123; Parker 1975).

Comparison of Quaternary nodules with Neogene ferruginous nodules

The major element compositions of Quaternary and Neogene nodules are given in Table 6 as a series of average compositions grouped ac-

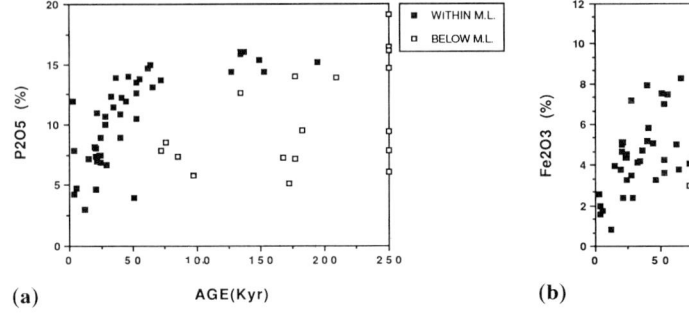
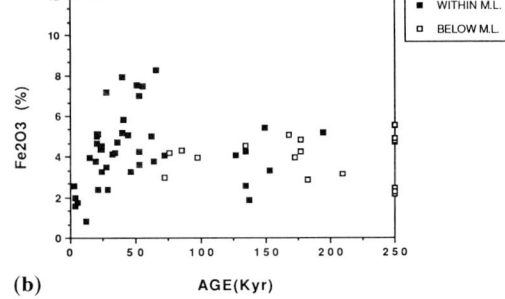

Fig. 7. (a) P_2O_5(%) content of Quaternary phosphate nodules which were located within and below the mixed layer (ML). The P_2O_5 content of nodules within the ML increases progressively with age, whereas P_2O_5 show little relationship to age in noodules buried below the ML. (b) Fe_2O_3(%) content of Quaternary phosphate nodules located within and below the mixed layer (ML). Increase in Fe_2O_3 in nodules within the ML is due to FeOOH precipitation (via Fe^{2+} oxidation) within the oxic zone. This process can enrich the nodules by several wt% Fe_2O_3 in 60 ka.

Table 6. *Average major element compositions of Quaternary non-ferruginous and Neogene ferruginous phosphate nodules, grouped according to Fe_2O_3 concentration*

Fe_2O_3(%)	0–8 (62)[§]	8–20 (8)	20–30 (2)	30–40 (4)	>40 (6)
Fe_2O_3*	4.32	12.03	28.00	34.11	45.22
FeO[†]	0.47(7)	0.38(3)	0.86(1)	0.44(2)	0.38(4)
MnO	0.02	0.04	0.09	0.08	0.09
TiO_2	0.28	0.29	0.38	0.28	0.27
CaO	32.9	27.9	18.1	17.8	11.9
K_2O	0.90	1.01	0.99	0.56	0.69
SO_3	0.71	0.71	0.56	0.44	0.39
P_2O_5	11.2	12.4	10.1	7.1	7.4
SiO_2	24.4	22.7	20.6	14.9	14.0
Al_2O_3	3.58	3.62	4.23	4.42	4.10
MgO	1.36	1.45	3.05	3.29	2.69
Na_2O	1.56	0.60	0.8	0.8	0.7
F	1.5 (16)	–	1.6(2)	1.3(3)	1.0(5)
F/P_2O_5	0.15(16)	–	0.17(2)	0.13(3)	0.14(5)
$CaCO_3$	–	–	–	–	6.5(2)
CaO/P_2O_5	2.94	2.25	1.79	2.51	1.62
TOC[‡]	0.55(14)	0.85(2)	0.17(1)	0.16(3)	0.11(5)
S_2	0.27(13)	0.05(2)	0.04(1)	0.04(3)	0.01(3)
HI	50(13)	3(2)	24(1)	23(3)	6(3)
LOI(%)	17.9	15.6	12.3	15.5	13.1

All analyses are in weight (%) except S_2 (kgHC/tonne) and HI (mgHC/gTOC).
* Total iron expressed as Fe_2O_3.
[†] Ferrous iron expressed as FeO.
[‡] Total organic carbon.
[§] Number of brackets below Fe_2O_3 range specifies total number of analyses used to calculate average for major elements (Average MA). Brackets beside components specify total number of analyses used to calculate average for that specific component, if different from Average MA.

Table 7. *Percentage change in elemental composition of the various Fe_2O_3 groups in Table 6, compared with the non-ferruginous (0–8% Fe_2O_3) Quaternary nodules*

Fe_2O_3(%)	0–8	8–20	20–30	30–40	>40
Fe_2O_3	4.32	+178	+548	+690	+947
FeO	0.47	−19	+83	−6	−19
MnO	0.02	+100	+350	+300	+350
TiO_2	0.28	+4	+36	0	−4
CaO	32.9	−15	−45	−46	−64
K_2O	0.90	+12	+10	−38	−23
SO_3	0.71	0	−21	−38	−45
P_2O_5	11.2	+11	−10	−37	−34
SiO_2	24.4	−7	−16	−39	−43
Al_2O_3	3.58	+1	+18	+23	+15
MgO	1.36	+7	+124	+142	+98
Na_2O	1.56	−62	−49	−49	−55
F	1.5	–	+7	−13	−33
F/P_2O_5	0.15	–	+13	−13	−7
CaO/P_2O_5	2.94	−23	−39	−15	−45
TOC	0.55	+55	−69	−71	−80
S_2	0.27	−81	−85	−85	−96
HI	50	−94	−52	−54	−88
LOI	17.9	−13	−31	−13	−27

Values for 0–8% Fe_2O_3 are in weight %. All other values are percent difference (Δ%) various groups relative to 0–8% Fe_2O_3 group. $\Delta\% = 100\ (x-y)$, where x = concentration in other group and y = concentration in 0–8% Fe_2O_3 group. −, No data cording to Fe_2O_3 concentration. This data is shown in Table 7 as the percentage change in composition (for the various Fe_2O_3 groups) from the composition of the average Quaternary nodule. It is clear from these data that the compositions of the nodules vary systematically with increasing Fe_2O_3 content, MnO and MgO are significantly enriched in the more Fe-rich nodules, while Al_2O_3 is slightly enriched. All other elements are depleted, although TiO_2 depletion is minor. Depletion is most pronounced in TOC, S_2, and HI, but CaO, SO_3 and the CaO/P_2O_5 ratio are also substantially reduced.

MgO and Al_2O_3 concentrations in nodules are plotted against total iron (as Fe_2O_3) in Figs 8a and 8b respectively. MgO increases progressively with increasing Fe_2O_3, although the increase in Al_2O_3 is more subtle. Nonetheless, the average Al_2O_3 content of Neogene nodules is consistently higher than that of Quaternary phosphorites. The positive relationship between Al_2O_3 and Fe_2O_3 could be explained by Al substitution in goethite. The relationship between MgO and Fe_2O_3 is more problematical. Electron microprobe analyses of the ferruginous rim on a Neogene nodule are given in Table 8. SiO_2, MgO and Al_2O_3 are all enriched with Fe_2O_3, perhaps indicating that an Si–Mg–Al (and Na–Ca?)-rich clay mineral is associated with the goethite in the Neogene nodules. Marshall (1983) reported berthierine, a form of chamosite, in iron-rich sediments from the same general area. The relatively high P_2O_5 and MnO concentrations in the goethite rim (Table 8) are possibly due to phosphate adsorption by goethite. Parker & Siesser (1972) reported that the goethite-rich phosphatized limestones from the Agulhas Bank are relatively enriched in Mg compared to nearby iron-poor phosphatized limestones and attributed this to the possible adsorption of Mg by goethite.

Both the CaO concentration and the CaO/P_2O_5 ratio of the nodules are strongly negatively correlated with Fe_2O_3 (Figs 9a & 9b). The CaO/P_2O_5 ratio is highly variable within the non-ferruginous nodules, a reflection of both the variable initial biogenic carbonate component of the sediments and also the variable P_2O_5 concentrations of the nodules. The decrease in CaO concentration and CaO/P_2O_5 ratio records the loss of biogenic carbonate from the nodules with time as a result of dissolution and/or replacement. SEM studies (O'Brien 1983) indicate that the fine-grained carbonate in coccolith tests is most susceptible

Table 8. *Electron probe analysis of a goethite rim surrounding a Neogene ferruginous nodule from the East Australian continental margin*

	Analysis#1	Analysis#2
Na_2O	0.37	0.37
MgO	2.76	2.90
Al_2O_3	1.52	1.42
SiO_2	3.23	2.71
P_2O_5	2.84	2.86
SO_3	0.02	0.10
K_2O	0.06	0.10
CaO	0.59	0.54
Fe_2O_3*	68.1	69.6
MnO	0.16	0.16
Cl_2O	0.01	0.00
Total	79.6	80.8

Location 71 RD/001, latitude : 30°44.886'S; longitude : 153°20.614'E; water depth 357 m
* Total iron expressed as Fe_2O_3

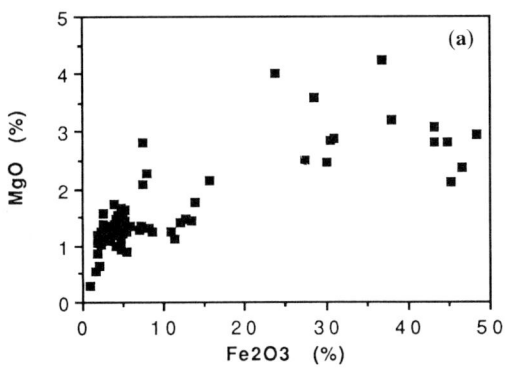 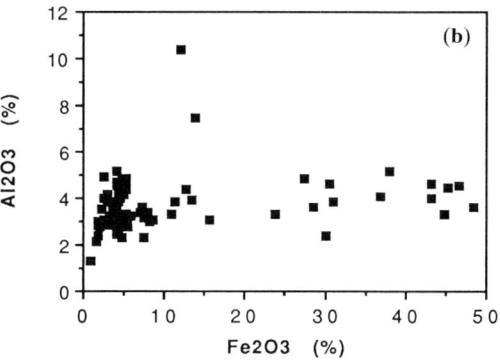

Fig. 8. (a) MgO versus Fe_2O_3 in phosphatic nodules from the East Australian continental margin. (b) Al_2O_3 versus Fe_2O_3 in phosphatic nodules from the East Australian continental margin.

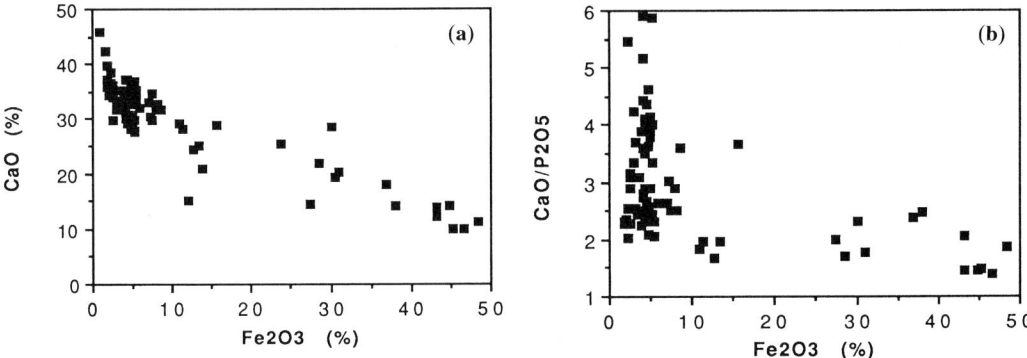

Fig. 9. (a) CaO versus Fe_2O_3 in phosphatic nodules. The pronounced decrease in CaO with increasing iron is due to the replacement or dissolution of $CaCO_3$ in the more Fe-rich nodules. (b) CaO/P_2O_5 versus Fe_2O_3 in phosphatic nodules.

to dissolution. In progressively more highly ferruginous nodules, foraminiferal tests are affected and the CaO/P_2O_5 ratio approaches that of carbonate fluorapatite. Apatite is probably also lost under conditions which favour dissolution of carbonate minerals, a proposition supported by the presence of remobilised apatite filling veins and fractures within some of the more iron-rich nodules (O'Brien 1983).

The relationships between Fe_2O_3 and TOC, S_2, HI, Na and SO_3 are shown in Figs 10a–e. TOC, S_2 and HI are progressively lost as ferruginization of the nodules proceeds, suggesting that the oxidative processes responsible for iron enrichment in the nodules has also oxidised the more labile organic matter. The most ferruginous nodules contain almost no labile organic carbon. The most likely explanation for the lower SO_3 concentrations in the Fe-enriched nodules is that pyrite originally present in the nodules was oxidized during ferruginization. Although McArthur (1978) proposed that subaerial weathering removes Na, Sr, CO_3^{2-} and SO_4 from carbonate fluorapatite, many of the Fe-rich nodules occur at water depths that suggest that they have never been subaerially exposed. Moreover, the CO_3^{2-} content of the apatite in both non-ferruginous and ferruginous nodules is very similar, which indicates differential behaviour of Na, SO_4 and CO_3^{2-}.

Comparison with nodules from other regions

Quaternary nodules from the East Australian continental margin tend to be enriched in the major oxides Fe_2O_3, K_2O, SiO_2, Al_2O_3 and MgO compared with phosphatic nodules from other regions (Table 9), due largely to the greater abundance of detrital quartz and clay and authigenic glauconite. SiO_2 concentrations in nodules from East Australia and Peru–Chile are similar, although the silica in Peru–Chile nodules is present largely as diatom frustules (Burnett 1977) which are rare in the East Australian material. P_2O_5 is lower in Quaternary East Australian nodules (average 11.2%) than in phosphatic nodules from Peru–Chile, Namibia, Morocco, Chatham Rise, southern California Borderland and the Phosphoria Formation (average >20%). The Agulhas Bank phosphatic nodules contain between 10–18% P_2O_5 and so are similar to the Australian phosphate nodules. The Australian and Agulhas Bank phosphatic nodules also have similar F concentrations. The average F/P_2O_5 ratio for the East Australian nodules (0.15) is twice that of the Namibian materials and about 1.5 times that of phosphatic nodules from Peru–Chile and the southern California Borderland. The Australian nodules are characterized by high concentrations of biogenic carbonate, resulting in high CaO/P_2O_5 ratios. Other phosphorites contain mainly siliceous biogenic material and little carbonate. The geochemical similarity between the Australian and Agulhas Bank nodules is paralleled by sedimentological features, as both are glauconitic, calcareous packstones (O'Brien 1983).

The Australian Neogene nodules are similar in some respects to Fe_2O_3-rich, glauco-conglomeratic Agulhas Bank nodules, Fe_2O_3-rich Moroccan samples and Fe_2O_3-rich nodules from Danois Bank (Table 9). There appear to be parallels between the association of non-

Table 9. Comparison of the average chemical composition of non-ferruginous and ferruginous phosphatic nodules from the East Australian continental margin with those of phosphatic rocks from other locations

	Eastern Australia					Agulhas Bank			Peru-Chile	Namibia	Morocco		Spain	Chatham Rise	Southern California Border-Land	Phosphoria Formation
	1A	1B	1C	1D	1E	Iron-poor 2A	Iron-rich 2B	2C	3	4	Iron-poor 5A	Iron-rich 5B	6	7	8	9
Fe_2O_3*	4.32	12.03	28.00	34.11	15.22	1.40	25.80	6.17	2.85	0.87	0.68	9.79	27.82	2.73	2.05	1.1
MnO	0.02	0.04	0.09	0.08	0.09	0.01	0.06	0.01	n.d.	n.d.	n.d.	n.d.	0.32	n.d.	n.d.	–
TiO_2	0.28	0.29	0.38	0.28	0.27	0.06	0.04	0.10	n.d.	0.06	n.d.	n.d.	0.17	n.d.	n.d.	–
CaO	32.9	27.9	18.1	17.8	11.9	47.02	33.42	36.87	33.93	51.3	49.44	40.20	25.35	44.3	42.8	44.0
K_2O	0.90	1.01	0.99	0.56	0.69	0.43	0.39	1.55	1.30	0.13	0.13	0.43	0.43	n.d.	n.d.	0.5
SO_3	0.71	0.71	0.56	0.44	0.39	0.77	0.52	1.12	0.40	n.d.	2.97	1.40	n.d.	n.d.	n.d.	1.75
P_2O_5	11.2	12.4	10.1	7.1	7.4	14.82	10.26	17.53	22.61	32.1	21.71	18.05	7.82	20.8	27.5	30.5
SiO_2	24.4	22.7	20.6	14.9	14.0	6.20	3.45	14.79	22.13	1.4	2.41	5.49	14.64	5.7	8.4	11.9
Al_2O_3	3.58	3.62	4.23	4.42	4.10	1.13	0.92	2.13	5.15	0.37	0.36	1.14	2.95	0.30	0.96	1.7
MgO	1.36	1.45	3.05	3.29	2.69	0.97	1.49	1.38	1.07	0.60	0.93	3.54	1.65	3.07	1.13	0.3
Na_2O	1.6	0.6	0.8	0.8	0.7	0.62	0.34	0.76	0.85	n.d.	2.31	1.73	0.56	n.d.	n.d.	0.6
F	1.5[†]	–	1.6[‡]	1.3[§]	1.0[‖]	2.12	1.42	2.15	2.22	2.39	n.d.	n.d.	0.7	1.9	3.17	3.1
Loss	17.9	15.6	12.3	15.5	13.1	25.52	23.24	16.40	8.78	n.d.	n.d.	n.d.	17.65	n.d.	14.9	6.5
CaO/P_2O_5	2.94	2.25	1.79	2.51	1.52	3.17	3.26	2.10	1.50	1.60	2.28	2.23	3.24	2.13	1.56	1.44
F/P_2O_5	0.15[†]	–	0.16[‡]	0.13[§]	0.14[‖]	0.143	0.138	0.123	0.098	0.074	–	–	0.090	0.091	0.115	0.100

All analyses are given in weight %.
* Total iron reported as Fe_2O_3.
[†] Average 2 analyses, (this study).
[‡] Average 2 analyses, (this study).
[§] Average 3 analyses, (this study).
[‖] Average 3 analyses, (this study).
nd = not detected or determined.

1A, Holocene-Pleistocene Nodules, 62 analyses, this study;
1B, Ferruginous nodules (8–20% Fe_2O_3), 8 analyses, this study;
1C, Ferruginous nodules (20–30% Fe_2O_3), 2 analyses, this study;
1D, Ferruginous nodules (30–40% Fe_2O_3), 2 analyses, this study;
1E, Ferruginous (40% Fe_2O_3), 6 analyses, this study;
2A, 'Iron-poor' phosphatized limestones, 3 analyses, Parker & Seisser (1972);
2B, 'Iron-rich' phosphatized limestones, 3 analyses, Parker & Seisser (1972);
2C, 'Glauco-conglomeratic phosphorite', 12 analyses, Parker (1975);
3, 'Phosphate nodules', 15 analyses, Burnett (1977);
4, 'Concretionary phosphates', 2 analyses (Price & Calvert 1978);
5A, 'Fe-poor phosphorites', 13 analyses, McArthur (1978);
5B, 'Fe-rich phosphorites', 22 analyses, McArthur (1978);
6, 'Nodules conglomeratique ferrugineaux', 13 analyses, Danois Bank, Lucas *et al.* (1978);
7, 'Phosphorite nodules', 5 analyses, Pasho (1972);
8, 'Phosphorite nodules', 22 analyses, Pasho (1972);
9, 'Phosphoria Formation', 60 analyses, Gulbrandsen (1966).

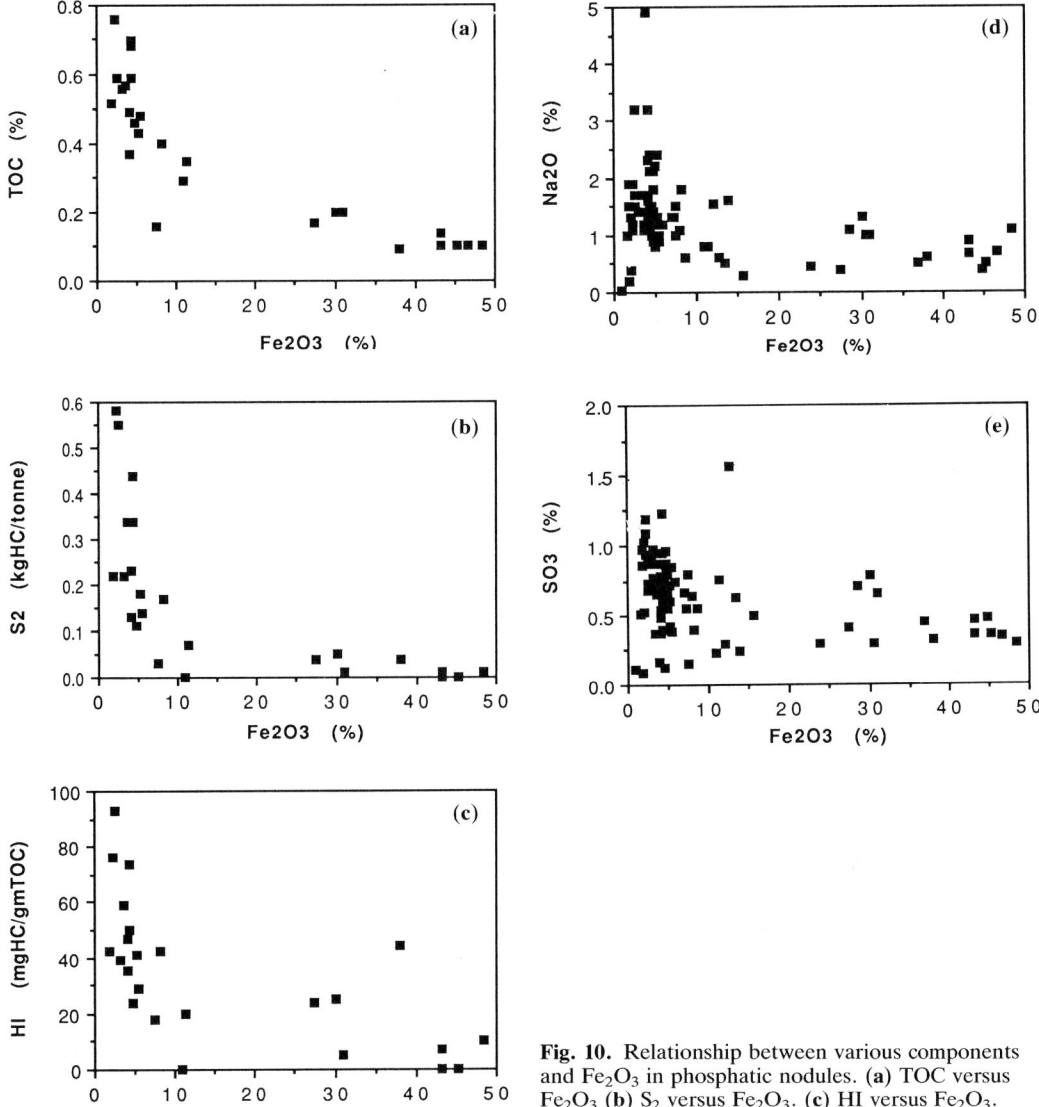

Fig. 10. Relationship between various components and Fe_2O_3 in phosphatic nodules. (a) TOC versus Fe_2O_3 (b) S_2 versus Fe_2O_3. (c) HI versus Fe_2O_3. (d) Na_2O versus Fe_2O_3. (e) SO_3 versus Fe_2O_3.

ferruginous and feruginous nodules in East Australian and iron-rich and iron-poor nodules on Agulhas Bank. On Agulhas Bank, the geochemistry of the Fe-rich and Fe-poor nodules is similar, except for greater concentrations of Fe_2O_3, MnO and MgO in the iron-rich phosphorites. Similarly, the CaO concentration and CaO/P_2O_5 ratio is much lower in the iron-rich Agulhas Bank samples. Parker & Siesser (1972) noted small amounts of goethite in the apatite matrix of iron-poor nodules, and suggested that a complete transition exists between the iron-poor and very iron-rich types. Such a transition also seems to exist off East Australia.

Discussion

The data presented in this paper provide an explanation for many aspects of phosphogenesis on the East Australian continental margin, including the genesis and diagenesis of the phosphatic nodules, the effect of sea-level variations on sedimentation and the origin of the iron-rich Facies 3 and 3' sediments.

Authigenic mineral formation

Phosphate nodules are presently forming within relatively organic-lean, glauconitic, foraminiferal sands (Facies 4) on the upper slope at water depths between 350–460 m. The nodules form at a transition-zone between Fe-P rich, organic carbon-poor shallow water sediments (Facies 3) and Fe-P deficient, organic carbon-rich deep basin sediments (Facies 5). The Facies 4 sediments become anoxic within about 1 cm of the sediment–seawater interface and hence the abundance of redox-sensitive phases, such as goethite, show strong vertical zonations. Fe-P cycling processes are critical to the formation of both apatite and glauconite within the sediments. Ferric oxy-hydroxides within the upper few centimetres adsorb PO_4^{3-} under oxic conditions; the PO_4^{3-} is then released to the pore waters when the FeOOH particles are shunted into the anoxic zone by sediment mixing processes and subsequently dissolve. Some of this released PO_4^{3-} is then taken up during apatite precipitation.

The phosphate nodules form as very friable layers or 'lumps' at depths of approximately 6–15 cm below the sediment surface, probably during periods of rapid sediment mixing; periods of slower mixing are characterized by minor, incipient apatite precipitation. Sediment mixing processes maintain many of the nodules within the mixed-layer for extended (>60 ka) periods after their formation. While in the mixed-layer, the nodules rapidly gain apatite and, to a lesser extent, ferric oxy-hydroxide; their degree of induration also increases rapidly. This apatite precipitates when the nodule is within the anoxic zone, whereas FeOOH precipitation occurs when the nodule is within the oxic zone very near the sediment–seawater interface. Nodules buried below the mixed layer soon after their formation gain little or no apatite and goethite. Sediment mixing processes may also explain the absence of nodules that contain <4% P_2O_5. Some nodules probably do initially form with only 2–4% P_2O_5 but they are so friable that they are disaggregated during sediment mixing into very fine, slightly phosphatic clasts. It appears that only nodules containing greater than about 5% P_2O_5 are sufficiently robust to be preserved. Glauconite forms contemporaneously with apatite within the sediments.

Sea-level and authigenic mineral formation

The importance of redox-controlled Fe-cycling in the genesis of both apatite and glauconite explains their co-existence on both the East Australian continental margin and in the geological record. Off Eastern Australia, both minerals are now forming within the sub-oxic to anoxic (Fe-cycling) zone in sediments at water depths between approximately 350–460 m. In slightly shallower water, Facies 3 sediments contain relict apatite and glauconite. The differences in depositional environment between Facies 3 and 4 are probably principally controlled by the organic carbon input to the sediments in the respective facies. Within Facies 3, the current velocities are sufficiently strong to restrict the organic carbon input to the sediment, so that the sediments are oxic, at least within the upper metre. In contrast, within Facies 4 the current velocities are lower and sufficient organic matter reaches the sediment to establish anoxic environments just below the sediment-seawater interface, and allow 'reduced' phases such as apatite and glauconite to form.

At the present time, the position and velocity of the EAC appear to control the sedimentary facies developed across the margin and presumably have done so throughout the Neogene. Thus, during low sea-level stands, when the current velocity was higher (O'Brien et al. 1986), the sedimentary facies described in this paper would be located further offshore. During periods of rising sea-level, the facies would migrate shorewards as current velocities fell. The now iron-rich Facies 3 and 3' sediments were probably originally deposited in a relatively iron-poor environment similar to that presently found in Facies 4. Sea-level was relatively high and consequently current velocities were relatively low, with the carbon input to the sediments sufficient to maintain anoxia just below the sediment–seawater interface. Fe-cycling processes within the sediments allowed the formation of both phosphate nodules and glauconite. Sea-level began to fall and the core of the EAC moved offshore, and current velocities over the outer shelf and upper slope rose. The carbon input to the sediments progressively decreased and eventually the sediments became oxic. With no oxic–anoxic transition in the sediments (or perhaps a very deep oxic–anoxic transition), the Fe-cycling processes stopped and neither apatite nor glauconite could form. As current velocities continued to rise, the environment became non-depositional and highly oxic and as a result, ferrous iron-bearing phases such as glauconite began to degrade. Continued alteration of glauconite liberated large amounts of goethite which was plastered over the associated phosphate nodules. The environment eventually became erosional, which resulted in both the concentration of the phosphate nodules into a 'lag' and the erosion of the finer grained sedimentary constituents. Repetition of these events through several sea-level oscillations would result in the production of a highly ferruginous conglomeratic facies identical to Facies 3′. This model is supported by the occurrence of Facies 3 and 3′ sediments below Facies 4 sediments on the outer shelf and slope at water depths between 350–425 m. These units were deposited during relatively low sea-level stands, and were progressively overlain by Facies 4 sediments as sea-level rose.

Iron enrichment of phosphate nodules thus occurs during at least two distinct phases on the East Australian continental margin. The first phase occurs soon after nodule formation and results from Fe-cycling processes within the mixed-layer. During this phase, the Fe_2O_3 content of the nodules may increase from approximately 2% to 8% within 60 ka of nodule formation. Some older (500–1000 ka old) nodules within Facies 4 contain as much as 12% Fe_2O_3, indicating that Fe-cycling processes can increase total Fe_2O_3 by a maximum of 10%.

The second iron-enrichment phase occurs during periods of low sea-level, and results in massive increases in the goethite contents of the nodules, probably via the alteration of glauconite associated with the nodules. Some highly conglomeratic nodules, which have undergone multiple erosional–depositional episodes, contain as much as 50% Fe_2O_3. Ferruginization results in depletion (by dilution and/or destruction) of most constituents, including $CaCO_3$, organic carbon, Na and SO_3 from the nodules. However, both Al_2O_3 and MgO show significant enrichments in the more Fe-rich nodules, perhaps because ferruginization is associated with the production of an Al-Mg clay mineral.

It is interesting to note that Baturin (1971) also proposed sea-level fluctuation as a major controlling factor in the genesis of phosphorite deposits on the Namibian shelf, with apatite nucleation in anoxic, diatomaceous muds during high sea-level stands, and lithification and enrichment during subsequent sea-level lows. Unfortunately, the very process which concentrates the originally widely dispersed phosphate nodules (i.e. erosion due to high current velocities) off Eastern Australia also results in the ferruginization of the nodules, making it unlikely that commercially exploitable phosphorite deposits are present on the margin.

Conclusions

Data from the East Australian continental margin indicate that modern phosphorite nodules are forming at a transition zone between Fe- and P-rich, TOC-poor, relict, outer shelf sediments and relatively rapidly accumulating Fe- and P-deficient, TOC-enriched upper slope sediments. Both apatite and glauconite form within the anoxic zone within the sediments, partly as a result of an interaction between two processes: Fe-cycling (between the oxic and anoxic sediment zones) and sediment mixing. Apatite and glauconite formation are favoured by periods of relatively high sea-level (such as during the late Holocene) when the velocity of the EAC over the outer shelf and upper slope is relatively low. Low current velocities allow sufficient organic matter to reach the sea-floor to maintain a shallow (i.e. within the mixed-layer) oxic–anoxic transition within the sediments, thereby allowing continuation of the Fe-cycling processes which appear critical to both apatite and glauconite formation. During sea-level lows, current velocities over the outer shelf and upper slope increase, and the organic carbon input to the sediments decreases. Consequently the sediments become oxic, or the oxic-anoxic transition zone occurs at a much greater depth, below the sediment mixed-layer. Fe-cycling either ceases or becomes much less effective, and the formation of both apatite and glauconite stops. Glauconite is progressively altered, resulting in the formation of goethite, and the phosphate nodules are concentrated into lags (by the strong bottom currents) and ferruginized. Repetition of this process results in the formation of the massive, highly ferruginous, conglomeratic phosphorites which characterize the Neogene sedimentary section off Eastern Australia.

The processes described in this paper may also explain many of the features reported from other phosphorite deposits. For example, there are strong similarities between the Eastern Australian phosphorites and both the iron-poor glauconitic packstones and the iron-rich glauco-conglomeratic phosphorites of Agulhas Bank (Parker & Siesser 1972). Similar parallels exist with the iron-rich phosphorites from Danois Bank (Lucas et al. 1978). Our model appears particularly applicable to unconformity or 'hardground-type' phosphorites which are common throughout the geological record. For example, there appears to be many similarities between the depositional settings, sedimentary facies and processes documented in the Eastern Australian continental margin phosphorites and those described from the Middle to Late Cretaceous nodular phosphorites of southeast England (Kennedy & Garrison 1975; Jarvis & Woodroof 1984; Jarvis & Tocher 1986).

The majority of samples used in this study were collected during cruises of the CSIRO and NZOI vessels R. V. 'Sprightly' and R. V. 'Tangaroa' respectively. The assistance of the CSIRO Division of Oceanography is gratefully acknowledged and we particularly thank the New Zealand Oceanographic Institute for making the 'Tangaroa' available for sampling. Global Marine samples were provided by C. C. von der Borch (Flinders University). Part of this study was completed while G. W. O'Brien was a recipient of a Commonwealth Post-Graduate Scholarship at the Flinders University of South Australia, while financial assistance for H. H. Veeh was provided by ARGS. We also thank the crew and scientific and technical staff of R. V. 'Rig Seismic' for their skill and application during and after the May 1987 PHOSREP cruise. This paper is a contribution to IGCP Project 156 (Phosphorites). G. W. O'B, D. T. H., J. F. M. and P. J. C. publish with the permission of the Director, BMR.

References

BATURIN, G. N. 1969. Authigenic phosphate concretions in Recent sediments of the Southwest African Shelf. *Doklady Academy of Sciences USSR (Translation, Earth Science Section)*, **189**, 227–230.
—— 1971. Formation of phosphate sediments and water dynamics. *Oceanology*, **11**, 372–376.
—— BATURIN, G. N., KOCHENOV, A. V. & PETELIN, V. P. 1970. Phosphorite formation on the shelf of southwest Africa. *Lithology and Mineral Resources*, **3**, 266–276.
——, MERKULOVA, K. I. & CHALOV, P. I. 1972. Radiometric evidence for recent formation of phosphatic nodules in marine shelf sediments. *Marine Geology*, **13**, M37–M41.
BERNER, R. A. 1973. Phosphate removal from seawater by adsorption on volcanogenic ferric oxides. *Earth and Planetary Science Letters*, **18**, 77–86.
BIRCH, G. F. 1975. *Surficial sediments on the continental margin off the west coast of South Africa.* SANCOR Marine Geology Programme, Cape Town, Bulletin 6.
—— 1979. Phosphatic rocks on the western margin of South Africa. *Journal of Sedimentary Petrology*, **49**, 93–110.
——, THOMSON, J., MCARTHUR, J. M. & BURNETT, W. C. 1983. Pleistocene phosphorites off the west coast of South Africa. *Nature*, **302**, 601–603.
BRINDLEY, G. W. & BROWN, G. (eds) 1984. *Crystal structures of clay minerals and their X-ray identification.* Mineralogical Society, Monograph No. 5, London.
BURNETT, W. C. 1977. Geochemistry and origin of phosphorite deposits from off Peru and Chile. *Geological Society of America Bulletin*, **88**, 813–823.
—— & VEEH, H. H. 1977. Uranium-series disequilibrium studies in phosphorite nodules from the west coast of South America. *Geochimica et Cosmochimica Acta*, **41**, 755–764.
COOK, P. J. & MARSHALL, J. F. 1981. Geochemistry of iron and phosphorus-rich nodules from the East Australian continental shelf. *Marine Geology*, **41**, 205–221.
DAVIES, P. J. 1979. *Marine geology of the continental shelf off southeast Australia.* Bureau of Mineral Resources, Australia, Bulletin 195.
ESPITALIÉ, J., MADEC, M. & TISSOT, B. 1979. Source rock characterization method for petroleum exploration. 9th *Annual Offshore Technology Conference, Houston*, 439–448.
FARRAH, H. & PICKERING, W. F. 1986. Interaction of dilute fluoride solutions with hydrous iron oxides. *Australian Journal of Soil Research*, **24**, 201–208.
GODFREY, J. S. 1973. Comparison of the East Australian Current with the western boundary flow in Bryan and Coxs' (1968) numerical model ocean. *Deep Sea Research*, **20**, 1059–1076.
——, CRESSWELL, G. R., GOLDING, T. J., PEARCE, A. P. & BOYD, R. 1980. The separation of the East Australian Current. *Journal of Physical Oceanography*, **10**, 429–440.
GULBRANDSEN, R. A. 1966. Chemical composition of phosphorites of the Phosphoria Formation. *Geochimica et Cosmochimica Acta*, **30**, 769–778.
—— 1970. Relation of carbon dioxide content of apatite of Phosphoria Formation to regional facies. *US Geological Survey Professional Paper*, **700**-B, B9–13.
HEGGIE, D. T., O'BRIEN, G. W., BLANKS, A., REIMERS, C. E., BURNETT, W. C. & MCARTHUR, J. M. 1988. E. Australian modern marine phosphorites: geochemical cycling through pore waters. AGU/ASLO Ocean Sciences Meeting, New Orleans, January 1988. *EOS*, **68**, 1711.
——, SKYRING, G. W., O'BRIEN, G. W. REIMERS, C., HERCZEG, A., MORIARTY, D. J. W., BURNETT, W. C. & MILNES, A. R. 1990. Organic Carbon cycling and modern phosphorite formation on the East Australian continental margin: an overview. *In*: NOTHOLT, A. J. G. & JARVIS, I. (eds). *Phosphorite Research and Developments*. Geological Society, London, Special Publication, **52**, 87–117.
JAHNKE, R. A., EMERSON, S. R., ROE, K. K. & BURNETT, W. C. 1983. The present day formation of apatite in Mexican continental margin sediments. *Geochimica et Cosmochimica Acta*, **47**, 259–266.
JACKSON, M. L. 1965. Free oxides, hydroxides and amorphous aluminosilicates *In*: BLACK, C. A. (ed.) *Methods of Soils Analysis. Part 1. Physical and mineralogical properties, including statistics of measurement and sampling.* American Society

of Agronomy Monograph No. 9. Madison, Wisconsin. 578–603.

JARVIS, I. & WOODROOF, P. B. 1984. Stratigraphy of the Cenomanian and basal Turonian (Upper Cretaceous) between Brancombe and Seaton, SE Devon, England. *Proceedings of the Geologists' Association*, **95**, 193–215.

—— & TOCHER, B. A. 1986. Field Meeting: the Cretaceous of SE Devon, 14–16th March, 1986. *Proceedings of the Geologists' Association*, **98**, 51–66.

KENNEDY, W. J. & GARRISON, R. E. 1975. Morphology and genesis of nodular phosphates in the Cenomanian Glauconitic Marl of south-east England. *Lethaia*, **8**, 339–360.

KIM, K. H. & BURNETT, W. C. 1986. Uranium-series growth history of a Quaternary phosphatic crust from the Peruvian continental margin. *Chemical Geology (Isotope Geoscience Section)*, **58**, 227–244.

KRESS, A. G. & VEEH, H. H. 1980. Geochemistry and radiometric ages of phosphatic nodules from the continental margin of northern New South Wales, Australia. *Marine Geology*, **36**, 143–157.

LUCAS, J., PRÉVÔT, J. & LAMBOY, M. 1978. Les phosphorites de la marge norde de l'Espagne. Chimie, minéralogie, génèse. *Oceanologica Acta*, **1(1)**, 55–72.

MARSHALL, J. F. 1979. The development of the continental shelf of northern New South Wales. *BMR Journal of Australian Geology and Geophysics*, **4**, 281–288.

—— 1980. *Continental shelf sediments: southern Queensland and northern New South Wales*. Bureau of Mineral Resources, Australia, Bulletin 207.

—— 1983. Geochemistry of iron-rich sediments on the outer continental shelf off northern New South Wales. *Marine Geology*, **51**, 163–175.

—— & COOK, P. J. 1980. Petrology of iron- and phosphorus-rich nodules from the E. Australian continental shelf. *Journal of the Geological Society, London*, **137**, 765–771.

MCARTHUR, J. M. 1978. Systematic variations in the contents of Na, Sr, CO_3 and SO_4 in marine carbonate fluorapatite and their relationship to weathering. *Chemical Geology*, **21**, 89–112.

MCCLELLAN, G. H. 1980. Mineralogy of carbonate fluorapatites. *Journal of the Geological Society, London*, **137**, 675–681.

NORRISH, K. & HUTTON, J. T. 1969. An accurate x-ray spectrographic method for the analysis of a wide range of geological samples. *Geochimica et Cosmochimica Acta*, **33**, 431–455.

O'BRIEN, G. W. 1983. *Geochemistry and origin of phosphatic nodules and sediments from the East Australian continental margin*. PhD Dissertation, Flinders University of South Australia.

—— & VEEH, H. H. 1980. Holocene phosphorite on the East Australian continental margin. *Nature*, **288**, 690–692.

—— & —— 1983. Are phosphorites reliable indicators of upwelling? In: SUESS, E. & THIEDE, J. (eds) *Coastal Upwelling — Its Sediment Record. Part A: Responses of the Sedimentary Regime to Present Coastal Upwelling*. Plenum, New York, N.Y. 399–419.

——, HARRIS, J. R., MILNES, A. R. & VEEH, H. H. 1981. Bacterial origin of East Australian continental margin phosphorites. *Nature*, **294**, 442–444.

——, HEGGIE, D. T. & HERCZEG, A. 1988. Influence of sediment mixing processes on modern phosphorite formation: Eastern Australian continental margin. *AGU/ASLO Ocean Sciences Meeting, New Orleans, January, 1988*.

——, VEEH, H. H., CULLEN, D. J. & MILNES, A. R. 1986. Uranium-series isotopic studies of marine phosphorites and associated sediments from the East Australian continental margin. *Earth and Planetary Science Letters*, **80**, 19–35.

——, ——, MILNES, A. R. & CULLEN, D. J. 1987. Sea-floor weathering of phosphate nodules off East Australia: Its effect on uranium oxidation state and isotopic composition. *Geochimica et Cosmochimica Acta*, **51**, 2051–2064.

PARKER, R. J. 1975. The petrology and origin of some glauconitic and glauco-conglomeratic phosphorites from the South African continental margin. *Journal of Sedimentary Petrology*, **45**, 230–242.

—— & SIESSER, W. G. 1972. Petrology and origin of some phosphorites from the South African continental margin. *Journal of Sedimentary Petrology*, **42**, 434–440.

PASHO, D. W. 1972. *Character and origin of marine phosphorites*. US Geological Survey Report USC-GEOL 72–5. Office of Marine Geology.

PRICE, N. B. & CALVERT, S. E. 1978. The geochemistry of phosphorites from the Nambian shelf. *Chemical Geology*, **23**, 151–170.

ROMANKEVICH, Ye. A. & BATURIN, G. N. 1972. Composition of the organic matter in phosphorites from the continental shelf of southwest Africa. *Geochemistry International*, **9**, 464–470.

RUTTENBURG, K. C. 1988. Fluoride adsorption onto ferric-oxyhydroxides in sediments. *Transactions of the American Geophysical Union*, **69**, 1235.

SENIN, Yu. M. 1970. Phosphorus in bottom sediments of the South West African Shelf. *Lithology and Mineral Resources*, **1**, 8–20.

SUESS, E. 1981. Phosphate regeneration from sediments of the Peru continental margin by dissolution of fish debris. *Geochimica et Cosmochimica Acta*, **45**, 577–588.

THOMSON, J., CALVERT, S. E., MUKHERJEE, S., BURNETT, W. C. & BREMNER, J. M. 1984. Further studies of the nature, composition and ages of contemporary phosphorite from the Namibian Shelf. *Earth and Planetary Science Letters*, **69**, 341–353.

VEEH, H. H., BURNETT, W. C. & SOUTAR, A. 1973. Contemporary phosphorites on the continental margin of Peru. *Science*, **181**, 844–845.

——, CALVERT, S. E. & PRICE, N. B. 1974. Accumulation of uranium in sediments and phosphorites on the South West African Shelf. *Marine Chemistry*, **2**, 188–202.

VON DER BORCH, C. C. 1970. Phosphatic concretions

and nodules from the upper continental slope, northern New South Wales. *Journal of the Geological Society of Australia*, **16(2)**, 755–759.

WALKLEY, A. & BLACK, I. A. 1934. An examination of the Degtjareff method for determining soil organic matter, and a proposed modification of the chromic acid titration method. *Soil Science*, **37**, 29–38.

ZUSSMAN, J. (ed.) 1977. *Physical Methods in Determinative Mineralogy*, 2nd edition. Academic Press, London.

Organic carbon cycling and modern phosphorite formation on the East Australian continental margin: an overview

D. T. HEGGIE[1], G. W. SKYRING[2], G. W. O'BRIEN[1], C. REIMERS[3],
A. HERCZEG[4], D. J. W. MORIARTY[5], W. C. BURNETT[6], A. R. MILNES[7]

[1] *Bureau of Mineral Resources, Division of Marine Geosciences and Petroleum Geology, PO Box 378, Canberra, ACT, Australia*
[2] *CSIRO, Division of Water Resources, GPO Box 1666 Canberra, ACT Australia*
[3] *Geological Research Division, Scripps Institution of Oceanography, La Jolla, California, 92093, USA*
[4] *CSIRO, Division of Water Resources, Private Mail Bag #2, Glen Osmond, South Australia, 5064, Australia*
[5] *CSIRO, Division of Fisheries, PO Box 120, Cleveland, Queensland, 4163, Australia*
[6] *Department of Oceanography, Florida State University, Tallahassee, Florida, 32306, USA*
[7] *CSIRO, Division of Soils, Private Bag #2, Glen Osmond, South Australia, 5064, Australia*

Abstract: During 1987, the Australian Bureau of Mineral Resources conducted a multi-disciplinary investigation of the modern phosphorites on the continental margin of southeastern Australia between 28 and 32°S. The objectives of the work were to examine the processes controlling the cycling of organic carbon and bioactive elements, nitrogen, phosphorus, sulphur and iron in the sediments, and to investigate the roles which these processes played in the formation of the modern phosphorites. Bacterial productivities, sulphate-reduction rates, sedimentary oxygen and pore-water concentrations of nitrate, ammonia, phosphate, iron, sulphate and fluoride were measured at sea. The highest rates of microbial productivity were found in the surficial (0–20 mm) sediments of the modern phosphorite zone in 350–460 m water depth. These rates were about double those in shallower shelf (<300 m) sediments and 3–4 fold those rates in mid-slope (600–1000 m) sediments. Aerobic and anaerobic oxidation rates of organic matter, calculated from sediment oxygen profiles and sulphate-reduction rates were highest in the surface sediments in the modern phosphorite zone. The recycling of sedimentary iron, via reductive dissolution of iron oxyhydroxides and reprecipitation at the oxic/anoxic boundary results in a near-surface sedimentary trap for iron in the phosphorite zone sediments. Phosphate released from organic matter in the interfacial sediments, and fluoride from seawater, are scavenged by iron oxyhydroxides in the top few centimetres of sediment. Phosphorus, in this way, is decoupled from organic carbon in the near-surface sediments and linked to the redox cycling of iron. Phosphate and fluoride scavenged onto iron oxyhydroxides, and concentrated in the surficial sediments, are subsequently released to pore waters in the anoxic sediments when iron oxyhydroxides are buried and dissolve. The recycling process releases phosphate and fluoride for incorporation into apatite; fluoride is depleted from pore waters at depths <18 cm, phosphorite nodules form within anoxic sediments at depths <18 cm and continue to accumulate iron and phosphorus while resident in the mixed layer. Combinations of rapid sediment mixing rates, a slow sedimentation rate and a mixed layer to about 18 cm result in an average particle residence time in the phosphorite zone sediments which is about ten-fold that of the mid-slope sediments. Long residence times and rapid mixing promote the oxidation of organic carbon and release of phosphate, while the continuous recycling of iron and phosphate concentrates the phosphorus for apatite precipitation and accumulation into phosphorite nodules. Phosphorite nodules are not found in mid-slope sediments probably because of combinations of relatively rapid sedimentation rates, ineffective iron, phosphorus and fluoride recycling and trapping mechanisms, plus dilution and dissemination of any incipient apatite.

Most deposits of modern marine sedimentary phosphate are associated with both the upwelling of cold nutrient-rich waters onto continental shelves and organic-rich diatomaceous sediments. These deposits are found on the continental margins of Peru−Chile (Veeh et al. 1973; Burnett & Veeh 1977; Burnett 1977; Burnett et al. 1982), Namibia (Baturin et al. 1972; Veeh et al. 1974) and Mexico (Jahnke et al. 1983). Studies conducted on these deposits have contributed to the development of the 'upwelling model' to explain the formation of marine sedimentary phosphate deposits ranging in age from Proterozoic to Holocene. Upwelling provides a continual supply of nutrients to surface waters and this results in high primary productivities, high organic carbon and phosphate fluxes into the sediments and the formation and maintenance of anoxic sediments. Apatite forms diagenetically within anoxic sediments when phosphate is released during bacterial degradation of organic matter (Baturin 1971; Burnett 1977; Jahnke et al. 1983) or dissolution of fish debris (Suess 1981).

However, there are ancient and modern phosphorite deposits which apparently were formed in areas with little coastal upwelling and low productivities (O'Brien & Veeh 1983). The marine phosphorite formations off western South Africa (Birch et al. 1983; McArthur et al. 1988) and East Australia (Kress & Veeh 1980; O'Brien & Veeh 1980; O'Brien et al. 1981) have Pleistocene−Holocene ages and have formed where upwelling is weak or absent, indicating that organic carbon and phosphate fluxes to the sediments are low. The organic carbon content of the East Australian sediments associated with the phosphorite nodules is typically <0.5 wt% (O'Brien et al. 1981), in contrast to the high (5−20 wt%) organic carbon contents of the Peru−Chile and Mexico sediments. The East Australian phosphorites may provide modern analogues for those deposits that do not easily fit the 'upwelling model'. The occurrences of modern phosphorites in both upwelling and non-upwelling environments suggest that different processes (and mechanisms) are operating which control the supply and concentration of phosphate, fluoride and other elements in the sediments for apatite precipitation and the formation of phosphorite nodules.

The Australian Bureau of Mineral Resources (BMR), in order to document the occurrences, and biogeochemical and sedimentological characteristics of a 'non-upwelling' modern phosphorite deposit, conducted an 18 day multidisciplinary research program aboard the R.V. Rig Seismic to the eastern Australian phosphorite deposits in 1987 (Fig. 1). A major objective of this program was to determine the processes which control the cycling of organic carbon (and associated elements) in continental margin sediments, to examine the role that these processes play in the formation of modern phosphorites, and hence, to contribute to an understanding of the distribution of marine phosphates in the geological record. The data and results presented in this overview are an

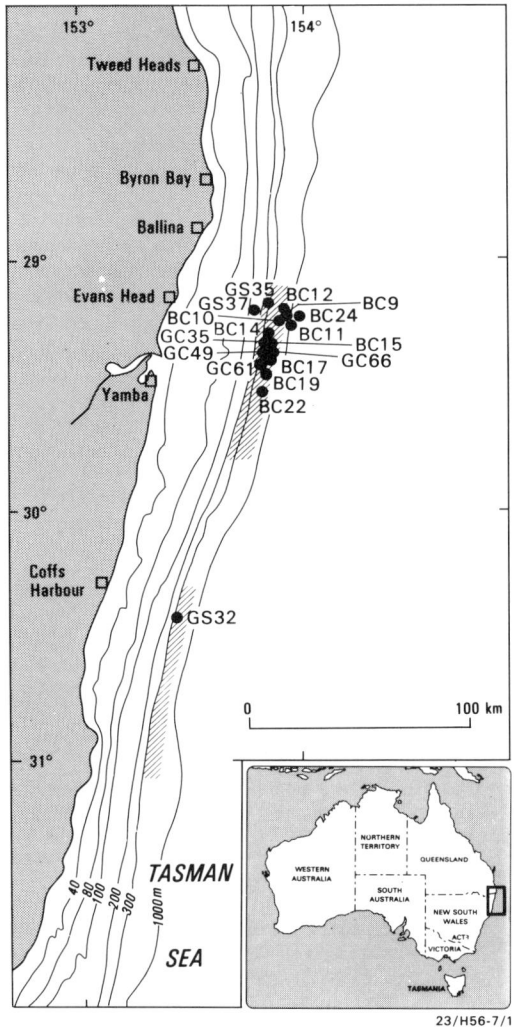

Fig. 1. Map of the East Australian continent showing the regional study area, bathymetry and locations of selected cores (Cross hatching indicates known areas of 'modern' phosphorite accumulations).

initial attempt to integrate several individual component studies. Detailed results of bacterial activities, oxygen and sulphate-reduction rates, pore-water geochemistry, sedimentology and mineralogy will be presented separately elsewhere. The data, results and observations reported here are generally confined to the top 20 cm of sediments where most biogenic activity occurs and where most nodules are found. Three processes are considered key factors in the formation of the modern phosphorites. These processes are:
(1) the oxidation and recycling of organic carbon by aerobic and anaerobic microbial processes in sediments;
(2) the recycling of iron, phosphate and fluoride between pore waters and solid phases during oxidation of organic matter;
(3) sediment mixing.

Study area: oceanography and sediments

The oceanography of this part of the Australian seaboard is dominated by the southward flowing East Australian Current (EAC). Meanderings of this current and its interaction with the continental shelf and slope result in weak and localized upwelling events which occur within 10–15 km of the coastline at water depths of less than 100 m (Rochford 1975; Tranter et al. 1986). The upwelling events are sporadic, of short (10 day) durations and the upwelled water apparently originates from shallow depths of 150–200 m. Because of the sporadic nature of the upwelling events and low nutrient concentrations (Fig. 2) in the upwelled waters from the shelf (PO_4^{3-} <0.5 μM; NO_3^- <5 μM; SiO_2^- <1 μM), primary productivity is low for most of the year, typically about 0.1 g C m^{-2} d^{-1}, but may reach 1.0 g C m^{-2} d^{-1} during upwelling events (Jitts 1965). The continental shelf, therefore, is typified by relatively low annual primary production, nearly continuous flushing of the shelf by oxygenated waters of the EAC, high concentrations of oxygen (200–300 μM) in bottom waters, and low concentrations of organic carbon (<0.5%) in sediments. The regional hydrography to about 1000 m is characterized by a weak oxygen-minimum layer and relatively low nutrient concentrations (Fig. 2), in contrast to an intense oxygen minimum and high nutrient concentrations off Peru–Chile.

Phosphorite nodules on the East Australian continental margin range in age from Neogene (probably Middle Miocene) to Holocene (von der Borch 1970; O'Brien & Veeh 1980; O'Brien et al. 1981), and are common between 28 and 32°S (Fig. 1). The Neogene phosphorites typically occur within massive hardgrounds on current swept areas of the shelf, are iron (goethite)-rich, often conglomeratic, and are concentrated at water depths between 200 and 350 m. The Quaternary phosphorites, in contrast, typically occur as small, discrete nodules within unconsolidated sands on the upper slope, usually at water depths between 350 and 460 m (O'Brien et al. 1986). The sedimentology has been discussed by O'Brien et al. (1990). The sediments between 28 and 32°S have been divided into five sedimentary facies types.
(1) Facies 1 sediments are found at water depths less than 50 m and are terrigenous quartz sands.
(2) Facies 2 sediments (water depths 50–200 m) are poorly sorted carbonate sands and gravels, the iron oxide content of these

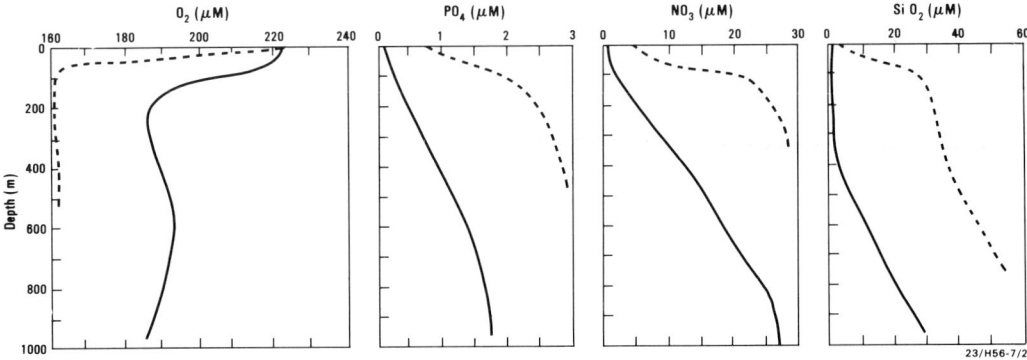

Fig. 2. Concentrations of oxygen, phosphate, nitrate and silicate in the water column of the continental margin of East Australian off Evans Head (solid lines, unpublished data from CSIRO, Division of Oceanography; compared to schematic oxygen and nutrient distributions off Peru–Chile (dashed lines).

sediments increases at water depths near 200 m and many of the shell fragments are iron stained.

(3) Facies 3 sediments (water depths 200–350 m) are goethite-rich unconsolidated glauconitic foraminiferal sands. Moderate to large amounts of goethite are distinctive features of Facies 3 sediments and glauconite comprises 5–30 wt%. The iron-rich Neogene phosphorites associated with the Facies 3 sediments occur as extensive hard grounds or as discrete nodules within thin layers of sediments. The goethite content of the sediments decreases at water depths of about 300 m and Facies 3 sediments grade into Facies 4 sediments at water depths near 350 m.

(4) Facies 4 sediments (water depths 350–460 m) are glauconitic foraminiferal sands which have a low goethite contents and contain the Quaternary phosphorites. These sediments are 7–30% glauconite, 20–35% quartz and 30–60% skeletal carbonate which is comprised mainly of planktonic foraminifera. The goethite content and abundance of oxidized glauconite decreases with increasing water depth. The upper 10–15 cm of these sediments are slightly iron-stained and contain more oxidized (red-brown) glauconite than do more deeply buried sediments.

(5) Facies 5 sediments (water depths >500 m) are compositionally similar to those of Facies 4 but are finer grained. Glauconite is less abundant but when present is green and unoxidized; goethite pellets are rare.

The data and diagenetic processes described in this paper are restricted to sediment Facies 4 and 5; emphasis is placed upon contrasting the spatial and temporal characteristics of biogeochemical processes in Facies 4 and 5, and within Facies 4 sediments, the formation of phosphate-rich sediments and phosphorite nodules.

Methods

Sampling

Sediments were obtained from an area of the continental shelf and slope between 28 and 32°S, and water depths between 150 and 1500 m. Combinations of a Soutar box core and a grab sampler were used to obtain surface sediment samples for interfacial studies, including pore-water analyses, dissolved oxygen measurements and microbiological (bacterial production and sulphate-reduction rate) experiments. Several gravity cores of 1–2 m were obtained and used for pore-water and sediment geochemical analyses. Pipe dredges were used to collect nodules and sediments from 'hardgrounds'. Seawater samples were obtained with 12 litre Niskin bottles.

Pore-water and sediment analyses

Cores were processed as soon as they were recovered. The oxygen contents of sub-cores of sediment from box-cores and selected grabs were measured at millimetre depth intervals with a microelectrode (Reimers et al. 1984; Reimers 1987). All sediment samples for pore-water analyses were processed under a nitrogen atmosphere, at in situ temperatures, in a cold-van aboard the ship. Sediments were extruded from cores, loaded into plastic centrifuge tubes and centrifuged at 15 000 rpm for 5-minutes in a Sorvall RC5B refrigerated centrifuge. Pore waters were syphoned from the centrifuge tubes and filtered through 0.45 μm filters. Ammonia, nitrate + nitrite, phosphate and silicate were measured at sea on small volume samples (typically <0.5 ml) colourimetrically with a Tecator flow-injection analyser. Samples for iron analyses were filtered directly into vials containing 25 μl of 10 N HCl and analysed spectrophotometrically (Stookey 1971). Fluoride was determined on 0.5 ml sub-samples of pore waters by a potentiometric technique similar to that described by Froelich et al. (1983). Sediment samples from the centrifuge tubes were dried and crushed in preparation for Al_2O_3, P_2O_5, Fe_2O_3 and CaO analyses by X-ray fluorescence (XRF).

Sulphate-reduction rates and sulphate-reducing bacteria in sediments

Sulphate reduction rates were determined by the ^{35}S radiometric method (Skyring et al. 1983; Skyring 1987b). This method measures rates of formation of acid-volatile (AVS) and non acid-volatile (NAVS) sulphur (and includes sedimentary sulphur concentrations) produced as a result of sulphate-reducing bacterial activity. Sulphate reduction rates in this paper are reported as the arithmetic sum of AVS and NAVS. Radioactive ^{35}S-sulphate was added to sediments (collected in 2 and 5 ml plastic syringes) on a glass rod immediately following sampling from the cores. The samples were then incubated at 7–12°C for between 30 and 60 days under O_2-free nitrogen. Zero-time control experiments were conducted using both upper and lower slope sediment samples. The sulphate concentration of core-top pore waters (28 mM) was used to calculate the specific activity of ^{35}S in incubated sediment samples. Two cores were sampled and extracted for gas chromatography-mass spectrometry determinations of the phospholipid, ester-linked fatty acids which are unique indicators of the sulphate-reducing bacteria (Nichols et al. 1987).

Bacterial productivities.

Bacterial productivities were calculated from measurements of rates of phospholipid (Moriarty et al. 1985), protein (Kirchman et al. 1986) and DNA

(Moriarty & Pollard 1981; Moriarty 1986; Moriarty 1987) synthesis. Most experiments with ^{3}H-thymidine (DNA) and ^{3}H-leucine (protein) and ^{32}P-PO$_4^{3-}$ were done in duplicate with core sections (8 mm diam × 2 mm) which were mixed with the radioactive substrate in tubes and incubated at in situ temperatures (7–12°C) for 1–2 hrs.

^{210}Pb in sediments

Sub-cores from box-cores were sealed and processed for ^{210}Pb measurements at sea, while gravity cores were processed ashore. Cores were extruded and sliced into 0.5–5 cm intervals and the peripheral 1–2 mm of sediment was removed from each slice to minimize contamination during core extrusion and processing. Most samples were oven-dried to determine water loss, porosity and bulk sediment density. Total ^{210}Pb activity was generally determined by analysis of the grandaughter ^{210}Po and isotope dilution techniques using ^{208}Po as a tracer. Total lead and polonium were extracted from the sediments by refluxing with nitric acid for 4-hours and coprecipitating the Pb and Po with iron hydroxide at pH 7. Po was liberated from the precipitate with both hydrochloric and ascorbic acids and then plated onto a silver disc (Flynn 1966). ^{210}Po activity was counted with an Ortec alpha spectrometer and a Tracor Northern pulse height analyser. Tracer recovery was consistently greater than 90%. One core (BC 17) was analysed by nondestructive gamma-ray spectroscopy using methods outlined in Kim & Burnett (1986).

Results

The locations of sampling sites, water depths and measurements carried out on most cores are summarized in Table 1.

Chemical indicators of microbial activity

The oxidation of organic carbon in marine sediments has been described as a series of sequential microbially-mediated reactions, in which oxidants are consecutively consumed according to the free energy released during oxidation (Froelich et al. 1979). The reactions important to the present work are shown in Table 2. Oxygen reduction is thermodynamically the most favoured reaction (reaction 1, Table 2). Oxygen depletion from sediments is accompanied by a rise in pore-water nitrate produced as a metabolic end-product of ammonia and nitrite oxidation. When oxygen is depleted, nitrate reduction (reaction 2, Table 2) proceeds. This reaction is indicated by decreasing nitrate concentrations from sediment pore-waters. Manganese oxyhydroxide reduction (not discussed in this work) and iron oxyhydroxide reduction are then favoured energetically and occur next. The characteristic indicator of sedimentary iron oxyhydroxide reduction (reaction 3, Table 2) is the release of soluble Fe^{2+} to pore waters. Sulphate is the next most favoured oxidant and the onset of this reaction is characterized by a release of ammonia and sulphide (reaction 4, Table 2) to sediment pore-water. The distribution of ammonia in the sediment pore-water is an indicator of the onset of sulphate reduction because oxidation of organic matter by anaerobic bacteria, including the sulphate-reducing bacteria, releases ammonia in predictable stoichiometric quantities (reaction 4, Table 2). A consideration of the stoichiometry of reaction 3, and the pore-water iron concentrations indicates that ammonia produced during sedimentary iron reduction is not quantitatively important.

The reactions in Table 2 may, in different sedimentary environments, extend over several metres of sediment, but the depth-dependence of these reactions is a complex function of several factors, including the organic carbon rain-rate to the sediment, the bulk sedimentation rate, biological mixing of organic carbon into the sediments, infaunal respiration and the bottom-water oxygen content (Emerson et al. 1985). The reaction sequences in Table 2 occur in the top few centimetres of sediment at locations where sediments receive a high flux of organic matter or underlie oxygen depleted bottom waters. Conversely, the reaction sequences in Table 2 are extended over several metres depth in sediments that receive a low flux of organic matter and that underlie bottom waters high in oxygen content (e.g. Grundmanis & Murray 1982; Jahnke et al. 1982; Bender & Heggie 1984). Oxygen is the major oxidant in the deep sea (Bender & Heggie 1984); sulphate reduction is quantitatively more important in continental margin sediments (e.g. Heggie et al. 1987) and also in coastal environments where the organic carbon flux may be particularly high (Skyring 1987a). The reactions in Table 2 all occur within the top few (<5 cm) centimetres of sediment in phosphorite-bearing sediments (Facies 4) of the East Australian margin, but are extended to depths >20 cm in Facies 5 sediments. Oxygen data from box cores taken across the continental shelf and slope are summarized in Appendix 1 and shown in Fig. 3. Pore-water metabolites are summarised in Appendices 2 and 3 and shown in Fig. 4.

Table 1. *Core locations, water depth and data reported here*

Core	Facies	Lat (deg. min S)	Long (deg. min E)	Depth (m)	O_2	NO_3	NH_3	PO_4^{3-}	Fe	F	SRR	BP	^{210}Pb	SPG
GS 35	2	29 13.56	153 45.94	149								X		
GS 37	3	29 13.19	153 50.10	294								X		
GS 40	4	29 19.35	153 49.78	371								X		
BC 22	4	29 30.84	153 50.12	375	X								X	
GC 77	4	29 18.82	153 51.07	368		X	X	X	X	X				
BC 14	4	29 17.23	153 51.38	377		X	X	X	X	X				
BC 15	4	29 17.96	153 51.73	375								X		
GC 49	4	29 18.24	153 50.68	381										
BC 17	4	29 23.05	153 49.99	412			X					X	X	X
GC 35	4	29 19.30	153 50.67	402		X		X	X	X	X		X	X
GS 32	4	30 25.29	153 26.14	431	X	X		X	X			X		X
BC 21	5	29 20.05	153 50.65	599	X	X	X	X	X	X	X	X	X	X
BC 19	5	29 24.34	153 49.73	604	X	X	X	X	X	X	X			
GC 16	5	29 20.98	153 51.75	595	X	X	X	X	X	X	X		X	X
BC 12	5	29 13.20	153 53.80	866										
BC 10	5	29 14.41	153 53.86	990	X								X	
BC 09	5	29 13.72	153 53.00	987		X	X	X	X	X	X	X		
BC 11	5	29 14.73	153 53.16	1071	X	X	X	X	X	X	X			X
BC 24	5	29 13.20	153 53.80	1484	X	X	X	X	X	X	X			X

BP, bacterial productivities measured by the DNA method. SRR, sulphate reduction rate measurements by ^{35}S. SPG, solid phase geochemistry, Al_2O_3, CaO, Fe_2O_3, P_2O_5.

Table 2. *Organic carbon oxidation reactions*

(1) $106(CH_2O)16(NH_3)(H_3PO_4) + 138O_2 \rightarrow 106 CO_2 + 16NH_3 + H_3PO_4 + 122H_2O$

(2) $106(CH_2O)16(NH_3)(H_3PO_4) + 94.4HNO_3 \rightarrow 106CO_2 + 55.2N_2 + H_3PO_4 + 177.2H_2O$

(3) $106(CH_2O)16(NH_3)(H_3PO_4) + 424FeOOH + 848 H \rightarrow 424Fe^{2+} + 106CO^2 + 16NH_3 + H_3PO_4 + 742H_2O$

(4) $106(CH_2O)16(NH_3)(H_3PO_4) + 53SO_4^{2-} \rightarrow 106 CO_2 + 16NH_3 + 53S^{2-} + H_3PO_4 + 106 H_2O$

(from Froelich *et al.* 1979)

Facies 5 (460–1500 m): Quaternary phosphorite nodules absent

Oxygen concentrations in interfacial sediments of core BC 24 (1484 m water depth) decrease rapidly from about 200 μM at the sediment–seawater interface (Fig. 3) to approximately 22 μM at around 3 cm depth; the depth distribution suggests that oxygen would be entirely depleted at around 5 cm depth. Nitrate is depleted because of denitrification at about 17 cm (Fig. 4) and there is no evidence for sulphate reduction (as indicated by pore water ammonia data) to 20 cm. Iron reduction (Fig. 4) begins at depths near 15 cm and pore-water phosphate rises from about 0.9 μM in interfacial sediments to near 2 μM at about 20 cm. The vertical distribution of each of these microbial indicators is consistent with free-energy considerations (Table 2). At shallower water depths (c.800–1100 m) on the mid-slope, oxygen penetrates to shallower depths in the sediment (Fig. 3; 987 m BC09), nitrate reduction occurs closer to the sediment–seawater interface (Fig. 4; 1071 m BC11, 866 m BC12) iron oxyhydroxide and sulphate reduction (indicated by rapidly increasing pore water Fe^{2+} and NH_3 concentrations respectively) begin at depths of 5–7 cm (more than 10 cm shallower in these sediments than at 1484 m water depth). These profiles are consistent with an increasing demand by the biota for oxygen and secondary oxidants at shallower water depths. Phosphate concentrations in pore waters increase from about 0.5 μM in surface pore waters to near 2 μM at about 10 cm. At water depths near 600 m (BC 19, BC21) oxygen is depleted at depths less than 2.5 cm (Fig. 3), nitrate is depleted at about 4 cm (Fig. 4) and sulphate reduction begins at about 4–5 cm. Fe^{2+} in pore waters (Fig. 4) rises above background at about 4–5 cm and increases to about 10–15 μM at depths of 8–16 cm. Phosphate concentrations rise from 0.7 μM in the near surface (0–2 cm) sediments to a broad maximum of 2.5–3.0 μM at about 8–16 cm (Fig. 4); thereafter phosphate decreases below the maximum at 8–16 cm, with increasing depth to concentrations <1 μM at depths of about 1 m in core GC 16. Pore water phosphate gradients in the top 10 cm appear concave-upward and suggest phosphate consumption in near surface sediments. Fluoride (Figs 4, 5a) shows a distinct increase in the upper parts of these cores from seawater values close to 70 μM to values near 80 μM and occasionally higher. These offshore cores tend to maintain these elevated fluoride concentrations to depths near 40 cm, but fluoride concentrations decrease at depth.

Facies 4 (350–460 m): Quaternary phosphorite nodules abundant

Oxygen concentrations at the sediment–seawater interface are close to that of the bottom water concentration (200 μM), but decrease rapidly to undetectable concentrations at depths of about 1 cm (BC 22, GS 32; Fig. 3). Nitrate (Fig. 4) is depleted at depths <5 cm and pore water ammonia concentrations begin to increase rapidly at depths near 3 cm. Fe^{2+} is detectable in pore waters at less than 3 cm and concen-

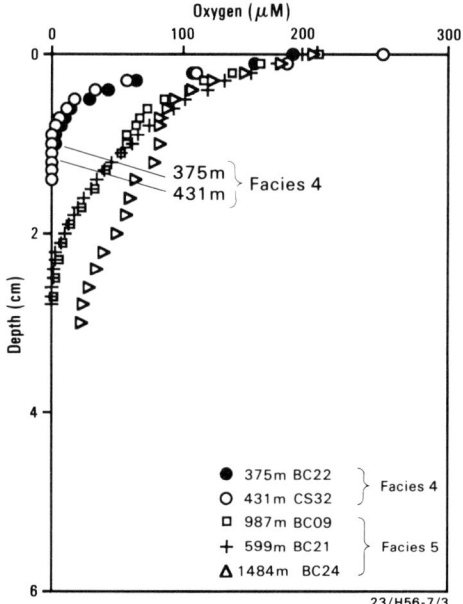

Fig. 3. Depth distributions of oxygen concentrations (μM) in phosphorite zone (Facies 4) and mid slope (Facies 5) sediments.

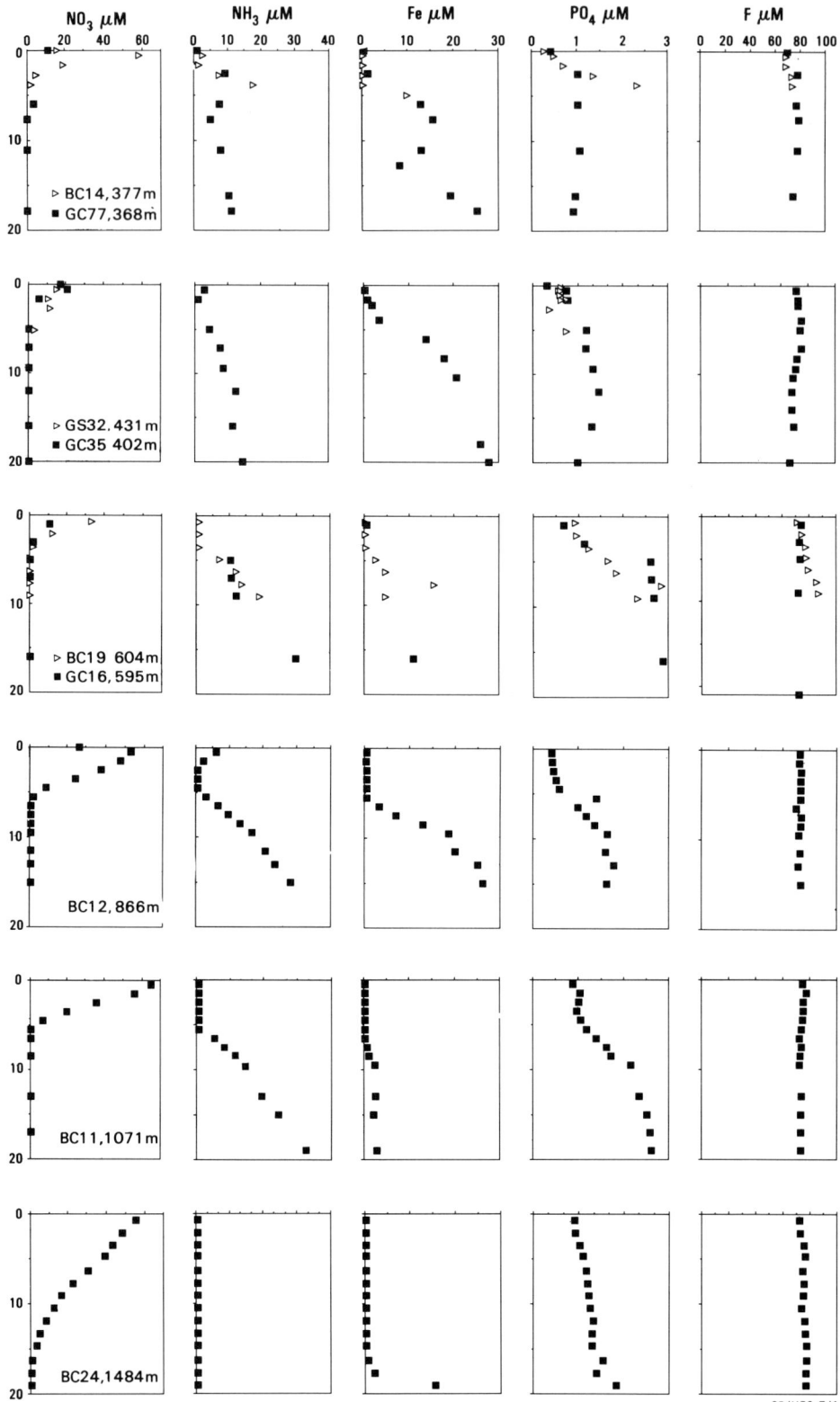

Fig. 4. Depth distributions of pore-water metabolites (NO_3^-, NH_3, PO_4^{3-}, Fe^{2+}, F^-) to depths of 20 cm for selected cores sampled from Facies 4 (350–460 m) and Facies 5 (>500 m) sediments across the continental margin.

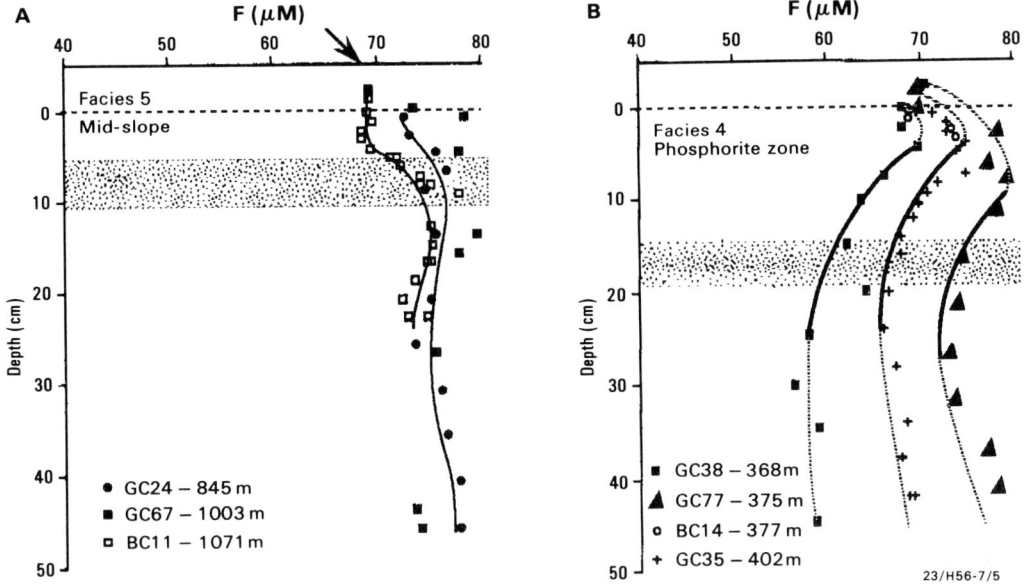

Fig. 5. Pore-water fluoride in the top 50 cm of selected cores from (**A**) Facies 5 and (**B**) Facies 4 sediments. Arrows on the upper scale represent bottom-water fluoride concentrations and the stippled areas indicate the approximate maximum depths of mixing in Facies 4 and 5 sediments.

trations increase to about 30 μM at 15–20 cm (Fig. 4). Phosphate concentrations increase above bottom-water values (0.3 μM) to concentrations between 0.5 and 0.8 μM within the upper 1.5 cm, but increase more gradually with increasing sediment depth to about 1.5 μM at depths <10 cm. One box core (BC14) from 377 m water depth indicated a shallow pore-water phosphate maximum of 2.4 μM at 4–5 cm. Phosphate concentrations in GC 35 (402 m water depth) were highest (1.5 μM) at 12 cm and decreased to <1 μM at 20 cm (Fig. 4).

There are three features of the pore-water phosphate data. First, pore water phosphate does not increase markedly in concentration within the interfacial sediments where sedimentary oxygen is depleted. Second, the depth of the concentration maxima fall within the depth ranges encompassing the rapidly increasing pore-water Fe^{2+} concentrations. Third, pore-water phosphate gradients, particularly that measured in the box core (BC 14) and within the top few centimetres appear concave-upward. These features suggest that phosphate consumption reactions are occurring and a sedimentary sink for phosphate exists within the top 10 cm.

Pore-water fluoride profiles (Fig. 5b) in the upper-slope phosphorite sediments display some unusual features. Fluoride increases above seawater values in the top few centimetres followed by a concave-downward trend and decreasing concentrations below this maximum (Figs 4 & 5b). These data suggest both release and uptake of fluoride are occurring within the same sediment, the mechanisms being separated by only a few centimetres.

Oxygen reduction rates

The sedimentary oxygen data presented in Appendix 1 and Fig. 3 have been used to calculate the oxygen flux into the sediments from the following equation, assuming that oxygen is transported within the sediments via molecular diffusion only:

$$J = -\phi D_s(d[O_2]/dx).$$

At the sediment−seawater interface, ϕ is porosity, D_s is the sediment molecular diffusion coefficient of oxygen ($D_s \approx D_{ionic} \times \phi$, Berner 1980), $d[O_2]/dx$ is the oxygen gradient. For each core ϕ was estimated after determining the water contents of known volumes of the surface 0.5 cm of sediment. The oxygen gradient was approximated as ($[O_2]$ bottom water $-[O_2]$ pore water at 5 mm)/0.5 cm. Oxygen fluxes are accurate within a factor of only two because oxygen exchange may not be solely a diffusive process, particularly where strong bottom currents impinge on the sea floor. Furthermore, the gradients measured were probably altered from an in situ state because of core disturbance and changes in the temperature and pressure of the sediment. The oxygen flux was converted to an equivalent organic carbon oxidation rate by multiplying the oxygen flux by the Redfield ratio (106/138). These results are summarized in Tables 3 and 4 (also see Fig. 6) and indicate about a two-fold higher organic carbon oxidation rate by oxygen in Facies 4 sediments (350–460 m) than in Facies 5 sediments. Two cores (BC 19, BC 21) on the upper slope (c.600 m) had organic carbon oxidation rates which were similar to those of the mid-slope (866–1484 m) cores (BC 12, BC 9, BC 24).

Sulphate reduction rates

Sulphate reduction rates measured in the top 20 cm of sediment are summarized in Appendix 4 and shown in Fig. 7 (for the top 6 cm of sediment only). There are five important points to be made from these data.

First, sulphate reduction was detected in the core-tops (0–3 cm) of most cores sampled (GS 32, BC 12, BC 19) even though oxygen was present. This result is significant because it indicates the existence of anoxic microniches and activity of sulphate-reducing bacteria in interfacial oxic sediments. Sulphate reduction rates were very low, but nevertheless detectable, in the sediments from 1484 m water depth (BC 24). Second, the highest rates of sulphate reduction in the core-tops were generally detected in the interfacial sediments (<2 cm) and rates generally decreased with increasing depth to minimum values at depths between about 3 and 8 cm. Sulphate reduction rates then increased with depth in the sediments, beyond the minimum zone, near where ammonia concentrations in the pore waters increase. This observation is in accord with the stoichiometry of equation 4, (Table 2) and the modelling of ammonia profiles to determine sulphate reduction rates (Bender

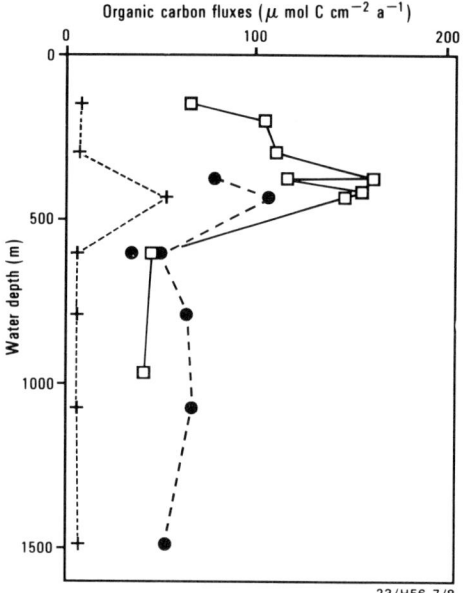

Fig. 6. Depth integrated bacterial productivities and carbon oxidation rates by oxygen and sulphate in the surface sediments of cores sampled at different water depths across the continental margin. Open squares are bacterial productivities. Crosses and filled circles are the organic carbon oxidation rates because of sulphate and oxygen reduction respectively. Bacterial productivities and organic carbon oxidation rates by oxygen plotted here include all data listed in Table 4. The purpose of the plot is to compare biogenic activities in surface sediments, consequently organic carbon oxidation rates by sulphate reduction include only data from the grabs and box-cores GS 37, 35, 32, BC 19, 12, 11, 24. Data from the gravity cores listed in Table 4 have not been plotted.

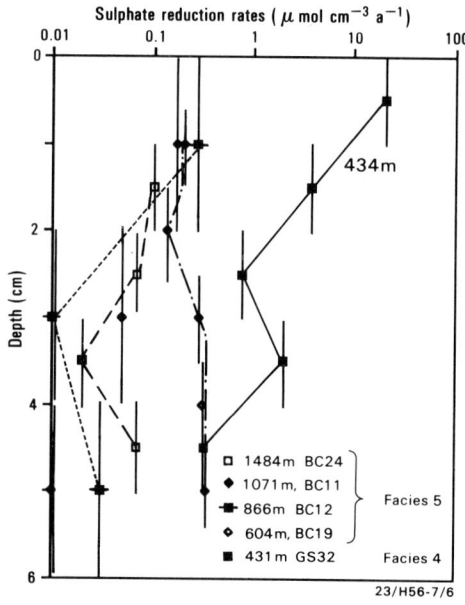

Fig. 7. Down-core distributions of surface-sediment sulphate-reduction rates in box-core samples only from Facies 4 (350–460 m) and Facies 5 (>500 m) sediments.

& Heggie 1984). Third, the depth-integrated rates in the top 5 cm of sediment are generally low compared to other coastal/shelf environments (Skyring 1987a). Fourth, the highest surface sediment sulphate reduction rates, about two orders of magnitude higher than rates measured in other cores, were measured in the phosphorite zone (Facies 4) sediments of core GS 32 (Fig. 7). Fifth, the depth distributions of the sulphate reduction rates were consistent with the downcore distributions, measured in two cores only (Glendenning & Nicholls, pers. comm.) of a phospholipid fatty acid (PLFA) biomarker for *Desulfobacter* sp. (Taylor & Parkes 1983). The biomarker abundance (up to 31%) for *Desulfobacter* sp. was higher in Facies 4 sediments than the abundances (<2%) estimated for deep-sea and estuarine sediments elsewhere (Nicholls, pers. comm.).

Anaerobic organic carbon oxidation rates were calculated from the depth-integrated sulphate reduction rates using the Redfield formula and the reaction stoichiometry of Table 2, (106C:53S) and these results are shown in Tables 3 and 4. The results of Table 3 are for sulphate reduction rates integrated to the same depth of oxygen penetration in sediments or the minimum in the depth distribution of sulphate reduction rates (whichever is shallowest), to enable a direct comparison to be made of the rates of carbon oxidation by sulphate and oxygen respectively in the surface sediments. Organic carbon oxidized during sulphate reduction in the top 5 cm of Facies 5 sediments, on the upper and mid-slope, is less than 3% of the carbon oxidized by oxygen reduction (Table 3, Fig. 6). However, in the shallower water Facies 4 sediments, carbon oxidized by sulphate reduction (where direct comparisons may be made e.g. in core GS 32) is about 42% of organic carbon oxidized during oxygen reduction (or about 30% of total organic carbon oxidized, Fig. 6).

The results of Table 4 are for sulphate reduction rates, depth-integrated over the distances indicated in brackets, to compare the relative importances of oxygen reduction and sulphate reduction on a 'whole-core' basis. When sulphate reduction rates are integrated with depth over the whole core (Table 4), sulphate becomes a relatively more important organic carbon oxidant in Facies 5 sediments. For example, where sulphate reduction rates were measured in box-cores (BC 24, 11, 12, 19) to depths up to about 20 cm, organic carbon oxidized during sulphate reduction is about 4−25% of organic carbon oxidized during oxygen reduction. Where sulphate reduction rates were measured in gravity cores (GC 24, 16) to depths near 1.5 m, carbon oxidized during sulphate reduction exceeds that oxidized during oxygen reduction. These results reflect the increasing importance of down-core sulphate-reducing activities and are probably a consequence of the combined effects of more rapid sedimentation rates (resulting in burial of metabolizable organic carbon below the aerobic zone) and slow sediment mixing rates (see below).

However, in Facies 4 sediments, when sulphate reduction rates were measured in gravity cores to depths also near 1.5 m (GC 42, 47 61, 36), organic carbon oxidized during sulphate reduction never exceeded that measured in the top 6 cm of sediments in GS 32. This result probably reflects, in part, the influences of slow sedimentation rates and rapid rates of sediment mixing, so that metabolizable organic matter is

Table 3. *Organic carbon oxidation rates by oxygen and sulphate in surface sediments*

Core	Water depth (m)	oxic/anoxic interface (cm)[1]	Rate of organic carbon oxidation	
			oxygen	sulphate[2]
			(μM C cm^{-2} a^{-1})	
BC 24	1484	c.5.0	48	0.5
BC 09	987	<3.0	62	NM
BC 12	866	<3.0	59	0.6
BC 19	604	NM	56	1.3
BC 21	599	2.6	54	NM
GS 32	431	1.0	107	41
BC 22	375	1.2	73	NM

[1] Depth in sediment where the pore-water oxygen concentrations were measured or estimated to fall to zero.
[2] Sulphate-reduction rates are integrated only to the same depth as the disappearance of oxygen.
NM, not measured.

Table 4. *Organic carbon oxidized during oxygen and sulphate reduction across the East Australian margin. Also shown are rates of organic carbon synthesis (productivities) by bacteria*

Core	Water depth (m)	C-OX(O_2)	C-OX(SO_4)	Bacterial Productivity
		(μM C cm^{-2} a^{-1})		
BC 24	1484	48	2 (0–12)	
BC 11	1071		15 (0–20)	
BC 09	987	62		50
GC 24	82		76 (10–95)	
BC 12	866	59	11 (0–12)	
GC 16	595		121 (0–150)	
BC 19	604	56	5 (0.5–8.5)	40
BC 21	599	54		
GS 32	431	107	54 (0–6)	142
GC 42	426		19 (14–150)	
GC 47	417		21 (1.5–8.1)	
GC 61	421		14 (0–95)	
BC 17	412			151
GC 36	417		19 (7–82)	
BC 15	375			112
BC 22	371	73		
GS 40				157
GS 37	294		1 (0–5)	106
GS 35	149		3 (0–5)	61

Data in brackets indicate depth in cm for integration of sulphate-reduction rates. All bacterial productivities (calculated from rates of DNA synthesis) were derived from experimental data gathered at 2mm intervals, and integrated to 2.0 cm depth in the sediment. Individual sulphate-reduction rate data for cores GS 35, 37, GC 36–61, 47, 42, 16, 24 are to be reported elsewhere. Data from GS 35, 37, 32, BC 19, 12, 11, 24 are plotted in Fig. 6.

recycled through the 'oxic' sediment zone and much of the organic matter flux to the sediments (see discussion below) is metabolized during oxygen reduction. Organic carbon oxidation via sulphate reduction is low in the core tops of Facies 3 sediments (Table 4; Fig. 6). Where comparisons are available (GS 32, BC 19), the bacterial productivities (approximations to the organic carbon fluxes required to sustain microbial activity) are comparable (within the combined errors of the measurements) to the organic carbon oxidized during both aerobic and anaerobic respiration (Table 4, Fig. 6).

Bacterial productivities

The bacterial productivities determined by rates of DNA synthesis are summarized in Appendix 5 and the down-core distributions of activities are shown in Fig. 8. Bacterial productivities calculated from synthesis rates of DNA, phospholipid and protein agreed within a factor of 2–3 (Moriarty, unpublished data). Two important points emerge from the productivity data. First, bacterial productivities were highest within the top 4 mm of all cores and decreased with increasing sediment depth. Most (>70%) of the depth-integrated bacterial growth occurred within the upper 2 cm of sediments, although detectable growth was evident (from DNA synthesis) to a depth of 70 cm in core GC 61 (412 m water depth). Second, the highest bacterial productivities in the top 20 mm of sediments were measured in Facies 4 sediments, the zone of modern phosphorite formation (Fig. 8). Depth-integrated bacterial productivities in facies 4 sediments were significantly higher than those measured in both Facies 3 and Facies 5 sediments (Table 4, Fig. 6).

^{210}Pb and sediment mixing

^{210}Pb is produced in the atmosphere from decay of ^{222}Rn and introduced into marine sediment following deposition and scavenging from seawater (Turekian *et al.* 1977). A small amount of

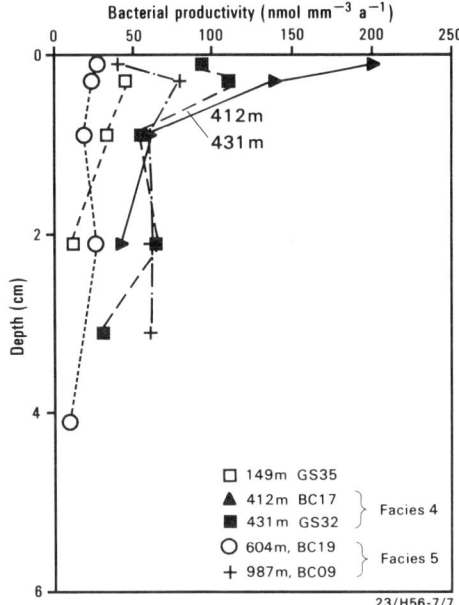

Fig. 8. Down-core distributions of bacterial productivities in Facies 3 (>200 m), Facies 4 (350–460 m) and Facies 5 (>500 m) sediments.

^{210}Pb is produced in sediments and bottom waters from decay of the grandparent ^{226}Ra. Unsupported ^{210}Pb (i.e., concentrations that are corrected for contributions from ^{226}Ra) have been used in the deep-sea (Peng *et al.* 1979; Cochran 1985) and coastal marine sediments (Benninger & Krishnaswami 1981; Carpenter *et al.* 1982; 1985) to calculate rates of sediment mixing at locations where sediment accumulation rates are low. The total ^{210}Pb data from one gravity core (GC 49) and two box-cores (BC 22, BC 17) collected from the Facies 4 sediments are summarized in Appendix 6 and shown in Fig. 9. The data indicate total ^{210}Pb of 5–9 dpm g^{-1} (disintegrations per minute per gram) in surface sediments and concentrations decrease with increasing depth. ^{210}Pb is a minimum at a depth of about 15 cm in Facies 4 sediments (GC 49). The increase in ^{210}Pb concentrations at depth below 20 cm observed in GC 49 (Appendix 6) is probably related to the in-growth of ^{210}Pb from uranium which is incorporated into the apatite crystal lattice during apatite precipitation in the mixed layer (O'Brien *et al.* 1990). The plots of log excess ^{210}Pb versus depth from the box cores of Facies 4 sediments, if extrapolated to an excess ^{210}Pb of about 0.2

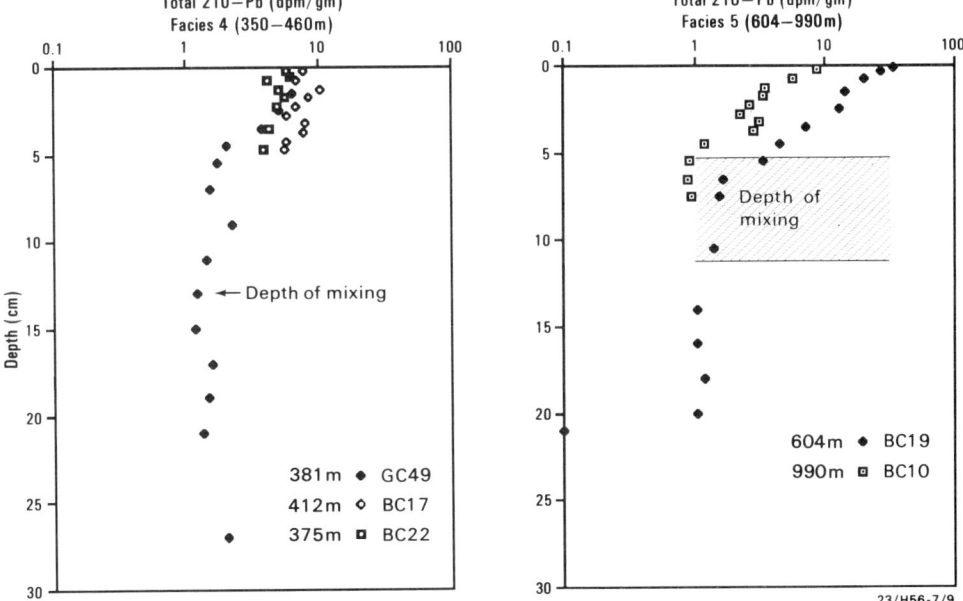

Fig. 9. Total ^{210}Pb activities in Facies 4 and 5 sediments. The estimated depths of mixing inferred from the minimum or the asymptotic concentrations are shown.

dpm g^{-1} (the probable detection limit) intercept the depth axis between about 15 and 20 cm. Total ^{210}Pb from two box-cores (BC 19, 604 m; BC 10, 990 m) collected from the mid-slope in Facies 5 sediments are summarized in Appendix 7 and also shown in Fig. 9. Data from these cores, in contrast to those from Facies 4 sediments, indicate minimum and asymptotic ^{210}Pb concentrations at depths between about 5 and 11 cm. These observations suggest that Facies 4 sediments, which contain the modern phosphorite nodules, are mixed to deeper (15–20 cm) depths than Facies 5 sediments (5–11 cm).

Discussion

Early diagenesis: iron and phosphorus cycling

According to equation 1, Table 2, one mole of phosphate is liberated from marine organic matter when 138 moles of O_2 are consumed. Pore water O_2 data from Fig. 3 indicate that O_2 is depleted from Facies 4 sediments (350–460 m) at depths of 1.1 cm but O_2 extends to nearly 5 cm in Facies 5 sediments at a water depth of 1484 m (core BC 24). The predicted increase in pore-water phosphate concentrations, above bottom-water values, during oxygen reduction was calculated from the following equation:

$$[PO_4]_{pw} - [PO_4]_{bw} = D_o/D_p[\triangle O_2]1/138.$$

$[PO_4]_{pw}$ and $[PO_4]_{bw}$ are the pore-water and bottom-water phosphate concentrations respectively, D_o and D_p are the oxygen and phosphate diffusion coefficients. Using diffusion coefficients from Krom & Berner (1980a) and Reimers et al. (1984) the ratio of the diffusion coefficients was calculated to be 3.2. From the above:

$$\triangle [PO_4] = 0.023 \triangle [O_2]$$

This relationship predicts that for bottom-water oxygen and phosphate concentrations of 200 μM and 0.5 μM respectively, pore-water phosphate will rise to 5.1 μM when all sedimentary oxygen is depleted. Measured pore-water phosphate concentrations are in the range 0.5–3.0 μM. Because the measured pore-water phosphate concentrations are lower than the predicted concentrations a sedimentary sink for pore-water phosphate probably exists in both Facies 4 and 5 sediment; this is supported by the observation that most phosphate pore-water profiles in the top 10 cm were nearly linear or concave-upward indicating that phosphate was consumed in the near-surface sediments.

Facies 4

The observations above, the coincidence of the pore-water Fe^{2+} and phosphate profiles (best illustrated from Facies 4 sediments in GC 35 and GC 77; Fig. 10), the correlations of phosphate and iron in the solid phase (O'Brien et al. 1990) and results from other studies (e.g. Krom & Berner 1980b, 1981; Sundby et al. 1986; Schaffer 1986; Lucotte & d'Anglejan 1987) suggest that phosphate released from organic matter near the sediment–seawater interface is scavenged by Fe^{3+} in the surficial oxic sediments. Sedimentary Fe^{3+} oxyhydroxides may be reduced in the sub-oxic zone according to equation 3 (Table 2); this reaction being mediated by the availability and concentration of pore-water nitrate (Sorensen 1982), thus releasing reduced soluble Fe^{2+} and phosphate to pore water. Fe^{2+} produced in this way within sub-oxic and anoxic sediments diffuses upwards where it is subsequently reoxidized to Fe^{3+} by oxygen or by nitrate diffusing down from oxic and sub-oxic near-surface sediments (Froelich et al. 1979; Klinkhammer 1980). Sedimentary iron (reported as Fe_2O_3) is high (>10 wt%) in the near surface (2–4 cm) sediments of Facies 4 (Fig. 10, Appendix 8). The recycling of iron between the oxic/anoxic zones of the sediments therefore results in a near surface trap for iron. Reprecipitated Fe^{3+} may again scavenge phosphate which has been released to pore waters from organic matter during aerobic oxidation. Similarly, sedimentary phosphorus (as P_2O_5) is enriched in the top 10 cm of Facies 4 sediments (Fig. 10, Appendix 8) with the maximum P_2O_5 concentration occurring in the same interval as the maximum Fe_2O_3 concentration. The iron oxyhydroxide phases (with adsorbed phosphate) may be again buried (by sedimentation and/or bioturbation) below the oxic zone where they subsequently undergo reduction and dissolution, liberating both iron and phosphate to pore-waters. Iron and phosphate therefore are continually recycled between the solid phase and pore-waters of oxic and anoxic sediments. The processes proposed here are summarized in Fig. 11, and are variations of the iron–phosphorus cycling scenarios described by Schaffer (1986) at the oxic/anoxic interface in the Black Sea and by Froelich et al. (1988) in interfacial sediments of the Peru continental margin. Phosphate adsorbed on iron oxyhydroxides and released to pore waters near the base of the mixed layer (15–18 cm) during dissolution of the iron oxyhydroxide carrier may:

(1) diffuse toward the sediment–seawater

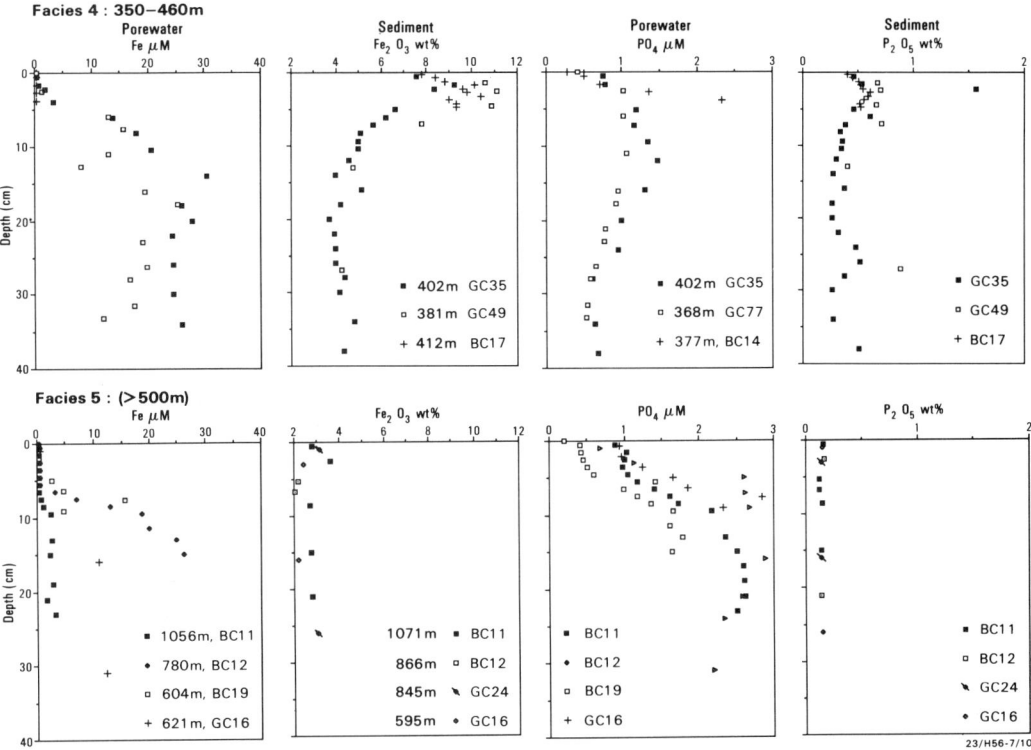

Fig. 10. Pore-water iron, phosphate and solid-phase Fe_2O_3 and P_2O_5 data from Facies 4 and Facies 5 sediments.

interface where it is probably rescavenged onto iron oxyhydroxides;
(2) be incorporated into apatite and/or nodules or
(3) diffuse to deeper depths in the sediments.

We suggest that apatite precipitation occurs in the mixed layer near the pore-water phosphate maximum and where pore-water fluoride concentrations begin to decrease. As an essential constituent of carbonate–fluorapatite, fluoride may serve as a useful tracer of authigenic phosphorite precipitation. The shapes of the pore-water fluoride profiles are quite different from those observed elsewhere. For example, simple concave-downward trends have been observed for sediments of the Peru shelf (Froelich et al. 1983, 1988) and Mexico (Jahnke et al. 1983) both areas where phosphorites are known to be presently forming. Slightly enriched F^- concentrations but generally featureless trends have been observed for pelagic sediments in the equatorial Pacific Ocean (Froelich et al. 1983). The main difference in the profiles from the East Australian margin is

the F^- maximum observed in the top several centimetres of sediments. The profiles of Facies 4 sediment, beyond the maximum, resemble the uptake profiles observed in other phosphorite-producing areas. The depth of the fluoride maximum and the onset of decreasing fluoride concentrations is variable, but always occurs within the mixed layer at depths < 18 cm (Fig. 5b).

We interpret these fluoride profiles as being a consequence of fluoride adsorption onto iron surfaces at or near the sediment-seawater interface, followed by burial and release of fluoride when iron is remobilized. The concave-downward portion of the profiles from the Facies 4 sediments are interpreted as representing uptake into apatite, just as shown for other environments. Fluoride adsorption onto ferric-oxyhydroxide surfaces in soils and marine sediments has recently been demonstrated experimentally (Farrah & Pickering 1986; Ruttenberg 1988). Furthermore, iron and fluoride are correlated with each other in the sediments of East Australia (O'Brien et al. 1990). Thus,

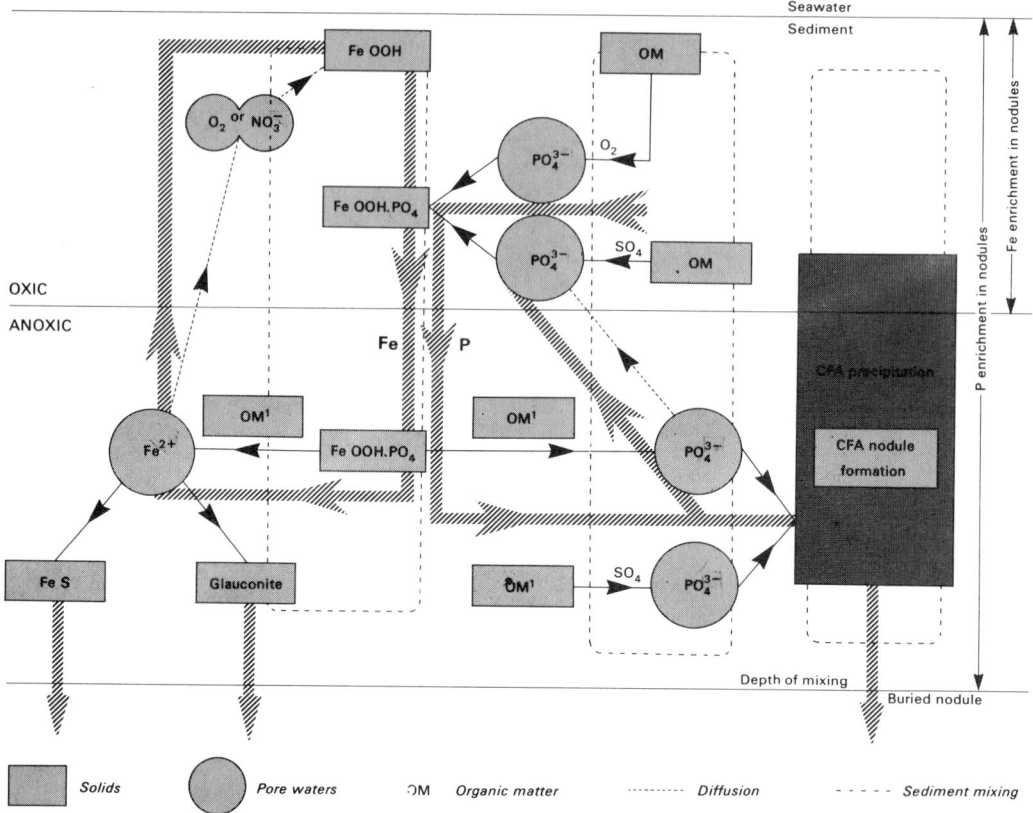

Fig. 11. Schematic of iron and phosphorus recycling within the mixed layer, between oxic and anoxic zones of Facies 4 sediments. The heavy arrowed lines indicate major pathways for phosphorus (on the right hand side of the diagram) and for iron (on the left). Rectangular boxes and circles represent solid and pore-water phases respectively. Dashed lines indicate all solid phases are recycled between oxic and anoxic zones. Light dotted lines indicate pore-water solute diffusion. On the far right of the diagram are indicated the proposed CFA (carbonate–fluorapatite) precipitation zone and the zones of nodule formation, and of subsequent phosphorus and iron enrichments in nodules.

recycling of iron in the mixed zone of Facies 4 sediments apparently is responsible for enhanced pore-water concentrations of phosphate and fluoride, two key elements required for apatite precipitation, near where apatite is proposed to precipitate.

Sedimentological and mineralogical observations support the pore-water evidence that the locus of apatite precipitation is within the mixed layer. First, Holocene nodules are always found within about 20 cm of the sediment-water interface (O'Brien et al. 1986), indicating that the nodules form somewhere within the mixed-layer. Secondly, solid phase geochemical data (O'Brien et al. 1990) show that the nodules contain 2–3 wt% Fe_2O_3 when they form. This is substantially less than the Fe_2O_3 content of the sediments within 10 cm of the sediment surface, but is similar to the Fe_2O_3 content of deeper (> 10 cm) sediments. Consequently, it appears that the nodules form towards the base of the mixed-layer at depths between approximately 10–20 cm. Thirdly, nodules which remain within the mixed layer for extended periods after they form continue to accumulate apatite, with the P_2O_5 content increasing from an initial 5–7 wt% to 15–18 wt% within 60 thousand years of formation (O'Brien et al. 1990). Nodules remaining within the mixed layer also accumulate iron oxyhydroxides at a similar rate as apatite. In contrast, nodules buried below the mixed layer soon after formation accumulate little apatite and no iron oxyhydroxides. These observations reflect the fact that when nodules are buried below the mixed layer they are removed from the zone of iron and phosphorus cycling and enrichment. The observations thus support the proposition that the locus of apatite precipitation is within the mixed layer.

Facies 5

Pore-water iron data from Facies 5 sediments (summarized in Figure 10) also indicate that iron undergoes a recycling between sediments and pore waters. However, the depth at which this process operates is deeper in Facies 5 sediments compared to Facies 4 sediments. Similarly, phosphate in pore waters of Facies 5 sediments is lower than predicted from the stoichiometric model and the concave-up profiles also suggest that phosphate released from organic matter is scavenged onto the sediments (probably by ferric oxyhydroxides). The solid phase data from Facies 5 sediments (a limited data set comprised of samples from several cores sampled in such a way to compare core-top concentrations with those deeper in the sediment) do not, however, indicate any enrichments in either iron of phosphorus in the core-tops. Iron concentrations in Facies 5 sediments (<4 wt% Fe_2O_3, Appendix 9) are about one-third those in the core tops of Facies 4 sediments. Similarly, phosphorus in Facies 5 sediments (<0.2 wt% F_2O_5) is only about one-third the concentrations in Facies 4 sediments. Significantly, there is no enrichment of either iron or phosphorus in the core-tops of Facies 5 sediments. These observations suggest that the recycling scenario outlined for Facies 4 sediments does not significantly control the solid phase concentrations in Facies 5.

The contrasts between Facies 4 and 5 sediments is schematically illustrated in Fig. 12. The diagram contrasts the spatial relationships of the key processes operating in the sediments and illustrates the spatial separation of the oxic sediments and the iron remobilization horizon in Facies 5 sediments as compared with Facies 4. Furthermore, the depth of sediment mixing in Facies 4 sediments encompasses the iron remobilization horizon, but in Facies 5, the iron remobilization horizon mostly lies beyond the depth of sediment mixing. These observations and the temporal properties of sedimentation

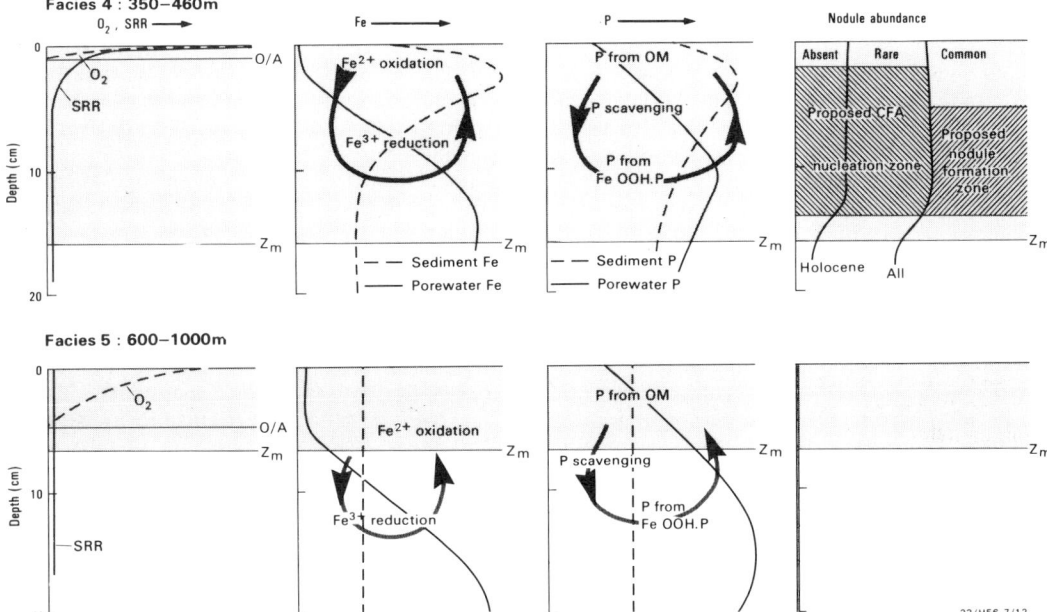

Fig. 12. Schematic of processes operating in Facies 4 and 5 sediments. The shaded areas indicate depths of sediment mixing. The dashed and solid lines on the P and Fe figures represent solid and pore-water distributions respectively. The heavier arrowed line in Facies 4 sediments represents more extensive Fe and P recycling between solids and pore waters than occurs in Facies 5. OM is organic matter.

and mixing (discussed below) are important in the recycling scenario and act to concentrate iron and phosphorus in the mixed layer of Facies 4 sediments.

^{210}Pb and particle mixing rates

If it is assumed that the deposition rate of ^{210}Pb is constant, that the sedimentation rate is small compared to the mixing rate and that post-depositional rearrangement of ^{210}Pb is controlled only by particle movement, then the excess ^{210}Pb distributions can be modelled to calculate sediment mixing rates (Peng et al. 1979; Berner 1980). The solution to the differential equation describing the down-core distribution of excess ^{210}Pb is:

$$C_z = C_o \exp - z(\lambda/D_b)^{\frac{1}{2}}$$

C_z and C_o are the excess ^{210}Pb concentrations respectively at any depth z and at the sediment–seawater interface, λ is the decay constant for ^{210}Pb and D_b is the sediment mixing rate. The sediment mixing rate is calculated from the slope of the line on a log excess ^{210}Pb versus depth plot and the expression:

$$D_b = \lambda \, [z/!\ln(C_o/C_z)]^2$$

Because the core tops of gravity cores may be lost, or at least disturbed during collection, gravity core ^{210}Pb data were not modelled. The box-core data are shown in Fig. 13. Modelling results indicate particle mixing rates of 60 cm^2 ka^{-1} (604 m) and 160 cm^2 ka^{-1} (990 m) in Facies 5 sediments and 1000 cm^2 ka^{-1} (412 m) and 2500 cm^2 ka^{-1} (376 m) in Facies 4 sediments. Patchiness and large variability characterize sediment mixing rate measurements, and variations of factors of 3–4 over relatively small areas are common (Cochran 1985). Nevertheless, data presented here indicate that the rate of sediment mixing in Facies 4 sediments is 5–40 fold more rapid than the rates measured in Facies 5 sediments, and the depth of sediment mixing in Facies 4 sediments is about twice that in Facies 5 sediments. The sediment mixing rates in Facies 4 sediments are comparable with, but generally less than those reported for open continental shelf environments and coastal embayments elsewhere (e.g. Carpenter et al. 1982, 1985). Mixing rates in Facies 5 sediments are, however, comparable to those measured in the deep sea (Cochran 1985). Most interpretations of these data suggest bioturbation by benthic infauna is the mixing mechanism and bioturbation has been found to extend typically to depths of 10–20 cm. Bioturbation may be the mixing agent on the East Australian margin but the East Australian Current, a major oceanographic feature of eastern Australia, impinges onto the shelf and slope between 28 and 32°S and has been suggested to be a major control on the development of the different sediment facies in this area (O'Brien et al. 1990). Current velocities on the shelf average 15–30 cm s^{-1} with maximum currents near 100 cm s^{-1} (Godfrey et al. 1980). High current velocities may induce bed-load transport, sediment resuspension and sediment mixing on the outer shelf and upper slope.

Mixing, sedimentation and particle residence times

Irrespective of the mechanism of mixing, this process is probably important in the cycling and trapping of phosphorus, iron and fluoride in sediments, the precipitation of apatite and the formation of phosphorite nodules. The following discussion compares and contrasts some of the consequences of the differences in sediment mixing and sedimentation rates of Facies 4 and 5 sediments. The average particle residence times in the mixed layer are calculated from:

$$t = z_m/S$$

z_m is the depth of the mixed layer, S is sedimentation rate and t is the particle residence time. These results and other characteristics of Facies 4 and 5 sediments are summarized in Table 5. Sedimentation rates have been reported by O'Brien et al. (1981). An average sedimentation rate for Facies 4 sediments, determined from ^{14}C analyses from several cores was found to be 1 cm ka^{-1}. The sedimentation rate determined from one core collected in Facies 5 sediments (550 m water depth) was significantly higher at 4 cm ka^{-1}. The calculations in Table 5 indicate average particle residence times in the mixed layer within Facies 4 sediments of about 15 ka (using $Z_m = 15$ cm), but only 1–3 ka for Facies 5 sediments.

The average time for a particle to transit the mixed layer distance via mixing processes is calculated from:

$$T_m = (z_m)^2/4D_b$$

D_b is the sediment mixing rate, z_m is the mixed layer distance and T_m is the mixed layer transit time. These times (using $z_m = 15$ cm) are about 23–56 years for Facies 4 and 190–510 years (using $z_m = 11$ cm) for Facies 5 sediments. A particle in Facies 4 sediments resides in the mixed layer for about 15 ka, but is transported across the mixed layer of 15 cm about once

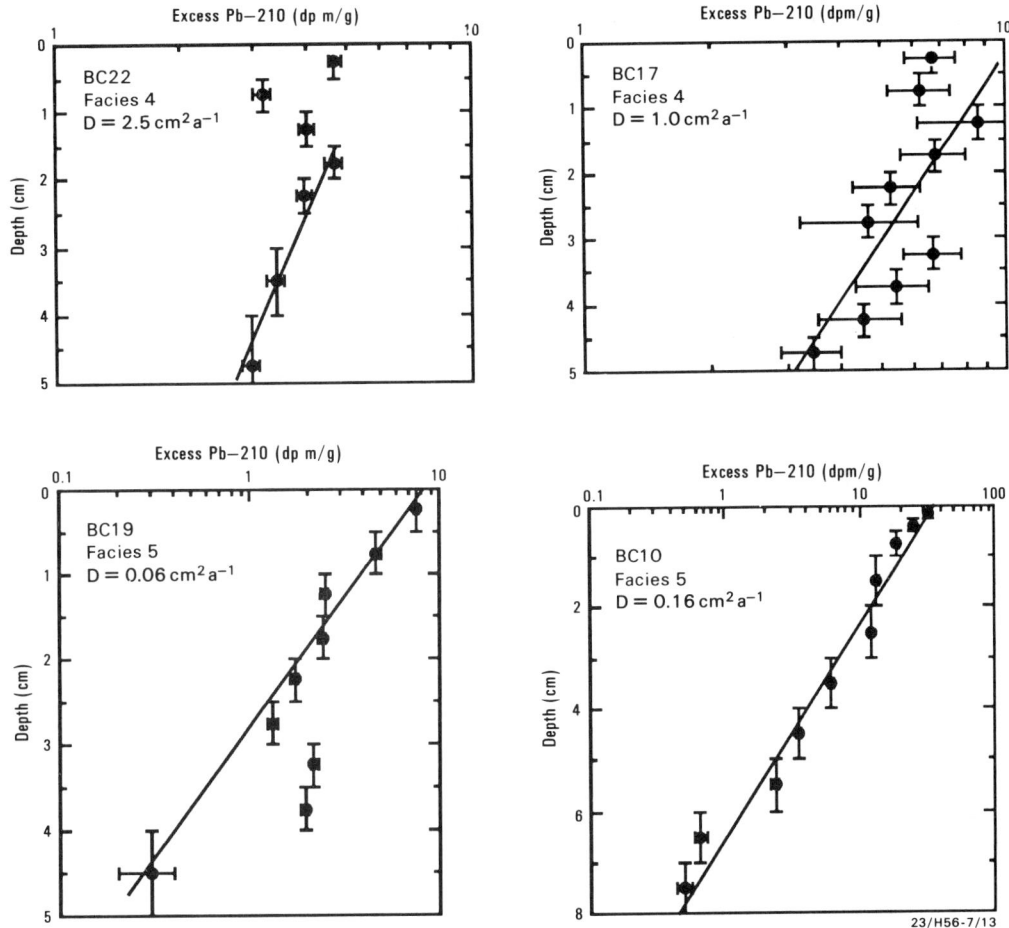

Fig. 13. ^{210}Pb distributions and least-squares fits to data from box-cores from Facies 4 (BC 17, BC 22) and Facies 5 (BC 10, BC 19) sediments.

Table 5. *Sedimentation and mixing characteristics of Facies 4 and Facies 5 sediments*

Characteristic	Facies 4 (phosphorite zone)	Facies 5 (midslope)
Sedimentation rate (S) cm ka^{-1}	<1.0	4
Depth of sediment mixing (Z_m) cm	c. 15	5–11
Sediment mixing rate (D_b) cm^2 ka^{-1}	1000–2500	60–160
Average particle residence time in mixed layer (t) ka	15	1–3
Average time for a particle to transit the mixed layer via mixing. (T_m) a	23–56	190–500
Average TOC content wt %	0.5	0.8

every 23–56 years. A sediment particle in Facies 5 sediments, in contrast, resides on average in the mixed layer only 1–3 ka, but takes approximately 190–510 years to transit a shallower mixed layer. Therefore, a particle in Facies 4 sediments will be recycled between the oxic/anoxic interface (1–2 cm) and through the iron and phosphorus remobilization horizons about (t/T_m) or 200–700 times before it is buried below the mixed layer. However, a particle in Facies 5 sediments will only be recycled (using t = 3000 years) about 6–15 times before being buried below the mixed layer. These temporal and spatial contrasts between Facies 4 and 5 sediments indicate that processes operating in Facies 4 sediments act as a pump that promotes the scavenging/dissolution processes across the oxic/anoxic interface to trap iron, phosphorus and fluoride in the near surface sediments, promote apatite precipitation and provide the reworking to accumulate apatite into nodules.

Organic carbon fluxes

The organic carbon oxidation rates by oxygen and sulphate presented in Tables 3 and 4 are summarized for the Facies 4 and 5 cores in Table 6. The total organic carbon oxidized in the sediments is the sum of carbon oxidized by oxygen and sulphate. Included in Table 6 are estimates of the organic carbon burial rate in the sediments. The latter have been calculated as the product of the TOC content in the sediments, the in situ dry bulk density estimated from the calcium carbonate content of the cores (Lyle & Dymond 1976) and the sedimentation rates (Table 5). The organic carbon contents of Facies 4 sediments (0.5 wt%) are about half those in Facies 5 sediments (0.8 wt %; O'Brien et al. 1990). The organic carbon fluxes are illustrated schematically in Fig. 14.

Despite the large uncertainties that result from grouping the data together according to

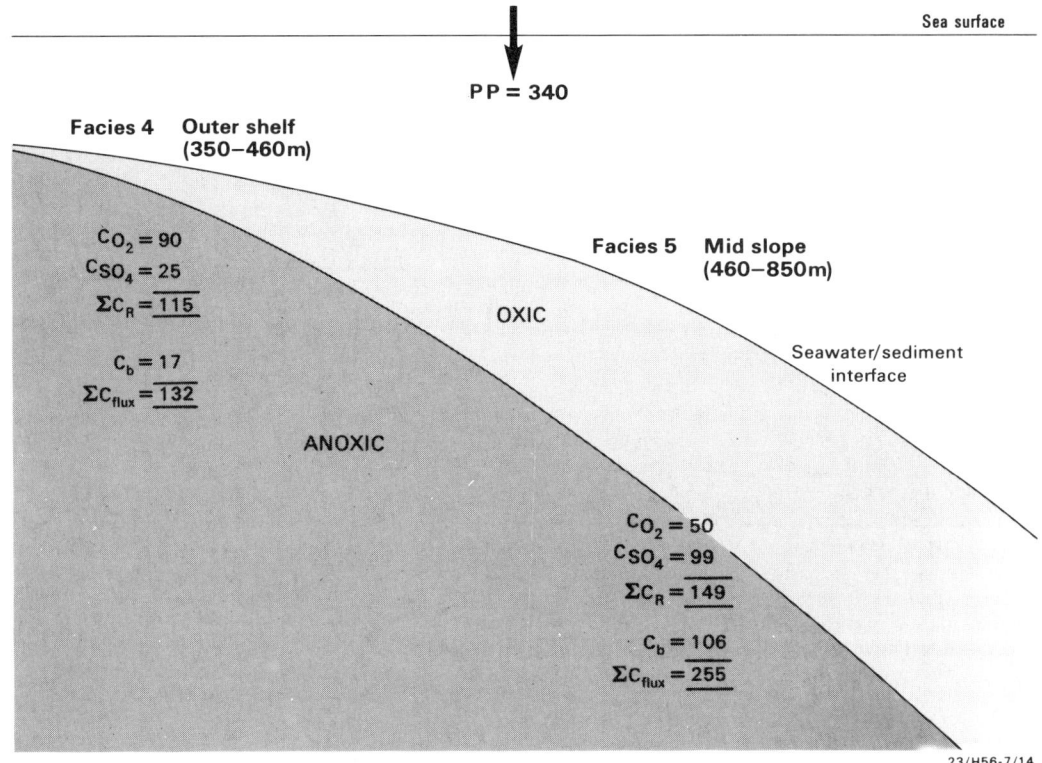

Fig. 14. Schematic representation of the organic carbon mass balance on this sector of the East Australian margin. All flux units are $\mu M\ C\ cm^{-2}\ a^{-1}$. C_{O_2} and C_{SO_4} are the organic carbon oxidized because of oxygen and sulphate reduction respectively. C_R is the total carbon respired, C_b is the organic carbon burial rate and C_{flux} is the total organic carbon flux to the sediment surface. PP is the estimated mean annual primary productivity.

Table 6. *Organic carbon fluxes.*

Facies	C_{O_2}	C_{SO_4}	Total C_{ox}	C_b	C_f
	(μM C cm^{-2} a^{-1})				
Facies 4					
(350–460 m)	90	25*	115	17	132
Facies 5					
(460–850 m)	50	99†	149	106	255
(850–1494 m)	55	?	?	?	?

C_{ox} is carbon oxidized by oxygen plus carbon oxidized by sulphate.
C_b is the carbon burial-rate in sediments.
C_f is the carbon flux to the sediments ($C_{ox} + C_b$)
* Calculated using all gravity core and grab sample data.
† Calculated using only the whole-core data from GC 16 and GC 24, recognizing the importance of down-core sulphate reduction. (Including BC data in the average would underestimate whole-core organic carbon oxidation rates).

facies type, several important observations can be made. First, oxygen is the most important oxidant in Facies 4 sediments. Sulphate reduction may account for about 25% of the total organic carbon oxidized in these sediments, although in one core where direct comparisons of aerobic and anaerobic oxidation can be made, sulphate reduction accounted for about 40% of carbon oxidized, (GS 32, Table 4). Furthermore, anaerobic oxidation found in the surface sediments, in the presence of oxygen, indicates the important existence of anoxic microniches for sulphate-reducing bacteria. Second, carbon oxidized during sulphate reduction in the core top of the grab sample (GS 32) was comparable to that oxidized when experiments were conducted on gravity cores (GC 42, 47, 61, 36) to depths near 1 m, suggesting that most sulphate-reducing microbial activity occurred in the top 10 cm of Facies 4 sediments.

A comparison of the results from Facies 4 and 5 sediments indicates the following. First, organic carbon oxidized by aerobic activity in Facies 5 sediments is about half that of Facies 4 sediments. Second, the organic carbon oxidized by anaerobic activity in Facies 5 sediments is comparable to that oxidized by oxygen. However, there is no evidence of significant interfacial anaerobic oxidation in Facies 5 sediments, and most of the anaerobic activity occurs below about 10 cm depth. This result is probably a consequence of the near five-fold higher sedimentation rates in Facies 5 sediments (compared to Facies 4) which bury metabolizable organic carbon below the oxic sediment zone where it is then metabolized by the sulphate-reducing bacteria. Third, an estimate of the total organic flux to the sediment (the sum of carbon oxidized by oxygen and sulphate plus the burial flux) indicates a higher flux of organic carbon to Facies 5 sediments; much of this is apparently accounted for buried in the sediments. There are insufficient data available on primary productivities from the area to comment upon spatial variability in productivity, but the winnowing effects of the East Australian Current, which impinges onto the outer shelf in this region and exerts a control on the chemical composition of the sediments, (O'Brien *et al.* 1990) apparently exports organic carbon associated with fine-grained resuspended sediments from the outer shelf (where current velocities are high) to the mid and lower-slope sediments, where surface current velocities are less.

Finally, an estimate of the amount of organic carbon recycled in the sediments, calculated by expressing the total organic carbon oxidized as a fraction of the organic flux to the sediments indicates that most (*c.*90%) of the organic carbon flux to Facies 4 sediments is oxidized therein, but only about 60% of the flux to Facies 5 sediments is oxidized. This result is, in part, a combination of the benthic infaunal activities, the nature of the organic matter, and the high sediment mixing rates (but low sedimentation rates) in Facies 4 sediments as contrasted to low sediment mixing but high sedimentation rates in Facies 5 sediments. The longer residence times of particles in Facies 4 sediments and the more rapid recycling of particles within the mixed layer indicates that organic carbon is recycled many times through 'oxic' sediments than in Facies 5 sediments.

The sediment fluxes in Fig. 14 suggest that a significant proportion of the estimated mean annual primary productivity is oxidized within the outer continental shelf and mid-slope sediments of this part of the Australian margin, although the mechanisms by which this organic carbon is recycled at various water depths are very different.

Bacterial activity

Previous workers (O'Brien *et al.* 1981) have suggested that the post-mortem alteration of the benthic bacterial populations to carbonate fluorapatite was responsible for formation and accumulation of phosphorite nodules within the sediments of Facies 4. They also suggested that bacteria gradually cemented the unconsolidated sediments and that the associated sediments were almost free of bacterial colonies. Experiments on three nodules conducted at sea indicated about a two-fold higher level of bacterial activity on the surfaces of nodules as compared to the adjacent sediments although the results were extremely variable, and data from one nodule indicated a lower level of activity than that in the sediments (Moriarty, unpublished data). The high rates of microbial activities (bacterial productivity, oxygen and sulphate reduction rates) in Facies 4 sediments support the idea that bacteria play an important role in phosphate diagenesis in these sediments. The sulphate-reducing bacteria are responsible for oxidizing about 20–40% of the organic matter flux into Facies 4 sediments, suggesting that there is a significant volume of anaerobic microniches in the surface sediments of the phosphorite zone. Phospholipids, specific to the sulphate reducing bacteria indicate that these organisms constitute 30–35% of the bacterial population (P. Nichols, pers. comm.). The shallow depths of oxygen penetration (1 cm), the presences of pyrite framboids (O'Brien et al. 1981) and glauconite in the phosphorite-bearing sediments are collectively indicative of anaerobic microbial activity in the near-surface sediments. These reducing conditions at or near the sediment-seawater interface are not characteristic of lower slope sediments.

Implications for phosphorite nodule formation in non-upwelling environments

The geochemical processes described in this paper help explain several features of both the East Australian and other marine phosphorite deposits. For example, in the East Australian setting, these processes explain: (1) why Holocene nodules are typically found within the mixed-layer, (2) why there is a zone of marked iron and phosphorus enrichments near the sediment-seawater interface in Facies 4 sediments; (3) the mechanism by which phosphate nodules become progressively enriched in phosphorus (Fig. 15) and iron after formation; (4) the reason why the phosphate nodules only form within a relatively narrow depth interval.

The processes described here also help to explain: (1) the origin of those phosphorite deposits in the geological record that did not apparently form within organic rich sediments of upwelling environments (Cook 1976; Bentor 1980) such as, the Agulhas Bank and other areas off South Africa (Birch et al. 1983 McArthur et al. 1988); (2) the well-known iron oxyhydroxide–phosphate association in the sedimentary rock record (e.g. the iron-rich glauco-conglomeratic Agulhas Bank phosphorites (Parker & Seisser 1972) and the iron oxide-rich Moroccan phosphates (McArthur 1978); (3) the formation of hardground and 'unconformity type' phosphorites, which are typically calcareous and may be ferruginous (Cook 1976; Jarvis 1980).

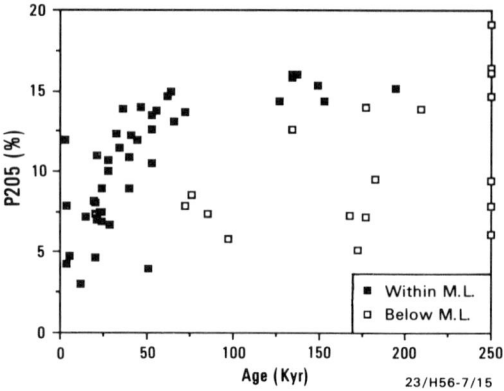

Fig. 15. Phosphorus content of nodules versus nodule age for nodules separated from within the mixed layer (filled symbols) and from below the mixed layer (open symbols; from O'Brien et al. 1990).

Summary

The data presented here show systematic spatial and rate differences between the geochemical processes operating in phosphorite-bearing Facies 4 (350–460 m) and Facies 5 phosphorite-free, mid-slope (>460 m) sediments, which provide new insights into the formation of the modern marine phosphorites on the East Australian margin.

(1) Facies 4 sediments are characterized by the highest rates of biogenic activity measured across the margin; oxygen is depleted at depths near 1 cm, sulphate reduction was measured in anoxic interfacial microniches, and bacterial productivities were highest in interfacial sediments.

(2) Estimates of carbon oxidized by the electron acceptors, oxygen and sulphate, indicated that in the top 10 cm of sediments, oxygen reduction accounted for about 60–80% of carbon oxidized in Facies 4 sediments, but >80% of carbon oxidized in Facies 5 sediments.

(3) Depth-integrated sulphate-reduction rates indicated that most anaerobic activity in Facies 4 sediments occurred in the near-

surface (<10 cm), where apatite is precipitating and most young nodules are found. In contrast, most of the anaerobic activity on Facies 5 sediments occurred beyond 10 cm and anaerobic oxidation of organic carbon was more significant 'at depth'.

(4) Much of the organic carbon flux to the sediments of Facies 4 is recycled within the mixed layer. However, a significant proportion of the organic carbon flux to Facies 5 sediments is recycled at depth during anaerobic oxidation, and an additional significant proportion of that flux is buried and apparently preserved.

(5) The efficient recycling of organic carbon in Facies 4 sediments promotes a demand for secondary oxidants and, in surface sediments, iron oxyhydroxides are recycled between the oxic and anoxic sediments. This process results in a near-surface trap for iron.

(6) Phosphate released from organic matter in interfacial sediments, and fluoride from sea-water, are scavenged by sedimentary iron oxyhydroxides. Phosphate and fluoride are subsequently recycled within the anoxic sediments, where the iron oxyhydroxides dissolve, liberating both phosphate and fluoride to pore waters. The liberated phosphorus may diffuse upwards to be re-scavenged and recycled again with the iron oxyhydroxides or be removed as apatite within the anoxic sediments of the mixed layer at depths between about 10 and 20 cm. The locus of fluoride depletion, an indicator of apatite precipitation, occurs at depths less than 18 cm in the sediments.

(7) The recycling of organic carbon in Facies 4 sediments is promoted by combinations of rapid sediment mixing and slow sedimentation rates which enhance the net flux of phosphorus into the surface sediments where apatite is precipitating and nodules are forming. Rapid sedimentation and slow mixing in Facies 5 sediments do not promote the oxidation and recycling of organic carbon, nor the trapping of iron and phosphorus in near surface sediments.

The mechanism of apatite precipitation remains elusive. However, anaerobic activity in surface sediments, phosphorus and fluoride cycling between oxic and anoxic sediments coupled with iron recycling and rapid sediment mixing but relatively slow sedimentation rates are key factors in the formation of phosphorite nodules on the East Australian margin.

We wish to thank: the BMR technical staff for navigation, maintenance of the geological equipment, their skills and assistance in sample collecting and processing; the Master of the *Rig Seismic*, H. Foreman, the Department of Transport engineers and deck-hands for their aid particularly in rebuilding the box-corer; A. Blank and S. Dibb for invaluable laboratory and on-board technical assistance and P. Nichols and L. Glendenning (CSIRO Division of Oceanography, Australia) for determining the phospholipid biomarkers for the sulphate reducing bacteria. Some of the radiochemistry was carried out at Florida State University by G. W. O'Brien and W. Burnett, with funding from the US-Australia Science and Technology Agreement. Funds for C. Reimers participation were provided by an award from the Academic Senate of the University of California, San Diego. W. Burnett received financial support from NSF grant INT86-13269. We thank P. J. Cook for support from the PHOSREP program. This manuscript is published with the permission of the Director of the Bureau of Mineral Resources, Australia and is a contribution to IGCP Project 156.

Appendix 1. Oxygen in sediments (concentrations in μM)

Depth (cm)	BC 22 375 m	GS 32 431 m	BC 21 599 m	BC 09 987 m	BC 24 1484 m	BC 12 866 m	BC 19 604 m
0	182.0	249.9	188.7	202.2	197.4	174.6	201.8
0.1	153.9	177.2	170.6	158.5	173.1	137.7	150.7
0.2	106.6	109.4	151.2	137.0	146.1	117.3	80.5
0.3	63.6	57.1	131.3	114.5	121.8	100.8	57.1
0.4	42.2	32.0	118.4	103.3	106.7	86.5	52.8
0.5	28.9	16.5	100.8	85.9	93.3		
0.6	14.8	10.7	92.0	72.6	87.6	68.9	42.6
0.7	10.4	5.8	83.3	66.5	82.9		
0.8	7.4	2.9	73.3	63.4	82.4	44.1	
0.9	3.3	0.7	65.6	57.3			
1.0	2.2	0	61.0	57.2	82.9	25.9	
1.1	0.7	0	53.3	53.2			
1.2	0	0	45.7		77.7	17.6	
1.3	0	0	39.9	40.9			
1.4		0	34.0		64.2	13.8	
1.5			29.3	32.7			
1.6			24.0		59.1	11.0	
1.7			20.5	22.5			
1.8			17.0		57.5	8.3	
1.9			13.5	14.3			
2.0			9.4		50.3		
2.1			7.0	8.2			
2.2			3.5		39.4		
2.3			2.9	5.1			
2.4			1.2		33.7		
2.5			1.8	3.1			
2.6			0		28.5		
2.7			0	2.0			
2.8			0		24.2		
2.9							
3.0					22.3		

Appendix 2. Facies 5 pore-water data (concentrations in μM)

Depth (cm)	NO$_3$	NH$_3$	PO$_4$	Fe	F	Depth (cm)	NO$_3$	NH$_3$	PO$_4$	Fe	F
BC-19, 604 m, GC-16, 595 m						BC-12, 604 m					
0–1.40	33.6	1.0	0.93	0.3	71.4	0–1.0	53.1	6.1	0.41	0.6	72.8
0–2.0	10.7		0.68	0.6	74.3	1.0–2.0	47.8	2.2	0.43	0.4	72.4
1.40–2.80	12.4	1.0	0.96	0.3	74.6	2.0–3.0	37.9	0.5	0.45	0.6	73.8
2.0–4.0	1.9		1.13		72.8	3.0–4.0	24.6	0.5	0.51	0.5	73.3
2.80–4.20	2.0	1.0	1.24	0.4	77.4	4.0–5.0	8.6	0.5	0.59	0.5	73.3
4.20–5.60	0.5	7.0	1.66	2.6	78.0	5.0–6.0	2.1	3.0	1.41	0.5	73.3
4.0–6.0	0.5	10.4	2.61		73.3	6.0–7.0	0.5	6.7	0.99	3.3	70.0
5.6–7.0	0.5	12.0	1.85	4.7	79.2	7.0–8.0	0.5	9.6	1.18	6.9	73.8
6.0–8.0	0.5	10.5	2.62			8.0–9.0	0.5	13.3	1.36	13.0	73.3
7.0–8.5	0.5	13.7	2.85	15.7	85.7	9.0–10.0	0.5	16.8	1.65	18.6	71.9
8.0–10.0		12.0	2.68		71.9	11.0–12.0	0.5	20.9	1.61	20.0	72.8
8.5–10.0	0.5	19.2	2.32	4.7	87.3	12.0–14.0	0.5	23.7	1.79	25.0	71.4
15.0–17.0	0.5	29.9	2.89	11.0		14.0–16.0	0.5	28.3	1.64	26.2	73.3
20.0–22.0		40.2	2.58		72.4						

Appendix 2. Continued

Depth (cm)	NO$_3$	NH$_3$	PO$_4$	Fe	F
BC 11 1071 m					
0–1.0	64.9	1.0	0.88	0.3	74.8
1.0–2.0	55.9	1.0	1.04	0.3	77.9
2.0–3.0	36.1	1.0	1.01	0.3	75.2
3.0–4.0	19.9	1.0	0.97	0.3	75.4
4.0–5.0	6.9	1.0	1.05	0.3	75.1
5.0–6.0	0.5	1.0	1.18	0.3	73.8
6.0–7.0	0.5	5.7	1.40	0.3	72.6
7.0–8.0		8.1	1.62	0.7	74.0
8.0–9.0	0.5	10.7	1.72	1.2	72.8
9.0–10.0		14.9	2.16	2.4	72.4
12.0–14.0	0.5	19.9	2.35	2.6	73.8
14.0–16.0		25.2	2.51	2.2	73.3
16.0–18.0	0.5		2.59		73.3
18.0–20.0		36.1	2.61	2.8	73.3
20.0–22.0	0.5	40.4	2.62	1.7	70.0
22.0–24.0		46.9	2.51	3.3	73.8
BC 24 1484 m					
0–1.4	55.3	0.5	0.91	0.3	72.7
1.4–2.8	48.6	0.5	0.92	0.3	73.2
2.8–4.2	43.3	0.5	1.02	0.3	75.8
4.2–5.6	39.6	0.5	1.09	0.3	76.8
5.6–7.0	30.9	0.5	1.17	0.3	74.7
7.0–8.4	22.9	0.5	1.19	0.3	75.8
8.4–9.8	16.5	0.5	1.22	0.3	75.2
9.8–11.2	12.6	0.5	1.25	0.3	73.7
11.2–12.6	8.4	0.5	1.32	0.3	76.3
12.6–14.0	5.0	0.5	1.29	0.3	76.8
14.0–15.6	3.3	0.5	1.29	0.3	77.9
15.6–17.0	1.1	0.5	1.53	0.7	77.9
17.0–18.4	0.5	0.5	1.40	2.2	77.4
18.4–19.8	0.5	0.6	1.83	15.7	77.4

Depth (cm)	NO$_3$	NH$_3$	PO$_4$	Fe	F
17.0–19.0				26	66.5
19.0–21.0	0.5	14.6	0.99	27.8	
21.0–23.0				24.3	66.1
23.0–25.0	0.5	15.1	0.95		
25.0–27.0				24.4	67.5
27.0–29.0	0.5	15.9	0.62		
29.0–31.0				24.4	66.5
33.0–35.0	0.5	16.1	0.65	26.0	67.9
37.0–39.0			0.69		69.1
41.0–43.0				20.7	61.8
43.0–45.0	0.5	12.9	0.61		
BC 14 377 m, GC 77 368 m					
0–1.10	58.3	2.6	0.50	0.3	68.9
1.10–2.20	19.2	1.6	0.71	0.3	68.9
1.70–3.40		9.2	1.03	1.3	78.1
2.20–3.30	4.5	7.5	1.36	0.3	73.4
3.30–4.40	1.9	17.8	2.33	0.3	74.0
3.40–5.10					
5.10–6.80	3.3	7.6	1.02	13.0	77.0
6.80–8.50	0.5	5.0		15.7	79.2
8.50–10.25					
10.25–12.10	0.5	8.0	1.06	13.1	78.1
12.0–13.7				8.2	
13.70–15.30					74.4
15.30–16.80		10.4	0.96	19.5	
16.80–18.6	0.5	11.2	0.92	25.3	73.9
20.40–22.1		10.8	0.79		
22.1–23.8		11.4	0.77	19.0	72.9
25.5–27.2	0.5	12.1	0.66	19.9	
27.2–28.9			0.59	16.8	73.4
30.6–32.3		11.2	0.55	17.6	
32.3–34.0			0.53		

Appendix 3. Facies 4 pore-water data (concentrations in μM)

Depth (cm)	NO$_3$	NH$_3$	PO$_4$	Fe	F
GC 35 402 m, GS 32 431 m					
0–1.10	21.2	2.8	0.75	0.3	71.4
0–1.30	15.5		0.58		
1.10–2.20	5.8	1.0	0.78	0.8	72.9
1.30–2.30	10.7		0.63		
2.20–3.30				1.9	72.9
2.20–3.40	11.7		0.37		
3.30–4.40				3.4	75.0
4.40–5.50	0.5	4.4	1.19		74.0
4.40–6.04	3.5		0.76		
5.50–6.60				13.9	
6.60–7.70	0.5	7.7	1.17		75.0
7.70–8.80				17.9	71.9
8.90–9.90	0.5	8.7	1.34		70.9
9.90–11.0	0.5	12.5	1.47		67.9
13.0–15.0				30.4	67.9
15.0–17.0	0.5	11.5	1.30		69.4

Appendix 4. Sulphate reduction rates (mM SO$_4$ m^{-2} cm^{-1} a^{-1})

	Sample				
Depth (cm)	GS 32 431 m	BC 19 604 m	BC 12 866 m	BC 11 1071 m	BC24 1484 m
0.5	202.8				
1.0		2.0	2.8	1.7	
1.5	37.4				1.0
2.0		1.4			
2.5	7.7				0.7
3.0		2.9	0.1	0.5	
3.5	19.6				0.2
4.0		3.1			
4.5	3.3				0.7
5.0		3.4	0.3	0.1	
6.0		2.3			
7.0		1.9	2.9		
7.5					0.6
8.0		5.8			
9.0			5.3	3.3	
11.0			21.2		1.6
12.5				0.5	
22.0				1.1	

Appendix 5. *Bacterial productivity (mM C m^{-2} mm^{-1} a^{-1})*

Depth (cm)	Sample								
	GS 35 149 m	GS 36 200 m	GS 37 294 m	GS 40 371 m	BC 15 375 m	BC 17 412 m	GS 32 431 m	BC 19 604 m	BC 09 987 m
0.1					22	201	94	27	40
0.3	45	81	114	155	64	140	110	24	79
0.9	33	24	36	61	116	61	55	19	61
2.0	12	53	24	37	49	43	64	26	61
3.0							31		61
4.0			24					9	
7.0								6	

Appendix 6. *Facies 4 ^{210}Pb data*
Table A. *Porosity, bulk density and ^{210}Pb data for box core BC22 375m*

Depth (cm)	porosity	density (g cm^{-3})	^{210}Pb (total) (dpm g^{-1})	excess^{210}Pb (dpm g^{-1})
0.0–0.5	0.620	1.592	5.71±0.18	4.71
0.5–1.0	0.554	1.688	4.15±0.14	3.15
1.0–1.5	0.538	1.712	5.02±0.17	4.02
1.5–2.0	0.523	1.734	5.66±0.20	4.66
2.0–2.5	0.515	1.746	4.96±0.17	3.96
3.0–4.0	0.508	1.756	4.37±0.15	3.37
4.0–5.5	0.504	1.762	3.93±0.13	2.93

Table B. *^{210}Pb and ^{226}Ra data for BC 17 412 m*

Depth (cm)	^{210}Pb (total) (dpm g^{-1})	^{226}Ra (dpm g^{-1})	excess ^{210}Pb (dpm g^{-1})
0–0.5	7.67±0.92	0.97±0.12	6.70
0.5–1.0	6.91±1.10	0.61±0.16	6.30
1.0–1.5	10.55±2.34	2.01±0.57	8.54
1.5–2.0	8.58±1.21	1.74±0.63	6.84
2.0–2.5	6.72±0.98	1.41±0.42	5.31
2.5–3.0	5.85±1.45	1.17±0.19	4.68
3.0–3.5	8.15±1.05	1.43±0.36	6.72
3.5–4.0	7.79±1.05	2.35±0.22	5.44
4.0–4.5	5.89±1.02	1.31±0.37	4.58
4.5–5.0	5.55±0.55	2.10±0.16	3.45

Table C. *^{210}Pb data for gravity core GC 49 381 m*

Depth (cm)	^{210}Pb (total) (dpm^{-1})	excess^{210}Pb (dpm^{-1})
0–1	6.28±0.20	5.12
1–2	6.33±0.21	5.17
2–3	5.05±0.18	3.89
3–4	3.83±0.14	2.67
4–5	2.00±0.06	0.84
5–6	1.74±0.06	0.58
6–8	1.53±0.05	0.37
8–10	2.26±0.07	1.10
10–12	1.44±0.05	0.28
12–14	1.21±0.03	0.05
16–18	1.58±0.05	0.42
18–20	1.48±0.04	0.32
20–22	1.33±0.04	0.17
26–28	2.03±0.06	0.87
32–34	1.49±0.05	0.33
38–40	2.17±0.07	1.01
44–46	2.55±0.07	1.39
50–52	2.57±0.07	1.41
56–58	3.07±0.08	1.91
68–70	6.06±0.15	4.90

Appendix 7. *Facies 5, ^{210}Pb data.*
Porosity, bulk density and ^{210}Pb data for box core BC 10 990 m

Depth (cm)	porosity	density (g cm^{-3})	^{210}Pb (total) (dpm g^{-1})	excess^{210}Pb (dpm g^{-1})
0.0–0.25	0.86	1.24	33.78±1.30	32.8
0.25–0.5	0.836	1.275	26.26±1.04	25.2
0.5–1.0	0.809	1.315	19.87±0.54	18.8
1.0–2.0	0.783	1.353	14.30±0.41	13.3
2.0–3.0	0.762	1.384	12.93±0.37	11.9
3.0–4.0	0.756	1.392	7.18±0.22	6.0
4.0–5.0	0.748	1.404	4.43±0.14	3.4
5.0–6.0	0.739	1.417	3.35±0.12	2.3
6.0–7.0	0.721	1.444	1.68±0.07	0.6
7.0–8.0	0.727	1.435	1.55±0.07	0.5
9.0–11.0	0.725		1.41±0.08	
12–14	0.703		1.05±0.05	
15–17	0.688		1.03±0.05	
17–19	0.699		1.21±0.05	
19–21	0.705		1.04±0.05	

Porosity, bulk density and ^{210}Pb data for box core BC 19 604 m

Depth (cm)	porosity	density (g cm^{-3})	^{210}Pb (total) (dpm g^{-1})	excess^{210}Pb (dpm g^{-1})
0.0–0.05	0.607	1.611	8.54±0.27	7.60
0.5–1.0	0.604	1.615	5.58±0.20	4.68
0.1–1.5	0.601	1.620	3.39±0.13	2.42
1.5–2.0	0.597	1.625	3.30±0.14	2.40
2.0–2.5	0.593	1.631	2.62±0.08	1.72
2.5–3.0	0.587	1.640	2.22±0.08	1.32
3.0–3.5	0.585	1.643	3.08±0.10	2.18
3.5–4.0	0.583	1.646	2.86±0.09	1.96
4.0–5.0	0.574	1.659	1.20±0.05	0.30
5.0–6.0	0.570		0.92	0.04
6.0–7.0	0.565		0.89	0.03
7.0–8.0	0.561		0.94	0.04

Appendix 8. *Solid-phase geochemistry, Facies 4*

Core	Depth (cm)	Al$_2$O$_3$	P$_2$O$_5$	Fe$_2$O$_3$	CaO
		(weight %)			
GC 35	0–0.5	3.96	0.40	7.79	29.29
GC 35	0–1.1	4.67	0.46	7.56	26.76
BC 17	0.5–1.0	3.95	0.45	8.40	28.82
BC 17	1.0–1.5	4.02	0.51	8.83	28.67
GC 49	1.0–2.0	4.72	0.68	10.61	23.74
GC 35	1.1–2.2	4.38	0.54	9.22	25.79
BC 17	1.5–2.0	4.28	0.53	10.11	26.62
BC 17	2.0–2.5	4.16	0.55	9.63	27.66
GC 35	2.2–3.3	4.31	1.57	8.36	27.00
GC 49	2.0–3.0	4.62	0.71	11.1	24.08
BC 17	2.5–3.0	4.05	0.61	9.81	27.92
BC 17	3.0–3.5	4.18	0.59	10.4	26.98
BC 17	3.5–4.0	4.01	0.56	9.01	28.83
BC 17	4.0–4.5	4.15	0.52	9.34	27.99
GC 49	4.0–5.0	4.57	0.67	10.83	23.82
BC 17	4.5–5.0	4.10	0.53	9.33	27.70
GC 35	4.4–5.5	4.37	0.46	6.62	26.90
GC 35	5.5–6.6	4.57	0.61	6.21	26.10
GC 49	6.0–8.0	4.15	0.72	7.78	27.36

Core	Depth				
GC 35	6.6–7.7	4.72	0.39	5.66	25.98
GC 35	7.7–8.8	4.97	0.34	5.07	25.92
GC 35	8.9–9.0	4.90	0.36	5.01	25.59
GC 35	9.9–11.0	5.08	0.35	4.98	25.81
GC 35	11.0–13.0	5.16	0.30	4.57	24.63
GC 49	12.0–14.0	4.98	0.41	4.77	26.81
GC 35	13.0–15.0	5.12	0.27	3.95	26.78
GC 35	15.0–17.0	4.84	0.38	5.14	25.42
GC 35	17.0–19.0	5.24	0.26	4.19	25.54
GC 35	19.0–21.0	5.48	0.26	3.69	26.16
GC 35	21.0–23.0	5.37	0.31	3.91	27.09
GC 35	23.0–25.0	5.04	0.48	3.94	27.59
GC 35	25.0–27.0	5.02	0.52	3.97	27.50
GC 49	26.0–28.0	4.71	0.86	4.22	27.69
GC 35	27.0–29.0	5.10	0.38	4.36	26.52
GC 35	29.0–31.0	4.94	0.26	4.16	26.84
GC 35	33.0–35.0	5.04	0.27	4.82	25.25
GC 35	37.0–39.0	5.52	0.51	4.35	26.00
GC 35	41.0–43.0	5.11	1.02	5.03	26.53
GC 35	43.0–45.0	5.13	0.38	5.17	25.77
GC 49	44.0–46.0	5.17	0.61	5.46	26.04

Appendix 9. *Solid-phase geochemistry, Facies 5*

Core	Depth (cm)	Al_2O_3	P_2O_5	Fe_2O_3	CaO
		(weight %)			
BC 12	0–1	6.82	0.16	2.81	27.43
GC 24	0–2	6.92	0.15	3.13	27.05
BC 24	1.4–2.8	6.51	0.14	2.50	30.95
BC 11	2.0–3.0	8.31	0.16	3.61	26.07
GC 16	2.0–4.0	5.00	0.14	2.44	26.00
BC 21	4.5–6.0	5.12	0.13	2.19	25.51
BC 21	6.0–7.5	5.01	0.12	2.05	25.52
BC 12	8.0–9.0	7.28	0.15	2.70	27.54
GC 67	11.0–13.0	7.53	0.14	2.75	27.16
BC 24	14.0–15.6	7.05	0.13	2.66	30.35
BC 12	14–16	7.29	0.14	2.77	27.45
GC 16	15–17	4.99	0.14	2.19	25.78
BC 24	18.4–19.8	7.03	0.15	3.24	29.58
BC 11	20–22	7.47	0.14	2.82	28.06
GC 24	25–27	8.05	0.15	3.05	26.02
GC 67	43–45	7.64	0.15	2.78	27.05
GC 16	60–62	6.81	0.12	2.85	24.26

References

BATURIN, G. N. 1971. Formation of phosphate sediments and water dynamics. *Oceanology*, **11**, 373–376.

——, MERKULOVA, K. I. & CHALOV, P. I. 1972. Radiometric evidence for recent formation of phosphatic nodules in marine shelf sediments. *Marine Geology*, **13**, M37–M41.

BENDER. M. L. & HEGGIE, D. T. 1984. Fate of organic carbon reaching the sea floor, a status report. *Geochimica et Cosmochimica Acta*, **48**, 977–986.

BENNINGER, L. K. & KRISHNASWAMI, S. 1981. Sedimentary processes in the inner New York Bight, evidence from excess ^{210}Pb and ^{239,240}Pu. *Earth and Planetary Science Letters*, **53**, 158–174.

BENTOR, Y. K. 1980. Phosphorites — the unsolved problems. *In*: BENTOR, Y. K. (ed.) *Marine Phosphorites–Geochemistry, Occurrence, Genesis*. Society of Economic Paleontologists and Mineralogists, Special Publication, **29**, 3–18.

BERNER, R. A. 1980. *Early Diagenesis*. Princeton University Press, Princeton, N.J.

BIRCH, G. F., THOMSON, J., MCARTHUR, J. M. & BURNETT, W. C. 1983. Pleistocene phosphorites off the west coast of South Africa. *Nature*, **302**, 601–603.

BURNETT, W. C. 1977. Geochemistry and origin of phosphorite deposits from off Peru and Chile. *Geological Society of America Bulletin*, **88**, 813–823.

—— & VEEH, H. H. 1977. Uranium series disequilibrium studies in phosphorite nodules from the west coast of South America. *Geochimica et Cosmochimica Acta*, **41**, 755–764.

——, BEERS, M. J. & ROE, K. K. 1982. Growth rates of phosphate nodules from the continental margin off Peru. *Science*, **215**, 1616–1618.

CARPENTER, R., PETERSON, M. L. & BENNETT, J. T. 1982. ^{210}Pb derived sediment accumulation and mixing rates for the Washington continental slope *Marine Geology*, **48**, 135–164.

——, —— & —— 1985. ^{210}Pb derived sediment accumulation and mixing rates for the greater Puget Sound region. *Marine Geology*, **64**, 291–312.

COCHRAN, J. K. 1985. Particle mixing rates in sediments of the eastern equatorial Pacific, evidence from ^{210}Pb, ^{239}Pu, ^{240}Pu and ^{137}Cs. *Geochimica et Cosmochimica Acta*, **49**, 1195–1210.

COOK, P. J. 1976. Sedimentary phosphatic deposits. *In*: WOLF K. H. (ed.) *Handbook of Strata-bound and Stratiform Ore Deposits*. Elsevier Scientific Publishing Co., Amsterdam, The Netherlands, 505–535.

EMERSON, S. R., REIMERS, C., FISCHER, K. & HEGGIE, D. T. 1985. Organic carbon dynamics and preservation in deep-sea sediments. *Deep Sea Research*, **32(1A)**, 1–22.

FARRAH, H. & PICKERING, W. F. 1986. Interaction of dilute fluoride solutions with hydrous iron oxides. *Australian Journal of Soil Science Research*, **24**, 201–208.

FLYNN, W. W. 1966. The determination of low levels of polonium-210 in environmental materials. *Analytica Chimica Acta*, **43**, 221–227.

FROELICH, P. N., KLINKHAMMER, G. P., BENDER, M. L., LUEDTKE, N. A., HEATH, D., CULLEN, P., DAUPHIN, D., HAMMOND, D., HARTMAN, B., & MAYNARD, V. 1979. Early oxidation of organic matter in pelagic sediments of the eastern equatorial Atlantic, suboxic diagenesis. *Geochimica et Cosmochimica Acta*, **43**, 1705–1090.

——, KIM, K. H., JAHNKE, R. A., BURNETT, W. C., SOUTAR, A. & DEAKIN, M. 1983. Pore water fluoride in Peru continental margin sediments, uptake from seawater. *Geochimica et Cosmochimica Acta*, **47**, 1605–1612.

——, ARTHUR, M. A., BURNETT, W. C., DEAKIN, M., HENSLEY, V., JAHNKE, R., KAUL, L., KIM, K.-H., ROE, K., SOUTAR, A. & VATHAKANON, C. 1988. Early diagenesis or organic matter in Peru continental margin sediments: phosphorite precipitation. *Marine Geology*, **80**, 309–343.

GODFREY, J. S., CRESSWELL, G. R., GOLDING, T. J., PEARCE, A. F. & R. BOYD. 1980. The separation of the East Australian Current. *Journal of Physical Oceanography*, **10**, 430–439.

GRUNDMANIS, V., & MURRAY, J. W. 1982. Aerobic respiration in pelagic marine sediments. *Geochimica et Cosmochimica Acta*, **46**, 1101–1120.

HEGGIE, D., MARIS, C., HUDSON, A., DYMOND, J., BEACH, R. & CULLEN, J. 1987. Organic carbon oxidation and preservation in NW Atlantic continental margin sediments. *In*: WEAVER, P. P. E. & THOMSON, J. (eds) *Geology and Geochemistry of Abyssal Plains*. Geological Society, London, Special Publication, **31**, 215–236.

JAHNKE, R. J., HEGGIE, D. T., EMERSON, S. R., & GRUNDMANIS, V. 1982. Pore waters of the central Pacific Ocean, nutrient results. *Earth and Planetary Science Letters*, **61**, 233–256.

——, EMERSON, S. R., ROE, K. K. & BURNETT, W. C. 1983. The present day formation of apatite in Mexican continental margin sediments. *Geochimica et Cosmochimica Acta*, **47**, 259–266.

JARVIS, I. 1980. The initiation of phosphatic chalk sedimentation in the Senonian (Cretaceous) of the Anglo-Paris Basin. *In*: Bentor, Y. K. (ed.) *Marine Phosphorites — Geochemistry, Occurrence, Genesis*. Society of Economic Paleontologists and Mineralogists, Special Publication, **29**, 167–192.

JITTS, H. R. 1965. The summer characteristics of primary productivity in the Tasman and Coral Seas. *Australian Journal of Marine and Freshwater Research*, **16**, 151–162.

KIM, K. H. & BURNETT, W. C. 1986. Uranium series growth history of Quaternary phosphatic crust from the Peruvian continental margin. *Chemical Geology*, **58**, 277–244.

KIRCHMAN, D. L., NEWELL, S. Y. & HODSON, R. E. 1986. Incorporation versus biosynthesis of leucine, implications for measuring rates of protein synthesis and biomass production by bacteria in marine systems. *Marine Ecology. Progress Series*, **32**, 47–59.

KLINKHAMMER, G. P. 1980. Early diagenesis in sediments from the Eastern Equatorial Pacific. II.

Porewater metal results. *Earth and Planetary Science Letters*, **49**, 81–101.

KRESS, A. G. & VEEH, H. H. 1980. Geochemistry and radiometric ages of phosphatic nodules from the continental margin of northern New South Wales, Australia. *Marine Geology*, **36**, 143–157.

KROM, M. D. & BERNER, R. A. 1980a. The diffusion coefficients of sulphate, ammonium and phosphate ions in anoxic marine sediments. *Limnology and Oceanography*, **25**, 327–337.

—— & BERNER, R. A. 1980b. Adsorption of phosphate in anoxic marine sediments. *Limnology and Oceanography*, **25**, 797–806.

—— & BERNER, R. A. 1981. The diagenesis of phosphorus in a nearshore marine sediment. *Geochimica et Cosmochimica Acta*, **45**, 207–216.

LUCOTTE, M. & D'ANGLEJAN. 1987. Processes controlling phosphate adsorption by iron hydroxides in estuaries. *Chemical Geology*, **67**, 75–83.

LYLE, M. W. & DYMOND, J. 1976. Metal accumulation rates in the southeast Pacific—errors introduced from assumed bulk densities. *Earth and Planetary Science Letters*, **30**, 164–168.

MCARTHUR, J. M. 1978. Systematic variations in the contents of Na, Sr, CO_3 and SO_4 in marine carbonate fluorapatite and their relationship to weathering. *Chemical Geology*, **21**, 89–112.

——, THOMSON, J., JARVIS, I., FALLICK, A. E. & BIRCH, G. F. 1988. Eocene to Pleistocene phosphogenesis off western South Africa. *Marine Geology*, **85**, 41–63.

MORIARTY, D. J. W. 1986. Measurement of bacterial growth rates in aquatic systems from nucleic acid synthesis. *Advances in Microbiological Ecology*, **9**, 245–292.

—— 1987. Accurate conversion factors for calculating bacterial growth rates from thymidine incorporation into DNA, elusive or illusive? *Archive fur Hydrobiologie, Ergebnisse der Limnologie*, **31**, 211–217.

—— & POLLARD, P. C. 1981. DNA synthesis as a measure of bacterial productivity in seagrass sediments. *Marine Ecology Progress Series*, **5**, 151–156.

——, BOON, P. I., HANSEN, J. A., HUNT, W. G., POINER, I. R., POLLARD, P. C., SKYRING, G. W. & WHITE, D. C. 1985. Microbial biomass and productivity in seagrass beds. *Geomicrobiological Journal*, **4**, 21–51.

NICHOLS, P. D., HANSON, J. M., ANTWORTH, C. P., PARSONS, J., WILSON, J. & WHITE, D. C. 1987. Detection of a microbial consortium, including type II methanotrophs, by use of phospholipid fatty acids in an anaerobic halogenated hydrocarbon degrading soil column enriched with natural gas. *Environmental Toxicological Chemistry*, **6**, 89–97.

O'BRIEN, G. W. & VEEH, H. H. 1980. Holocene phosphorite on the East Australian continental margin. *Nature*, **288**, 690–692.

—— & —— 1983. Are phosphorites reliable indicators of upwelling? *In*: SUESS E. & THIEDE, J. (eds) *Coastal Upwelling – Its sediment Record. Part A: Responses of the Sedimentary Regime to Present Coastal Upwelling*. NATO Advanced Research Institute, (Plenum, New York, N.Y.), 399–419.

——, HARRIS, J. R., MILNES, A. R. & VEEH, H. H. 1981. Bacterial origin of East Australian continental margin phosphorites. *Nature*, **294**, 442–444.

——, MILNES, A. R., VEEH, H. H., HEGGIE, D. T., RIGGS, S. R., CULLEN, D. J., MARSHALL, J. F. & COOK, P. J. 1990. Sedimentation dynamics and redox iron-cycling: controlling factors for the apatite–glauconite association on the East Australian continental margin. *In*: NOTHOLT, A. J. G. & JARVIS, I. (eds) *Phosphorite Research and Development*. Geological Society, London, Special Publication, **52**, 61–68.

——, VEEH, H. H., CULLEN, D. J., & MILNES, A. R. 1986. Uranium-series isotopic studies of marine phosphorites and associated sediments from the East Australian continental margin. *Earth and Planetary Science Letters*, **80**, 19–35.

PARKER, R. J. & SEISSER, W. G. 1972. Petrology and origin of some phosphorites from the South African continental margin. *Journal of Sedimentary Petrology*, **42**, 434–440.

PENG, T.-H., BROECKER, W. S. & BERGER. W. H. 1979. Rates of benthic mixing in deep-sea cores from project FAMOUS. *Earth and Planetary Science Letters*, **34**, 167–173.

REIMERS, C. E., 1987. An in-situ microprofiling instrument for measuring interfacial pore water gradients: methods and oxygen profiles from the North Pacific Ocean. *Deep Sea Research*, **34**, 2019–2035.

—— KALHORN, S., EMERSON, S. R. & NEALSON, K. H. 1984. Oxygen consumption rates in pelagic sediments from the Central Pacific, first estimates from microelectrode profiles. *Geochimica et Cosmochimica Acta*, **48**, 903–910.

ROCHFORD, D. J. 1975. Nutrient enrichment of East Australian coastal waters. II. Laurieton upwelling. *Australian Journal of Marine and Freshwater Research*, **26**, 233–243.

RUTTENBERG, K. C. 1988. Fluoride adsorption onto ferric-oxyhydroxides in sediments (abs.). *Transactions of the American Geophysical Union*, **69**, 1235.

SCHAFFER, G. 1986. Phosphate pumps and shuttles in the Black Sea. *Nature*, **321**, 515–517.

SKYRING, G. W. 1987a. Sulphate reduction in coastal ecosystems. *Geomicrobiological Journal*, **5**, 295–374.

—— 1987b. Acetate as the main energy substrate for the sulphate reducing bacteria in Lake Eliza South Australia hypersaline sediments. *FEMS Microbial Ecology*, **53**, 87–94.

——, CHAMBERS, L. A. & BAULD, J. 1983. Sulphate reduction in sediments colonized by cyanobacteria, Spencer Gulf, South Australia. *Australian Journal of Marine and Freshwater Research*, **34**, 359–374.

SORENSEN, J. 1982. Reduction of ferric iron in anaerobic, marine sediment and interaction with reduction of nitrate and sulphate. *Applied and*

Environmental Microbiology, **43**, 319–324.

STOOKEY, L. 1970. Ferrozine — a new spectophotometric reagent for iron. *Analytical Chemistry*, **42**, 779–781.

SUESS, E. 1981. Phosphate regeneration from sediments of the Peru continental margin by dissolution of fish debris. *Geochimica et Cosmochimica Acta*, **45**, 577–588.

SUNDBY, B., ANDERSON, L. G., HALL, PER. O. J., IVERFELDT, A., RUTGERS VAN DER LOEFF, M. M. & WESTERLUND, STIG. F. G. 1986. The effect of oxygen on release and uptake of cobalt, manganese, iron and phosphate at the sediment-water interface. *Geochimica et Cosmochimica Acta*, **50**, 1281–1288.

TAYLOR, J. & PARKES, R. J. 1983. The cellar fatty acids of the sulphate-reducing bacteria, *Desulfobacter* sp., *Desulfobulbus* sp., and *Desulfovibrio desulfuricans*. *Journal of General Microbiology*, **129**, 3303–3309.

TRANTER, D. J., CARPENTER, D. J. & LEECH, G. S. 1986. The coastal enrichment effect of the East Australian Current eddy field. *Deep-Sea Research*, **33**, 1705–1728.

TUREKIAN, K. K., NOZAKI, Y. & BENNINGER, L. K. 1977. Geochemistry of atmospheric radon and radon products. *Annual Reviews of Earth Science*, **5**, 227–255.

VEEH, H. J., BURNETT, W. C. & SOUTER, A. 1973. Contemporary phosphorites on the continental margin of Peru. *Science*, **181**, 844–824.

——, CALVERT, S. E. & PRICE, N. B. 1974. Accumulation of uranium in sediments and phosphorites on the south west African shelf. *Marine Chemistry*, **2**, 189–202.

VON DER BORCH, C. C. 1970. Phosphatic concretions and nodules from the upper continental slope, northern New South Wales. *Journal Geological Society of Australia*, **16**, 755–759.

Diagenetic stability of the isotopic composition of phosphate-oxygen: palaeoenvironmental implications

J. M. McARTHUR[1] & A. HERCZEG[2]

[1] *Department of Geological Sciences, University College London, Gower Street, London WC1E 6BT, UK*

[2] *Research School of Earth Science, Australian National University, ACT 2601, Canberra, Australia. Present address, Division of Water Resources, CSIRO, Private Bag 2, Glen Osmond, SA 5064, Australia*

Abstract: Meteoric diagenesis of the Lower Pliocene Varswater Formation phosphate deposit, Cape Province, South Africa, has altered the chemistry of the francolite component. The isotopic composition of the phosphate-oxygen has been changed, as have other chemical and isotopic parameters that are known to be susceptible to meteoric diagenesis. Reduced major axis regressions between $\delta^{18}O_{PO_4}$ and Sr/Ca, $^{87}Sr/^{86}Sr$ and francolite-$\delta^{13}C_{CO_3}$, and the correlation of francolite-$\delta^{18}O_{CO_3}$ and francolite-$\delta^{13}C_{CO_3}$, predict an original isotopic composition of +30.7‰ (SMOW) for francolite-$\delta^{18}O_{CO_3}$ and +22.7‰ (SMOW) for francolite-$\delta^{18}O_{PO_4}$. These data fix the temperature of phosphogenesis at 16°C and the $\delta^{18}O$ of the phosphatizing fluid as +0.3‰ (SMOW).

The isotopic composition of phosphate-oxygen in living and fossil apatite, such as teeth, bones and early-diagenetic francolite in phosphorites, has excited much interest because of its potential for palaeoenvironmental analysis. For example, Longinelli (1984) has suggested that, because mammalian body temperatures are constant and known within narrow limits, the $\delta^{18}O$ of phosphate-oxygen in mammal bone may permit calculation of the isotopic composition of meteoric water in the mammal's environment. This isotopic composition can be interpreted in terms of palaeoclimate. In addition, oxygen isotope data for pairs of chemical species or minerals, e.g. chert–phosphate and calcite–apatite, may permit the calculation of both the temperature and the isotopic composition of the equilibrating water from which the species or mineral precipitated (Karhu & Epstein 1986; Shemesh *et al.* 1988).

Longinelli & Nutti (1968, 1973*a, b*), who were early exponents of the use of phosphate-oxygen in environmental studies, drew attention to the possible role of post-depositional alteration in controlling the isotopic composition of phosphate-oxygen (Longinelli & Nutti 1968), and they gave examples of such alteration (examples not invalidated by later correction of the data; Longinelli & Nutti 1973*a*). More recent exponents (Kolodny *et al.* 1983; Shemesh *et al.* 1983) have stressed how stable the phosphate-oxygen bond should be. They have suggested that it preserves its initial isotopic composition even under conditions that are geologically extreme, although this view has recently been modified slightly (Shemesh *et al.* 1988).

Palaeoenvironmental indicators have no value unless they preserve their original environmental signal, in this case the isotopic composition of oxygen in phosphate. Francolite is susceptible to chemical and isotopic alteration during weathering, diagenesis and metamorphism (Lucas *et al.* 1980; Flicoteaux & Lucas 1984; McArthur 1985; McArthur *et al.* 1987). Biogenic apatite and early-diagenetic francolite will be altered by such processes to similar extents. We show here that francolite showing clear evidence of meteoric alteration (petrographically and chemically) shows equally clear evidence of accompanying changes in the isotopic composition of its phosphate-oxygen. The alteration has caused a systematic change in francolite chemistry so that clear relationships have been induced between such chemical and isotopic parameters as Sr/Ca, Na/Ca, S/P and francolite-$^{87}Sr/^{86}Sr$ ratios, francolite-$\delta^{18}O_{CO_3}$, francolite-$\delta^{13}C_{CO_3}$ and francolite-$\delta^{18}O_{PO_4}$ (McArthur *et al.* 1987; see below). The clarity of the evidence for alteration of $\delta^{18}O_{PO_4}$ decreases confidence in palaeoenvironmental reconstructions based on $\delta^{18}O_{PO_4}$. In favourable cases, however, there are ways to see through the alteration veil and deduce the original isotopic composition.

Samples and results

The Lower Pliocene Upper Varswater Formation of Cape Province, South Africa is well exposed in the New Varswater Quarry, on Langebaan Road, which lies 20 km NE of Saldanha Bay and 130 km NNW of Capetown. The Formation has been described elsewhere (Dingle et al. 1979; Hendey 1981; McArthur et al. 1987), so the description given here is brief. The Upper Varswater Formation in the locality is about 10 m thick. It consists of poorly consolidated quartz sand intimately mixed with peloidal francolite. Locally, cementation by francolite has produced discontinuous lithification in zones that are commonly 10–30 cm thick. Quartz sand comprises 55 ± 15% of the poorly-consolidated units. The remainder is peloidal francolite with < 2% of other minerals. The francolite is predominantly phosphatized bone fragments (70 ± 10%) with a subordinate fraction of peloidal authigenic francolite. Both types show varying degrees of meteoric alteration that is visible petrographically (McArthur et al. 1987). The Upper Varswater Formation is capped by over 2 m of calcrete deposited from upward moving fluids that presumably caused the alteration.

We present here data on the isotopic composition of phosphate-oxygen and the content and isotopic composition of uranium ($^{234}U/^{238}U$ activity ratio) in 13 samples of francolite from the Upper Varswater Formation. The samples are the same as those studied by McArthur et al. (1987). They represent three stratigraphic levels in the formation: near the top (V3), middle (V7) and bottom (V8).

The method of analysis for $\delta^{18}O_{PO_4}$ is given by Shemesh et al. (1988) and that for $^{234}U/^{238}U$ is given by Ku et al. (1965). The results of the new analyses are shown in Table 1. Correlations between data are shown in Fig. 1 and Table 2. Most of the correlations are significant at the 95% confidence level or better. Reduced major axis correlation is used as all data are controlled by a master influence, meteoric alteration. There are no strictly dependent or independent variables so least-squares regression is inappropriate.

Discussion

Uranium contents and U/P ratios are variable (Table 1, Fig. 1). The U/P ratios correlate with other parameters related to alteration, such as francolite-$\delta^{13}C_{CO_3}$ (Fig. 1, Table 2). These correlations show that uranium is affected by meteoric processes and provide confirmation of the exchangeable nature of U in francolite. Samples cemented with francolite (H in Table 1) are at equilibrium with respect to their $^{234}U/^{238}U$ activity ratios. It takes 800 ka for such equilibrium to be established, so these samples must have been closed to exchange of uranium for at least this length of time, indicating that they must have been cemented by secondary francolite prior to 800 ka. For unconsolidated samples, activity ratios for $^{234}U/^{238}U$ are not at equilibrium and vary between 0.95 and 1.20. Uranium mobilization has continued after local cementation, but over what period cannot be determined without a detailed knowledge of the $^{234}U/^{238}U$ activity ratios in ground water during the last 800 ka.

The isotopic composition of the phosphate-oxygen correlates with variables that are clearly affected by the meteoric alteration, such as $^{87}Sr/^{86}Sr$, Sr/Ca and U/P ratios, francolite-$\delta^{13}C_{CO_3}$ and francolite-$\delta^{18}O_{CO_3}$ (Fig. 1). The scatter in the $\delta^{18}O_{PO_4}$ data is large because the total range of its isotopic composition (1.5‰) is small compared to the analytical error in its measurement (± 0.3‰). Most of the correlations are, nevertheless, significant at levels better than 95% (Table 2). The slope of the reduced major axis correlation between francolite-$\delta^{18}O_{CO_3}$ and francolite-$\delta^{18}O_{PO_4}$ is 0.57 ± 0.19, which is within error of the value of 0.65 ± 0.06 for the reduced major axis correlation calculated from the 55 data pairs on world-wide phosphates presented by Shemesh et al. (1988). The theoretical slope is 0.76, for isotopic equilibrium with water of constant isotopic composition (Shemesh et al. 1988).

The $\delta^{18}O_{PO_4}$ and $\delta^{18}O_{CO_3}$ in phosphorites trend towards lighter isotopic compositions with increasing age (Longinelli & Nutti 1968; Shemesh et al. 1983; Karhu & Epstein 1986; Shemesh et al. 1988). Such trends were interpreted by Karhu & Epstein (1986), and by Shemesh et al. (1988) as showing higher temperatures in past surficial environments. Here we suggest that they are an expression of the premise that old rocks are more likely to have been deeply buried, and heated more, than young rocks. During burial, francolites continuously equilibrate their isotopic compositions in response to increasing temperature and for most phosphorites, this process approaches equilibrium with formation waters of marine origin and isotopic composition.

On Fig. 1g ($\delta^{18}O_{PO_4}$ v. $\delta^{18}O_{CO_3}$) are super-

Table 1. *Isotopic composition of francolite from the Varswater Quarry, Cape Province, South Africa*

Sample	Whole rock ppm U	Activity ratio $^{234}U/^{238}U$	‰ SMOW $\delta^{18}O$-PO_4	‰ SMOW *$\delta^{18}O$-CO_3	‰ PDB *$\delta^{13}C$-CO_3
V3 (A)	4.9 ± 0.2	0.99 ± 0.02	21.9 ± 0.3	28.8	−6.7
V3 (B)H	4.6 ± 0.2	1.02 ± 0.02	21.2 ± 0.4	28.2	−9.5
V3 (C)H	4.6 ± 0.1	1.00 ± 0.04	–	28.3	−9.5
V3 (D)H	4.7 ± 0.1	1.01 ± 0.02	21.8 ± 0.3	28.6	−9.1
V3 (E)	4.1 ± 0.2	1.08 ± 0.03	21.5 ± 0.3	28.3	−8.9
V3 (F)	5.1 ± 0.2	1.20 ± 0.03	22.1 ± 0.3	28.4	−9.5
V7 (A)	8.1 ± 0.3	0.97 ± 0.01	21.8 ± 0.3	29.9	−4.7
V7 (B)	7.5 ± 0.3	0.97 ± 0.01	22.1 ± 0.3	30.2	−4.5
V7 (C)H	6.3 ± 0.3	0.99 ± 0.02	21.3 ± 0.3	28.9	−7.7
V7 (D)	13.1 ± 0.4	0.98 ± 0.01	22.6 ± 0.3	29.8	−5.0
V8 (A)	8.1 ± 0.2	1.03 ± 0.01	21.7 ± 0.3	29.1	−7.0
V8 (B)	10.8 ± 0.4	1.02 ± 0.02	21.9 ± 0.3	30.0	−6.3
V8 (C)	13.3 ± 0.6	0.95 ± 0.02	22.5 ± 0.3	29.8	−5.3
V8 (D)H	14.8 ± 0.4	1.00 ± 0.02	22.1 ± 0.3	29.4	−6.7

* Data from McArthur *et al.* (1987). Errors are ± 0.25‰ for $\delta^{18}O$, ± 0.1‰ for $\delta^{13}C$.

Table 2. *Pearson product-moment coefficients of correlation and reduced major axis regressions*

Variables y	Variables x	† Slope of regression	Correlation coefficient	Level of significance
$\delta^{18}O_{CO_3}$	$\delta^{13}C_{CO_3}$	0.382 ± 0.062	0.95	>99%
$\delta^{13}C_{CO_3}$	$\delta^{18}O_{PO_4}$	4.63 ± 1.93	0.61	>95%
$\delta^{18}O_{PO_4}$	$\delta^{18}O_{CO_3}$	0.565 ± 0.193	0.63	>95%
$^{87}Sr/^{86}Sr$	$\delta^{18}O_{PO_4}$	−160 ± 76 × 10^{-6}	0.42	<90%
* $^{87}Sr/^{86}Sr$	$\delta^{18}O_{PO_4}$	−186 ± 66 × 10^{-6}	0.64	>94%
Sr/Ca	$\delta^{18}O_{PO_4}$	18.0 ± 7.3	0.67	>98%
Sr/Ca	$\delta^{13}C_{CO_3}$	3.89 ± 0.64	0.95	>99%
U/P	$\delta^{18}O_{PO_4}$	227 ± 63	0.74	>99%
U/P	$\delta^{13}C_{CO_3}$	49.1 ± 12.9	0.75	>99%
$^{87}Sr/^{86}Sr$	U/P	−0.71 ± 0.34 × 10^{-6}	0.37	<90%

Variable	Mean value	Standard deviation
$\delta^{13}C_{CO_3}$	−7.17	1.880
$\delta^{18}O_{PO_4}$	21.87	0.4059
$\delta^{18}O_{CO_3}$	29.12	0.7186
$^{87}Sr/^{86}Sr$	0.70920	6.50 × 10^{-5}
*$^{87}Sr/^{86}Sr$	0.70919	7.55 × 10^{-5}
Sr/Ca	21.23	7.32
U/P	168.9	92.3

Data from Table 1 and McArthur *et al.* 1987.
* $^{87}Sr/^{86}Sr$: omitting V8 samples
† error on slope is ± 1.96 standard deviations

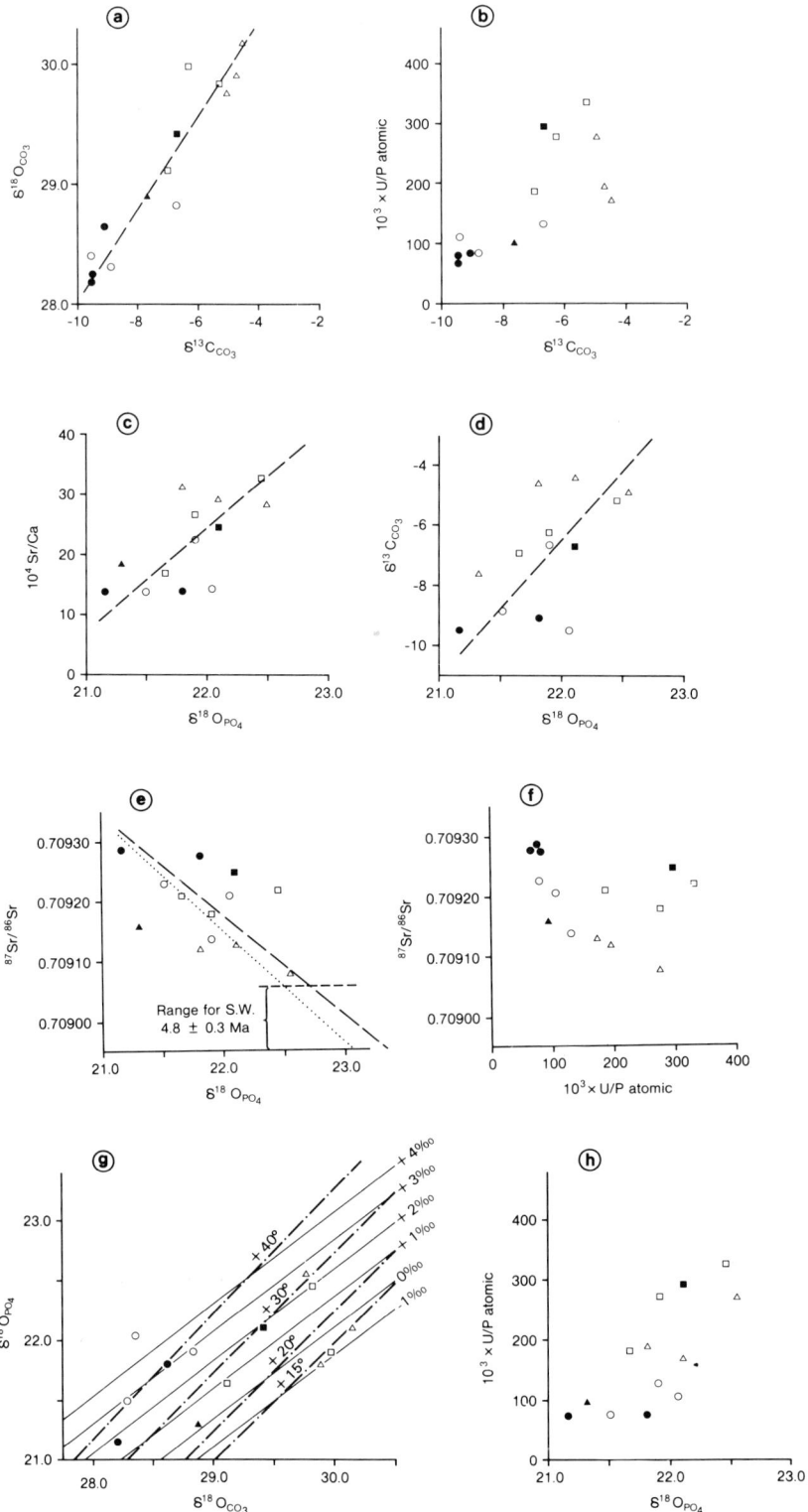

Fig. 1. Chemical and isotopic correlations in Varswater francolite. Symbols are: ○, V3; △, V7; □, V8; representing sample groups from different stratigraphic levels in the deposit. Filled symbols represent cemented samples (H in Table 1). Lines for the reduced major axis regression are shown for data used to calculate the original $\delta^{18}O_{PO_4}$ and $\delta^{18}O_{CO_3}$ in the francolite. The dotted line in (e) represents the reduced major axis regression omitting anomalous V8 samples; S.W., Seawater Sr. Lines on (g) represent equilibrium with waters of various isotopic compositions at various temperatures. Regression data are given in Table 2.

imposed lines representing theoretical equilibrium for waters of various temperatures and isotopic compositions. Comparison of the plotted data with these lines shows that for some samples (e.g. V3(f)), extreme temperatures and isotopic compositions (50°C and +5‰) for meteoric waters must be invoked if alteration is assumed to take place under equilibrium conditions. Such temperatures seem extreme as the deposit has never been buried to depths greater than a few hundred metres (Dingle et al. 1979; Hendey 1981; Dingle pers. comm. 1988). Isotopic compositions of +5‰ for meteoric water also seem implausible. A more likely explanation for these data is that PO_4-O exchanges more slowly than CO_3-O and the scatter of the data for $\delta^{18}O_{PO_4}$ represents an approximation to equilibrium.

The results presented here provide a striking example of the instability of the phosphorus-oxygen bond under conditions of meteoric alteration. The results reduce confidence in the use of the isotopic composition of phosphate-oxygen as a palaeoenvironmental indicator. There is, nevertheless, a way to see through the 'weathering veil' in order to derive useful palaeoenvironmental information, despite the difficulties introduced by incomplete equilibration and the resulting large scatter of the data. If alteration is suspected in a sample set, and sufficient samples are analysed, it may prove possible to define alteration trends of the type shown in Fig. 1. If these are well defined they may be extrapolated to the point of zero alteration thereby providing a usable palaeoenvironmental signal. For the Varswater samples the original Sr/Ca atom ratio was 37 ± 2. This is the characteristic value for minimally altered offshore francolite. It is also the value for peloidal francolite from the nearby Namibian shelf (McArthur 1985; McArthur et al. 1987). The excellent correlation of Sr/Ca with $\delta^{13}C_{CO_3}$ (McArthur et al. 1987; Table 2) allows the original $\delta^{13}C_{CO3}$ of -3.1 ± 0.7‰ to be derived. The original $^{87}Sr/^{86}Sr$ ratio was 0.70901 ± 0.00006, a value derived from an age of 4.8 ± 0.3 Ma for the deposit given by Hendey (1981) and data for the isotopic evolution of Neogene marine strontium given by McKenzie et al. (1988).

From these values, and the regression analysis in Table 2 and Fig. 1, it can be deduced that the original $\delta^{18}O_{PO_4}$ was: +22.8‰ from the correlation with $\delta^{13}C_{CO_3}$, +22.7‰ from the correlation with Sr/Ca ratios; and +23.1‰ from the correlation with $^{87}Sr/^{86}Sr$ (an alternative value of +22.7‰ is obtained if the anomalous data for V8 samples are omitted; see Fig. 1, and fig. 8 of McArthur et al. 1987). The original $\delta^{18}O_{CO_3}$ was +30.7‰ from the excellent correlation with $\delta^{13}C_{CO_3}$ (Fig. 1, Table 2). These original values of +22.7‰ and +30.7‰ are determined independently of the correlation between $\delta^{18}O_{CO_3}$ and $\delta^{18}O_{PO_4}$ in francolite. The regression line for this correlation, nevertheless, passes exactly though these co-ordinates (for $\delta^{18}O_{CO_3}$ = 30.7‰ the regression predicts $\delta^{18}O_{PO_4}$ = 22.8‰). Applying the PO_4-CO_3 and PO_4–water palaeothermometers of Shemesh et al. (1988) to these isotopic compositions gives an average temperature of 16°C for phosphogenesis (there were, presumably, diurnal and seasonal variations) and an isotopic composition of +0.3‰ for the sea water in which it occurred.

The maximum alteration to the isotopic composition of PO_4-O is -1.5‰ in sample V3(B). Precipitation on the west coast of southern Africa has a $\delta^{18}O$ of -3.5‰ (Dansgaard 1964). This value is not far from the value for sea water and necessarily restricts the degree of alteration that can occur. Greater alteration might be found in phosphates existing at higher latitudes than the Varswater deposits because the difference between the isotopic composition of oxygen in sea water and meteoric water increases with increasing latitude.

Conclusions

The isotopic composition of $\delta^{18}O_{PO_4}$ in francolite from the Varswater Formation has been altered by meteoric diagenesis. Palaeoenvironmental interpretation based on the analysis of the isotopic composition of phosphate-oxygen in fossil apatite (eg. teeth, bones and authigenic francolite) must allow for this effect of diagenetic alteration. Analyses of many samples from one site, however, may permit alteration trends to be defined, as has been done for the samples reported on here, and useful depositional signals may then be obtained.

The isotopic data for oxygen were provided by R. Nissan and Y. Kolodny, Hebrew University of Jerusalem. We express our sincere thanks for their assistance.

References

DANSGAARD, W. 1964. Stable isotopes in precipitation. Tellus, **XVI**, 436–468.

DINGLE, R. V., LORD, A. R. & HENDEY, Q.B. 1979. New sections in the Varswater Formation (Neogene) of Langebaan Road, Southwestern Cape, South Africa. Annals of the South African Museum, **78**, 81–92.

FLICOTEAUX, R. & LUCAS, J. 1984. Weathering of phosphate minerals. *In*: NRIAGU, J. O. & MOORE, P. B. (eds) *Phosphate Minerals*. Springer-Verlag, 292–317.

HENDEY, Q. B. 1981. Geological succession at Langebaanweg, Cape Province, and global events of the Late Tertiary. *South African Journal of Science*, **77**, 33–38.

KARHU, J. & EPSTEIN, S. 1986. The implication of the oxygen isotope records in coexisting cherts and phosphates. *Geochimica et Cosmochimica Acta*, **50**, 1745–1756.

KOLODNY, Y., LUZ, B. & NAVON, O. 1983. Oxygen isotope variations in phosphate of biogenic apatites, I. Fish bone apatite – rechecking the rules of the game. *Earth and Planetary Science Letters*, **64**, 398–404.

KU, T.-L. 1965. An evaluation of the $^{234}U/^{238}U$ method as a tool for dating pelagic sediments. *Journal of Geophysical Research*, **70**, 3457–3474.

LONGINELLI, A. 1984. Oxygen isotopes in mammal bone phosphate: a new tool for paleohydrological and paleoclimatological research? *Geochimica et Cosmochimica Acta*, **48**, 385–390.

—— & NUTTI, S. 1968. Oxygen isotopic composition of phosphorites from marine formations. *Earth and Planetary Science Letters*, **5**, 13–16.

—— & —— 1973a. Revised phosphate-water isotopic temperature scale. *Earth and Planetary Science Letters*, **19**, 373–376.

—— & —— 1973b. Oxygen isotope measurements of phosphate from fish teeth and bones. *Earth and Planetary Science Letters*, **20**, 337–340.

LUCAS, J., FLICOTEAUX, R., NATHAN, Y., PREVOT, L. & SHAHAR, Y. 1980. Different aspects of phosphorite, weathering. *Society of Economic Paleontologists and Mineralogists, Special Publication*, **29**, 41–51.

MCARTHUR, J. M. 1985. Francolite geochemistry – compositional controls during formation, diagenesis, metamorphism and weathering. *Geochimica et Cosmochimica Acta*, **49**, 23–35.

——, HAMILTON, P. J., GREENSMITH, J. T., WALSH, J. N., BOYCE, A. B., FALLICK, A. E., BIRCH, G., BENMORE, R. A. & COLEMAN, M. L. 1987. Francolite geochemistry – meteoric alteration on a local scale. *Chemical Geology (Isotope Geoscience)*, **65**, 415–425.

MCKENZIE, J. A., HODELL, D. A., MUELLER, P. A. & MUELLER, D. W. 1988. Application of strontium isotopes to late Miocene–early Pliocene stratigraphy. *Geology*, **16**, 1022–1025.

SHEMESH, A., KOLODNY, Y. & LUZ, B. 1983. Oxygen isotope variations in phosphate of biogenic apatites, II. Phosphorite rocks. *Earth and Planetary Science Letters*, **64**, 405–416.

——, —— & —— 1988. Isotope geochemistry of oxygen and carbon in phosphate and carbonate of phosphorite francolite. *Geochimica et Cosmochimica Acta*, **52**, 2565–2572.

Episodes of phosphogenesis and phosphorite concretion formation in the North Sea Tertiary

PETER S. BALSON

British Geological Survey, Keyworth, Nottingham NG12 5GG, UK

Abstract: It is well established that certain periods through geological time have been more favourable to the formation of phosphorite deposits than others. These 'phosphogenic episodes' have been correlated with periods of eustatic high sea-levels, global warm temperatures and increased marine productivity, amongst other factors.

The development of phosphorite concretions in marine sediments of the southern part of the North Sea Tertiary Basin reflects episodicity corresponding to times of sea-level high-stands. In particular, concretions in Early Eocene marine mudstones make their appearance just above levels which indicate initiation of a more open connection between the Atlantic Ocean and the relatively enclosed North Sea Basin. Similarly, during the Middle Miocene, phosphorite concretions occur associated with indicators of oceanic influence. Elsewhere in the world, remote from the North Sea, phosphate-rich sediments also formed during these stratigraphically brief episodes. Occurrences of relatively sparse diagenetic phosphorite concretions in the North Sea Basin therefore appear to be linked to large scale, possibly global, cycles and not solely to the local diagenetic environment.

Most sedimentary rocks contain some phosphate but some rocks contain anomalously high concentrations. Phosphate-rich sediments have probably been deposited during all major sea-level transgressions during the Tertiary. However, some periods have been more important than others and resulted in the formation of large phosphorite deposits (Riggs 1987). Many authors have discussed the occurrence and causes of episodicity in the formation of phosphorites (e.g. Cook & McElhinny 1979; Bentor 1980; Sheldon 1980; 1981; Cook 1984; Yanshin & Zharkov 1986) but there is still no general agreement as to the conditions which are responsible (see also Cook et al. 1990). Among the factors suggested are sea-floor spreading rates, orogenic and volcanic episodes, climate, oceanic circulation and sea-level changes. Most of these studies concentrate on the abundance and volume of phosphorite deposits through time. Usually only major phosphorite occurrences are considered which represent phosphogenic episodes of millions of years in duration (e.g. Cook & McElhinny 1979). This paper, on the other hand, is concerned with volumetrically very small phosphorite occurrences within the Tertiary sequence of the North Sea Basin, and examines the possible causes of their episodic distribution.

Upwelling of deep ocean water onto continental shelves is often cited as a prerequisite to the formation of phosphorite deposits. Upwelling replenishes the nutrient supply in surface waters leading to localized areas of enhanced productivity and at the present time occurs in open ocean, continental margin and some continental shelf settings. Only two areas of modern continental margin upwelling are known to be connected with the deposition of phosphate-rich sediments, off southwest Africa and Peru–Chile. Upwelling is probably not involved in the formation of the only other known Recent phosphorites in an area off eastern Australia (O'Brien & Veeh 1983). Thus upwelling is a process operating on a local or regional scale which in certain areas *may* lead to the deposition of phosphate-rich sediments. Upwelling processes may have been more vigorous at certain times in the past (Riggs & Sheldon 1990) but upwelling is not *per se* the cause of the observed episodicity of phosphogenesis on a global scale.

The occurrences of phosphorite concretions in the North Sea Tertiary are depicted on Fig. 1, which represents a revised and more accurate version of Fig. 3 of Balson (1987). In this paper, the stratigraphy of each concretion-bearing horizon has been re-examined and correlated with the standard nannoplankton zones of Martini (1971) which have been widely applied in the North Sea Tertiary and which give a fair degree of stratigraphic resolution: 25 zones in the Palaeogene giving an average of 1.65 Ma per zone; 18 zones in the Neogene giving an average of 1.31 Ma per zone, although resolution is better in certain parts of the sequence. The width of the stippled bands in Fig. 1 represents the maximum age range within which concretion development, possibly during a relatively brief

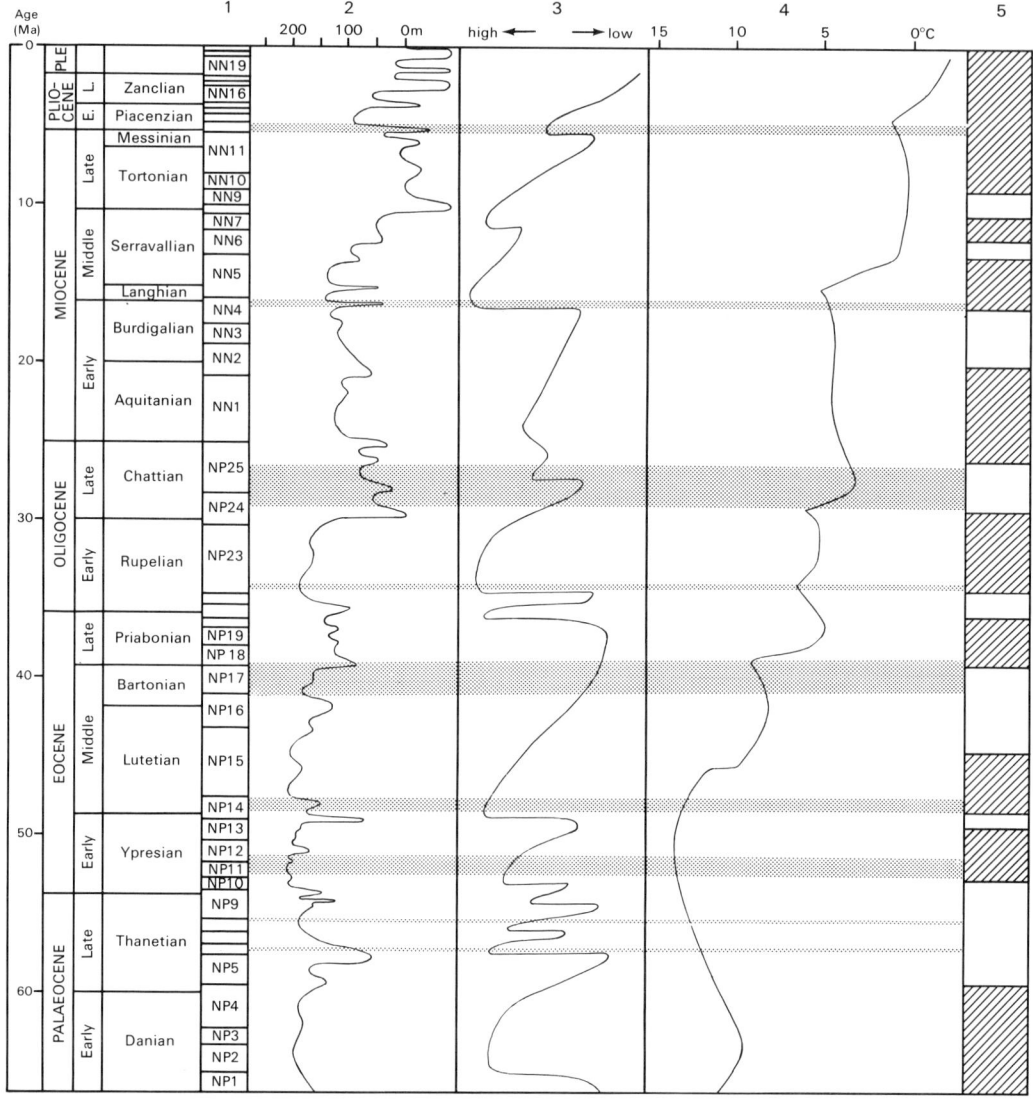

Fig. 1. Tertiary stratigraphy, sea-level, temperature and North Sea phosphorite concretion occurrences. Horizons bearing phosphorite concretions shown by horizontal stippled bands. Time scale and stratigraphy from Haq *et al.* (1987).
1, standard nannoplankton zones (Martini 1971); 2, eustatic sea-level curve (Haq *et al.* 1987); 3, North Sea sea-level (after Kockel 1988*b*); 4, isotopic palaeotemperature data for Tertiary benthic foraminifera (after Savin 1977). Temperature scale calculated assuming water $\delta^{18}O$ values of -1.00 per mil. 5, periods (shaded) of open connection between the North Sea and North Atlantic, as indicated by planktonic foraminifera. (After King 1989 and pers. comm.).

episode, occurred. The sea-level and temperature curves have been drawn to correlate with the nannoplankton stratigraphy and the geochronology and stage stratigraphy of Haq *et al.* (1987).

Some discrepancies will, however, continue to exist with data compiled from a variety of sources where the primary correlation of the data may be unknown or tied to a different standard stratigraphy, as for instance Savin's (1975) temperature data, which were plotted against the planktonic foraminifera stratigraphy and the geochronology of Berggren & Van Couvering (1974). By adopting a policy of correlation to a single standard stratigraphy, as in this paper, it is hoped to minimize such discrepancies.

The North Sea during the Tertiary

The North Sea Basin has undergone a long and complex evolution. A number of genetically different basins have occupied the area at various times. During late Cretaceous times an extensive shallow sea covered most of the area presently occupied by the North Sea and the British Isles. The British Isles became an emergent land area at the close of the Cretaceous and since that time the North Sea Basin has contained a semi-enclosed, epicontinental sea (Fig. 2). The geographical extent of the North Sea varied through the Tertiary but has occupied the same general area as at present. The palaeolatitude of the southern North Sea Basin was approximately 44°N at the start of the Tertiary (Scotese *et al.* 1988) and has generally moved northwards to its present position of approximately 52°N.

At the onset of seafloor spreading in the northern Atlantic Ocean during the Palaeogene the Mesozoic rift systems of the North Sea were already inactive. The Tertiary development of the North Sea Basin has been characterized by regional subsidence that is, in part, still continuing. This subsidence has allowed deposition of over 3000 m of sediments in the centre of the basin (Fig. 3) which are dominated by clastic shelf deposits with influxes of thick clastic fan deposits into a subsiding Central Graben, particularly in the central and northern areas. Minimum water depths of 300–500 m have been proposed for the Central Graben area during the Palaeogene (Lovell 1986). The southern end of the basin by contrast has been more tectonically stable with the development of a sequence of shallow marine shelf deposits deposited in <200 m water depth. The sequences in the southern North Sea record a succession of transgressive cycles periodically inundating this area. Connection with the Norwegian Sea to the north was probably maintained throughout the Tertiary. Connection between the cold waters of the Norwegian Sea and the progressively opening North Atlantic was prevented by a volcanic ridge between the Faeroe Islands and Iceland until mid-Miocene times.

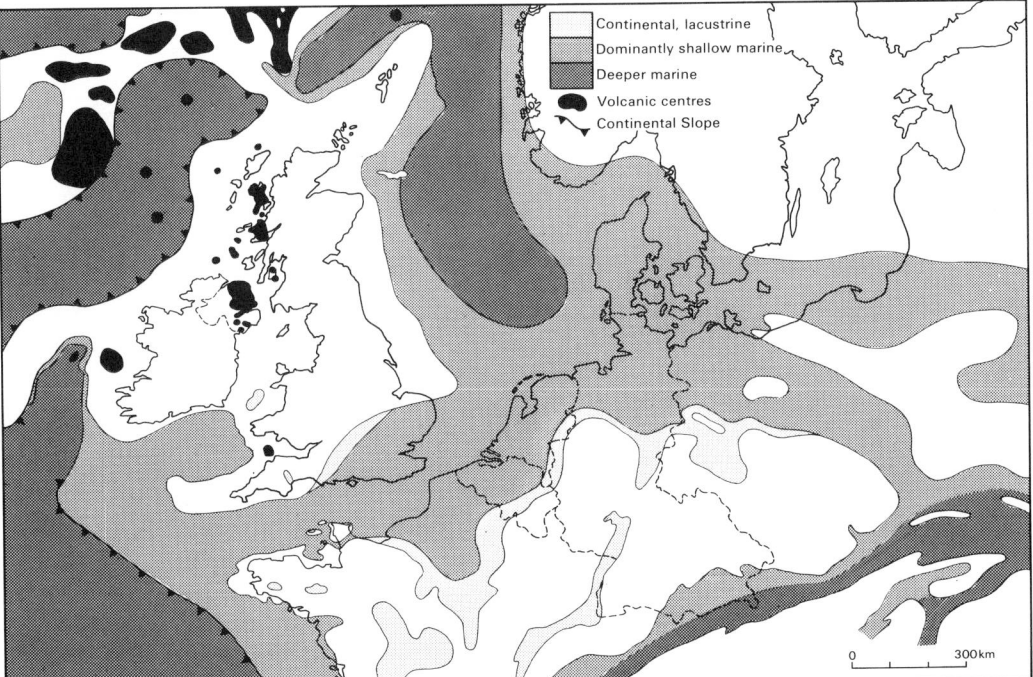

Fig. 2. Palaeogeography of northwest Europe during the Palaeocene–Eocene (modified after Ziegler 1982).

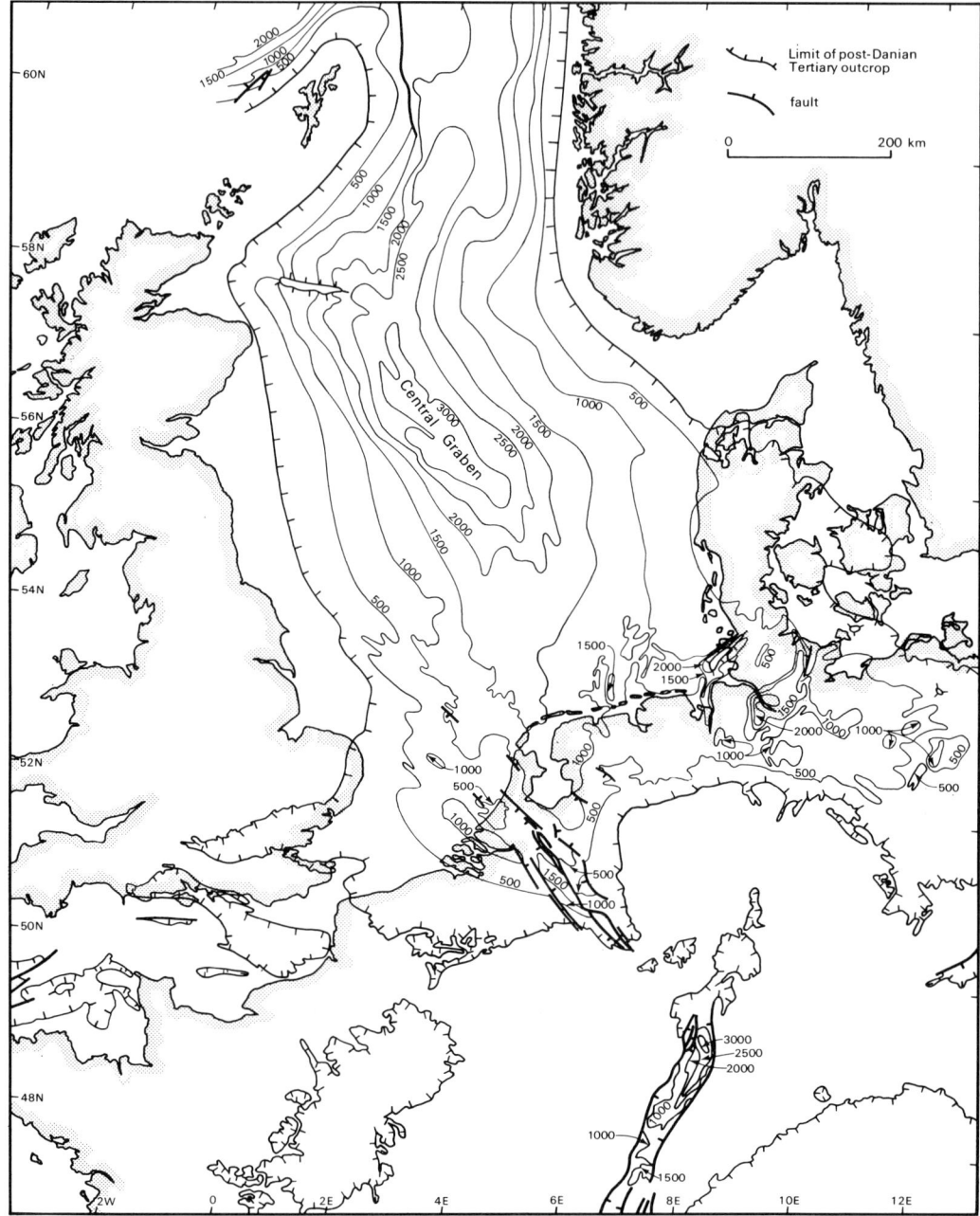

Fig. 3. Structural contour map on the base of the post-Danian Tertiary, contours in metres below sea-level. (after Kockel 1988a).

At the southern end of the North Sea connection was made periodically during periods of high eustatic sea level with the warmer waters of the North Atlantic through the area presently occupied by the English Channel. A marine connection may also have existed towards the east from time to time. Thus, for most of the Tertiary the southern North Sea was probably a semi-enclosed embayment with only a remote connection to oceanic waters.

Phosphorite concretions

Phosphorite deposits are found at numerous horizons within the Tertiary of the southern North Sea Basin. These deposits occur largely as pebbly lag deposits on unconformity planes. A deposit of this type is found at the base of the Late Pliocene Red Crag Formation in eastern England which was exploited as a source of phosphates for fertiliser manufacture as early as 1843. These deposits formed by the winnowing of marine formations containing authigenic phosphorite concretions (Balson 1980, 1989). The ages of these reworked deposits do not reflect the age of phosphogenesis, therefore, and these deposits are not considered in this paper.

The in situ concretions are composed of carbonate fluorapatite (francolite), the stable form of apatite in seawater (Nathan & Sass 1981), and are commonly between 2 and 10 cm in size although some may reach over 30 cm. Their stratigraphic distribution is shown in Fig. 1. They are found within marine muds, silts and muddy sands which may be extensively bioturbated and contain rich benthic faunas. Many originated by growth around organic nuclei. The concretions may be spherical or ellipsoid, particularly where they have grown around an obvious nucleus, or may be elongate or irregularly shaped. The concretions grew authigenically within the sediment but probably at very shallow depth. Those concretions which formed within mud-rich sediments may contain pyrite-filled shrinkage cracks. Some of the concretion-bearing deposits are conspicuously glauconitic, particularly those of Late Oligocene and Neogene age. The mid-Miocene Antwerp Sands of Belgium contain abundant glauconite which composes up to 80% of the sediment in parts. Usually the concretions are decalcified and only moulds of molluscan shells are seen. However shell material (including primary aragonite) is occasionally preserved. In some Early to mid-Miocene concretions from the Netherlands shell preservation is the rule rather than the exception.

Factors controlling distribution of phosphorite concretions

The occurrence of francolite concretions within the Tertiary sequence may be controlled by a variety of factors. These factors can be considered under three headings: (1) local sedimentary environment; (2) regional setting and (3) global environment.

(1) Local sedimentary environment

The supply of phosphate to the pore waters is probably the limiting factor in controlling francolite precipitation. Phosphate is fixed from sea-water primarily by phytoplankton which, upon death, may become deposited on the sea bed. The activity of other scavenging organisms within the water column and on the sea-floor will be important in determining the extent to which organic material becomes buried within the sediment and becomes available for diagenetic reactions within the porewaters. Organic matter is generally the only reducing agent to be buried with the sediment and thus is of fundamental importance in these reactions. Once buried the organic matter is degraded, initially by microbial processes, and phosphate is released. Organic matter degradation reactions strongly influence the pH of sediment porewaters and almost all authigenic mineral precipitation/dissolution reactions are pH sensitive (Curtis 1987). In general precipitation of apatite within the sediment appears to be favoured by conditions of low pH (Coleman 1985). This may help to explain the observation that the majority of the North Sea Tertiary phosphorite concretions which enclosed shells of calcite and aragonite preserve only moulds. However in some rare examples aragonitic shell material may be preserved. It is believed that the optimum site for precipitation of authigenic francolite will be under suboxic conditions at shallow depths within the sediment (Coleman 1985) and where the bottom waters are oxic (Baturin & Savenko 1985). Slightly acidic conditions may arise from reactions with H_2S diffusing upwards from the sulphate reduction zone below (Curtis 1987). Glauconite precipitation may also be favoured by slightly acid conditions within suboxic sediments (Coleman 1985) where there is a sufficient supply of iron.

The production and maintenance of suitable suboxic conditions will depend, in turn, on the sedimentation rate, the activities of burrowing organisms and the porosity/permeability of the sediment. Sedimentation rate will affect residence time of reactive organic matter within any particular diagenetic zone within the sediment. If the rate is too slow, all the organic matter may be oxidized and eliminated from the sediment before burial. If the rate is too fast, rapid burial takes sediment quickly to the zone of microbial methanogenesis where organic matter degradation is least intensive.

Many of the concretions are seen to have formed around organic nuclei. These nuclei include vertebrate remains (particularly teeth and bones of fish but occasionally also of cetaceans), crustaceans, molluscs and invertebrate burrows. Vertebrate skeletal remains are composed primarily of apatite as is the enclosing concretion. Crustacean carapaces consist of high-Mg calcite (Morrison & Brand 1986) but with relatively high phosphate content (Clarke & Wheeler 1922) compared with other invertebrates. Molluscs and detritus-filled burrows represent sites of organic matter concentration within the sediments. However, the majority of the concretions studied have no obvious nucleus. Furthermore, concretions do not form around every available potentially suitable nucleus. It would appear, therefore, that the composition or mineralogy of a nucleus, where present, is not a factor controlling concretion formation.

Benthic faunal evidence indicates that the sea-bed was oxic. This contrasts strongly with areas of modern phosphorite formation where the decomposition of micro-organisms descending from highly productive surface waters may lead to anoxia on the sea-bed (e.g. Burnett et al. 1980). Off eastern Australia, however, phosphorite concretions are presently forming in shelf areas associated with benthic faunas (O'Brien et al. 1990). It is possible that periodic anoxia occurred within high productivity areas in the Tertiary North Sea allowing burial of organic matter. It also seems possible that the activities of burrowing organisms may be important in burying the necessary organic matter in the sediment. In fine-grained sediments, where the oxic layer is generally thin, perhaps only a few mm thick, the infauna will burrow into the suboxic or anoxic sediments below. Indeed, the death of members of the infauna within burrows may in itself represent a mechanism for the transport of organic matter into the sediment. Large infaunal molluscs are commonly found to be articulated within phosphorite concretions indicating that they died in situ within the sediment. Bioturbation also serves to increase the residence time of the concretion within the suboxic conditions necessary for growth (O'Brien et al. 1990).

It is clear, however, that the organic matter of the mollusc alone is not sufficient to provide the required PO_4^{3-} to produce the concretion. Phosphorite concretions in the North Sea Basin Tertiary generally contain between 35 and 44% PO_4^{3-}, equivalent to between 26 and 33% P_2O_5. Organic matter containing approximately 3% PO_4^{3-} represents an enrichment by a factor of 140 000 with respect to sea-water concentration but a further concentration by a factor of 10–15 is required for the formation of the phosphorite concretion (Bentor 1980) and therefore additional PO_4^{3-} must diffuse laterally or vertically towards the concretion site.

The mineralogy of the host sediments has been thought to be connected with the formation of phosphorite sediments. A correlation with sediments rich in montmorillonite (smectite) was observed by Riggs (1979) and an association with illite was proposed by Weaver & Wampler (1972). In the North Sea Tertiary, smectite is the dominant clay mineral in many Palaeogene sediments associated with Palaeocene–Early Eocene volcanism and subsequent weathering of ash-bearing sediments, but in general the smectite content decreases and illite and kaolinite increase upward through the Tertiary (Nielsen 1988) and there is no clear relation to the occurrence of phosphorite concretions. Several of the concretion horizons are not associated with clay-rich sediments but have formed within relatively sandy deposits. These sands are often extremely glauconitic. An association between phosphorites and glauconite has often been noted (e.g. Odin & Letolle 1980; Burnett 1980). However, not all concretion horizons in the North Sea Tertiary are notably glauconitic.

(2) Regional setting

Regional or basin-wide factors which directly affect the local sedimentary environment may determine when and where phosphorite concretions would have formed in the Tertiary North Sea. Such factors include sea-level, basin circulation, regional volcanism and palaeolatitude (climate).

Sea-level in the North Sea Basin has shown pronounced fluctuations throughout the Tertiary (see Fig. 1) some of which may have been caused by the pattern of regional subsidence within the North Sea Basin. Comparison with the eustatic sea-level curve of Haq et al. (1987) shows a fair correlation with that for the North Sea determined by Kockel (1988b) indicating that the majority of the inferred sea-level changes may be attributed to global eustatic sea-level changes (see below) rather than basin subsidence and regional factors.

Changes in sea-level within a semi-enclosed basin will affect not only the geographical extent of the basin and the distribution and types of marine facies but will also have an effect on the size and nature of the basin's connection with the open ocean and in turn on the circulation pattern established in the basin. The clearest example of the effect of rising sea-level on the distribution of phosphorite concretions

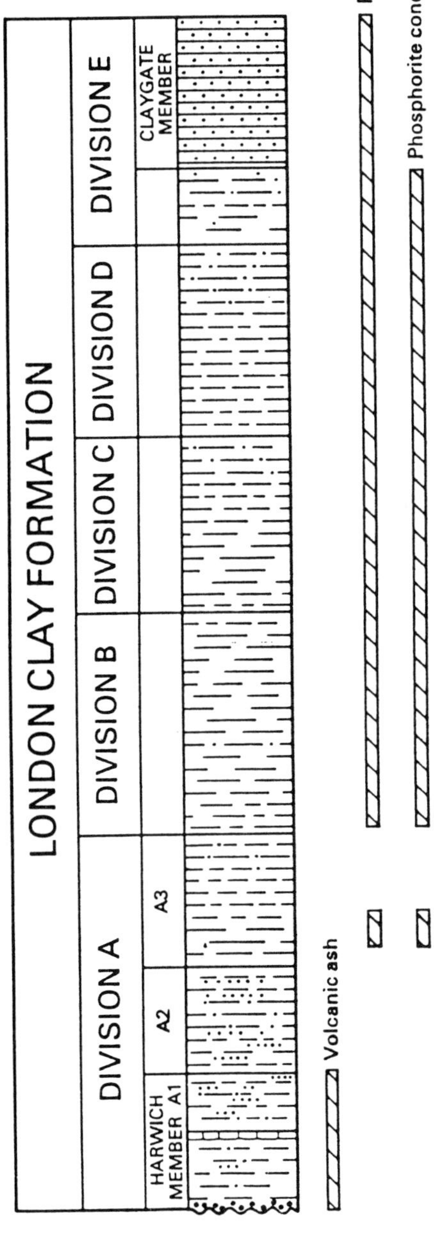

Fig. 4. Composite stratigraphy of the London Clay Formation (Late Palaeocene–Early Eocene) in eastern England (modified after King 1981) showing range of phosphorite concretion occurrences. Thickness represented: approximately 134 m.

within the North Sea Basin is seen in the Late Palaeocene–Early Eocene. This time period is represented by the London Clay Formation in southeastern England and the equivalent Ieper (Ypres) Clay Formation in western Belgium and northwestern France. These formations consist of a *c.* 150 m thick sequence of marine muds (Fig. 4) deposited under subtropical conditions.

The earliest sediments, Division A1 (the Harwich Member), of late Palaeocene age, contain many layers of argillized volcanic ash deposited under shallow marine conditions. At this time the basin was probably open only to the Norwegian Sea to the north. As the transgression proceeded and water depths increased a connection was initiated with the North Atlantic to the west either through the area presently occupied by the English Channel or via a more remote route around the north of the British Isles. This connection resulted in a major influx of planktonic foraminifera into the southern North Sea forming a 'planktonic datum' within the sediments at the base of division B of Fig. 4. Water depth reached a maximum of more than 100 m in division B and the lower part of division C (King 1981, 1984) followed by a gradual regression terminating with shallow water, tidally-influenced sediments (Claygate Member).

Phosphorite concretions within the London Clay Formation occur exclusively after the 'planktonic datum' and before shallower conditions returned. A brief influx of planktonic foraminifera occurs below the planktonic datum and within division A3 of Fig. 4 and persist only for 1–2 m of the sequence. Phosphorite concretions are not known from the London Clay at this level but within equivalent sediments of the Ypres Formation of northern France phosphorite concretions are found within this stratigraphic interval and are absent immediately above and below. There appears to be therefore a very clear correlation between rising sea-level, the influx of planktonic foraminifera, and the formation of diagenetic phosphorite concretions within the sediments.

Evidence of intermittent connection between the southern North Sea and the North Atlantic comes also from the stratigraphic distribution of *Nummulites*. *Nummulites* is a large benthic foraminiferid characteristic of the Tethyan faunal province. The first appearance of Nummulites in the North Sea Tertiary occurs in late Ypresian sediments (Blondeau 1972) equivalent to NP11 (C. King, pers. comm., 1989). A second influx occurs in Lutetian times (NP14/NP15) before a barrier in the late Lutetian prevented further northward migrations. Late Eocene *Nummulites* in the southern North Sea evolved in a basin separated from the Atlantic Ocean by the closure of this western route (Blondeau 1972).

In the Netherlands the Early–Middle Miocene Aalten Member (Miste Bed) contains abundant phosphorite concretions, and vertebrate material including cetacean bones and teeth of the shark *Carcharodon megalodon* (Van den Bosch *et al.* 1975). A similar fauna from the central part of the Antwerp Sands sequence of equivalent age in Belgium is also associated with phosphorite concretions. The fauna indicates a subtropical climate, high productivity, shallow marine conditions and an open connection with the Atlantic Ocean (Van den Bosch *et al.* 1975).

Fig. 1 shows a clear correlation between the observed phosphorite concretion occurrences and times of high sea-level in the North Sea. Changes in sea-level and marine connections with the Atlantic Ocean will have a consequent effect on the pattern of currents within the basin that may also have had an effect on the siting and occurrence of phosphate-rich sediments.

One apparent exception to the correlation of phosphorite concretions to high sea level in Fig. 1 is due to the previously uncertain assignation of certain marine concretion-bearing sediments of Bartonian age in Federal Germany which would imply an additional transgressive peak should be added to Kockel's (1988b) curve.

Local volcanism may have an effect in increasing the input of PO_4^{3-} either directly from ash falls or indirectly through the continental weathering of ashes and transport of phosphates into the marine basin by rivers. Volcanism, as already mentioned, was extensive during the late Palaeocene–early Eocene to the north and west of the British Isles (Fig. 2). This volcanicity was related to the opening of the North Atlantic and occurred during two phases resulting in the formation of over 200 thin ash layers within the North Sea Tertiary sequence (Knox & Morton 1988). These ashes, however, originate from tholeitic magma which is not especially phosphorus-rich. More P-rich alkaline basalts were produced from volcanic centres in the Rhine Graben area later in the Tertiary but these did not result in widespread ash falls. The volcanic episodes apparently do not correlate with episodes of phosphorite concretion formation. For example, concretions in the Palaeocene–Early Eocene London Clay Formation are present only in beds higher in the sequence than the ash layers.

Palaeolatitude has been thought to have an influence on phosphogenesis (Sheldon 1964, 1980) in so much as it will affect the local climate of an area. Whilst it may be true that most major phosphorite deposits formed at low palaeolatitudes, other factors are involved, such as the pattern of global climate and ocean circulation, which override the importance of palaeolatitude.

(3) Global environment

Global factors which may control the distribution of phosphorites in time include global climate, eustatic sea-level changes, changes in oceanic circulation and sea-water chemistry. A correlation between phosphorite formation and eustatic high sea-levels has been noted by many authors (e.g. Arthur & Jenkyns 1981; Riggs 1984). Eustatic sea-level rises have the effect of inundating greater areas of epicontinental shelf creating the necessary sites for phosphorite development. In the case of the North Sea eustatic sea-level rise has the effect of increasing the size and number of connections between this semi-enclosed basin and oceanic waters. However, correlations between phosphorite formation and eustatic sea-level noted in continental margin areas which had continuous open connection to the ocean imply that some other factors, linked to eustatic sea-level fluctuations, control the observed episodicity in phosphogenesis. The North Sea, because of its semi-enclosed nature can only reflect this underlying episodicity when it is sufficiently 'connected' to the global system. Hence the phosphogenic episodes in the North Sea Basin are probably briefer and are associated with volumetrically much smaller deposits than those occurring on open continental margins, as can be shown by a comparison with the stratigraphic distribution of phosphorite sediments in the Neogene of the southeastern United States and adjacent continental shelf.

In North Carolina and the adjacent shelf areas, Neogene phosphorite sediments have also been correlated with high eustatic sea levels (Riggs 1984, Riggs et al. 1985). Two periods of phosphorite development are represented by the Pungo River Formation (Early–Middle Miocene) and Lower Yorktown Formation (Early Pliocene). A third, lesser episode of phosphorite formation occurred during the late Pleistocene (Riggs 1984). The Early–Middle Miocene episode, which correlates with a second order (Haq et al. 1987) sea-level high-stand, can be further subdivided into three phosphate-rich units which were believed to have been deposited during third order sea-level high-stands. The oldest of these was assigned to nannoplankton zone NN3, and the youngest to NN6 (Snyder 1988). The middle unit was assigned to latest NN4–NN5, and thus is of very similar age to some phosphorite concretion-bearing sediments in the Southern North Sea (Fig. 1). Within each of the sedimentary sequences deposited during these third order transgressive-regressive cycles it was found that phosphogenesis was most intense during the midstages of each transgression with carbonate sedimentation during the high-stand.

An indication of global temperature can be obtained from the isotopic composition of benthic foraminifera in the deep ocean which reflect bottom water temperatures. These bottom water temperatures should be indicative of surface water temperatures at high latitudes where dense cold water sinks to form bottom water (Savin 1977). A general cooling trend is evident throughout the Tertiary punctuated by a number of warmer peaks. A comparison of Tertiary palaeotemperature data and North Sea phosphorite concretion occurrences can be seen in Fig. 1. A good correlation exists between the occurrences of North Sea phosphorite concretions and peaks in the palaeotemperature curve.

Discussion

Phosphorite concretions are not randomly distributed within the Tertiary sequence of the Southern North Sea Basin. They occur only during specific relatively narrow intervals within marine sequences and are associated with other features diagnostic of open connections with the oceanic realm, e.g. planktonic foraminifera, cetaceans and, during the Miocene, the giant oceanic shark *Carcharodon megalodon*.

The episodes, within which the North Sea phosphorite concretions formed, seem to correspond well with those given by other authors for periods of global phosphogenesis, even allowing for differences in geochronologic scales. Cook & McElhinny (1979) cite 54 Ma as a peak in Palaeogene phosphogenesis (cf. Fig. 1). For the later Tertiary, Riggs (1987) lists peaks at 29–25 Ma, 19–13 Ma and 5–4 Ma. Phosphorite concretions formed in the North Sea Basin sequence during each of these time intervals (see Fig. 1).

Many of the concretions are associated with glauconite, particularly those of the Late Oligocene and Neogene. Glauconitic sediments also occupy relatively narrow time intervals although, as in the case of phosphorite occurrences, some may be reworked. Authigenic glauconitic sediments are known to be genetically associated with phosphorites (e.g. Odin & Letolle 1980; Burnett 1980).

Within the North Sea Basin certain time intervals are characterized by a dominance of non-calcareous agglutinated foraminifera within the microfaunas. Such assemblages may reflect basin circulation patterns (Gradstein & Berggren 1981). Assemblages of agglutinated foraminifera dominate in areas undersaturated with $CaCO_3$ and with low pH. Abundance of fine-grained organic-rich and carbonate-poor sediments will lower pH on the sea bed and in the uppermost part of the sediment column and may lead to the dissolution of calcareous benthic foraminifera or inhibit calcification of their tests. Gradstein & Berggren (1981) believed that such conditions existed at times of restricted bottom water circulation in the deeper part of the basin, particularly during the Palaeocene–Eocene.

The development of silled basins may lead to a restricted circulation below the level of the sill with a normal circulation in surface waters. This situation will be accentuated during periods of low sea-level. King (1989) noted a general inverse relationship between the stratigraphic distribution of assemblages of non-calcareous agglutinating foraminifera and assemblages of planktonic foraminifera, and attributed the variations as being controlled primarily by eustatic changes in sea-level. These changes resulted in an alternation of 'open' and 'closed' connections with oceanic environments. Data from King (1989, and pers. comm., 1989) has been used in fig. 1 to indicate periods of open connection with the North Atlantic Ocean as determined from study of foraminiferal assemblages. A good correlation is seen with phosphorite concretion occurrences in the Early Eocene, Middle Eocene and Middle Miocene but not in the Oligocene. The stratigraphy and microfaunas of the Late Oligocene, in particular, are not well constrained and require further study. The development of a silled basin and the resulting restricted circulation are regional factors which would not account for the observed correlations with episodicity in areas outside the North Sea Basin.

The process of upwelling of ocean water from beneath productive surface waters onto continental shelves is often cited in models of phosphorite formation either as coastal upwelling caused by longshore winds as in the case of the modern southwest African shelf phosphorites or dynamic upwelling caused by currents over sea-bed topographic irregularities as in the case of Neogene phosphorites in the southeastern USA (Riggs 1984). In surface waters phosphorus is generally completely used by phytoplankton and concentrations in seawater are low, of the order of a few ppb (Bentor 1980).

Upwelling brings fresh supplies of phosphate and other nutrients from water masses beneath the photic zone which can then stimulate enhanced productivity in the surface waters. These areas of enchanced productivity tend to be geographically restricted to the locus of the upwelling. In the North Sea Basin significant upwelling could only have occurred from the relatively deeper waters of the Central Graben area where water depths may have reached 300–500 m during the Palaeogene (Lovell 1986). By late Neogene times sedimentation had filled the Central Graben so that only shallow marine conditions existed (Gradstein & Berggren 1981) and upwelling of nutrients could not occur. The southern areas of the North Sea and the intermittent seaways through the English Channel have also been relatively shallow water areas throughout the Tertiary. Thus, any upwelling from the North Atlantic to the west would not have been able to affect productivity in the Southern North Sea area.

Bonde (1979) believed an extensive zone of coastal upwelling existed on the eastern side of the North Sea Basin in Palaeocene/Eocene times leading to the formation of a 60 m thick formation of diatomites. Bonde's upwelling model has been disputed by Pedersen & Surlyk (1983) who believe the diatomite to reflect very local environmental conditions. In any case this diatomite is of similar age to the ash-bearing Harwich Member of the London Clay Formation and therefore predates the Early Eocene occurrences of phosphorite concretions.

Upwelling is a factor which would not account for the observed correlations between phosphorite concretion occurrences in the North Sea Tertiary and the episodicity of phosphogenesis elsewhere, although upwelling intensity may also be episodic with more vigorous upwelling at certain times. These periods of enhanced upwelling would have occurred during sea-level low-stands and during the late Tertiary correlate with periods of polar glaciation (Riggs & Sheldon 1990). At such times the ocean may have withdrawn to a level below the outer edge of shallow continental shelves and therefore phosphorite development in the southern North Sea Basin would have been impossible. In the Tertiary most major periods of phosphogenesis appear to correlate with high sea levels, periods when upwelling intensity was relatively reduced (Fischer & Arthur 1977; Riggs & Sheldon 1990).

Conclusions

The episodicity of phosphogenesis and associated authigenic minerals such as glauconite within the North Sea Tertiary appears to reflect a larger-scale episodicity observed elsewhere during this period. This same episodicity exists even though the nature of the phosphatic sediments differs. A strong correlation is seen between the times of occurrence of phosphorite concretions and peaks of ocean bottom water temperature which requires further investigation. Local environmental factors, whilst important in providing the correct milieu for precipitation of these authigenic minerals, do not in themselves determine at what time they will form. Thus, at times favourable to phosphogenesis, concretions may form given appropriate local sedimentary conditions. Upwelling, although an important mechanism in production of some major Tertiary phosphorite deposits, was probably not a factor in the production of phosphorite concretions in the southern North Sea Basin.

I would like to thank C. King (Paleoservices Ltd.) and R. P. Sheldon (consultant) for constructive comments. The paper is published with the permission of the Director, British Geological Survey (NERC).

References

ARTHUR, M. A. & JENKYNS, H. C. 1981. Phosphorites and paleoceanography. *Oceanologica Acta*, SP, 83–96.

BALSON, P. S. 1980. The origin and evolution of Tertiary phosphorites from eastern England. *Journal of the Geological Society of London*, **137**, 723–729.

—— 1987. Authigenic phosphorite concretions in the Tertiary of the Southern North Sea Basin: An event stratigraphy. *Mededelingen van de Werkgroep voor Tertiaire en Kwartaire Geologie*, **24**, 79–94.

—— 1989. Tertiary phosphorites in the southern North Sea Basin: Origin, evolution and stratigraphic correlation. *In*: HENRIET, J. P. & DE MOOR, G. (eds) *The Quaternary and Tertiary geology of the Southern Bight, North Sea*. Belgian Geological Surrey Brussels, 51–70.

BATURIN, G. N. & SAVENKO, V. S. 1985. Mechanism of formation of phosphorite nodules. *Oceanology*, **25**, 747–750.

BENTOR, Y. K. 1980. Phosphorites – the unsolved problems. *Special Publication of the Society of Economic Paleontologists and Mineralogists*, **29**, 3–18.

BERGGREN, W. A. & VAN COUVERING, J. A. 1974. The late Neogene: Biostratigraphy, geochronology and palaeoclimatology of the last 15 million years in marine and continental sequences, *Palaeogeography, Palaeoclimatology, Palaeoecology*, **16**, 1–216.

BLONDEAU, A. 1972. *Les Nummulites*. Librairie Vuibert, Paris.

BONDE, N. 1979. Palaeoenvironment in the 'North Sea' as indicated by the fish bearing Mo-Clay deposit (Paleocene/Eocene), Denmark. *Mededelingen van de Werkgroep voor Tertiaire en Kwartaire Geologie*, **16**, 3–16.

BURNETT, W. C. 1980. Apatite-glauconite association off Peru and Chile: palaeo-oceanographic implications. *Journal of the Geological Society, London*, **137**, 757–764.

——, VEEH, H. H. & SOUTAR, A. 1980. U-series, oceanographic and sedimentary evidence in support of Recent formation of phosphate nodules off Peru. *Special Publication of the Society of Economic Paleontologists and Mineralogists*, **29**, 61–71.

CLARKE, F. W. & WHEELER, W. C. 1922. The inorganic constituents of marine invertebrates. US Geological Survey Professional Paper, No. 124.

COLEMAN, M. L. 1985. Geochemistry of diagenetic non-silicate minerals: kinetic considerations. *Philosophical Transactions of the Royal Society London*, Series A, **315**, 39–56.

COOK, P. J. 1984. Spatial and temporal controls on the formation of phosphate deposits – a review. *In*: NRIAGU, J. O. & MOORE, P. B. (eds) *Phosphate Minerals*, Springer-Verlag, New York, 242–274.

COOK, P. J. & MCELHINNY, M. W. 1979. A reevaluation of the spatial and temporal distribution of sedimentary phosphate deposits in the light of plate tectonics, *Economic Geology*, **74**, 315–330.

——, SHERGOLD, J. H., BURNETT, W. C. & RIGGS, S. R. 1990. Phosphorite research: a historical overview. *In*: NOTHOLT, A. G. & JARVIS, I. (eds) *Phosphorite Research and Development*. Geological Society, London, Special Publication, **52**, 1–22.

CURTIS, C. 1987. Mineralogical consequences of organic matter degradation in sediments: Inorganic/Organic diagenesis. *In*: LEGGETT, J. K. & ZUFFA, G. G. (eds) *Marine Clastic Sedimentology*, Graham and Trotman, London, 108–123.

FISCHER, A. G. & ARTHUR, M. A. 1977. Secular variations in the pelagic realm. *Special Publication of the Society of Economic Paleontologists and Mineralogists*, **25**, 19–50.

GRADSTEIN, F. M. & BERGGREN, W. A. 1981. Flysch-type agglutinated foraminifera and the Maestrichtian to Paleogene history of the Labrador and North Seas. *Marine Micropaleontology*, **6**, 211–268.

HAQ, B. U., HARDENBOL, J. & VAIL, P. R. 1987. Chronology of fluctuating sea levels since the Triassic. *Science*, **235**, 1156–1167.

KING, C. 1981. *The stratigraphy of the London Clay and associated deposits*. Tertiary Research Special Paper, No. 6.

—— 1984. The stratigraphy of the London Clay Formation and Virginia Water Formation in the coastal sections of the Isle of Sheppey (Kent, England). *Tertiary Research*, **5**, 121–160.

—— 1989. Cenozoic of the North Sea. *In*: JENKINS, D. G. & MURRAY, J. W. (eds) *Stratigraphical Atlas of Fossil Foraminifera*, 2nd edition, Ellis Horwood Ltd, Chichester. 418–489.

KNOX, R. W. O'B & MORTON, A. C. 1988. The record of early Tertiary N. Atlantic volcanism in sediments of the North Sea Basin. *In*: MORTON, A. C. & PARSON, L. M. (eds) *Early Tertiary Volcanism and the Opening of the NE Atlantic*. Geological Society, London, Special Publication, **39**, 407–419.

KOCKEL, F. 1988*a*. Base post-Danian Tertiary structural contour map. *In*: VINKEN, R. (compiler). *The Northwest European Tertiary Basin*. Geologisches Jahrbuch, Reihe A, **100**.

—— 1988*b*. Assumed basin-wide relative sea-level changes in the NW European Tertiary Basin. *In*: VINKEN, R. (compiler). *The Northwest European Tertiary Basin*, Fig. 35a, Geologisches Jahrbuch, Reihe A, **100**.

LOVELL, J. P. B. 1986. Cenozoic. *In*: GLENNIE, K. W. (ed.) *Introduction to the Petroleum Geology of the North Sea*, Blackwell Scientific Publications, London, 179–196.

MARTINI, E. 1971. Standard Tertiary and Quaternary calcareous nannoplankton zonation. *In*: FARINACCI, A. (ed.), *Proceedings of the Second Planktonic Conference, Rome 1970*, **2**, Edizioni Tecnoscienza, Rome, 739–785.

MORRISON, J. O. & BRAND, V. 1986. Geochemistry of Recent Marine Invertebrates. *Geoscience Canada*, **13**, 237–254.

NATHAN, Y. & SASS, E. 1981. Stability relations of apatites and calcium carbonates. *Chemical Geology*, **34**, 103–111.

NIELSEN, O. B. 1988. The clay mineralogy in the eastern and southeastern part of the Tertiary North Sea Basin. *In*: VINKEN, R. (compiler). *The Northwest European Tertiary Basin*, Geologisches Jahrbuch, Reihe A, **100**, 141–142.

O'BRIEN, G. W. & VEEH, H. H. 1983. Are phosphorite reliable indicators of upwelling? *In*: SUESS, E. & THIEDE, J. (eds) *Coastal Upwelling, its sediment record. Part A, Responses of the sedimentary regime to present coastal upwelling*, NATO Conference Series IV, marine sciences, *10A*. Plenum Press, New York, 399–419.

——, MILNES, A. R., VEEH, H. H., HEGGIE, D. T., RIGGS, S. R., CULLEN, D. J., MARSHALL, J. F. & COOK, P. J. 1990. Sedimentary dynamics and redox iron-cycling: controlling factors for the apatite–glauconite association on the East Australian continental margin. *In*: NOTHOLT, A. J. G. & JARVIS, I. (eds) *Phosphorite Research and Development*. Geological Society, London, Special Publication, **00**, 000–000.

ODIN, G. S. & LETOLLE, R. 1980. Glauconitization and phosphatization environments: a tentative comparison, *Special Publication of the Society of Economic Paleontologists and Mineralogists*, **29**, 227–237.

PEDERSEN, G. K. & SURLYK, F. 1983. The Fur Formation, a late Paleocene ash-bearing diatomite from northern Denmark, *Bulletin of the Geological Society of Denmark*, **32**, 43–65.

RIGGS, S. R. 1979. Phosphorite sedimentation in Florida — A model phosphogenic system, *Economic Geology*, **74**, 285–314.

—— 1984. Paleoceanographic model of Neogene phosphorite deposition, U.S. Atlantic continental margin, *Science*, **223**, 123–131.

—— 1986. Proterozoic and Cambrian phosphorites — specialist studies: phosphogenesis and its relationship to exploration for Proterozoic and Cambrian phosphorites. *In*: COOK, P. J. & SHERGOLD, J. H. (eds) *Phosphate Deposits of the World. Volume 1. Proterozoic and Cambrian phosphorites*, Cambridge University Press, 352–368.

—— 1987. Model of Tertiary phosphorites on the world's continental margins. *In*: TELEKI, P. G., DOBSON, M. R., MOORE, J. R. & VON STACKELBERG, U. (eds) *Marine Minerals. Advances in research and resource assessment*, NATO ASI series, Series C, Mathematical and Physical Sciences, **194**, 99–118.

—— & SHELDON, R. P. 1990. Paleoceanographic and paleoclimatic controls of the temporal and geographic distribution of Upper Cenozoic Continental Margin phosphorites. *In*: BURNETT, W. C. & RIGGS, S. R. (eds) *Phosphate Deposits of the World*, Vol. 3, *Neogene to Modern Phosphorites*, Cambridge University Press. (in press)

——, LEWIS, D. W., SCARBOROUGH, A. K. & SNYDER, S. W. 1982. Cyclic deposition of Neogene phosphorites in the Aurora area, North Carolina, and their possible relationship to global sea-level fluctuations, *Southeastern Geology*, **23**, 189–204.

——, SNYDER, S. W. P., HINE, A. C., SNYDER, S. W., ELLINGTON, M. D. & MALLETTE, P. M. 1985. Geologic framework of phosphate resources in Onslow Bay, North Carolina Continental Shelf, *Economic Geology*, **80**, 716–738.

SAVIN, S. M. 1977. The history of the earth's surface temperature during the past 100 million years. *Annual Review of Earth and Planetary Sciences*, **5**, 319–355.

SCOTESE, C. R., GAHAGAN, L. M. & LARSON, R. L. 1988. Plate tectonic reconstructions of the Cretaceous and Cenozoic ocean basins. *Tectonophysics*, **155**, 27–48.

SHELDON, R. P. 1964. Paleolatitudinal and Paleogeographic distribution of phosphorite. *U.S. Geological Survey Professional Paper*, **501-C**, C106–C113.

—— 1980. Episodicity of phosphate deposition and deep ocean circulation — a hypothesis. *Special Publication of the Society of Economic Paleontologists and Mineralogists*, **29**, 239–247.

—— 1981. Ancient Marine Phosphorites. *Annual Review of Earth and Planetary Sciences*, **9**, 251–284.

SNYDER, S. W. 1988. Synthesis of biostratigraphic and paleoenvironmental interpretations of Miocene sediments from the shallow subsurface of Onslow Bay, North Carolina Continental Shelf. *Cushman Foundation for Foraminiferal Research Special Publication*, **25**, 179–189.

VAN DEN BOSCH, M., CADÉE, M. C. & JANSSEN, A. W. 1975. Lithostratigraphical and biostratigraphical subdivision of Tertiary deposits (Oligocene-Pliocene) in the Winterswijk–Almelo region (eastern part of the Netherlands), *Scripta Geologica*, **29**, 1–167.

WEAVER, C. E. & WAMPLER, J. M. 1972. The illite-phosphate association, *Geochimica et Cosmochimica Acta*, **36**, 1–13.

YANSHIN, A. L. & ZHARKOV, M. A. 1986. Epochs and evolution of phosphate-deposition through geological time. *International Geology Review*, **28**, 390–401. [Translated from *Byulleten' Moskovskogo Obshchestva Ispytateley Prirody, Otdeleniye Geologii*, **61**, 7–19].

ZIEGLER, P. A. 1982. *Geological Atlas of Western and Central Europe*. The Hague: Shell Internationale Petroleum Maatschappij B.V.

Comparative geology and mineralogy of the southeastern United States and Togo phosphorites

S. J. VAN KAUWENBERGH[1] & G. H McCLELLAN[2]

[1] *Fertilizer Technology Division, International Fertilizer Development Center (IFDC), Muscle Shoals, Alabama 35662, USA*
[2] *Department of Geology, University of Florida, Gainesville, Florida 32611, USA*

Abstract: The physical and mineralogical characteristics of the southeastern United States and Togo phosphorite deposits are the result of both primary depositional factors and secondary diagenetic processes. The North Carolina phosphorites appear to be 'unaltered' or are in the first stages of alteration. Both the Florida and Togo phosphorites have resulted in the removal of mineral species (primarily carbonates), systematic decarbonatization of the carbonate-fluorapatite, the development of iron and aluminium phosphates, and clay mineral alteration profiles.

In sedimentary phosphate deposits the most important mineral groups are carbonate-fluorapatite (francolite), carbonates, clays, and quartz. They are also the most important from an economic viewpoint. Francolite is generally the only economically important phosphate mineral present. Carbonates, which are difficult to remove both technically and economically, dilute the ore, contribute impurities, and increase acid consumption during processing. Clays are a significant source of chemical impurities and often present serious disposal problems. Quartz, which is inert in most processes, dilutes the ore and may cause excessive wear of equipment. Hydrous forms of silica (opal) can form troublesome precipitates during processing.

Variations in mineralogy over a region of sedimentary phosphorite accumulation are the result of primary compositional differences and postdepositional alterations. The latter category includes diagenetic processes that result in changes in the chemical and physical properties of primary minerals or the formation of authigenic minerals. Postdepositional alteration may result in the removal of some species (for example, carbonates) or the formation of new minerals. Reworking of deposits can concentrate minerals with a higher specific gravity by winnowing action (for example, phosphates and heavy minerals).

Some examples of the geological and mineralogical variations that occur in phosphate deposits are presented using data from Togo and the southeastern United States. These deposits are relatively young (Eocene and younger) and are mainly unconsolidated. Older phosphate deposits which have been subjected to more intense alteration may not exhibit the variations found in these deposits. North Carolina marine phosphorite has either been preserved as it was deposited or is in the first stages of diagenetic alteration. Postdepositional processes have altered both the Florida and Togo deposits, removing undesirable mineral components and concentrating phosphate in minerable ore zones.

The Neogene phosphorite deposits of the southeastern United States extend along the Atlantic Coastal Plain from North Carolina to the center of the Florida Peninsula. The deposits form a large phosphogenic province (Fig. 1) which has been subdivided into the Carolina Phosphogenic Province and the Florida Phosphogenic Province (Riggs 1984). The identified resources of these vast deposits are estimated at approximately 7×10^9 tons of phosphate rock (Cathcart *et al.* 1984).

Mining in North Carolina is undertaken at the Lee Creek mine in the Aurora area on the south shore of the Pamlico River (Fig. 2). The geological section in the Aurora area consists of four formations (adapted from Crowson *et al.* 1985).

(1) *Croatan Formation* (Pleistocene). Marginal to marine clayey sand filling and covering a highly irregular erosion surface at the top of the Yorktown Formation (2−24 m thick).

(2) *Yorktown Formation* (Pliocene). Grey clayey fossiliferous sand burrowed by marine organisms. A weakly cemented carbonate zone separates the Upper and Lower Yorktown Formation (maximum of about 36 m thick in the Aurora area). The contact between the Lower Yorktown and Pungo

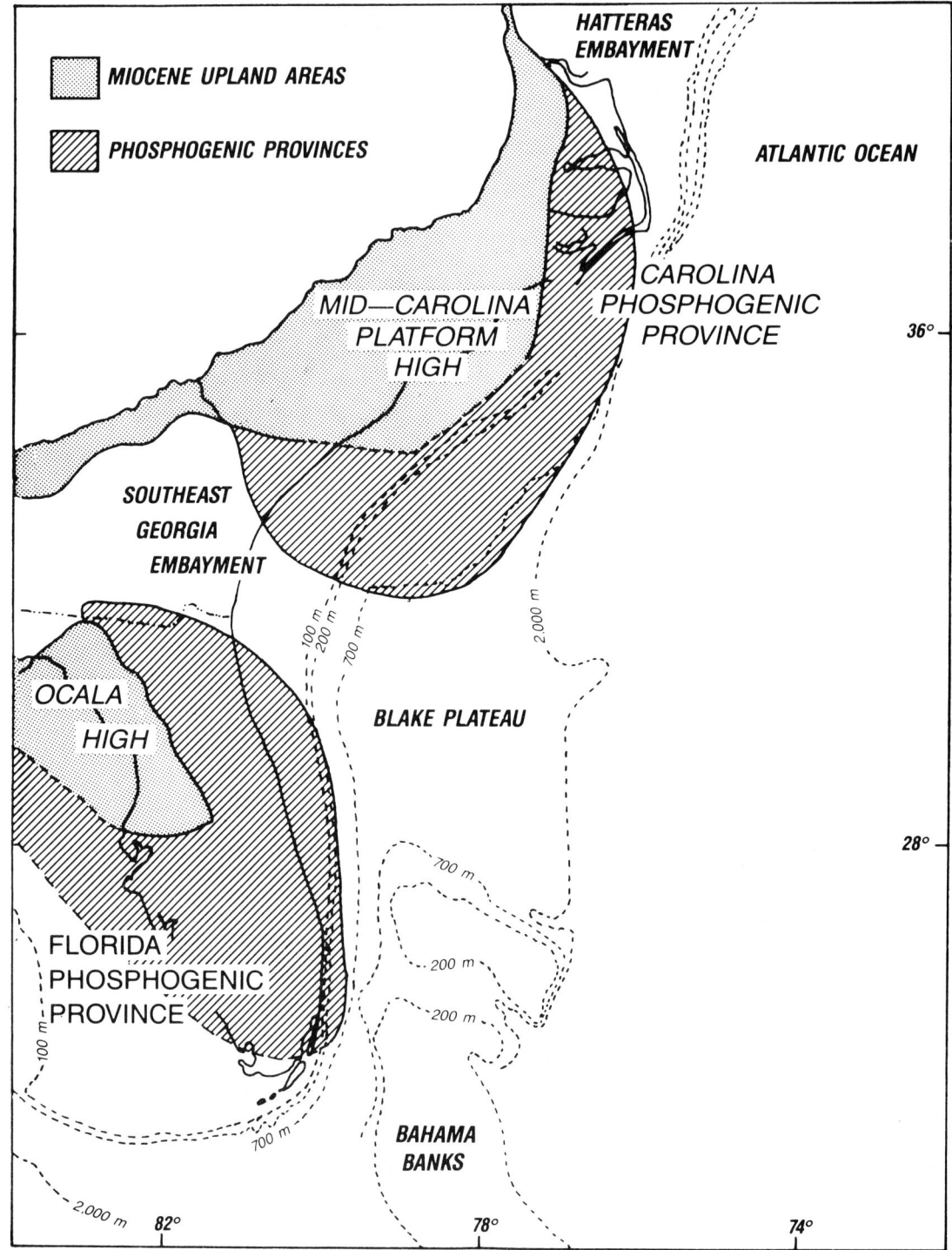

Fig. 1. Southeastern United States phosphogenic provinces (adapted from Riggs 1984).

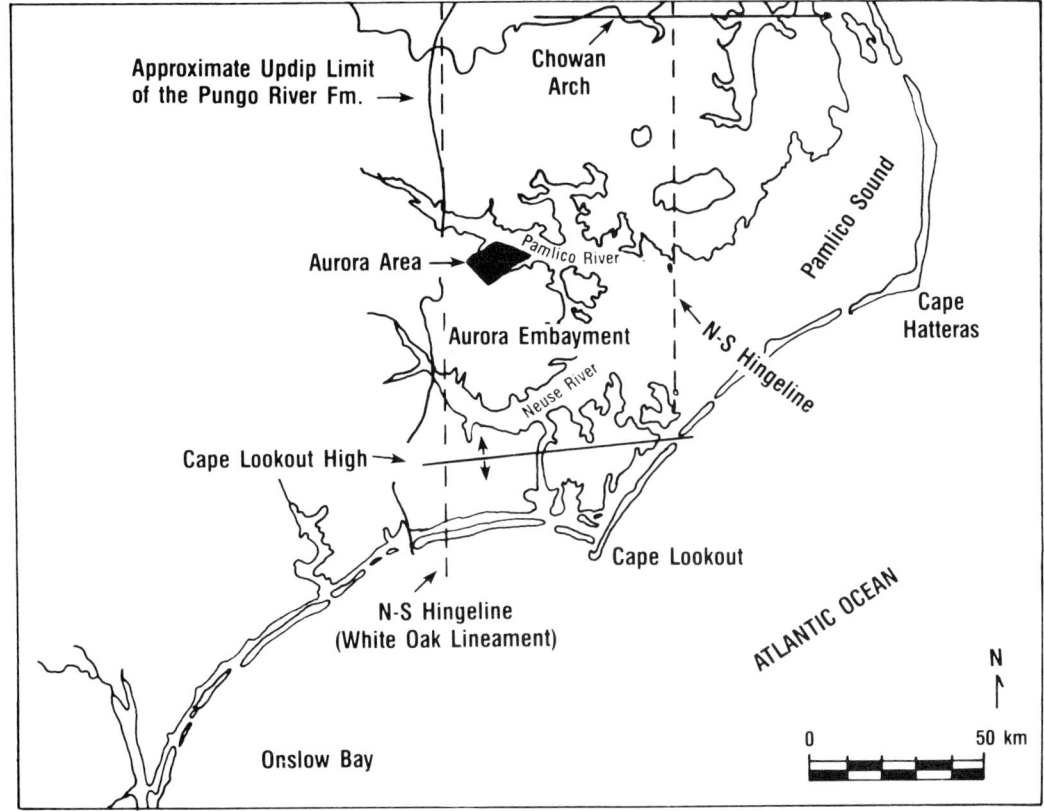

Fig. 2. Major structural features of eastern North Carolina and location of Aurora area (after Riggs *et al.* 1982).

River Formation is a major unconformity.
(3) *Pungo River Formation.* Pelletal Miocene phosphorite that has been divided by Riggs *et al.* (1982) into a number of informal units (from the base units A, B, C, and D) (12–37 m thick).
(4) *Castle Hayne Formation.* Eocene limestone separated by a sharp disconformity from the overlying Pungo River Formation.

The phosphatic Lower Yorktown and low-grade (<10% P_2O_5) upper portion of the Pungo River Formation are removed as overburden. The middle portion of the Pungo River Formation, the ore zone, is relatively thick (12–13 m) and homogeneous. Average ore grade of the economic zone is approximately 14.7% P_2O_5 (Redeker & Wright 1979). The low-grade (<10% P_2O_5) lower portion of the Pungo River Formation is left as an impermeable cap over the underlying Castle Hayne limestone aquifer.

Mining in Florida is concentrated in the areas known as the central Florida phosphate district and the southern extension (Fig. 3). A typical section in the central Florida district can be described as follows, using nomenclature commonly employed by geologists in the area (adapted from Cathcart 1985):
(1) Pleistocene — bedded loose to slightly clayey sand; some channel deposits of coarse phosphate at base (3–20 m thick);
(2) Upper Bone Valley Formation — Pliocene to Upper Miocene; clay and sand, minor phosphate, and Fe–Al phosphates where leached (0–6 m thick);
(3) Lower Bone Valley Formation — Upper Miocene; sandy clayey to coarse phosphorite at the base, abundant Fe–Al phosphates where leached (10–15 m thick);
(4) Hawthorn Formation — Miocene; upper clastic unit, clayey sand-silt, some coarse phosphate beds, dolomitic (6–30 m thick);
(5) Hawthorn Formation — Miocene; lower carbonate unit, dolomite with interbeds of

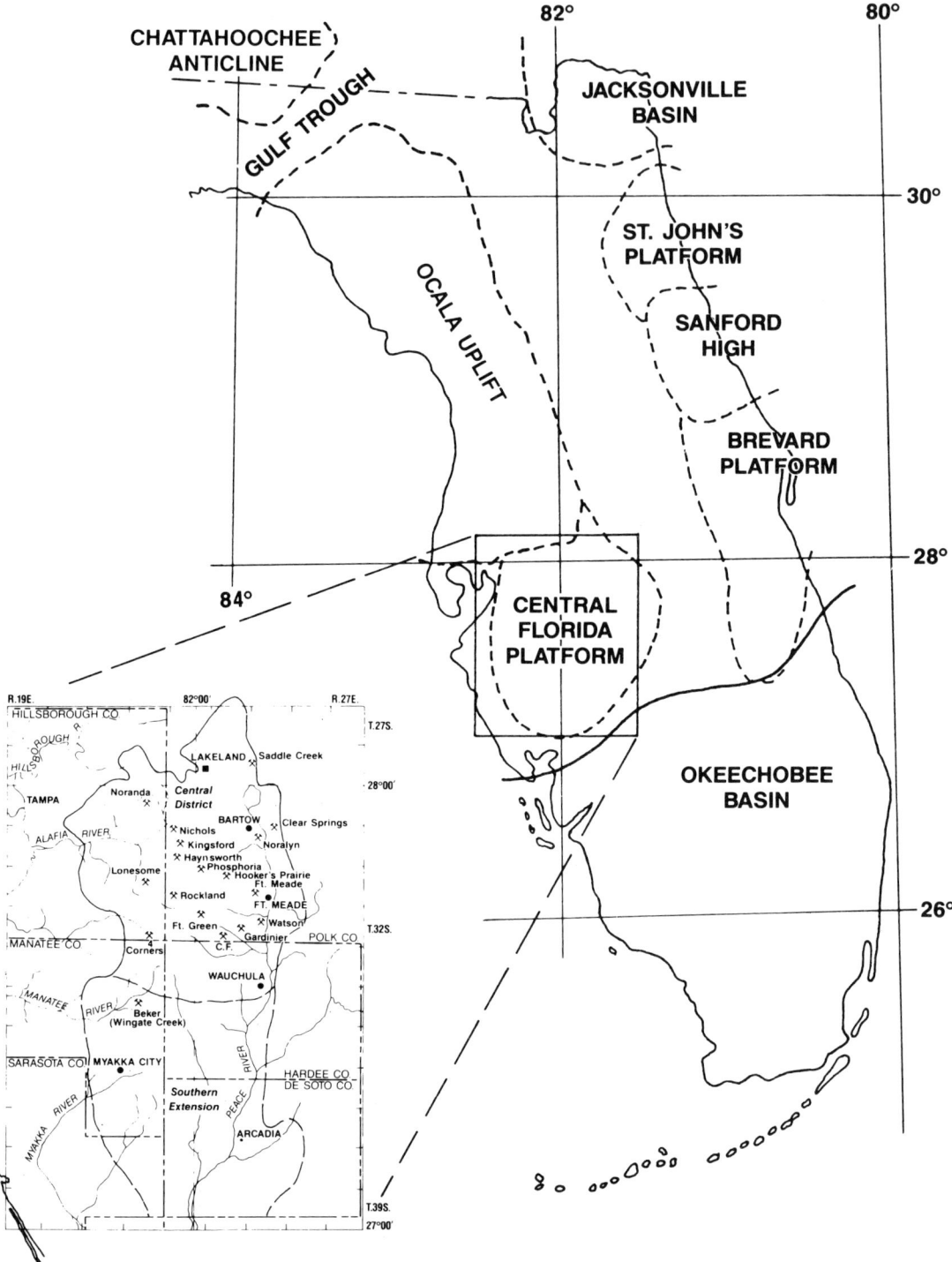

Fig. 3. Major structural features of Florida and location of central Florida phosphate district and its southern extension (adapted from Van Kauwenbergh *et al.* 1990).

clay and phosphorite (15- 30 m thick).

It should be noted that this is a description of a maximum section found in the central Florida district. In the southern extension, the Bone Valley pebbly phosphorite beds are thin or not present, and the Hawthorn Formation thickens downdip and becomes more dolomitic.

Pleistocene sand and the non-economic aluminophosphate zone at the top of the section are removed as overburden. An average of about 5 m of overburden is removed to mine an average of about 5 m of combined Bone Valley and Hawthorn Formations. In the central district the combined Bone Valley and Hawthorn Formation ore averages 14.5% P_2O_5 (Van Kauwenbergh & McClellan 1985). In this district, mining stops at the transition from the upper clastic unit of the Hawthorn Formation to the lower carbonate unit.

The Togo phosphate deposit occurs in a palaeogeographic coastal basin that extends from the southeastern corner of Ghana through Togo and Benin into Nigeria (Fig. 4). The deposit is intersected by a number of regional north-south trending faults. The phosphatic formations are of Lower−Middle Eocene age (Upper Ypresian−Lower Lutetian) and have a gentle southeasterly dip. These Eocene formations are overlain by a continental sequence of sand, gravel, and clay. The entire sequence is covered by a ferruginous lateritized top soil ('terre de barre').

Two areas favourable for mining occur in a 35 km by 2.5 km southwest to northeast strip between Avéta and Dagbati. The present mining is at Hahotoé and Kpogamé. A typical lithological sequence in this area, based on the nomenclature used by the Office Togolaise des Phosphates (OTP), is as follows:

(1) top soil − Overburden (2−17 thick);
(2) continental terminal − partially indurated continental sand, gravel, and clay (up to 22 m thick);
(3) Bed 0 − Eocene; phosphorite, kaolinitic with occasional pelletal phosphate stringers, usually rejected but sometimes mixed with Bed 1 (0−2 m thick);
(4) Bed 1 − Eocene; clayey phosphorite, the high-grade pelletal ore now being mined (0−6 m thick);
(5) Bed CSC− phosphatic limestone and marl, the lateral equivalent of Bed 1;
(6) bicoloured clay − marker bed (0−2 m thick);
(7) Bed 2/3 − Eocene; phosphatic bed consisting of marl, clay, and limestone containing thin high-grade beds of phosphate as coprolites, francolite pellets, and organic debris (2−8 m thick);
(8) palygorskite bed.

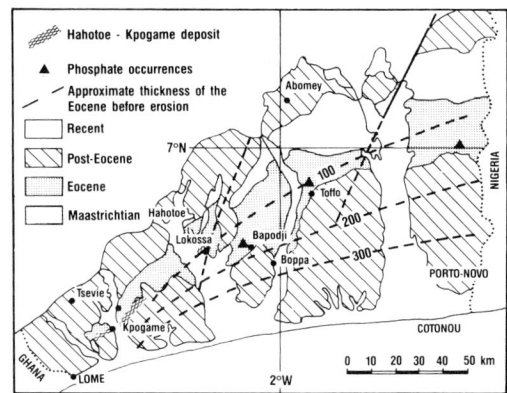

Fig. 4. Geology of the Togo−Benin coastal basin (after Slansky 1986).

One significant primary compositional difference between the Togo and southeastern United States phosphorites is the lack of sandy detrital sediments in the Togo deposit. The section is dominated by phosphate, clay, and carbonate. The Togo phosphate deposits are much higher in grade than the North Carolina or Florida deposits. Samples of Bed 1 from 15 cores in the Kpogamé area contain an average of 26.1% P_2O_5. Four samples from the noneconomic carbonate Bed CSC contain an average of 13.7% P_2O_5.

Deposits of the southeastern United States and Togo, although widely separated temporally and geographically, have a common ancestry in their marine origins. Phosphorites, in general, form from seawater although the mechanisms for formation and significant accumulation are a matter of continuing research and hypothesis. The present state of the deposits is the result of differing postdepositional histories.

One interpretation (Riggs 1984) associates the deposition of the North Carolina phosphorites with a major transgression of the sea during the Early to Middle Miocene (Fig. 5). Gulf Stream dynamics and palaeotopographic features controlled upwelling currents and the deposition of the phosphate. Palaeobathymetric estimates indicate deposition at depths of 80 m or less (Snyder et al. 1982). The Pungo River sediments were truncated by an extensive erosional episode that produced an apparent stratigraphic offlap pattern and a major unconformity with the overlying Yorktown Formation (Riggs et al. 1982). This erosional episode and depositional hiatus lasted approximately 7

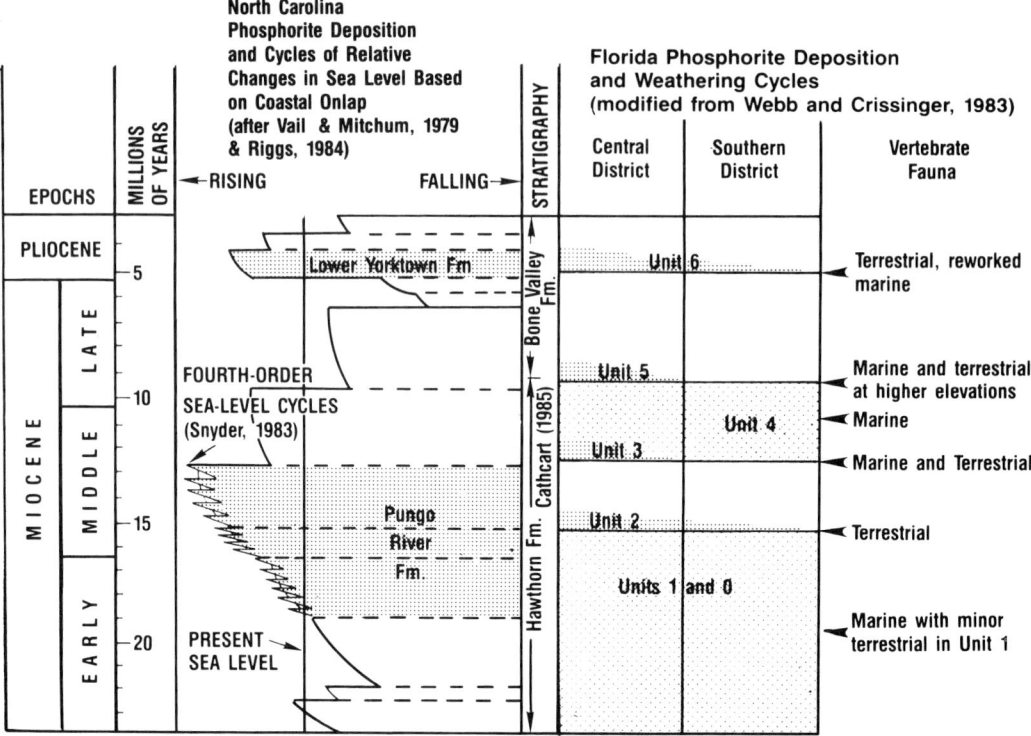

Fig. 5. Contrasts in the depositional histories of the North Carolina and Florida phosphorite deposits (adapted from Van Kauwenbergh *et al.* 1990).

million years.

Some evidence of reworking is found within the Pungo River Formation. Rooney & Kerr (1967) interpreted the phosphorites to have been reworked on a local scale. Evidence of reworking includes the presence of mixtures of light- and dark-coloured phosphate pellets in some samples; light-coloured pellets were oxidized by exposure on bars and beaches and then transported and redeposited in deeper water. More recently, Snyder *et al.* (1982) have interpreted the lowermost zones of the Pungo River Formation (Units A and B) as in situ accumulations. Unit C was transported from adjacent areas in the embayment or derived from the underlying units.

Deposition of the Florida phosphate deposits began in the Early Miocene. Currents upwelling from the Gulf of Mexico crossed the Florida peninsula and turned northward much like the modern Gulf Stream, depositing phosphate on the flanks of structural highs and behind a barrier of clastic sediments built by longshore currents from the north (Freas & Riggs 1968). Freas & Riggs (1968) proposed that the phosphate was initially deposited as an orthochemical phosphorite mud (microsphorite). The phosphorite muds were then indurated, possibly by periodic intertidal exposure, and broken up to form a high-energy inner interclastic phosphogenic belt (Riggs 1979). Fine-grained pelletal phosphorite was deposited in deeper water offshore. Whereas the North Carolina phosphorites appear to have been deposited continuously in a marine environment, the Florida phosphorites were affected by numerous episodes of subaerial exposure, weathering, and reworking (Fig. 5). Changes in characteristics from the central Florida district to the southern extension have been attributed to certain geological factors. The southern deposits are less intensively altered, and there is a transition in the depositional environment from nearshore to offshore conditions (Riggs 1979; Bernardi & Hall 1980; Cathcart 1985).

Much of the Bone Valley Formation is the product of reworking of the preexisting Hawthorn Formation (Cathcart 1963; Altschuler 1965). Furthermore, the Hawthorn Formation itself contains evidence of cycles of

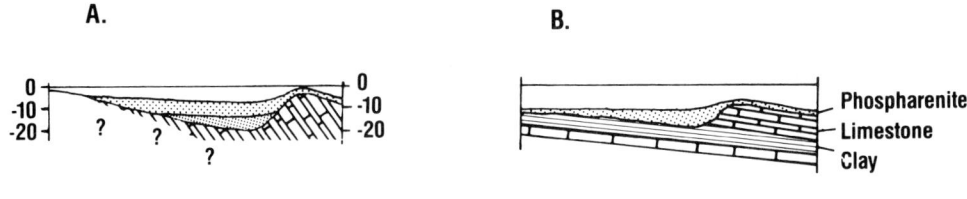

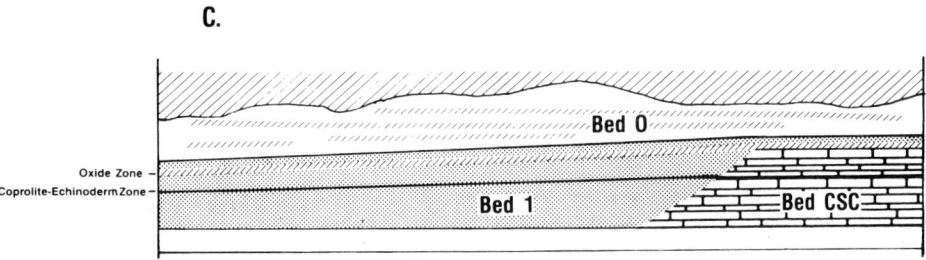

Fig. 6. Evolution of theories in formation of Togo phosphate deposit: (**A**) carbonate barrier and entrapment basin (Kilinc & Cotillon 1977); (**B**) bedded carbonate barrier and entrapment of the phospharenite (Slansky 1986); (**C**) lateral continuity of Bed 1 and the carbonate Bed CSC (Johnson 1987).

reworking and subaerial exposure. Replacement and/or leaching of preexisting phosphatic limestone or marl (Altschuler et al. 1964) with secondary recycling and reprecipitation (Altschuler 1965) has been proposed as an origin of the deposits. Riggs (1979) does not believe that replacement of calcite by phosphate played a major role in the formation of the Florida deposits, and he stresses the primary nature of phosphorite in this system. Although the origin of the phosphorite remains problematical, recent investigations have shown that the Florida deposits contain material in primary, altered, and mixed states (Van Kauwenbergh & McClellan 1985).

Kilinc & Cotillon (1977) proposed that the Togo phosphorite and underclays accumulated in a confined trap behind a carbonate barrier subjected to high-energy conditions (Fig. 6). Modifying this model, Slansky (1986) proposed a bedded carbonate (Bed CSC) as a limestone barrier. Johnson (1987) indicated that carbonate Bed CSC is the lateral equivalent of Bed 1 and described the transition in detail. Johnson proposed that phosphate was initially deposited during rising sea level with upwelling and high organic productivity at water depths of 15–50 m; the phosphatic sediments were then reworked and concentrated during a regressive phase and modified by postdepositional diagenetic processes.

Examination of five sets of drill cores from the mining area (unpublished IFDC work) and results presented in this paper indicate that a substantial portion of Bed 1 was derived from carbonate Bed CSC. The section appears to be partially condensed by chemical weathering (Fig. 7) that resulted in the concentration of phosphate at the top of Bed CSC and at its termination. Evidence of extensive reworking after subaerial weathering, as found in the Florida deposits, is not apparent in the Togo deposits. A conglomeratic zone found at the interface of Bed 2/3 and the palygorskite bed may indicate that some reworking has occurred.

Francolite

Variations in the crystallographic and chemical properties of francolites are among the criteria

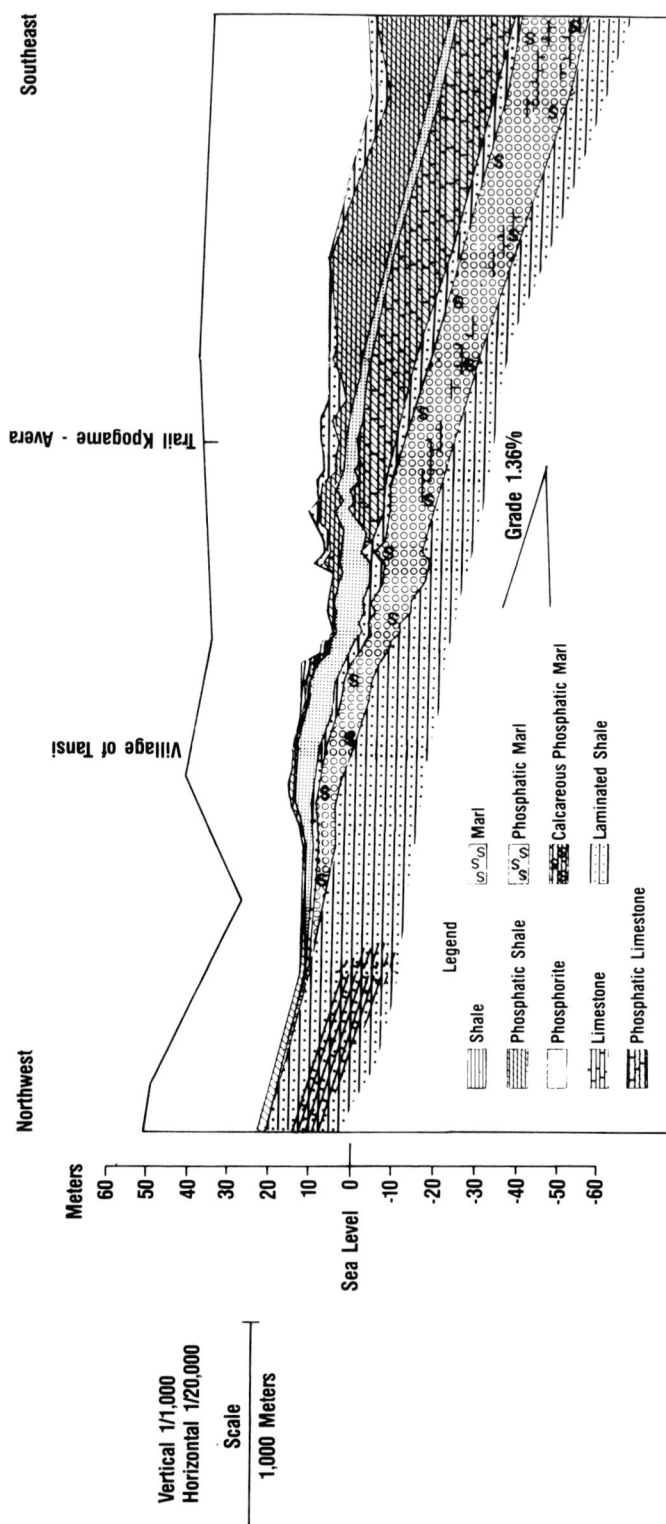

Fig. 7. Cross section of the Togo phosphorite deposit.

that can be used to analyse phosphorite deposits. A basic understanding of francolites is required to evaluate the variations within each deposit and the similarities and differences between the phosphorite deposits of the southeastern United States and Togo.

Previous work has established a series of systematic relationships between francolites (Lehr et al. 1967; McClellan & Lehr 1969; McClellan 1980; McClellan & Van Kauwenbergh 1990a). The results of these studies indicate that only six (Ca, Na, Mg, P, CO_2, and F) of the more than 25 chemical elements reported to occur in carbonate-fluorapatites are required to adequately describe francolites. Carbonate substitutes for phosphate in a 1:1 ratio, and the maximum amount of substitution is limited to about 6% CO_2. Similarly, Na^{1+} and Mg^{2+} substitute for Ca^{2+} within the structure and cannot be removed by physical beneficiation. Fluorine contents also increase with increasing carbonate substitution to compensate for charge imbalances.

These substitutions have been shown to be represented by a series with the following end-member empirical formula (McClellan 1980):

$$\text{Fluorapatite} = Ca_{10}(PO_4)_6F_2$$
$$\text{Francolite} = Ca_{10-x-y}Na_xMg_y(PO_4)_{6-z}(CO_3)_zF_{0.4z}F_2$$

An important economic result of these substitutions is the reduction of the maximum P_2O_5 content of a 100% francolite concentrate by as much as 20% when compared with the pure end-member fluorapatite (34.8% P_2O_5 in maximum substituted francolite compared with 42.2% in fluorapatite).

Crystal chemical variations in francolite compositions result in changes in the a and c unit-cell dimensions. These unit-cell dimensions are calculated from data collected by precise measurement of the d-spacings of francolite diffraction patterns. The a unit-cell dimension (hereafter referred to as the a-value) shows the greatest variability and can be used as an indicator of isomorphic substitutions (CO_3, Na, Mg, F, etc.) in the francolite structure. The range in francolite to fluorapatite chemical compositions with changes in the a-value, based on the McClellan (1980) model, are given in Table 1.

This model predicts that the minimum a-value for francolite is 9.320Å although a few samples have been measured that are slightly smaller than the predicted limit. As the table indicates, the lower the a-value the greater the amount of carbonate and impurity substitution in the francolite. Thermodynamic studies indicate that francolites with low a-values and high carbonate contents are metastable with respect to fluorapatite (Chien & Black 1976). Francolites in geologic systems alter from highly substituted forms with low a-values to forms approaching a fluorapatite composition with higher a-values (McClellan 1980). In theory, the lower the a-value the less altered the francolite.

Table 1. *Range in francolite chemical composition and changes in the a-value**

a-value	wt%					
	CaO	P_2O_5	F	CO_2	Na_2O	MgO
9.320	55.1	34.0	5.04	6.3	1.4	0.70
9.330	55.2	35.7	4.78	5.0	1.1	0.56
9.340	55.3	37.4	4.52	3.7	0.83	0.41
9.350	55.4	39.0	4.26	2.4	0.54	0.27
9.360	55.5	40.7	4.00	1.2	0.26	0.13
9.369	55.6	42.2	3.77	0.0	0.00	0.00

* Calculated from data in McClellan (1980).

The alteration of francolite is influenced by the presence of other minerals. Carbonates have been reported to increase the field of stability of francolite (Nathan & Sass 1981). When carbonates are present in the sediment, they buffer the pore water and prevent the alteration of the associated francolite to varieties containing less carbonate. Altschuler (1973) and Lucas et al. (1980) state that the carbonate must be dissolved before the francolite begins to change and/or disappear. In the absence of clays, the francolite becomes a less carbonated variety. When clays are present, decarbonation may be followed by the formation of iron and aluminium phosphate minerals.

Mineralogy

The general mineralogical variations of the North Carolina, Florida, and Togo deposits are listed in Tables 2, 3, and 4. The minerals, shown in the tables as 'common' minerals and clay minerals, are listed in the approximate order of decreasing abundance and importance. Although individual mine sections or beds in the deposits may exhibit differences, these tables represent the sequences commonly found in the deposits.

The North Carolina deposit contains a suite of minerals that may be primary (francolite, quartz, and calcite) or may have been formed by diagenetic processes (dolomite, clinoptilolite, and sepiolite). The distribution of carbonate, terrigeneous (primarily quartz), and

Table 2. *General mineralogy of the North Carolina deposit*

	Common minerals	Clays and zeolites
Pungo River Formation	Francolite Quartz Dolomite Calcite Opal Microcline (minor) Plagioclase (minor) Pyrite (minor)	Smectite Illite Sepiolite Clinoptilolite

phosphate minerals is one of the criteria used by Riggs *et al.* (1982) to differentiate the Pungo River Formation into various units. The top of each unit is marked by an increase in carbonate content and a relative decrease in phosphorite and clastic sedimentation. These variations are primary features.

Carbonates are found throughout the section in North Carolina. Calcite is the common carbonate in the formations overlying the Pungo River Formation. At the top of the Pungo River Formation, portions of Unit D are calcareous (Unit DD of Scarborough *et al.* 1982). Dolomite becomes the most common carbonate with depth in the central part of the Aurora Embayment and occurs as dispersed grains and indurated lenses. To the south and east, and downdip, these dolomitic phosphorites grade laterally into calcareous facies (Units BB and CC).

With the exception of the variation in carbonate species, the overall mineralogy is rather homogeneous. The clay fraction throughout the phosphorite section is dominated by smectite with minor illite. Kaolinite is present in the overlying formations but has not been identified in the phosphorite zones. Clinoptilolite can be found throughout the phosphorite section. Sepiolite has been identified in Unit C in the Aurora area (McClellan & Van Kauwenbergh 1990*b* and in the Upper and Lower Pungo River Formation in the Onslow Bay area (Lyle 1984).

Textural relationships observed in the scanning electron microscope (SEM) suggest that the minor amounts of sepiolite, clinoptilolite, and opal present are authigenic. Whether these minerals formed shortly after the deposition of the sediment or at some later time is a matter of conjecture. Miller (1971) proposed that calcareous Pungo River phosphorites were dolomitized by the movement of magnesium-bearing groundwater. Magnesium-bearing authigenic minerals may have formed during dolomitization.

The *a*-values of francolites from the Pungo

Table 3. *General mineralogy of the Florida phosphate deposits*

Formation	Common minerals	Clays
Bone Valley	Francolite Quartz Al-phosphates Fe-phosphates Microcline Dolomite (minor) Calcite (very minor)	Kaolinite Smectite Illite
Hawthorn	Quartz Francolite Dolomite Al-phosphates Microcline Albite (minor) Calcite (very minor)	Smectite Palygorskite Illite Kaolinite (very minor) Sepiolite (very minor)

Table 4. *General mineralogy of the Togo deposit*

	Common minerals	Clays
Bed 0	Francolite Al-phosphates Fe-phosphates	Kaolinite Smectite
Bed 1	Francolite Quartz (minor) Fe-phosphates (minor)	Kaolinite Smectite
Bed CSC	Calcite Francolite	Smectite Kaolinite (minor)
Bed 2/3	Calcite Francolite Quartz (minor)	Smectite Kaolinite Palygorskite (abundant at base)

River and Yorktown Formations are at the lower end of the francolite series (Table 5). Eleven samples of Pungo River concentrates (the processed −0.87 to +0.106 mm fraction) ranged from 9.318 to 9.332Å with an average of 9.324Å. There is not any significant vertical variation although some variability can be seen if samples from the various zones are broken down into size fractions. The larger size fractions contain the most highly substituted francolites, indicative of primary formation, near the lower limit of the francolite model (9.320Å).

The Florida deposits are more complex mineralogically than those of North Carolina. These complexities are the result of a more complicated geologic history and significant post-depositional alteration of the deposit.

Aluminophosphates (primary wavellite and crandallite) and some iron phosphates are common in the 'leached zone ore' overlying the commercial phosphate ore matrix (Altschuler et al. 1955). Wavellite and iron phosphates are primarily confined to the Bone Valley Formation while traces of crandallite may be found at the base of some producing zones in the Hawthorn Formation. Quartz and microline are ubiquitous in the sections. The finer grained portions of the Bone Valley Formation in the Central Florida district are devoid of dolomite; minor amounts of dolomite may be found in the coarse pebble fractions. Dolomite becomes common in the Hawthorn Formation as indurated lenses, dispersed grains, and inclusions in phosphate particles. Calcite is rarely detected in the mineable sections.

The crystallographic properties and compositions of francolite are highly variable in the Florida deposits. These variations can be demonstrated by examining data from selected sections of the Clear Springs, Fort Meade, and Four Corners mines (Table 6). The Clear Springs and Fort Meade mines are located within the central district where the Bone Valley Formation is best developed and alteration has been most intense. The Four Corners mine is located in the transition zone to the southern extension where alteration is less pronounced.

Data from the Clear Springs Mine show that francolite a-values increase with decreasing particle size and decrease with increasing depth and changing stratigraphy. The Bone Valley Formation at the top of the section does not contain dolomite. Dolomite is found in the Hawthorn Formation and increases in abundance to the base of the section. An alteration profile has developed; highly altered francolites are found at the top of the section and the more highly substituted francolites are found at depth. This is the most common type of alteration found in the central district when mine sections are deep enough so that the basal samples contain dolomite.

Francolites in the section from the Fort Meade mine exhibit increasing a-values with decreasing particle size and little change with stratigraphy or depth. Dolomite was not detected in this section. The francolites in both formations are altered in a similar fashion and to a similar extent.

Dolomite is present throughout the section from the Four Corners mine. The Bone Valley Formation at the top of the section contains francolites with low a-values and does not show the progression of higher a-values with de-

Table 5. Unit-cell a-values of North Carolina francolites*

	Number of Samples	Range	Average
Concentrates	11	9.318–9.332	9.324
Size fractions and vertical variation			
Formation or bed		Size (mm)	
		+0.87	−0.87 + 0.106
Upper Yorktown		9.321	9.330
Lower Yorktown		9.321	na[†]
Pungo River			
Upper Unit C		9.323	9.324
Pungo River			
Lower Unit B		9.317	9.329

* ±0.001 Å.
[†] Not analysed; francolite content below limits for reliable x-ray diffraction measurement.

Table 6. *Unit-cell a-values of selected mines in Florida**

Mine	Formation	Depth (m)		a-Values, Å† Size (mm)	
			+0.87	−0.87 +0.106	−0.106
Clear Springs	BV‡	4.6−9.6	9.353	9.353	9.361
	H	9.2−11.7	9.337	9.333	9.344
	H	12.9−14.1	9.338	9.334	9.337
Fort Mead	BV	5.8−8.9	9.328	9.332	9.344
	H	8.9−13.2	9.328	9.334	9.345
	H	13.2−15.4	9.328	9.332	9.343
Four Corners	BV	4.3−7.0	9.332	9.330	9.327
	H	7.0−10.9	9.324	9.332	9.333
	H	11.3−13.2	9.328	9.335	9.349

* Data from Van Kauwenbergh & McClellan (1985).
† ±0.003 Å.
‡ BV, Bone Valley Formation; H, Hawthorn Formation.

creasing particle size typical of mines in the central district. Higher a-value francolites in the section beneath lower a-value francolites indicate that a previous weathering and reworking event has been preserved.

The range of a-values observed in the Bone Valley Formation (9.327−9.361Å) and the Hawthorn Formation (9.324−9.349Å) in this data set almost spans the range of the francolite model (9.320−9.369Å). In general, the francolites of Bone Valley Formation at the top of the section in the central district are more highly altered. Downdip and into the Hawthorn Formation, the francolites are more highly substituted. Highly substituted francolites may occur stratigraphically above altered francolites, thus indicating the complexity of primary formation, subaerial exposure, weathering, and reworking. With reworking, altered francolites are mixed with less altered francolites and the resulting a-values of such samples vary in proportion with the compositions of the constituents.

The general sequence in the clay mineralogy of the economic Florida deposits, from top to base, is kaolinite ← smectite ← smectite + palygorskite (Altschuler et al. 1964). Kaolinite may be the only clay mineral present in the leached zone; in the less weathered portions of the deposit, smectite has been shown to be altering progressively to kaolinite (Altschuler et al. 1963). Minor palygorskite is associated with the coarse pebble fractions in the top of the section; it becomes common in the lower Hawthorn. This palygorskite has been postulated to have formed by alteration of preexisting smectite (Altschuler et al. 1964), by neoformation (Isphording 1973), by a combination of alteration and primary formation in perimarine environments (Weaver & Beck 1977) or in perialkaline lakes (Upchurch et al. 1982).

The Togolese phosphorite deposits exhibit many of the mineralogical characteristics of the deposits of the southeastern United States. In the Togo deposits, as in the Florida deposits, there are vertical zonations in the phosphate and clay mineralogy. Bed 0, at the top of the section, contains significant amounts of iron and aluminum phosphates. Iron and aluminum phosphates are not a significant component of the Bed 1 ore zone. At the top of the section, francolite a-values are similar to those of the most weathered zones in Florida (Table 7, Fig. 8). Analysis of francolite a-values from size fractions of a sample of Bed 1 indicates that the least altered francolites occur in the larger size fractions. Deeper in the section and in Bed CSC where carbonates are present, francolites are found with a-values similar to those of the less altered samples in the Florida sections as well as the material from North Carolina.

The Togo deposits are also similar to the Florida deposits in the vertical variation of clay mineralogy (Fig. 9). Kaolinite dominates the upper portion of the section and is replaced by smectite lower in the section. Although a portion of the kaolinite component may be detrital, this mineralogical sequence is probably the result of the low temperature alteration of smectite to kaolinite that has been demonstrated in Florida (Altschuler et al. 1963). Palygorskite at the base of the section may have formed as the result of neoformation (Millot 1970) or one or more of the processes proposed for the Florida deposits.

The two major differences between the Togo

Table 7. Unit-cell a-values of Togo francolites*

	Number of Samples	Range of a-values	Average a-values
Concentrates†	12	9.345–9.353	9.349
Vertical variation			
Bed 0	1		9.355
Bed 1	6	9.354–9.345	9.350
Bed CSC	2	9.333	9.333
Bed 2/3	6	9.324–9.333	9.328
Size fractions:			
Run-of-mine ore, Bed 1 (26% P_2O_5)		Size (mm)	a-values
		+0.3	9.345
		−0.3 +0.15	9.348
		−0.15 +0.106	9.354
		−0.106 +0.075	9.354
		−0.075 +0.038	9.353
		−0.038	9.354

* ±0.001 Å.
† All samples from Bed 1, the producing bed.

and southeastern United States phosphorite deposits are the lack of sand-size clastic components in the Togo deposit and the nature of the carbonate minerals. Unlike the southeastern United States deposits, the Togo deposits exhibit a lack of dolomitization within or immediately below the phosphate-bearing beds. In Bed CSC and in Formation 2/3, calcite is the only carbonate mineral. Johnson (1987) has noted minor dolomitization in the underlying Tabligbo Formation.

The interpretation of Bed 1 as the lateral

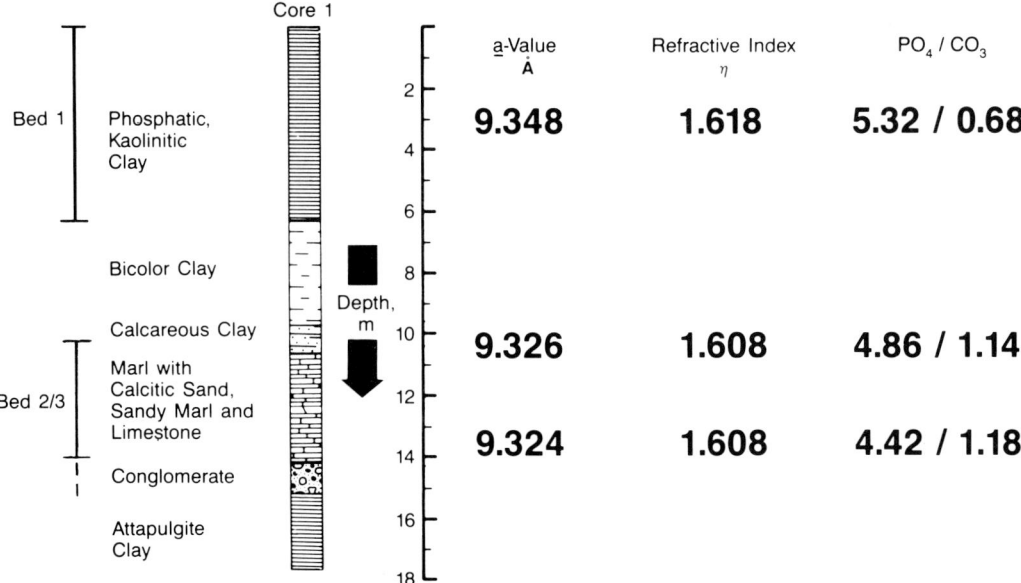

Fig. 8. Variation of francolite properties in a core from the Togo deposit.

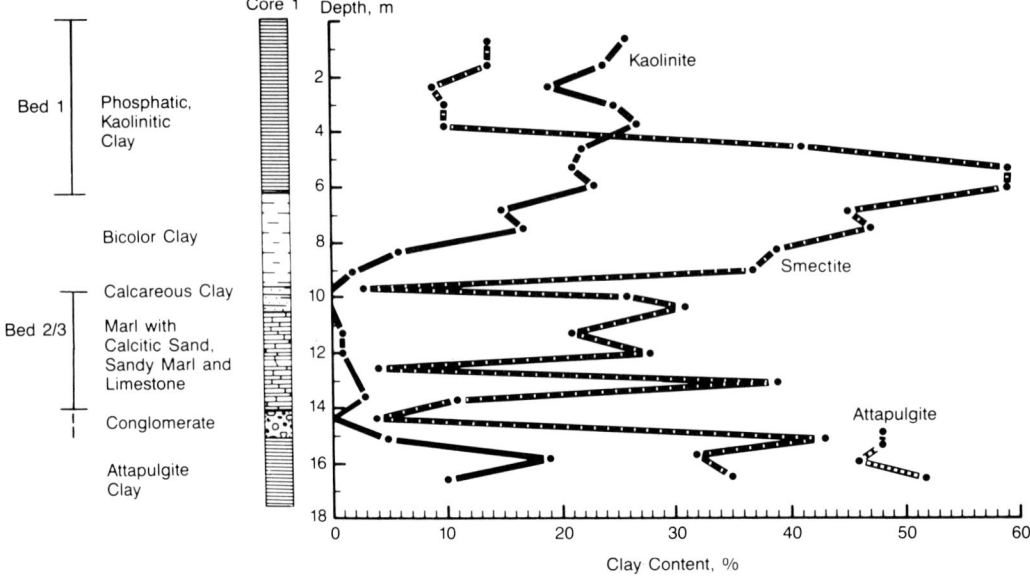

Fig. 9. Variations in clay mineralogy in a core from the Togo deposit.

equivalent of Bed CSC and the variations in phosphate and clay mineralogy suggest that weathering is the main source of alteration of these sediments. Table 8 gives a mineralogical comparison of samples of Bed CSC and Bed 1. Calcite has been leached from Bed CSC and is essentially zero in Bed 1, and the reduction in calcite content is accompanied by a two- to threefold increase in the amounts of francolite, quartz, and goethite. The francolite has been decarbonated, and a-values have increased. The total clay content is roughly the same in Bed 1 and Bed CSC; however, kaolinite, the dominant clay mineral in Bed 1, replaces the dominant smectite of Bed CSC.

Conclusions

The differences in mineralogy of the Togo and southeastern United States phosphate deposits, summarized in Table 9, are the result of primary compositional and depositional features as well as superimposed postdeposition alterations. The differences in primary composition may be the result of differing source areas, palaeoenvironmental conditions, or mixing of primary francolites with altered or recycled francolites. Postdepositional alterations are the result of all the physical and chemical factors that have modified the sediments. These modifications may be in their first stages, as in the North Carolina deposits, or highly advanced, as in Togo, and/or the system may have been influenced by numerous episodes of alteration that have left multiple imprints on the deposits, as in Florida.

The North Carolina deposit is an example of a marine phosphorite that has been mineralogically preserved as it was deposited or is in the first stages of alteration. The francolite is highly carbonate substituted, which indicates its primary formation. Carbonates are present throughout the phosphorite and overlying sediments, buffering the pore water and protecting the francolite. With the exception of variations in the distribution of carbonate species, the mineralogical character of the deposit is rela-

Table 8. *Mineralogical comparison of samples from Bed CSC and Bed 1*

	Bed CSC*	Bed 1[†]
	(%)	(%)
Calcite	49	0
Francolite	36	79
Smectite	12	4
Kaolinite	1	11
Quartz	<1	3
Goethite	1	3

* 12.2% P_2O_5, a-value = 9.333 Å ± 0.001.
[†] High-grade ore, 30.0% P_2O_5, a-value = 9.350 Å ± 0.001.

Table 9. *Comparison of the phosphate deposits of the southeastern United States and Togo*

	North Carolina	Florida	Togo
Depositional environment	Offshore marine	Marine to nearshore marine	Offshore marine
Primary composition	Terrigenous sands, clays, carbonates, phosphates	Terrigenous sands, clays, carbonates, phosphates	Carbonates, clays, phosphates
Subaerial exposure and reworking	Minor	Multiple exposure, extensive reworking	Deeply weathered, reworking not evident
Francolite *a*-values	Low	Low to high	Low to high
Variability in francolite *a*-values	Minor	Vertically and laterally variable, size variations, mixed character	Vertically and laterally variable
Carbonates	Present throughout the section, dolomitic in producing beds	Dolomitic, leached from upper altered zones	Calcareous, below producing zones
Al- and Fe-phosphate zone	No	Extensive	Well developed
Clay-weathering profile	No	Well developed	Well developed

tively homogeneous. Some minerals found in the deposit may indicate the influence of postdepositional diagenetic processes (dolomite, clinoptilolite, sepiolite); however, the location of the deposit near sea level and the relatively impermeable nature of the beds have preserved the mineralogic character.

The central Florida deposits have been subjected to numerous cycles of subaerial exposure, weathering, and reworking. The deposits have a well-developed aluminophosphate zone and a systematically altered clay mineral profile. Carbonates have been leached from the upper parts of sections. Francolite compositions range from highly substituted francolite varieties to varieties approaching a fluorapatite composition. The more highly altered francolites are found at the extensively altered tops of sections. Francolites are more highly substituted at depth where protected by impermeable beds and/or the buffering influence of carbonates. The smaller size fractions contain decarbonated francolites. Altered fracolites found below more highly substituted francolites point out the complexity of sedimentation, weathering, and reworking. Reworking has resulted in a mixing of francolites in various states of alteration.

The Togo deposit has been extensively altered, and the section appears to have been condensed by chemical weathering. Reworking of the deposit is not evident. The deposit has an iron aluminium phosphate zone at the top of the section and a well-developed clay alteration profile. The francolite compositions vary both vertically and laterally. Francolite composition at the top of the deposit resembles that of the altered francolites from Florida. The francolites from Bed CSC and deeper in the section resemble less-altered francolites from Florida and North Carolina. The enrichment of the Bed 1 ore zone is primarily caused by the leaching (decarbonation) of its lateral equivalent Bed CSC.

The alteration of phosphorite deposits may be evaluated by studying the gross physical and mineralogical characteristics and the subtle differences in francolite composition. Low *a*-value francolites and a homogeneous mineral distribution are indicative of a general lack of alteration. Iron and aluminium phosphate zones, clay mineral profiles, and a wide range of francolite compositions indicate that deposits are altered. Francolite composition profiles may develop in response to weathering and the removal of carbonates. Reworking of deposits can result in a mixing of francolites. Studies of phosphorite deposits should include evaluations of the extent to which the deposits may have been affected by postdepositional alterations. Sedimentological, stratigraphic, and geochemical interpretations may be confused by unperceived diagenetic effects.

References

ALTSCHULER, Z. C. 1965. Precipitation and recycling of phosphate in the Florida land-pebble phosphate deposits. *United States Geological Survey Professional Paper*, **525-B**, B91–B95.

—— 1973. The weathering of phosphate deposits —

geochemical and environmental aspects. *In*: GRIFFITH, E. J. (ed.), *Environmental Phosphorus Handbook*. John Wiley & Sons, New York, 33–96.

——, CATHCART, J. B. & YOUNG, E. J. 1964. *Geology and geochemistry of the Bone Valley Formation and its phosphate deposits, West Central Florida*. Geological Society of America, Annual Convention.

——, DWORNIK, E. J. & KRAMER, H. 1963. Transformation of montmorillonite to kaolinite during weathering. *Science*, **141**, 148–152.

——, JAFFEE, E. B. & CUTTITTA, F. 1955. The aluminum phosphate zone of the Bone Valley Formation, Florida, and its uranium deposits. *In*: *Contributions to the Geology of Uranium and Thorium*. United States Geological Survey and Atomic Energy Commission for the United Nations International Conference on the Peaceful Uses of Atomic Energy, Geneva, Switzerland, 495–504.

BERNARDI, J. P. & HALL, R. B. 1980. Comparative analysis of the Central Florida Phosphate District and its southern extension. *Mining Engineering*, 1256–1261.

CATHCART, J. B. 1963. Economic Geology of the Plant City Quadrangle, Florida. *United States Geological Survey Bulletin*, **1142-D**, D1–D56.

—— 1985. Economic geology of the Land Pebble District of Florida and its southern extension. *In*: SNYDER, S. (ed.) *IGCP Project 156 — Phosphorites, Eight International Field Workshop and Symposium (Southeastern United States) Guidebook*. Greenville, North Carolina, USA 127–137.

——, SHELDON, R. P. & GULBRANDSEN, G. A. 1984. *Phosphate rock resources of the United States*. United States Geological Survey Circular **888**.

CHIEN, S. H. & BLACK, C. A. 1976. Free energy of formation of carbonate apatites in some phosphate rocks. *Soil Science Society of America Journal*, **40**, 234–239.

CROWSON, R. A., SNYDER, S. W., RIGGS, S. R. & MALLETTE, P. M. 1985. Geology of the Aurora Phosphate District. *In*: SNYDER, S. (ed.) *IGCP Project 156 — Phosphorites, Eighth International Field Workshop and Symposium (Southeastern United States) Guidebook*, Greenville, North Carolina, USA, 17–32.

FREAS, D. H. & RIGGS, S. R. 1968. Environments of phosphorite deposition in the Central Florida Phosphate District. *In*: *4th Forum on Geology of Industrial Minerals*, Austin, Texas, 117–128.

ISPHORDING, W. C. 1973. Discussion of the occurrence and origin of sedimentary palygorskite/sepiolite deposits. *Clays and Clay Minerals*, **21**, 391–401.

JOHNSON, A. K. C. 1987. *Le Bassin côtier à Phosphates du Togo (Maastrichtien — Eocène moyen)*. PhD thesis, Université de Bourgogne, Centre des Sciences de la Terre-Sedipal, Université du Benin (Togo).

KILINC, M. & COTILLON, P. 1977. Le gisement d'Hahotoe-Kpogame (Tertiaire du Sud Togo), exemple de piège sédimentaire à sables phosphatés. *Bulletin du Bureau de Recherches Geologiques et Minieres*, **II(1)**, 43–63.

LEHR, J. R., MCCLELLAN, G. H., SMITH, J. P. & FRAZIER, A. W. 1967. Characterization of apatites in commercial phosphate rocks. *Colloque international sur les phosphates minéraux solides*, Vol. 2, Toulouse, France, 16–20 May, 29–44.

LYLE, M. E. 1984. *Clay mineralogy of the Pungo River Formation, Onslow Bay, North Carolina Shelf*. Master's thesis, East Carolina University.

LUCAS, J., FLICOTEAUX, R., NATHAN, Y., PRÉVÔT, L. & SHABAR, Y. 1980. *Differential aspects of phosphorite weathering*. Society of Economic Paleontologists and Mineralogists, Special Publication, **29**, 41–51.

MCCLELLAN, G. H. 1980. Mineralogy of carbonate fluorapatites. *Journal of the Geological Society, London*, **137**, 675–681.

—— & LEHR, J. R. 1969. Crystal chemical investigation of natural apatites. *American Mineralogy*, **54**, 1374–1391.

—— & VAN KAUWENBERGH, S. J. 1990a. Mineralogy of sedimentary apatites. *In*: NOTHOLT, A. J. G. & JARVIS, I. (eds) *Phosphorite Research and Development*. Geological Society, London, Special Publication, **52**, 23–31.

—— & —— 1990b. Clay mineralogy of the phosphorites of the Southeastern United States. *In*: RIGGS, S. R. & BURNETT, W. L. (eds) *Phosphate Deposits of the World: Volume 3, Genesis of Neogene to Recent Phosphorites*. Cambridge University Press, Cambridge. (in press).

MILLER, J. A. 1971. *Stratigraphic and structural setting of the Middle Miocene Pungo River Formation of North Carolina*. PhD dissertation, University of North Carolina, Chapel Hill, North Carolina, USA.

MILLOT, G. 1970. *Geology of Clays*, Springer-Verlag, New York, USA.

NATHAN, Y. & SASS, E. 1981. Stability relations of apatites and calcium carbonates. *Chemical Geology*, **34**, 103–111.

REDEKER, I. H. & WRIGHT, T. J. 1979. Phosphate mining in North Carolina. *Sonderdruck aus Zeitschrift ERZMETALL*, **32**(2), 77–85.

RIGGS, S. R. 1979. Petrology of the Tertiary phosphate system of Florida. *Economic Geology*, **74**(2), 195–220.

—— 1984. Palaoceanographic model of Neogene phosphorite deposition, U.S. Atlantic Continental Margin. *Science*, **223**(4632), 123–131.

——, LEWIS, D. W., SCARBOROUGH, A. K. & SNYDER, S. W. 1982. Cyclic deposition of the Upper Tertiary phosphorites of the Aurora Area, North Carolina, and their possible relationship to global sea level fluctuations. *Southeastern Geology*, **23**(4), 189–204.

ROONEY, T. P. & KERR, P. F. 1967. Mineralogic nature and origin of phosphorite, Beaufort County, North Carolina. *Geological Society of America Bulletin*, **78**, 731–748.

SCARBOROUGH, A. K., RIGGS, S. R. & SNYDER, S. W. 1982. Stratigraphy and Petrology of the Pungo River Formation, Central Coastal Plain of

North Carolina. *Southeastern Geology*, **23**(4), 205–215.

SLANSKY, M. 1986. *Geology of Sedimentary Phosphates*, Elsevier Science Publishing Co., New York.

SNYDER, S. W. P. 1983. *Seismic Stratigraphy Within the Miocene Carolina Phosphogenic Province: Chronostratigraphy, Paleotopographic Controls, Sea Level Clyclicity, Gulf Stream Dynamics, and the Resulting Depositional Framework*. MS thesis, University of North Carolina, Chapel Hill.

SNYDER, S. W., RIGGS, S. R., KATROSH, M. R., LEWIS, D. W. & SCARBOROUGH, A. K. 1982. Synthesis of phosphatic sediment-faunal relationships within the Pungo River Formation: Paleoenvironmental implications. *Southeastern Geology*, **23**(4), 233–246.

UPCHURCH, S. B., STROM, R. N. & NUCKELS, M. G. 1982. Silicification of Miocene rocks from Central Florida. *In*: SCOTT, M. & UPCHURCH, S. B. (eds) *Miocene of the Southeastern United States*, T. Florida Bureau of Geology, Special Publication, 25, 251–284.

VAIL, P. R. & MITCHUM, R. M. 1979. Global cycles of relative changes of sea level from seismic stratigraphy. *In*: WATKINS, J. S. MONTADERT, L. & DICKERSON, P. W. (eds) *Geological and Geophysical Investigations of Continental Margins: American Association of Petroleum Geologists Memoir*, **29**, 469–472.

VAN KAUWENBERGH, S. J. & MCCLELLAN, G. H. 1985. Variations in the Mineralogy of the Florida Phosphate District. *In*: *Florida Land-Pebble Phosphate District Guidebook*, Geological Society of America Annual Meeting, Orlando, Florida, 38–68.

——, CATHCART, J. B. & MCCLELLAN, G. H. 1990. Mineralogy and alteration of the phosphate deposits of Florida. *United States Geological Survey Bulletin*. (In press).

WEAVER, C. E. & BECK, K. C. 1977. *Miocene of the S. E. United States: A Model for Chemical Sedimentation in a Perimarine Environment*, Elsevier Publishing Co., New York, USA.

WEBB, S. D. & CRISSINGER, D. B. 1983. Stratigraphy and vertebrate paleontology of the Central and Southern Phosphate District of Florida. *In*: *The Central Florida Phosphate District Field Trip Guidebook*. Geological Society of America, Southeastern Section, 28–72.

Microbial mediation in phosphatogenesis: new data from the Cretaceous phosphatic chalks of northern France

MICHEL LAMBOY

Department of Geology, University of Rouen, BP 118, 76134 Mont Saint Aignan, France

Abstract: Detailed petrographic study of the Upper Cretaceous phosphatic chalks of northern France, particularly by scanning electron microscopy, shows that phosphate occurs predominantly as phosphatized bacterial remains. Shiny phosphatic crusts on hardgrounds and anisotropic phosphate coatings on grains are interpreted as phosphate mineralized microbial ('microstromatolitic') structures. Phosphatic remains of ovoid bacteria and of microbial colonies with botryoidal surfaces have been observed in chambers of foraminiferans and bryozoans, and within other bioclasts, and occur as pore linings in peloids, coprolites and lenses of phosphatic chalk incorporated into hardgrounds. The microbial community which formed the crusts and grain coatings developed by surface accretion, whereas the other community grew solely in the intragranular porosity of grains and in the extragranular porosity of semilithified sediments. Analysis of these two types of microbial growth, which have also been documented from other phosphorite deposits, permits a better but as yet incomplete understanding of the mechanisms of phosphatization.

The Upper Cretaceous phosphatic chalks of northern France were exploited commercially during the late nineteenth and early twentieth centuries and they provide the thickest and most concentrated reserves of economic phosphate in northwest Europe (Jarvis 1980a, b). The deposits, however, are no longer commercially viable and, recently, extraction has ceased. The phosphatic chalks occur as metre to decimetre-thick beds deposited in small (1 km long or less) shallow, elongate, erosional troughs. Deposits are generally floored by an extensive hardground surface, termed the basal hardground by Jarvis (1980a, b). Detailed lithostratigraphic and biostratigraphic study (Jarvis 1980a, b) has demonstrated the petrographic diversity of these Santonian to Campanian deposits. Facies include strongly lithified and phosphatized hardgrounds, white chalks containing large phosphatic chalk-filled burrows, bioturbated phosphatic chalks, and coarse grained phosphatic gravels overlying complex erosional surfaces. In these rocks, several distinct types of phosphatic material can be recognized: (1) shiny phosphatic crusts and veneers, coating in situ or reworked hardgrounds, (2) phosphatized chalk, in the form of lithoclasts or occuring as lenses plastered onto hardground surfaces, and (3) a variety of phosphatic grains, which dominate the sand-sized component of the phosphatic chalk, and which are also found cemented into some hardgrounds. The geochemistry of these deposits has been investigated by Jarvis (1980c, 1984).

Numerous recent studies have demonstrated the presence of phosphatic microbial remains inside both granular and nodular ancient and modern phosphorites (Bignot 1980; O'Brien et al. 1981; Soudry 1983, 1987; Soudry & Champetier 1983; Dahanayake & Krumbein 1985; Zanin et al. 1985, 1987; Southgate 1986; Lamboy 1987a; Lamboy & Monty 1987, 1989; Soudry & Lewy 1988; Purnachandra Rao & Nair 1988; Soudry & Southgate 1989; Bréhéret 1989; Lewy 1990); the synthesis of apatite by microbial processes has also been achieved experimentally (Lucas & Prévôt 1981, 1985; El Faleh 1988; Lucas et al. 1990). In this study, detailed petrographic analysis of Cretaceous phosphatic chalks was undertaken in an attempt to document the presence of microbial remains in the different kinds of phosphatic material and to compare phosphatogenic processes in grains and hardgrounds.

Studied samples were collected from Beauval (Somme, x 599 89 y 266 80 z 130) and Hardivillers Quarries (Oise, x 593 22 y 213 74 z 120). No exposure of the basal hardground now remains at Hardivillers, and the sample of this level came from the collection of G. Bignot. Detailed observations were made on washed samples, polished slabs and corresponding thin sections, using binocular and petrographic microscopes; studies were undertaken using reflected light (RL), plane polarized light (PPL) and crossed polarized light (CPL). Lastly, chips of representative specimens were examined using scanning electron microscopy (SEM).

Chips were first cleaned in an ultrasonic tank, dried and coated immediately with gold/palladium. No etching was undertaken.

Petrographic study of phosphatic components

Shiny phosphatic crusts and veneers

The hardgrounds which commonly underlie these deposits of phosphatic chalk contain irregular phosphatic surfaces which were termed phosphatic skins by Jarvis (1980a, b). In section, they correspond to thin veneers or crusts (Fig. 1a) up to 1 mm in thickness. Generally, these crusts show an alternation of dark and light laminae (Fig. 1b). Laminae which are dark under RL, are transparent under PPL and anisotrophic under CPL. SEM studies indicate that they consist of massive to more or less sheet-like layers of compact anhedral nannograins (Figs 1c & d). Laminae which are light-coloured under RL are porous but have little permeability. This produces a dark appearance under PPL due to lack of penetration of the Canada balsam mounting medium, and consequently the total reflection of the light (Lamboy 1988a; Fikri et al. 1989). When these light-coloured laminae are thin, they consist entirely of layers of calcareous nannofossil tests which have been coated by phosphate and then dissolved (this texture is illustrated in Fig. 2b, see below). Thicker laminae contain fewer dissolved tests, and the thickest layers include thin chalk lenses which remain unphosphatized or contain only scattered phosphatic grains (Fig. 1b). In some areas, laminae form small squat columns, up to 100 μm high (Figs 1b & c). Hollows between these columns are filled by partly decalcified chalk. The ultrastructure of the columns (Fig. 1c) is similar to that of single layers in sheet-like morphologies.

Jarvis (1980a, b) proposed an algal or microbial origin for the phosphatic skins. The observed columns confirm that these skins are biosedimentary structures with layered cryptalgal fabrics (cf. Monty 1976). Some probable algal filaments have been observed during optical analysis, but these have not yet been confirmed by SEM studies. Since the ultrastructure is composed of nannograins, it might be suggested that the stromatolitic layers originally consisted mainly of cocci, as proposed by Soudry & Lewy (1988) for phosphate coatings in phoshate nodules and megafossil moulds from the Negev. However, other phosphatic microbial remains exist (see below) whose ultrastructure is also composed of nannograins (Figs 2c & f), and in these cases nannograins do not correspond to phosphatized cocci. So, we believe that the stromatolitic layers was gener-

Fig. 1. (a) Photograph of a polished slab of the basal hardground from Hardivillers Quarry. Phosphate is dark, unmineralized chalk is white. Numerous cross sections through laminated phosphatic crusts can be seen. Individual crusts delimit irregular lenses of sediment which vary in size, contain different numbers and sizes of enclosed particles, and exhibit variable degrees of phosphatization. (b) Photograph of the same slab of basal hardground as (a) showing detail of a crust. Note the alternating dark and light-coloured laminae and the occurrence locally of 'microstromatolitic' columns. Note also the finely laminated phosphatic coatings on grains and the planktonic foraminiferan which forms the nucleus of the central grain. (c) SEM photomicrograph of a phosphatic 'microstromatolite'; same slab as (a) & (b). The layers of the microcolumns are separated by a thin sheet-like lamina. (d) Detail of (c) (outlined area), showing the compact structure of tightly packed anhedral phosphate nannograins. (e) SEM photomicrograph of a fractured cross-section through a sediment lens containing grains with microlaminated phosphatic cortices (compare with (b) same slab). Note the dissolution of foraminiferal tests and bioclasts inside the nuclei. The arrow points to a phosphatic crust which separates a lower sediment lens (the matrix of which is lithified and phosphatized) from an upper lens (the matrix of which is unlithified, and has been partially removed during ultrasonic cleaning). (f) Detail, under SEM, of a thick laminated cortex around a chalk intraclast; this cortex can be compared with those shown on (b) & (e) which came from the same sample. The surface of the intraclast is visible at the bottom of the photomicrograph. The laminae which are dark under reflected light (RL) show a compact structure. The beginning of a microcolumn can also be seen (black arrow) in one of these areas. The laminae which are light-coloured under RL show a slightly porous structure (white arrows); the pores correspond to decalcified layers of nannofossils tests (see also Fig. 2b). (g) SEM photomicrograph of the contact between the cortex and the nucleus of the grain shown in (f). At the bottom, a fragment of a coccolith is visible in the nucleus. In the middle, laminae display a typical compact structure composed of anhedral nannograins (cf. (d)). (h) SEM view of a fractured phosphatic grain from a phosphatic chalk (Hardivillers Quarry). A foraminiferal test 'a' forms the nucleus of the grain. A thin phosphatic cortex 'b' coats the test and penetrates through its pores to cover the internal mould 'c'. The chamber is mainly filled by unmineralized chalk 'd'. In some areas, however, a microbial phosphate 'e' has developed in an empty part of the chamber (under higher magnification, this phosphate has an identical texture to that displayed in Fig. 2c).

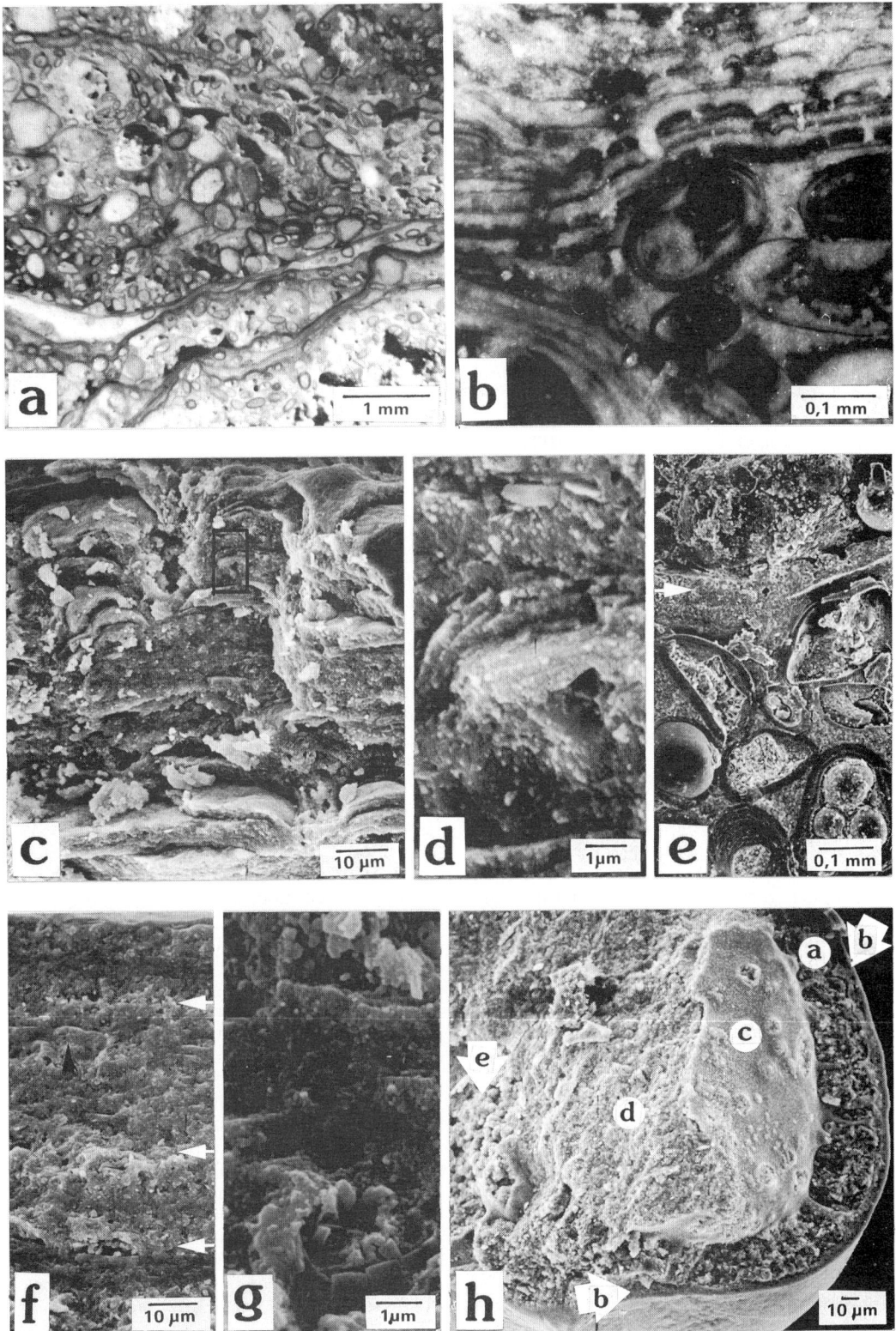

ated by microbes, the type of which remains unknown. According to the small size of the columns, the structures can be classified as microstromatolites (cf. Lanier 1988). The similarity between the ultrastructure of the columns and that of the compact laminae indicates a probable microbial origin for all of the shiny phosphatic veneers and crusts.

Phosphatic crusts commonly constitute the surface of the hardground sediments. They are also observed on the walls of burrows and boring associated with the hardgrounds, as noted by Jarvis (1980a, b). In some complex sedimentary structures, the direction of growth of the columns indicates their development on the walls of small caverns or galleries; in these cases, they may be termed endostromatolites (cf. Monty 1982; Delamette 1988).

The role, whether active or passive, of microorganisms in apatite precipitation is still under discussion (e.g. Choudhury & Roy 1986); this study does not provide any new evidence concerning this matter.

Anisotropic phosphatic cortices

The great majority of phosphatic particles from both hardgrounds (Figs 1a & b) and phosphatic chalks consist of coated grains. Their cortex is invariably composed of anisotropic phosphate, whatever the nucleus (e.g. foraminiferan, bioclast, tooth, bone fragment, chalk intraclast), whether phosphatized or not. Such grains from the phosphatic chalks were regarded as phosphatic ooids by Tabatabaï (1977) and Jarvis (1980a, b).

These cortices are composed of one or several laminae (Fig. 1b). The number of laminae is commonly fewer than in the crusts and their thickness is generally less, particularly those laminae which are light-coloured under RL. With optical microscopy, the structure of the laminae in the cortices is similar to that of the laminae in the crusts. Using SEM, their microstructures and ultrastructures are also comparable (Figs 1e-g, 2a & b). Sometimes, under the SEM (Figs 1f & g) and in thin section, the initiation of column structures can be seen. When the cortex surrounds a foraminiferal test, it occasionally extends through the pores and coats the internal part of the shell (Fig. 1h).

Jarvis (1980a, b) suggested that these cortices may have the same algal or other organic origin as the crusts. The similarity of the microstructures and the presence of initial growth stages of 'microstromatolitic' columns indicate that the anisotropic phosphatic cortices probably did indeed have the same microbial origin as the crusts. Those grains with an ooid-like structure, therefore, may be micro-oncoids although they do not display typical oncoid microstructures (cf. 'oolites' from the Cambrian Karatau phosphorites; Eganov 1988). They are different, however, from the phosphate micro-oncoids of microbial origin described by Dahanayake & Krumbein (1985) which contain identifiable microbial filaments.

Phosphatized chalk

Using optical microscopy, texture, colour, anisotropy and birefringence allow areas of phosphatized chalk to be distinguished. Such areas can be observed in lithoclasts, in the chalk nuclei of phosphatic ooids, and in hardgrounds, particularly inside lenses delimited by phosphatic crusts (Figs 1a & b). The size and shape of these phosphatized areas vary considerably. Light microscopy yields little information concerning their origin.

Using SEM, complete and fragmentary nannofossil tests can be seen to constitute the main component of unmineralized chalks. In phosphatized areas, however, complete and partial dissolution of carbonate particles make their recognition more difficult, although nannofossil material may be preserved as composite moulds in phosphate (Fig. 2b). Here, pores commonly display linings of botryoidal phosphate with radialfibrous internal structures (Fig. 2c). These are similar to structures described by Lamboy & Monty (1987, 1989) in phosphatic grains from several other phosphorite ore-deposits, and by Soudry & Lewy (1988) and Lewy (1990) in phosphate nodules and megafossil moulds from phosphorites of the Negev. The former authors interpreted the structures as mineralized microbial colonies while the latter authors regarded them as representing phosphatized communities of cocci. Ovoid phosphatic globules, 1 to 3 μm long, are present both between chalk particles (Fig. 2f) and inside microfossil tests (Fig. 2e). These globules are similar to the phosphatic 'rod-like' bacteria described in other phosphorites (O'Brien et al. 1981; Zanin et al. 1985, 1987; Lamboy & Monty 1987, 1989; Purnachandra Rao & Nair 1989; Bréhéret 1989), and to those observed experimentally in the microbial precipitation of apatite (Lucas & Prévôt 1981, 1985; El Faleh 1988). These morphological similarities suggest that the phosphatic globules may also be mineralized bacteria. Figures 2e and f show that the bacteria had grown against carbonate particles prior to their dissolution, flattening themselves against the particles and producing composite phosphatic moulds; the lateral coalescence of bacterial bodies is often incomplete and leaves some gaps. Phosphatization of the bacteria and subsequent dissolution of the carbonate grains has preserved the morphology of the original bodies.

Phosphate and carbonate particles are in contact where the phosphate laminae of crusts or cortices cover chalk (Figs 1f & g). In some cases, carbonate particles are dissolved, particularly where they are sandwiched between adjacent phosphatic laminae (Fig. 2b). In these cases, a phosphatic composite mould of the particles is produced which is composed of nannograins similar to those forming the phosphatic laminae. The small size (< 0.1 μm) of the phosphatic nannograins make them difficult to identify between individual phosphatic particles, but it appears that the microbes of the mat covering the chalk were unable to penetrate pore spaces more than a few microns deep.

It is proposed, therefore, that phosphatization of the chalk was not simply a surface-related replacive process, as has been suggested previously (Jarvis 1980a, b, c), but was caused by the invasion of pore-spaces in the chalk by microbes which were subsequently phosphatized. In other phosphate deposits, phosphatic microbial fills occur in pores formed by dissolution of carbonate (Lamboy 1987a); this type of fill has been observed in some bioclasts such as fragments of echinoderm skeletons, but cannot be proven in phosphatized chalks where they are composed solely of nannofossil tests.

Other phosphatized components

Several types of the more common phosphatic or phosphatized components in the sediments have also been examined, without an exhaustive study being made. Fragments of bones and teeth, which are of primary phosphatic composition, are common. These generally exhibit a compact structure, similar to that observed for comparable material from other phosphorite deposits (Lamboy 1982a; Soudry 1983); some of them display peripheral microborings, the fills of which have been phosphatized (phosphomicritization of Soudry, 1979).

Foraminiferal tests commonly form the nuclei of phosphatic ooids (Figs 1b, e & h) and they are also present, without cortices, within areas of phosphatized chalk (Figs 2d & e). Here, the tests are either carbonate (Fig. 1h), or has been dissolved (Figs 1e, 2d & e) and are preserved as composite moulds of phosphate. Sometimes, fungal or microbial phosphatic filaments are present within the mouldic porosity. These micro-organisms have either bored into the test or have entered the cavity after test dissolution. The replacement of carbonate tests by phosphate has never been observed in these deposits. The chambers are generally filled with chalk, which is rarely phosphatized. When the chambers were partially or totally empty, microbial colonies (Fig. 1h) and ovoidal bacteria (Fig. 2d) have grown inside. Similar observations have been made in other phosphorites (Lamboy & Monty 1987, 1989).

In some areas, bryozoans have grown on the hardgrounds and have then been covered by microstromatolitic crusts. Here, their chambers are filled with phosphatized microbial colonies displaying radialfibrous internal structures and botryoidal surfaces (cf. Fig. 2c). Similarly, fragments of echinoderm skeletons are present in both hardgrounds and phosphatic chalks. In both cases, the calcitic stereome is still visible but may be partially dissolved. Botryoidal radialfibrous phosphate fills both primary and secondary (dissolution) cavities within the stereome, indicating that phosphatization and dissolution were early diagenetic processes. Calcitic bioclasts commonly constitute the nuclei of phosphatic ooids (Fig. 1e). In many cases, they have been intensively bored by micro-organisms and phosphate is localized in the borings, which appear in relief within the mouldic porosity formed after dissolution of the primary carbonate.

In phosphatic chalks and hardgrounds, some grains are totally phosphatic and are isotropic in thin section. Their size and morphology enable their identification as faecal pellets or coprolites as proposed by Jarvis (1980a, b). Some of these are also surrounded by anisotropic cortices. Using SEM, this isotropic phosphate can be seen to contain numerous remains of phosphatic ovoid bacteria and microbial colonies similar to those described above. Such remains are also common in the grains of the Tunisian phosphorites (Lamboy & Monty 1987, 1989) which have also been interpreted as fossil faecal pellets (Lamboy 1982a, 1987b, 1988b, Ben Amar 1985).

Discussion

Occurrence and significance of microbially mediated phosphatization

Inorganic chemical precipitation have been advocated as the main mechanism for the formation of phosphate in phosphatic chalks (Kennedy & Garrison 1975; Tabatabaï 1977) and other phosphorite deposits (Slansky 1986;

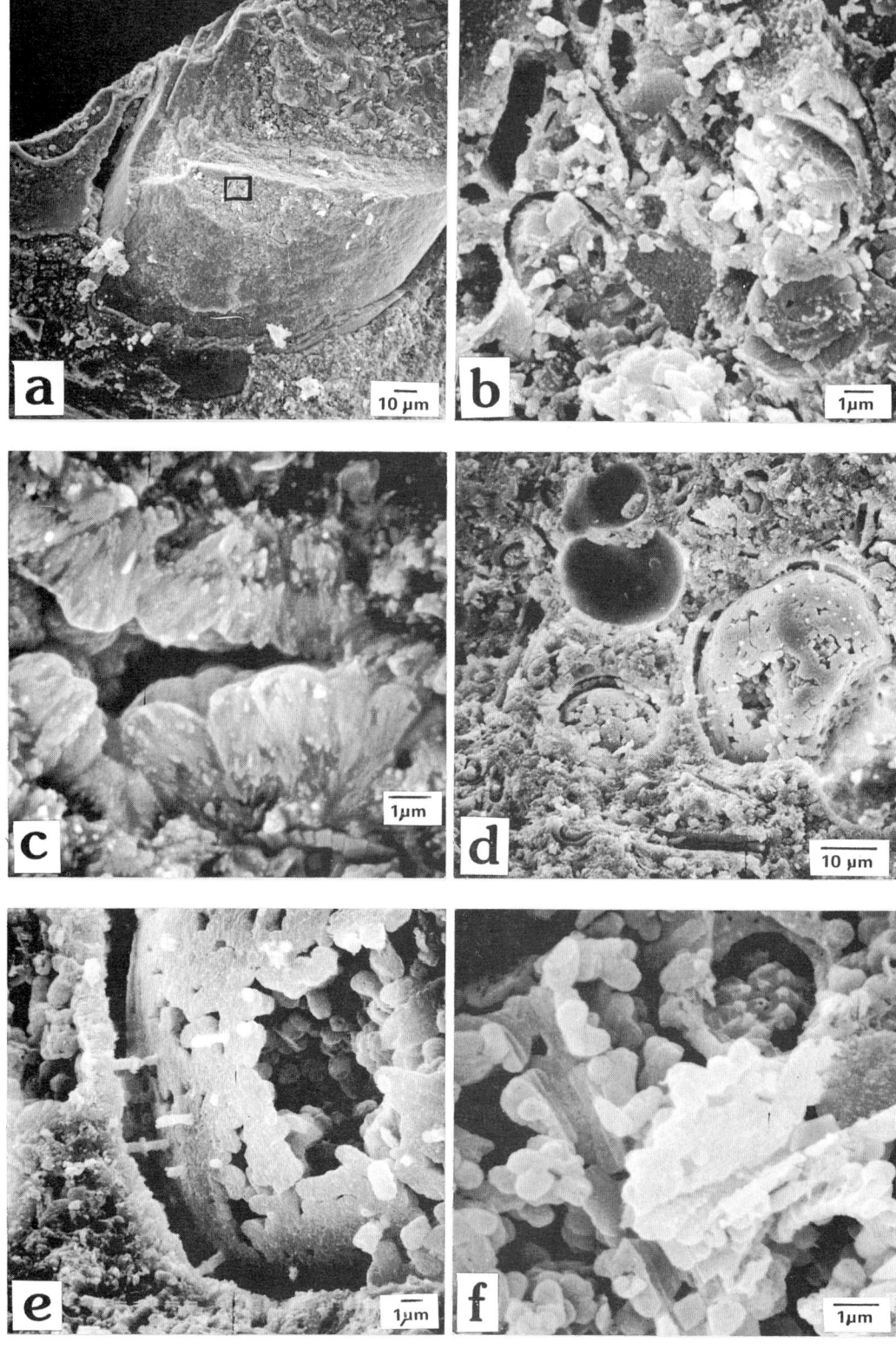

Lamboy 1986; Poels & Robaszynski 1988; Van Cappellen & Berner 1988). Many authors have suggested that phosphate can be form from carbonate by direct replacement (Ames 1959; Slansky 1986); phosphatization of chalks has been quoted as an example of this process (Cayeux 1939; Tabatabaï 1977; Jarvis 1980a, b, c).

Our observations indicate a more complex process. SEM studies clearly indicate the dissolution of carbonate (Figs 1e, 2b & d) and formation of phosphatic composite moulds of microbial origin (formed prior to dissolution) which preserved the morphology of the primary grains. However, in thin section, these structures appear to have a simple replacive origin. In French Upper Cretaceous phosphatic chalks, phosphate is localized in two kinds of structures which are interpreted as having microbial origins: (1) Laminae within phosphatic crusts and forming the cortices of coated grains. The laminated structure and the anisotropy of the resulting phosphate allow its identification in thin section. (2) Individual ovoid bacteria and small microbial colonies having botryoidal surfaces and radialfibrous internal structures. These are observed in phosphatized chalk, chambers of foraminiferans and bryozoans, fragments of echinoderm tests, and coprolites and faecal pellets. The resulting phosphate is isotropic in thin section, although in other phosphorites similar microbial colonies developed in large enough cavities to form anisotropic phosphate (Lamboy 1987b; Lamboy & Monty 1989). Recognition of these microbial remains and their relationships to non-phosphatic material is difficult and often not possible in thin section.

In the absence of recognisable microstructures, it is often difficult to demonstrate (even using SEM) a microbial origin for phosphate within any specific area. In spite of this, evidence of microbial structures and the absence of definite chemical precipitates or replacement textures lead us to believe that microbial processes are a major mechanism of phosphatogenesis. The precise role of microbes in the nucleation and precipitation of phosphate remains to be clarified. However, the fact that the nearly all of the apatite is located in microbial remains allows us to envisage an active role for microbes in the apatite nucleation, as demonstrated by experimental studies. On the other hand, the phosphatized microbial bodies are now solid, indicating that some of the apatite must have precipitated *post mortem*, and that the source of this phosphate cannot be from fossilized microbes.

It is estimated that the phosphatic microbial laminae, which form both crusts and grain cortices within the phosphatic chalks and associated sediments, represent the volumetrically dominant phosphate phase in the deposits studied here. This conclusion is based on the observation that a large proportion of phosphatic coated grains have non-phosphatic nuclei.

Fig. 2. (a) SEM photomicrograph of a grain with a thick laminated phosphatic cortex surrounded by a matrix of phosphatized chalk (same slab as Fig. 1a). The cortex is composed of alternations of more or less compact laminae (compare with Figs 1e−g). (b) Detail of central area outlined in (a). Surface view of a porous laminae showing the remains of numerous nannofossil tests which have been preserved as composite moulds in phosphate. The phosphate consists of nannograins the size of which is similar to those in the compact laminae (cf. Figs 1d & g). The fine grain size of the phosphate leads to preservation of the minutest detail (cf. (c) & (f)). (c) Detail of the area outlined at the bottom left of (a), showing the typical appearance of pore-filling phosphate. A fragment of calcareous nannofossil is visible in the bottom right corner. The radialfibrous structure of the phosphate and the botryoidal surface are both typical of phosphatized microbial colonies observed in numerous phosphorites (Lamboy & Monty 1987, 1989; Soudry & Lewy 1988). (d) SEM photomicrograph of a phosphatized lens in a hardground (same slab as Fig. 1a). The biogenic nature of the sediment is clearly visible. Major cavities have resulted from the dissolution of calcareous microfossil and nannofossil tests. (e) Detail of (d) showing a close-up of a dissolved test of a planktonic foraminiferan. To the right, the phosphatic internal mould of a chamber is composed of ovoid phosphate globules up to 2 μm long. These globules are interpreted as being phosphatized bacteria. Some of them have grown in the chamber before test dissolution, flattening themselves against the test wall and locally coalescing. A few have extended through the test pores. The bacteria which have grown inside the chamber have a more regular ovoid shape because their growth has not been constrained by the test walls. These bacterial remains are similar to those observed in numerous other phosphorites (see text for details). To the left, the external part of the test has been preserved as an external mould by a thin layer of microbial phosphate. (f) SEM photomicrograph showing detail of phosphatized chalk matrix in a lens from the same hardground. The tests of calcareous nannofossils are preserved as composite moulds in phosphate. These phosphatic moulds have been produced by ovoid bacteria whose shape is only fully developed within primary pore spaces.

Microbial morphology and phosphatogenesis

Whatever the exact mechanism of fixation may be, from the above evidence it is concluded that the phosphate in phosphatic chalks is essentially localized within microbial remains. In the absence of evidence to the contrary, we must assume that the mineralogy of the phosphate (carbonate-fluorapatite was the only phosphate mineral identified by Jarvis 1980a, b, c) is the same in both types of microbial structure. The two types probably correspond to two different kinds of microbial communities, the ecologies and modes of growth of which were different.

The microbial communities which were initiated as thin crusts on hardgrounds surfaces or occurred as coatings on grains, grew on a substrate. The microbes could penetrate a short distance into their substrate (Fig. 1h), but this kind of growth seems to have been very limited, and their growth was essentially a surface-related process. Growth was able to continue despite intermittent occlusion of the surface by sediment particles which were subsequently included into the crusts. Crusts are absent on unlithified sediments (both phosphatic and white chalks), indicating that the microbes required a surface compact and stable enough to permit colonization. However, many of coated grains in these same chalks have a rounded chalk nucleus. These nuclei must correspond to granules compact enough for the microbes to have fixed onto them; some of these nuclei were undoubted rounded intraclasts, others were probably faecal pellets. In some coated grains, the nucleus consists of a foraminiferal test or other bioclast surrounded by chalk and then coated by microbial laminae; such a structure is also best explained as having a faecal origin. These faecal pellets were probably produced by the infauna responsible for the intense bioturbation of the chalks.

Ovoid bacteria and microbial colonies with botryoidal surfaces correspond to another kind of community which grew in various cavity structures including chambers of foraminiferans and bryozoans, primary and secondary porosity developed within echinoderm stereome, microborings within bioclasts, primary porosity within coprolites and faecal pellets. The soft chalky matrix of the phosphatic chalks does not contain such bacterial remains, whereas they are common in the lenses deliminated by crusts on hardgrounds and in the chalky infills of foraminiferal chambers. Thus again these microbes only appear to have be able to colonize pore-spaces of semilithified material. In phosphorites of Tunisia, the same kind of microbial community has transformed faecal pellets into isotropic phosphate grains, the fine matrix between which has not been phosphatized (Lamboy 1982b, 1988b). Surficial phosphatization of peloids, whether or not surrounded by anisotropic phosphate cortices, has also been described in the Maastrichtian phosphatic chalk of Ciply, Belgium (Poels & Robaszynski 1988).

The existence of a microbial community on the surface of a substrate seems to have influenced the growth of the microbes forming the other community inside the material. The microstromatolitic laminae, by trapping lenses of sediment on hardground surfaces, have permitted the growth of ovoid bacteria and microbial colonies within the accreted chalk. Numerous grains with anisotropic phosphate cortices contain chalk nuclei which display little or no phosphatization, whereas faecal pellets without cortices have been transformed into isotropic phosphate by the intragranular microbial community. This suggests that the presence of microbial laminae around a nucleus does not necessarily lead to, and may in fact impede, its phosphatization. Dissolution of skeletal carbonate inside coated grains indicates the cortex is chemically permeable, but the growth of microbial laminae around a grain probably limits the ability of the other microbial community to penetrate and develop inside the nucleus.

It is concluded, therefore, that the two kinds of microbial communities developed in phosphatic chalks had different habitats and modes of development, although interactions between the two occurred. These differences allow a better understanding of the location of the phosphate and its relation to the non-phosphatic particles in the sediment in which the microbes grew.

Phosphatogenesis in phosphatic chalks compared to other phosphorites

Each of the two phosphatogenic processes observed in phosphatic chalks involving the surficial and intragranular development of microbial communities can be evoked in the formation of other phosphate deposits.

Major phosphate deposits of Precambrian age are commonly composed of stromatolites (e.g. papers in Cook & Shergold 1986; Sisodia & Chauhan 1990). In the Cambrian phosphorite deposits from the Karatau Basin, Eganov (1988) described typical stromatolites, thin flat-laminated phosphatic crusts with microstromatolitic columns up to 1 mm high, and phosphate ooids whose cortices have the same origin as the crusts. The latter structures seem similar to those described here.

Phosphatic stromatolites and microstromatolites are also recorded in the mid-Cretaceous condensed deposits of the northern Tethyan margin (Krajewski 1981a, b, 1983, 1984; Delamette 1988; Föllmi 1990). Phosphate laminated crusts have been described from the recent phosphatic and glauconitic nodules from the continental margin of northern Spain (Lamboy 1976). The microstructure of these crusts is similar to those described in this paper, supporting the microbial origin proposed by Delamette (1988). Moreover, the presence of microbial crusts in recent phosphatic nodules indicates that microbial processes are concentrating phosphorus in modern sediments.

Phosphatic grains with anisotropic cortices exist in the Middle Eocene granular phosphorites of Senegal (Flicoteaux 1982) and Guinea Bissau (Prian et al. 1987), but their percentage is much lower than that of totally isotropic grains. The structure of these cortices is similar to that of the coated grains in the Cretaceous phosphatic chalks described in this paper, so a microbial origin may be envisaged.

Numerous authors have observed phosphatic remains of ovoid bacteria (O'Brien et al. 1981; Zanin et al. 1985, 1987; Lamboy & Monty 1987, 1989; Purnachandra Rao & Nair 1989; Bréhéret 1989) and of microbial colonies with a botryoidal surface (Lamboy & Monty 1987, 1989; Soudry & Lewy 1988) in various phosphorites. In numerous cases, microbial remains have been observed inside phosphatic components, corresponding to the intragranular microbial community described here. Composite phosphate moulds and evidence of carbonate dissolution have also been frequently recorded. In Tunisian phosphorites, for example, microbially mediated phosphatization inside grains appears to be the dominant phosphatogenic process (Lamboy 1988a, b). In these phosphorites, foraminiferans and diatoms are also preserved as composite phosphatic moulds, confirming that dissolution of carbonate and also of biogenic opal postdated phosphatization.

On the other hand, the tortuous hollow phosphatic tubules, 5 to 15 μm external diameter, described from several phosphorites and interpreted as cyanobacterial remains (Soudry 1983, 1987; Soudry & Champetier 1983; Dahanayake & Krumbein 1985; Southgate 1986; Soudry & Southgate 1989; Abed & Fakhouri 1990) have not been observed in the phosphatic chalks studied here.

Conclusions

Detailed petrographic and SEM study of the phosphatic chalks of northern France has resulted in a better understanding of their genesis. Textural evidence suggests that phosphate occurs solely as mineralized microbial remains. Two kinds of microbial populations are distinguished according to their occurrence and morphology. Lamellar microbial mats grew on both hardground and grain surfaces producing complex phosphatic crusts and anisotropic phosphate coatings respectively. A second kind of community, composed of dispersed ovoid bacteria and of void-filling microbial colonies, developed in intragranular and intergranular pore-spaces within semilithified sediment, and resulted in the precipitation of radialfibrous phosphate with a botryoidal surface. Phosphatization of microbial bodies is commonly accompanied by dissolution of chalk and other carbonate trapped within the mineralized area.

Comparison with other phosphorites indicates that the microbial structures observed in phosphatic chalks may also be identified in other deposits. It is concluded that microbial activity exerts a major control on phosphatogenesis in many phosphorites.

I thank G. Bignot who kindly provided a sample of the basal hardground from Hardivillers Quarry. I also thank warmly I. Jarvis for critical reading and for English improvements of the first manuscript. This paper is a contribution of the Project 156 (Phosphorites) of the International Geological Correlation Program.

References

ABED, A. M. & FAKHOURI, K. 1990. Role of microbial processes in the genesis of Jordanian Upper Cretaceous phosphorites. *In*: NOTHOLT, A. J. G. & JARVIS, I. (eds) *Phosphorite Research and Development*. Geological Society, London, Special Publication, **52**, 193–203.

AMES, L. L. 1959. The genesis of carbonate apatites. *Economic Geology*, **54**, 829–841.

BEN AMAR, Z. 1985. *Contribution à l'étude des phosphates de Jellabia (Bassin de Gafsa, Tunisie)*. Unpublished Thesis, University of Rouen.

BIGNOT, G. 1980. A la recherche des bactéries fossiles. *Bulletin trimestriel de la Société Geologique de Normandie*, Le Havre, **67**, 15–41.

BREHERET, J. G. in press. Phosphatic concretions in black facies of the mid-Cretaceous marls of the Vocontian Basin (SE France) and in Site 369 D.S.D.P.: witnesses of benthic microbial activity. *Marine Geology*, Special Issue.

CAYEUX, L. 1939. *Les phosphates de chaux sédimentaires de France, I: France métropolitaine*. Etude des gites minéraux de la France, Imprimerie Nationale, Paris.

CHOUDHURI, R. & ROY, A. B. 1986. Proterozoic and Cambrian phosphorites deposits. Jhamarkotra, Rajasthan, India. *In*: COOK, P. J. & SHERGOLD, J. H. (eds). *Phosphate Deposits of the World, Vol. 1*, Cambridge University Press, 202–219.

COOK, P. J. & SHERGOLD, J. H. (eds) 1986. *Phosphate Deposits of the World: Vol. 1: Proterozoic and Cambrian Phosphorites*. Cambridge University Press.

DAHANAYAKE, K. & KRUMBEIN, W. E. 1985. Ultrastructure of a microbial mat-generated phosphorite. *Mineralium Deposita*, **20**, 260–265.

DELAMETTE, M. 1988. L'évolution du domaine helvétique entre Bauges et Morcles de l'Aptien supérieur au Turonien. Séries condensées, phosphorites et circulations océaniques. *In*: *Publications du Département de Géologie et de Paléontologie de l'Université de Genève*, **5**.

EGANOV, E. A. 1988. Phosphate deposition and stromatolites. *Publication of the Academy of Sciences of the USSR, Siberian Division, Institute of Geology and Geophysics, Novosibirsk*, 56–89 (in Russian).

EL FALEH, E. 1988. *Les mécanismes de synthèse de l'apatite par activité bactérienne; rôle et comportement de quelques éléments minéraux. Application aux phosphates sédimentaires*. Unpublished Thesis, University of Strasbourg.

FIKRI, A., LAMBOY, M., BENALIOULHAJ, S., TRICHET, J. & BELAYOUNI, H. 1989. Contribution à l'étude pétrologique de la matière organique dans les phosphates naturels. Nouvelles approches méthodologiques. *Bulletin de la Société géologique de France*, **5**, 979–987.

FLICOTEAUX, R. 1982. *Genèse des phosphates alumineux du Sénégal occidental. Etapes et guides de l'altération*. Sciences Géologiques, Strasbourg, Mémoire 67.

FOLLMI, K. B. 1990. Condensation and phosphogenesis: example of the Helvetic mid-Cretaceous (northern Tethyan margin). *In*: NOTHOLT, A. J. G. & JARVIS, I. (eds) *Phosphorite Research and Development*. Geological Society, London, Special Publication, **52**, 237–252.

GIOT, D. 1988. Granular phosphate ores: a proposal for the definition of particles. *Chronique de la recherche minière, Special Issue Phosphates, B.R.G.M. Orléans*, 51–66.

JARVIS, I. 1980a. *Genesis and diagenesis of Santonian to early Campanian (Cretaceous) phosphatic chalks of the Anglo-Paris Basin*. PhD Thesis, University of Oxford.

—— 1980b. The initiation of phosphatic chalk sedimentation — the Senonian (Cretaceous) of the Anglo-Paris Basin. *In*: BENTOR, Y. K. (ed.) *Marine Phosphorites*. SEPM Special Publication, **29**, 167–192.

—— 1980c. Geochemistry of phosphatic chalks and hardgrounds from the Santonian to early Campanian (Cretaceous) of northern France. *Journal of the Geological Society, London*, **137**, 705–721.

—— 1984. Rare-earth element geochemistry of late Cretaceous chalks and phosphorites from northern France. *Special Publication of the Geological Survey of India*, **17**, 179–190.

KENNEDY, W. J. & GARRISON, R. E. 1975. Morphology and genesis of nodular chalks and hardgrounds in the Upper Cretaceous of southern England. *Sedimentology*, **22**, 311–386.

KRAJEWSKI, K. 1981a. Phosphate microstromatolites in the High-Tatric Albian limestones in the Polish Tatra Mountains. *Bulletin de Académie polonaise des Sciences, Série des Sciences de la Terre*, **29**, 175–183.

—— 1981b. Phosphate pisolite structures from condensed limestones of the High-Tatric Albian (Tatra Mts). *Annales de la Société Géologique de Pologne*, **51**, 339–352.

—— 1983. Albian pelagic phosphate-rich macrooncoids from the Tatra Mts, Poland. *In*: PERYT, T. M. (ed.) *Coated Grains*. Springer Verlag, Berlin, 344–357.

—— 1984. Early diagenetic phosphate cements in the Albian condensed glauconitic limestone of the Tatra Mountains, Western Carpathians. *Sedimentology*, **31**, 443–470.

LAMBOY, M. 1976. *Géologie marine et sous-marine du plateau continental au Nord-Ouest de l'Espagne. Genèse des glauconies et des phosphorites*. Thesis, University of Rouen.

—— 1982a. Importance des pelotes fécales comme origine des grains de phosphate: l'exemple du gisement de Gafsa (Tunisie). *Comptes endus de l'Académie des Sciences*, Paris, **295**, 595–600.

—— 1982b. La phosphatization en micro-milieu granulaire d'après un exemple tunisien. *Comptes rendus de l'Académie des Sciences*, Paris, **295**, 799–802.

—— 1986. Relations entre propriétés optiques et nannostructures des grains de phosphate. Implications génétiques. *Revue de Géologie dynamique et Géographie physique*, Paris, **27**, 311–318.

—— 1987a. Genèse de grains de phosphate à partir de débris de squelette d'échinodermes: les processus et leur signification. *Bulletin de la Société géologique de France*, **8**, 759–768.

—— 1987b. Genèse des phosphates granulaires. Enseignements des grains centrés sur des foraminifères. Importance et modalités de la précipitation. *Comptes rendus de l'Académie des Sciences*, Paris, **304**, 435–440.

—— 1988a. New data related to the petrography and the genesis of granular phosphorites. *IGCP 156, Regional meeting on Cretaceous and Tertiary phosphorites. Abstracts*, Amman, 15.

—— 1988b. The role of bioturbation in the genesis of granular phosphorites. *IGCP 156, Regional meeting on Cretaceous and Tertiary phosphorites, Abstracts*, Amman, 16.

—— & MONTY, C. 1987. Bacterial origin of phosphatized grains. *Terra Cognita*, **7**, 207.

—— & —— in press. Observations on microbial phosphorites. *Marine Geology*, Special Issue.

LANIER, W. 1988. Structure and morphogenesis of microstromatolites from Transvaal supergroup,

South Africa. *Journal of Sedimentary Petrology*, **58**, 89–99.

LEWY, Z. 1990. Pebbly phosphate and granular phosphorite (Late Cretaceous, southern Israel) and their bearing on phosphatization processes. *In*: NOTHOLT, A. J. G. & JARVIS, I. (eds) *Phosphorite Research and Development*. Geological Society, London, Special Publication, **52**, 167–178.

LUCAS, J. & PRÉVÔT, L. 1981. Synthèse d'apatite à partir de matière organique phosphorée (ARN) et de calcite par voie bactérienne. *Comptes rendus de l'Académie des Sciences*, Paris, **292**, 1203–1208.

—— & —— 1985. The synthesis of apatite by bacterial activity: mechanism. *Sciences Géologiques, Strasbourg, Mémoire*, **77**, 83–92.

——, EL FALEH, E. M. & PRÉVÔT, L. 1990. Experimental study of the substitution of Ca by Sr and Ba in synthetic apatites. *In*: NOTHOLT, A. J. G. & JARVIS, I. (eds) *Phosphorite Research and Development*. Geological Society, London, Special Publication, **52**, 33–47.

MONTY, C. 1976. The origin and development of cryptalgal fabrics. *In*: WALTER, M. R. (ed.) *Stromatolites*, Developments in Sedimentology, **20**, 193–249.

—— 1982. Cavity or fissure dwelling stromatolites (endostromatolites) from Belgian Devonian mud mounds. *Annales de la Société géologique de Belgique*, **105**, 343–344.

O'BRIEN, G. W., HARRIS, J. R., MILNES, A. R. & VEEH, H. H. 1981. Bacterial origin of East Australian continental margin phosphorites. *Nature*, **294**, 442–444.

POELS, J. P. & ROBASZYNSKI, F. 1988. Les grains phosphatés de la craie de Ciply (Maastrichtien, Belgique). Eléments d'interprétation pour la phosphatogenèse. *Mededelingen Rijks Geologische Dienst*, Roermond, The Netherlands, **42**, 51–75.

PRIAN, J. P., GAMA, P., BOURDILLON, C. & ROGER, J. 1987. Le gisement de phosphate éocène de Farim-Saliquinhé (République de Guinée-Bissau). *Chronique de la Recherche minière*, **486**, 25–54.

PURNACHANDRA RAO, V. & NAIR, R. R. 1988. Microbial origin of the phosphorites of the Western continental shelf of India. *Marine Geology*, **84**, 105–110.

SISODIA, M. S. & CHAUHAN, D. S. 1990. The influence of magnesium ions during the formation of stromatolitic phosphorites of Udaipur, Rajasthan, India. *In*: NOTHOLT, A. J. G. & JARVIS, I. (eds) *Phosphorite Research and Development*. Geological Society, London, Special Publication, **52**, 313–320.

SLANSKY, M. 1986. *Geology of Sedimentary Phosphate*. North Oxford Academic, Oxford, and B.R.G.M., Orléans.

SOUDRY, D. 1979. Intervention de schizophytes dans la phosphomicritisation des débris osseux. *Comptes rendus de l'Académie des Sciences*, Paris, **288**, 669–671.

—— 1983. Etude de la série phosphatée de la région d'Ein Yahav (Negev central, Israel). Logique séquentielle, pétrologie, approche de la phosphatogenèse. Unpublished Thesis, University of Nancy.

—— 1987. Ultra-fine structures and genesis of the Campanian Negev high-grade phosphorites (southern Israel). *Sedimentology*, **34**, 641–660.

—— & CHAMPETIER, Y. 1983. Microbial processes in the Negev phosphorites (southern Israel). *Sedimentology*, **30**, 411–423.

—— & LEWY, Z. 1988. Microbially influenced formation of phosphate nodules and megafossil moulds (Negev, southern Israel). *Palaeogeography, Palaeoclimatology, Palaeoecology*, **64**, 15–34.

—— & SOUTHGATE, P. N. 1989. Ultrastructure of a middle Cambrian primary nonpelletal phosphorite and its early transformation into phosphate vadoids: Georgina Basin, Australia. *Journal of Sedimentary Petrology*, **59**, 53–64.

SOUTHGATE, P. N. 1986. Cambrian phoscrete profiles, coated grains, and microbial processes in phosphogenesis: Georgina Basin, Australia. *Journal of Sedimentary Petrology*, **56**, 429–441.

TABATABAI, C. M. 1977. *La sédimentation phosphatée (ses modalités). Pétrographie et sédimentologie des craies phosphatées du Nord du bassin de Paris*. Thesis, Université Pierre et Marie Curie, Paris.

VAN CAPPELLEN, P. & BERNER, R. A. 1988. A mathematical model for the early diagenesis of phosphorus and fluorine in marine sediments: apatite precipitation. *American Journal of Science*, **288**, 289–33.

ZANIN, Y., LETOV, S. V., KRASIL'NIKOVA, N. A. & MIRTOV, Y. V. 1985. Phosphatized bacteria from Cretaceous phosphorites of European platform and Palaeocene phosphorites of Morocco. *Sciences Géologiques, Strasbourg, Mémoire*, **77**, 79–81.

——, GORLENKI, V., MIRTOV, Y., KRASIL'NIKOVA, N. & LETOV, S. 1987. Bacteriomorphic formations in nodular and granular phosphorites. *Soviet Geology and Geophysics, Novosibirsk*, **28**, 39–44.

Pebbly phosphate and granular phosphorite (Late Cretaceous, southern Israel) and their bearing on phosphatization processes

ZEEV LEWY

Geological Survey of Israel, 30 Malkhe Yisrael Street, Jerusalem 95501, Israel

Abstract: Phosphate nodules, including common mineralized internal moulds of macrofauna, are found only at certain levels in the chalky, locally bituminous Ghareb Formation (uppermost Campanian–Maastrichtian) of southern Israel. The microstructure of these nodules suggests that they consist of marine sediment cemented by a microbially-mediated apatite, precipitated locally in protected microenvironments, e.g., within partly closed burrows and shells, or within the upper sediment layer. The globular microstructures commonly recognized in the phosphate cements of the pebbly phosphates are regarded as mineralized cells of endobenthic micro-organisms. The global occurrence of pebbly phosphate in diverse lithofacies throughout the geological column indicates the tolerance of these globular micro-organisms to a wide range of ecological conditions.

The grains in the Upper Cretaceous phosphorites of the Negev (southern Israel) consist of bone fragments, phosphatized faecal pellets and ovular grains of different types. The commonest variety of ovular grain consists of microcrystalline phosphate (microphosphate) containing relics of globular microstructures, either forming the whole grain, coating a foraminifera or bone fragment, or cementing small clasts. The latter types resemble the pebbly phosphates and can be regarded as micronodules. The matrix of these granular phosphorites is mainly micritic, and in places argillaceous. In some Negev phosphorites the ovular grains and their matrix consist mainly of loosely packed tubules. These tubular microstructures resemble mineralized filaments of microphytes which form microbial mats on modern shallow seafloors lying within the photic zone. The origin of these granular phosphorites is discussed, and a comparison with the pebbly phosphates indicates that different ecological controls and microbial mediators resulted in the different kinds of phosphate bodies.

Since most of the phosphate matter can be related to microbial activity, it is doubtful whether direct replacement of solid carbonate by phosphate contributes significantly to phosphogenesis in sediments. It is suggested, however, that in many deposits phosphate-precipitating micro-organisms have penetrated and mineralized the borings of algae and other endolithic groups, or have filled dissolution cavities within lithified carbonates.

The Upper Campanian–Maastrichtian sequence in southern Israel comprises the siliceous–phosphatic Mishash Formation and the overlying chalk of the Ghareb Formation. The Mishash Formation consists of a lower Chert Member made of alternating beds of chert, porcelanite and some carbonate (Soudry *et al.* 1985), and an upper Phosphate Member consisting of alternating beds of granular phosphorite, argillaceous carbonates (dominated by buliminid bethonic foraminifera) and some chert and porcelanite. Towards the upper part of the latter member high-grade granular phosphorite layers become more abundant in the sequence, forming the Phosphorite Unit of the uppermost part of the Mishash Formation. The contact with the overlying argillaceous chalk of the Ghareb Formation may be sharp, suggesting a paraconformity. Nevertheless, the lower part of the Ghareb Formation is locally bituminous and contains phosphate grains reaching a P_2O_5 content of 1–4% (e.g., in the Oron region, Fig. 1; Shahar 1968). Higher in the sequence the formation grades into almost pure coccolith-foraminiferal chalk of a rather monotonous appearance. The Ghareb Formation contains very few macrofossils; records are generally restricted to a few forms possessing calcitic shells, such as pectinid and oyster bivalves, the brachiopod *Gyrosoria gracilis* (Schlotheim) and serpulid worm tubes. Local concentrations of phosphatized internal moulds representing a diverse macrofaunal assemblage are, therefore, of great biostratigraphic and palaeoecological significance. Recently, Soudry & Lewy (1988) described the petrography and distribution of spherical phosphatic nodules and invertebrate internal moulds in the lower part of the formation, and related these to the palaeogeography and tectonic history of the shallow-marine palaeoenvironments which characterize the Late Cretaceous of southern Israel. They suggested that the spherical nodules were portions of an uppermost sediment layer from a lagoonal

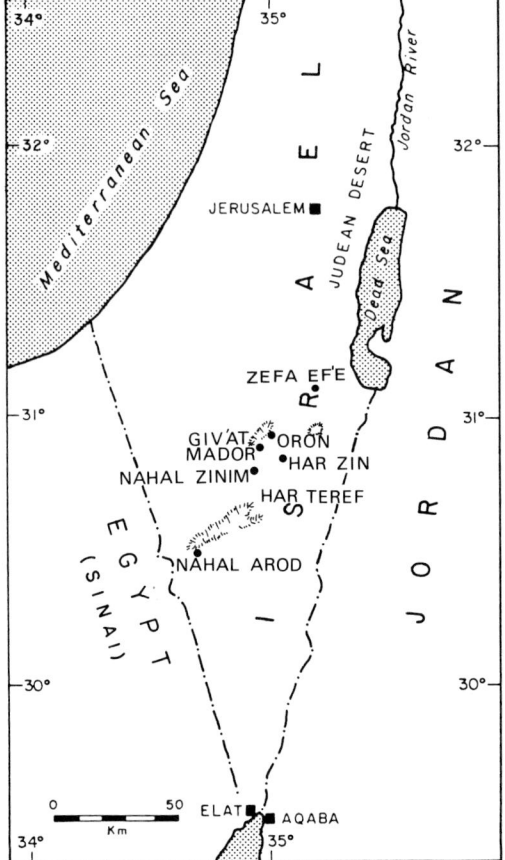

Fig. 1. Map of southern Israel showing the location of the phosphate deposits discussed in the text.

environment, densely colonized by microorganisms and forming an organic mat at the top. Soudry & Lewy (1988) suggested that this cohesive, yet plastic sediment was disrupted mainly by bioturbation, rolled by currents and redeposited as mineralized lag deposits. Lithification resulted (Soudry & Lewy 1988) from microbially-mediated precipitation of apatite cement on, and between sedimentary particles, and did not affect the composition of the enclosed calcareous material.

Additional nodules and moulds collected from Late Cretaceous basinal palaeoenvironments of the northeastern Negev has yielded additional forms of nodules, the nature of which suggests a mode of formation different from that of the shallow-marine nodules. Both the basinal and shallow-marine types of pebbly phosphate, however, display common textural features, completely different from those ob-

served by Soudry (1987) in the high-grade granular phosphorites of the upper part of the Mishash Formation. A comparison between the pebbly phosphates and the granular phosphorites of southern Israel is presented, and it will be argued that different palaeoenvironmental controls and microbial assemblages are responsible for the formation of the two types of phosphate.

Pebbly phosphates in pelagic sediments

Occurrence

The Ghareb Formation has been divided in its type region at Giv'at Mador (Oron Syncline, Fig. 1; Shahar 1968) into three members (Fig. 2): the lower Oil Shale Member (19.8 m), the middle Marly Member (18 m), and the upper Chalky Member (37.5 m). The lower 10 m of the Oil Shale Member consist of dark grey, bituminous (c. 6–13% total organic carbon, TOC; Shahar 1968) argillaceous chalk, overlain by a 3–4 m thick light grey-pink unit of lower organic matter content (c. 4% TOC; Shahar 1968), indicating a zone of oxidation. Phosphate nodules occur close below the oxidized zone (Fig. 2) and are associated with scattered oysters, vertebrate remains and a few phosphatic moulds of baculitid ammonites. The overlying upper part of the Oil Shale Member (c. 7 m) contains 2.6–1.5% TOC (Shahar 1968) but its ochreous colour distinguishes it from the highly bituminous, dark grey lower part. Rare, small (<1 cm) phosphate nodules are present in the lower part of this less bituminous unit (Fig. 2).

The Ghareb Formation exposed in the Nahal Zin Syncline (southeast of Oron; Roded 1982) at Har Zin (Fig. 1) consists of a 14 m thick Oil Shale Member and a 35 m thick chalk and argillaceous chalk sequence, which can be correlated with both the Marly and the Chalky Members recognized in the Oron region (Fig. 2). The lower 10 m of the Oil Shale Member is highly bituminous (dark grey) and is topped by a 5 cm thick bed of pebbly phosphate, phosphatic invertebrate moulds, oyster shells and vertebrate remains (Fig. 2). Nodules over 1 cm are uncommon and most phosphate particles are faecal pellets, bone fragments and grains (tiny nodules) 0.5–1.5 mm in size. This condensation of biogenic components at the top of the bituminous sequence indicates an increase in water-energy (winnowing) and improved bottom oxygenation (benthonic activity). This latter trend is reflected also by the associated

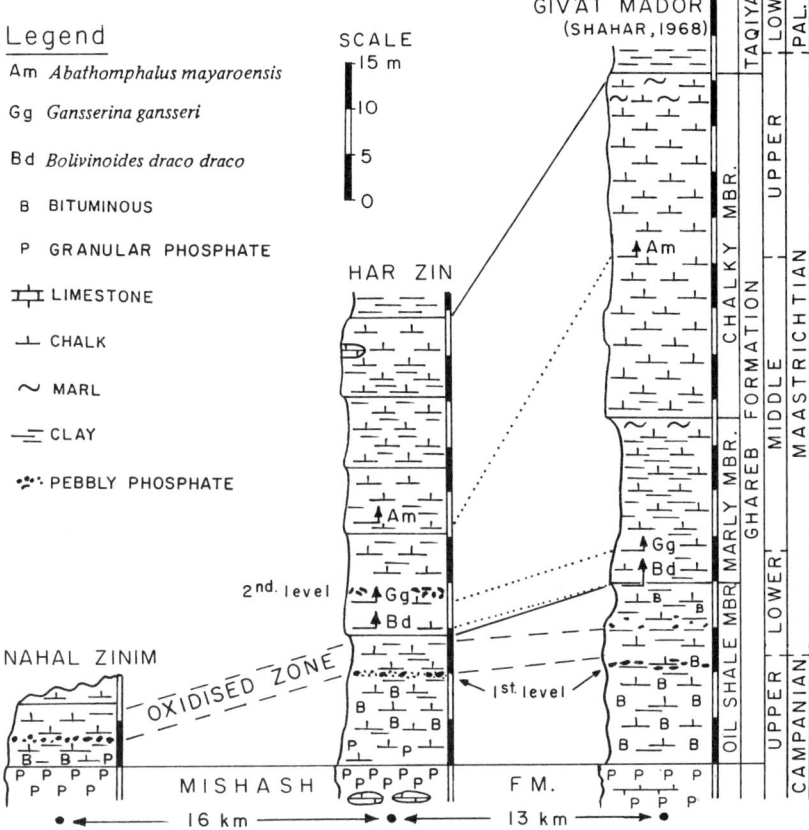

Fig. 2. Litho- and biostratigraphic correlation of the Ghareb Formation from its type area at Giv'at Mador (Oron Syncline) to Har Zin (Zin Syncline) and Nahal Zinim. Stratigraphic data from Magaritz et al. (1985), Reiss et al. (1985), Shahar (1968) and Lewy (this paper).

lithological change to the overlying ochreous-pinkish coloured zone of oxidation (c. 4 m thick) which terminates the Oil Shale Member. Approximately 5 m above the base of the 'Marly Chalky' Member at Har Zin, phosphate nodules and invertebrate moulds occur, scattered throughout 45 cm of white chalk (Fig. 2), in association with calcitic bivalves, brachiopods and serpulids.

At Nahal Zinim (Figs 1 & 2) the bituminous lower part of the Oil Shale Member is about 3 m thick. It is overlain by 3–4 m of a grey-pinkish coloured zone with phosphatic nodules and invertebrate moulds at its base. Higher parts of the Ghareb sequence are not preserved at this site (Fig. 2).

Tentative correlation of the lower pebbly phosphate level in the upper part of the highly bituminous sequence, as well as the overlying oxidized zone in all three sites (described above) is supported by the first occurrence of *Bolivinoides draco draco* (Marsson) at the top of the Oil Shale Member in the Oron region and in the lowermost part of the 'Marly Chalky' Member at Har Zin (Fig. 2). The oxidation zone and the associated pebbly phosphate in the Oron, Har Zin and Nahal Zinim regions reflect, therefore, a late Early Maastrichtian event rather than a much younger late diagenetic oxidation (Shiloni 1988).

Nodule shape and texture

Nodules in the lower pebble bed of the Oron region have rounded, irregular tabloid shapes, 1.5 to 2.5 cm across and 0.8 to 1.2 cm thick. Their surfaces are commonly covered by lines of single or paired scratch marks (produced by crustacean claws?), and incised by tangential and perpendicular burrows (1–3 mm in diameter), and tiny (0.5 mm) shafts of burrowing or boring organisms (Fig. 3A–G). Similar structures are commonly developed on comparable nodular phosphates elsewhere, such as those from the Albian Gault Clay of southern England (I. Jarvis pers. comm. 1989).

Nodules from the upper phosphatic bed at Har Zin [which is coincident in the first occurrence of *Gansserina gansseri* (Gandolfi) of Middle Maastrichtian age, Figs 1 & 2] consist of cemented granular phosphate, which produces a rough surface due to the protrusion of hard bone fragments and peloids from the interior. Ovoid faecal pellets are quite abundant. Many nodules have irregular subcylindrical to elongated shapes, 0.9 to 1.3 cm in diameter. Broken fragments exhibit a composite internal structure consisting of clasts of several different lithologies. Nodules differ mainly in the concentrations of the granular fraction and the colour of the enclosing microcrystalline phosphate (microphosphate). A dark zone defines the boundaries of individual clasts, indicating deformed contacts between semi-lithified material. These nodules do not exhibit any clear burrows, borings or ?crustacean scratches.

Large phosphate nodules, 5–7 cm in diameter, were also collected from the lower part of the Ghareb Formation at Nahal Arod (Fig. 1), situated on the southern flank of the Ramon Anticline, which developed in Senonian–Eocene times. These nodules consist of cemented aggregates of mineralized clasts (Fig. 3O & P). Deformation of adjacent clasts indicates that amalgamation occurred while most clasts remained plastic, although a few had indurated rims. The compound nodules are coated by thin apatitic veneers and are easily differentiated from the surrounding friable, argillaceous chalk, which contains some granular phosphate.

In addition to the phosphatic internal moulds of macrofauna, abundant mineralized infills of *Rhizocorallium* (Fig. 3J) and a few *Ophiomorpha* burrows are also present in most of the pebbly phosphate levels. When well-preserved, the latter exhibit finely nodose surfaces (Fig. 3H–I) consisting of a thin layer of apatite coating small faecal pellets and other phosphate grains (similar, but coarser features are described in *O. nodosa*; Kennedy & Sellwood 1970). The diameter of the shafts is about 1 cm, suggesting that some of the subcylindrical nodules (second level at Har Zin) are probably poorly preserved mineralized infills of similar *Ophiomorpha* burrows.

The tabloid nodules from the lower bed of the Oron region consist of microphosphate in a very dense texture. In places moulds recalling coccolith plates (Fig. 3Q) or dissolved fragments of foraminifera can be seen, as well as a rare dolomite and pyrite crystals. A few voids preserve a globular and cauliflower-like microstructure (Fig. 3R) similar to those observed in the phosphate cement in the spherical nodules from the Late Cretaceous shallow marine facies in southern Israel (Soudry & Lewy 1988). The complete infilling of porosity and of the globular structures themselves, by a microbially-mediated microphosphate cement obliterates any evidence of microbial involvement in mineralization (Soudry & Lewy 1988). Such a dense microphosphate cement incorporates bone fragments, phosphatic peloids and faecal pellets (Fig. 4) in the subcylindrical nodules from the second level at Har Zin (see above).

Origin of the phosphate nodules

The Late Cretaceous phosphate nodules described here accumulated in intra-shelf basins characterized by pelagic regimes, as indicated by the associated planktonic foraminifera and nannoplankton in the Oil Shale Member in the Oron region (Reiss *et al.* 1985) and at Har Zin. Rates of sedimentation of the Ghareb chalk were around 1 cm per 1000 years (the Ghareb Formation, which is 76 m thick in the type section at Giv'at Mador and 49 m at Har Zin, accumulated between 73 Ma and 66.5 Ma, Fig. 2; Lewy 1986, Shahar 1968). The initial high porosity of the upper layer of this Late Cretaceous pelagic sediment can be evaluated by comparison with recent deep-sea analogues, where calcareous oozes reach porosities of 70–90% (Cook & Egbert 1983). The relative low volume of primary sedimentary particles within the original Oil Shale mud may have been locally reduced even further by the growth and expansion of the microbial colonies in the intergranular spaces. This may explain the low content of particles (or their moulds) recognized in the tabloid phosphate nodules (Oron region), compared to the phosphatized burrow fills in the second pebbly level at Har Zin and to the spherical, shallow-marine nodules described by Soudry & Lewy (1988).

It can be concluded that Late Cretaceous–Early Tertiary pebbly phosphates (including invertebrate phosphatic moulds), in the Negev all consist of marine sediment, whose particles have been cemented by apatite, probably precipitated on degrading globular microbial structures (cf. Soudry & Champetier 1983; Soudry & Lewy 1988; Lamboy 1990). These microorganisms must have lived within the muddy

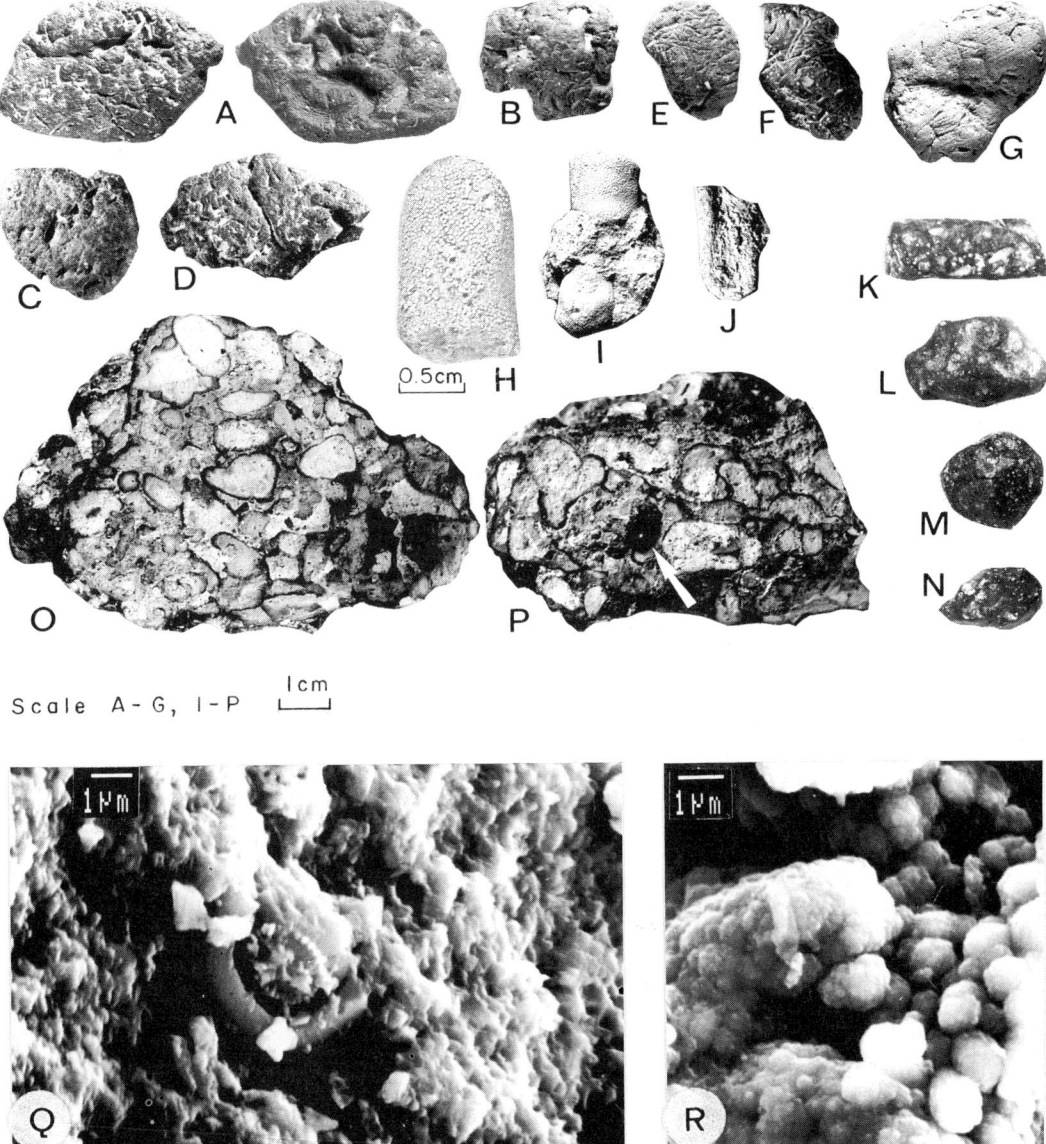

Fig. 3. Pebbly phosphates from the Ghareb Formation (Upper Campanian–Maastrichtian) of southern Israel. (A–G) nodules from the lower phosphate bed in the Oron region: (A–D) tabloid, intensively burrowed (A: both sides); (E–F) coarse scratches (by crustaceans?); (G) sets of fine scratches (gastrolith?). (H–I) well-preserved moulds of *Ophiomorpha* burrows; (H), Har Zin, upper phosphate bed; (I) Har Teref. (J) *Rhizocorallum* from the lower level at Har Zin. (K–N) abraded-polished nodules, probably gastroliths, lower phosphate bed, Oron region (K may be an abraded internal mould of a baculitid ammonite). (O–P) conglomeratic phosphates (fractured surfaces) from Nahal Arod; arrow in P points at the protruding surface of an individual nodule, indicating earlier lithification than the surrounding clasts. (Q–R) SEM photomicrographs of a phosphate nodule, lower nodule bed at Giv'at Mador; (Q) dense microphosphate with a composite mould (centre) which resembles a coccolith plate; (R) cluster of globular microbial structures preserved in a void.

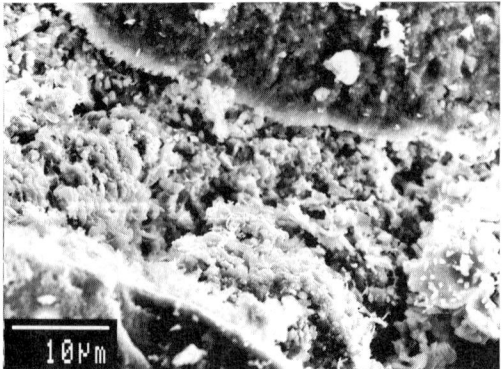

Fig. 4. SEM micrograph of a coarse-grained, subcylindrical nodule, (upper bed at Har Zin), showing microphosphate cement in between two phosphatic particles.

sediment in the uppermost sediment layer in lagoons (Soudry & Lewy 1988), and in marine settings both in isolated sheltered microenvironments (e.g. within mollusc shells or in burrows), and at discrete levels within the sediment where they developed around scattered nuclei of colonization which were in contact with the surrounding pore-water. These latter microbial colonies produced indurated patches which where within the reach of burrowers, and in some cases were temporarily exhumed and exposed on the sea floor, where grazers and other organisms scratched the surfaces of semi-lithified nodules and bored into cemented ones. Benthonic activity during the development of the phosphate nodules in the upper part of the highly bituminous (9–13% TOC), pelagic chalk in the Oron region, as well as in the succeeding zone of oxidation (<4% TOC, Shahar 1968), indicates gradual improvement in oxygenation of the bottom and probably periodic current activity associated with bioturbation, as deduced for the coeval lower pebbly phosphate at Har Zin.

Well-cemented nodules exposed on the sea bottom may have served as gastroliths ('stomach stones'), used by certain marine reptiles (e.g., Plesiosauria; Swinton 1973) to improve their digestive processes, as suggested by the few abraded and polished nodules (Fig. 3K–N, G?) found together with the non-abraded ones. The truncated edges of the peripheral coarse sedimentary particles in this rare type of nodule could hardly have resulted merely from a more intensive or longer history of reworking, which, however, did not affect the majority of the phosphate nodules.

The subcylindrical phosphate nodules of the second level at Har Zin, are regarded as infills of burrows (*Ophiomorpha*?) by phosphatic particles (including partly mineralized faecal pellets) and organic matter. These phosphatic-organic components probably stimulated microbial colonization which led to local phosphate cementation of the burrow fills only, without affecting the surrounding chalk, the white non-bituminous nature of which suggests deposition under normal aerated pelagic bottom-water conditions.

Thin phosphate crusts on calcareous concretions embedded in Upper Campanian granular phosphorite of the Mishash Formation at Nahal Zinim (Fig. 1) also contain microbial globular microstructures in the phosphatic matter similar to those cementing the pebbly phosphates (Soudry & Lewy 1988). SEM analyses indicate a well-phosphatized, narrow, peripheral zone followed inward by a zone of phosphatic dendroid meshwork, which penetrates a few millimetres into the micrite. Trace fossils and sedimentary features suggest that here the phosphatization proceeded when the concretions were temporarily exposed or buried at very shallow depth, while their peripheral zone was not completely lithified. Here, it is suggested that the micro-organisms penetrated the intergranular spaces and were stopped at the indurated core.

Granular phosphorites of southern Israel

Nathan *et al.* (1979) reviewed the mineralogy, petrography and distribution pattern of the Upper Cretaceous phosphorite deposits of the Negev (southern Israel). The granular fraction in these phosphorites consists of various kinds of ovular grains and bone fragments; the former include phosphatized oval and cylindrical bodies which resemble faecal pellets (coprolites). The matrix is mainly micritic, and in places dolomitic. Some clay and secondary gypsum or silica (silicified phosphate) also occur.

The phosphatic ovules (0.1–0.5 mm in size) consist predominantly of microphosphate, either making up the whole grain or enveloping foraminifera, bone fragments or other kinds of small clasts (Soudry & Nathan 1980). A detailed SEM study of these coated grains (Soudry & Champetier 1983) distinguished globular bodies 15–20 μm in diameter which resemble colonies of coccoid unicells in the microphosphate. Further SEM analyses of the microphosphate cement in both grains and pebbles (Soudry & Lewy 1988) has shown that these globules are themselves aggregates of much smaller ones

(<1 μm), indicating that the size of the globules cannot be used as a means of classification. Accordingly, the coated grains or cemented aggregates in the granular phosphorites can be regarded as micronodules formed in a way similar to that discussed above for the phosphate nodules (*viz.*, phosphate cement mediated by endobenthic globular microorganisms). The small size of these micronodules indicates short periods of microbial activity, probably terminated by the exhumation and redeposition of these sand-sized grains outside the microbial habitat.

Bone fragments are a common constituent in the Phosphorite Unit at the Zefa Ef'e field (Fig. 1), locally forming the whole granular fraction (Würzburger *et al.* 1963). The coeval lateral equivalent of the unit at Har Zin (Nathan *et al.* 1979) consists of black laminae of bituminous micrite interbedded with fish bones and scales, some grey, friable phosphate nodules (2–7 mm in size) and tiny phosphatic ovules. Locally this rock is penetrated by crustacean burrows (*c.* 1 cm in diameter) filled with a grey granular phosphorite, composed of phosphatic ovules and fragmentary bones. It seems that bioturbation reworked the initially laminated bituminous sediment, fragmenting the fragile bones. This reworking was associated with the oxidation of the organic matter, winnowing of the fine fraction and concentration of the grains on the sea-floor before they were redeposited in burrows. Although there need not be any direct relation between nekton productivity and endobenthic microbial activity, it seems that the preservation of unusual quantities of vertebrate remains (relative to their presence in other marine sediments in Israel) probably requires oxygen-depleted burial conditions (cf. Arthur & Jenkyns 1981, p. 90), such as those favoured by the endobenthic micro-organisms. The original pores of the bones in the Upper Cretaceous phosphorites of Israel are filled with apatite, and additionally many have undergone micro-phosphatization, probably in association with boring by endolithic algae (Soudry & Nathan 1980). Consequently the bones became brittle and were easily fragmented into sharp-edged, angular particles. The dense microstructure of the mineralized bone fragments, however, gives them a high resistivity to chemical degradation and allows them to form a major part of the granular constituent.

In addition to the types of phosphate grains described above, Soudry (1987) described and illustrated the microstructure of a kind of phosphatic ovule which consisted mainly of clusters of tiny phosphate tubules, 4–8 μm in diameter. These grains are embedded in a matrix consisting mainly of similar phosphate tubules, oriented subparallel to the bedding, and contributing to the overall high-grade of the whole phosphate rock. It has been suggested (Soudry 1987) that the tubular structures of both the grains and the matrix are phosphatized filamentous microphytes, derived from microbial mats. The excellent SEM photomicrographs clearly show the phosphate tubules, forming both the grains and the matrix, merely touching each other, resulting in a friable and fragile tubular meshwork. Soudry (1987) proposed that this kind of high-grade phosphorite resulted from repeated cycles of colonization by filamentous microphytes (associated with other types of micro-organisms), mineralization and thereafter partial destruction and redeposition of the phosphatic detrital grains in a newly formed microbial mat. These periods of high-energy and probable bioturbation, which destroyed and reworked the mineralized mat structure, must have been short, basically retaining the general ecological regime favoured by these filamentous microphytes (low-energy, oxygen depleted, shallow seafloors; Soudry 1987).

Discussion

Calcite replacement by phosphate

Ames (1959) was the first worker to demonstrate calcite replacement by phosphate in an alkaline phosphate solution in the laboratory, and suggested that large marine phosphorite deposits may have formed in a similar way (*ibid.* p. 839). Kennedy & Garrison (1975) adopted this replacement mechanism for the phosphatization of omission surfaces and burrows in Upper Cretaceous chalks in England, and their work has been extensively referred to in succeeding publications on phosphatization. Recent studies on phosphates, however, have emphasized the large variety of phosphatic sediments which range from scattered grains and nodules to extensive mineralized omission surfaces. Mineraliation has generally taken place within, or at the sediment surface, at various bathymetries, and under normal, dysaerobic or anaerobic bottom conditions, associated in some cases with upwelling currents. In all of these phosphate types, relics of microbial microstructures have been recognized. It is unlikely that the replacement mechanism (Ames 1959) can operate under such a broad variety of bottom conditions. The increasing number of examples of recent and fossil phosphatized microbial structures (e.g., O'Brien et al. 1981; Soudry & Lewy 1988; Lamboy 1990) have convinced many authors of the importance of microbially-mediated apatite precipitation. The recent publication of Glenn & Arthur (1988, p. 232 and fig. 6f) for example, mentioned and illustrated such a microbial involvement in phosphatization; nevertheless, these authors still regarded the replacement mechanism as the main phosphatization process.

Oligocene phosphatized hardground surfaces and nodular phosphates of New Zealand were studied by Carter et al. (1982). On the basis of detailed petrographic analysis they reconstructed the initial sedimentary and the early and late diagenetic processes which had resulted in various phases of phosphate mineralization. They concluded that their phosphate nodules 'generally resulted from direct cementation by phosphate on and just below the seafloor' (ibid. p. 40). Their illustrations (ibid. fig. 20b, e) clearly show a cauliflower-like development of a phosphate cement and infill of foraminiferal tests, which closely resembles the clusters of mineralized globular microstructures in the Israeli pebbly phosphates and the phosphate crusts. Of great significance is their observation of boring by invertebrates and endolithic algae into lithified rhodoliths and calcareous fragments, through which phosphate cement (or other minerals) has been introduced from the periphery inwards, until the fragments have been totally altered (ibid. p. 20).

Kennedy & Garrison (1975) noticed that 'away from the marginal zones of total replacement, irregular mixtures of phosphatized micrite and unreplaced micrite give the rock a blotchy or patchy appearance in thin section' (ibid. p. 358). This observation recalls the mode of penetration of the phosphate through the endolithic algal borings into the Oligocene rhodolith nodules (Carter et al. 1982), and the microbially-mediated phosphatization of the intergranular spaces in the unconsolidated periphery of the Mishash calcareous concretions (Soudry & Lewy, in prep.)

It seems that the intensive destructive activity of endolithic algae, which bore into calcareous marine substrates at various bathymetries (e.g., Golubic et al. 1984), may also have taken place in many calcareous hardground surfaces. Apatite-precipitating micro-organisms, penetrating the tiny tunnel systems of the borings, would have gradually changed the composition of this bored zone into phosphate. Peripheral phosphatization of both lithified and unconsolidated sediments in which similar microbial globular micro-structure have been recognized, suggests that the micro-organism involved was the same in both types of phosphatization. This means that the pebbly phosphates, the phosphate crusts and the phosphatized surfaces, all of which now contain well-cemented phosphate bodies, share a common microbial generator. This micro-organism prefers living within the unconsolidated uppermost sediment layer, or within pores in lithified bodies, and will live in aerobic, dysaerobic or anaerobic conditions, in shallow to deep marine environments not necessarily connected to upwelling currents or high productivity zones (O'Brien et al. 1981, 1990). Nevertheless, this ecologically highly tolerant micro-organism prefers dark (endobenthic and 'deep' marine) sites. Therefore, pebbly phosphates, phosphate crusts and phosphatized surfaces are quite common throughout the geological record, and occur in various palaeoenvironments worldwide.

Phosphate in bituminous sediments

Bituminous phosphorites (e.g. Mishash Formation, southern Israel) preserved mainly in the subsurface, suggest that some phosphorites were originally bituminous, and subsequently have been partly oxidized by late diagenetic processes (Shiloni 1988). The primary accumulations of phosphate sediment in Egypt (Germann et al. 1984) and northwest Jordan (Abed & Al-Agha 1989) are locally represented by organic-rich, laminated, calcareous shales, and it is suggested that these shales represent the initial sediment which has been reworked, oxidized and winnowed soon after deposition, thereby concentrating the phosphate grains. These apparent relations between phosphate and bituminous sediments apply in Israel only to relative shallow marine, Upper Cretaceous palaeoenvironments; no significant phosphate occurrences have been observed in the bituminous outer shelf/continental slope facies of the Ghareb Formation. It is noteworthy that Arthur & Jenkyns (1981) indicated that 'oceanic anoxic events (OAEs: episodes of worldwide deposition and preservation of organic carbon) ... generally do *not* coincide with major phosphorogenic episodes' (ibid. p. 83; see p. 88).

Bottom physiography and water energy

If the pebbly phosphate is disregarded as a marker for phosphogenic episodes, then the physiography of the environment of deposition of large deposits of granular phosphorite becomes most significant. These deposits occur in southern Israel, Jordan and central Egypt in synclinal basins partly sheltered from the open Tethys Ocean by submarine to temporarily subaerially exposed anticlinal structures (e.g., Abed & Al-Agha 1989; Schröter 1986, p. 65; Soudry et al. 1985). These structural barriers must have significantly affected the water circulation and energy of the lower, and to a lesser extent the upper part of the water column. Surface to intermediate waters, rich in nutrients fluxed (upwelled) into these semi-closed basins, enhancing biogenic productivity. Within bottom waters, periods of low-energy or stagnation resulting in depletion in dissolved oxygen, alternated with short periods of high-energy (winnowing) and improved bottom aeration (bioturbation). Longer periods of improved water circulation and bottom aeration are evidenced by thin beds of highly fossiliferous limestone in the Phosphorite Unit of the Oron region. By contrast, carbonates which accumulated under dysaerobic bottom conditions consist of micrite densely packed with foraminiferid tests of a single type of endobenthic buliminid, which favoured such bottom conditions (Reiss 1988, p. 326); phosphatic particles are absent. Many of the contacts between phosphorites and the underlying carbonate beds are sharp however, commonly with crustacean burrows in the carbonate filled with granular phosphorite, suggesting lateral transport of the phosphorite sediment onto a burrowed surface. Such transport is also evident from the heterogenous grain composition of these phosphorites, indicating a mixture of particles (including relics of the nektonic community and faecal pellets of nektonic and benthonic organisms) formed under various ecological controls and mineralized by different microbial assemblages.

Bathymetry

In the attempt to reconstruct the environmental setting of phosphogenesis, many authors have regarded pebbly phosphates, phosphatized hardground surfaces and granular phosphorites as having been mineralized in the same way. However, the occurrence of recently formed (or nearly so) pebbly phosphates in depths between 100–500 m (e.g., Glenn & Arthur 1988) contrasts with the shallow-marine palaeoenvironments indicated by ancient granular phosphorites. Soudry (1987) emphasized this shallowness, showing that mats of filamentous microphytes (probably cyanobacteria) played a significant role in the formation of phosphate particles (both grains and cement). The deduced restricted ecological conditions required by these filamentous micro-organisms, however, contrasts, with the ecologically much more tolerant globular varieties (forming the phosphate cements in grains, pebbles, crusts and void fills), and clarifies the apparent bathymetric discrepancy between ancient and recent phosphatization sites (see Reiss 1988 and Soudry 1987 for further discussion on palaeoecological criteria).

Conclusions

The present study focuses on the genetic differentiation between granular phosphorite deposits and pebbly phosphates in the Upper Cretaceous sequence of southern Israel, and discusses their relationship to 'phosphatization' phenomena in ancient and recent sediments elsewhere. It is suggested that the globular micro-organisms identified in the Israeli deposits have also 'phosphatized' carbonate particles and hardground surfaces by depositing microbially precipitated apatite in voids produced in the carbonate by endolithic algae and other boring organisms, or by dissolution. Thus the significance of replacement of carbonate by phosphate as a major phosphatization process is in doubt.

The wide ecological tolerances of the globular micro-organisms discussed here, indicates that they and their phosphatic products are poor palaeoenvironmental indicators in contrast to the restricted ecological setting demonstrated by Soudry (1987) for the filamentous microphytes, which form the tubular-microstructure phosphate grains and matrix of high-grade granular phosphorites in southern Israel.

The global oceanic—climatic regime had a minor affect on phosphogenesis (cf. Arthur & Jenkyns 1981, p. 93) compared to that of the local, shallow-marine bottom physiography. In southern Israel, late Cretaceous tectonic movements elevated anticlinal structures which affected local water energy and circulation, and produced intervening synclinal basins. These changes led to repeated cycles of varying water energies and oxygenation, without any significant sea-level changes, which were crucial for the formation of the phosphorite deposits.

This study is part of a more comprehensive research project (No. 28891): Senonian biostratigraphy and palaeoecology in Israel, carried out at the Geological Survey of Israel (GSI). D. Soudry and Y. Nathan (GSI), Z. Reiss (Hebrew University, Jerusalem), I. Jarvis (Kingston Polytechnic, England) and M. Lamboy (University of Rouen, France) are acknowledged for their constructive remarks. B. Katz, M. Dvorachek and Y. Levy (GSI) are acknowledged for their technical assistance.

References

ABED, A. M. & AL-AGHA M. R. 1989. Petrography, geochemistry and origin of the NW Jordan phosphorites. *Journal of the Geological Society, London*, **146**, 499–506.

AMES, L. L. JR. 1959. The genesis of carbonate apatites. *Economic Geology*, **54**, 829–841.

ARTHUR, M. A. & JENKYNS, H. C. 1981. Phosphorites and paleoceanography. *Oceanologica Acta*, **45**, 83–96.

CARTER, R. M., LINDQVIST, J. K. & NORRIS, R. J. 1982. Oligocene unconformities and nodular phosphate-hardground horizons in western Southland and northern West Coast. *Journal of the Royal Society of New Zealand*, **12**, 1, 11–46.

COOK, H. E. & EGBERT, R. M. 1983. Diagenesis of deep-sea carbonates. *In*: LARSEN, G. & CHILINGAR, G. V. (eds) *Diagenesis in sediments and sedimentary rocks*, 2. Developments in Sedimentology 25 B, Elsevier Scientific Publishing Company, 213–293.

GERMANN, K., BOCK, W.-D., SCHRÖTER, T. 1984. Facies development of Upper Cretaceous phosphorites in Egypt: sedimentological and geochemical aspects. *Berliner Geowissenschaftliche Abhandlungen*, A 50, 345–361.

GLENN, C. R. & ARTHUR, M. A. 1988. Petrology and major element geochemistry of Peru margin phosphorites and associated diagenetic minerals: authigenesis in modern organic-rich sediments. *Marine Geology*, **80**, 231–267.

GOLUBIC, S., CAMPBELL, S. E., DROBNE, K., CAMERON, B., BALSAM, W. L., CIMERMAN, F. & DUBOIS, L. 1984. Microbial endoliths: a benthic overprint in the sedimentary record, and a paleobathymetric cross-reference with foraminifera. *Journal of Paleontology*, **58**, 2, 351–361.

KENNEDY, W. J. & GARRISON, R. E. 1975. Morphology and genesis of nodular chalks and hardgrounds in the Upper Cretaceous of southern England. *Sedimentology*, **22**, 311–386.

—— & SELLWOOD, B. W. 1970. *Ophiomorpha nodosa* Lundgren, a marine indicator from the Sparnacian of south-east England. *Proceedings of the Geologists' Association*, **81**, 1, 99–110.

LAMBOY, M. 1990. Microbial mediation in phosphatogenesis: new data from the Cretaceous phosphatic chalks, northern France. *In*: NOTHOLT, A. J. G. & JARVIS, I. (eds) *Phosphorite Research and Development*. Geological Society, London, Special Publication, **52**, 157–167.

LEWY, Z. 1986. *Anaklinoceras reflexum* Stephenson in Israel and its stratigraphic significance. *Newsletters on Stratigraphy*, **16**, 1–8.

MAGARITZ, M., MOSHKOVITZ, S., BENJAMINI, C., HANSEN, H. J., HAKENSSON, E. & RASMUSSEN, K. L. 1985. Carbon isotope-, bio- and magnetostratigraphy across the Cretaceous-Tertiary boundary in the Zin Valley, Negev, Israel. *Newsletters on Stratigraphy*, **15**, 100–113.

NATHAN, Y., SHILONI, Y., RODED, R., GAL, I. & DEUTSCH, Y. 1979. *The geochemistry of the Northern and Central Negev phosphorites (Southern Israel)*. Geological Survey of Israel, Bulletin 73.

O'BRIEN, G. W., HARRIS, J. R., MILNES, A. R. & VEEH, H. H. 1981. Bacterial origin of East Australian continental margin phosphorites. *Nature*, **294**, 442–444.

——, MILNES, A. R., VEEH, H. H., HEGGIE, D. T., RIGGS, S. R., CULLEN, D. J., MARSHALL, J. F. & COOK, P. J. 1990. Sedimentation dynamics and redox iron-cycling: controlling factors for the apatite glauconite association on the East Australian continental margin. *In*: NOTHOLT, A. J. G. & JARVIS, I. (eds) *Phosphorite Research and Development*. Geological Society, London, Special Publication, **52**, 61–86.

REISS, Z. 1988. Assemblages from a Senonian high-productivity sea. *Revue de Paléobiologie*, Special No. 2, Benthos'86, 323–332.

——, ALMOGI-LABIN, A., HONIGSTEIN, A., LEWY, Z., LIPSON-BENITAH, S., MOSHKOVITZ, S. & ZAKS, Y. 1985. Late Cretaceous multiple stratigraphic framework of Israel. *Israel Journal of Earth Science*, **34**, 147–166.

RODED, R. 1982. *The geological and structural map of Israel on a 1:50 000 scale, Sheet 19 III: Oron*. Geological Survey of Israel.

SCHRÖTER, T. 1986. Die lithofazielle Entwicklung der oberkretazischen Phosphatgestein Ägyptens — ein Beitrag zur Genese der Tethys-Phosphorite der Ostsahara. *Berliner Geowissenschaftliche Abhandlungen*, **A 67**, 1–105.

SHAHAR, Y. 1968. *Type section of the Campanian–Maastrichtian Ghareb Formation in the Oron syncline (northern Negev)*. Geological Survey of Israel, Stratigraphic Section, 6.

SHILONI, Y. 1988. Variations of organic matter content of the Mishash and Ghareb explained by late oxidation processes. *Israel Geological Society, Annual Meeting, En Boqeq*, 1988, 105–106.

SOUDRY, D. 1987. Ultra-fine structures and genesis of the Campanian Negev high-grade phosphorites (southern Israel). *Sedimentology*, **34**, 641–660.

—— & CHAMPETIER, Y. 1983. Microbial processes in the Negev phosphorites (southern Israel). *Sedimentology*, **30**, 411–423.

—— & LEWY, Z. 1988. Microbially influenced formation of phosphate nodules and megafossil moulds (Negev, southern Israel). *Palaeogeography, Palaeoclimatology, Palaeoecology*, **64**, 15–34.

—— & NATHAN, Y. 1980. Phosphate peloids from the Negev phosphorites. *Journal of the Geological Society, London*, **137**, 749–755.

——, —— & RODED, R. 1985. The Ashosh-Haroz facies and their significance for the Mishash palaeogeography and phosphorite accumulation in the northern and central Negev, southern Israel. *Israel Journal of Earth Sciences*, **34**, 211–220.

SWINTON, W. E. 1973. *Fossil Amphibians and Reptiles*. British Museum (Natural History), 5th edition.

WÜRZBURGER, U., LASMAN, N., GROSS, S. & SHILONI, Y. 1963. *The Zefa Ef'e phosphate deposits*. Geological Survey of Israel, Report MP 126/62, Parts 1 and 2.

Geological significance of carbonate substitution in apatites: Israeli phosphorites as an example

Y. NATHAN[1], D. SOUDRY[1] & A. AVIGOUR[2]

[1] *Geological Survey of Israel, 30 Malkhei Israel Street, 95501, Jerusalem, Israel*
[2] *Negev Phosphate Company, Tel Aviv 6100, Israel*

Abstract: The extent of carbonate for phosphate substitution in francolite (carbonate-fluorapatite) was measured in 177 selected phosphorite samples of Santonian to Eocene age from Israel. The results show that unaltered apatites from different stratigraphic positions have different CO_2 concentrations while those from the same stratigraphic level have a relatively narrow range of concentrations. As a rule phosphorite samples from rocks with a low P_2O_5 content have a higher carbonate substitution than those from rocks with a high P_2O_5 content. Apatites from the carbonate-rich Santonian–Campanian Menuha Formation contain on average 5.5% CO_2, while the average content of the apatites from the phosphatic Campanian–Maastrichtian Mishash Formation is 3.7%. These changes are probably a reflection of the early diagenetic environment and, more specifically, are a function of the varying composition of the interstitial waters.

Epigenesis and/or weathering bring about a significant decrease in the CO_2 concentrations causing an overprint which masks the above picture. Therefore, it is possible to use a CO_2 value for geological interpretation only after making a petrographic analysis, in order to distinguish between early diagenesis and late epigenetic processes.

The most characteristic and important substitution in francolite (carbonate-fluorapatite) is the carbonate for phosphate (Lehr *et al.* 1968; McClellan & Lehr 1969; McClellan & Van Kauwenbergh 1990). It is also the substitution which has the greatest range of variability and is also a key factor in determining the chemical reactivity of phosphate rock and its agronomic effectiveness (Lehr & McClellan 1972). The increase of chemical reactivity with increasing carbonate for phosphate substitution, has been also demonstrated thermodynamically by Chien (1977). Two main proposals have been made to explain the degree of substitution. The first relates carbonate concentration to the environment of deposition (Gulbrandsen 1970) and the second relates it to weathering (McArthur 1978). The subject has been reviewed by Nathan (1984).

Samples and methods

The samples represent most of the variation in francolite composition that occurs in Israeli phosphorites. Most of the samples come from the Negev Desert (Fig. 1), collected from rocks ranging in age from the Santonian (Menuha Formation) to the Lower Eocene (Mor Formation) (Fig. 2). In two stratigraphic horizons (base of the Chert Member and of the Nodular Phosphorite Unit) where altered phosphorites (together with iron and aluminum phosphates) are known to occur, seven altered samples (two from the Chert Member and five from the Nodular Phosphorite Unit; Table 1) were deliberately chosen. In all other cases, the aim was to obtain fresh samples. As far as possible, the samples were chosen to represent the different petrographic types of phosphorites and phosphatic rocks.

The method used to evaluate the CO_2 content is the peak-pair diffraction method proposed by Gulbrandsen (1970). Only samples with relatively high apatite content can be accurately measured by this method. Thus only rocks relatively rich in P_2O_5 (> 15%) or which contained discrete particles enriched in P_2O_5 were sampled (for these the phosphate particles were hand-picked before analysis). The Gulbrandsen method embodies a certain degree of uncertainty, because of its statistical nature, therefore each sample was analysed in triplicate and the average value was used. The CO_2 content of apatite samples with known CO_2 content (samples free of carbonate which were analysed chemically) was determined in this manner. The results showed on average an absolute + 0.2% from the chemical analyses; the extreme samples (very low and very high CO_2 content) gave the largest deviations. In all, 177 samples were analysed, 117 from outcrops and 60 from cores.

Petrography

Three main phosphate lithotypes, differentially distributed stratigraphically, are distinguished in the examined Santonian–Eocene succession (Fig. 2): (1) coarse biodetrital phosphorites, either coarse calcareous biodetrital phosphorites, (CCBP), or coarse siliceous biodetrital phosphorites, (CSBP); (2) sandy peloidal phos-

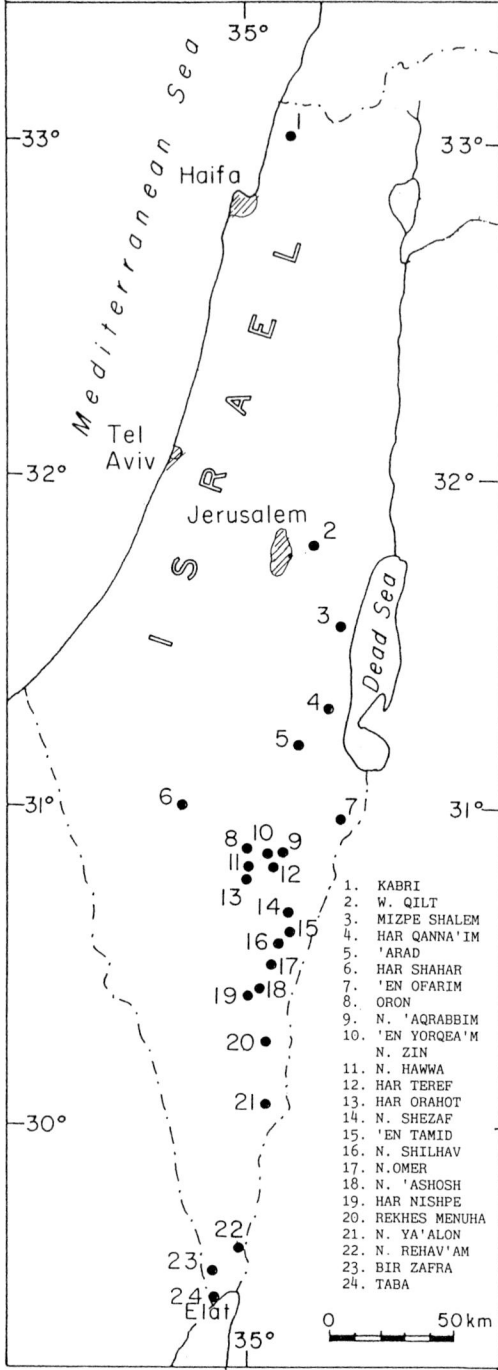

Fig. 1. Location map.

1. KABRI
2. W. QILT
3. MIZPE SHALEM
4. HAR QANNA'IM
5. 'ARAD
6. HAR SHAHAR
7. 'EN OFARIM
8. ORON
9. N. 'AQRABBIM
10. 'EN YORQEA'M N. ZIN
11. N. HAWWA
12. HAR TEREF
13. HAR ORAHOT
14. N. SHEZAF
15. 'EN TAMID
16. N. SHILHAV
17. N. OMER
18. N. 'ASHOSH
19. HAR NISHPE
20. REKHES MENUHA
21. N. YA'ALON
22. N. REHAV'AM
23. BIR ZAFRA
24. TABA

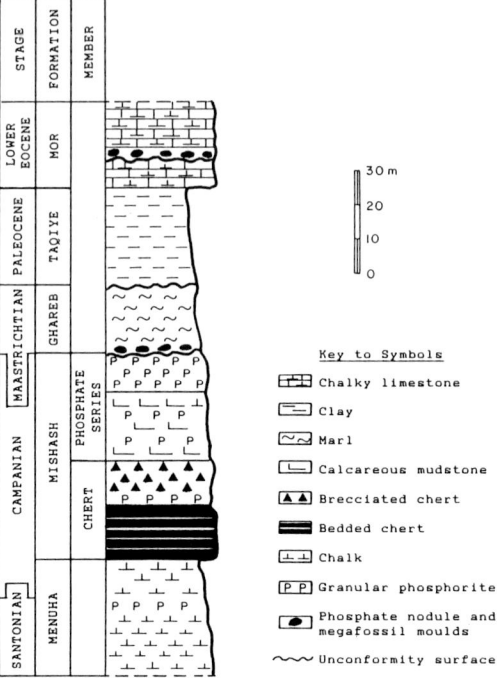

Fig. 2. Schematic composite section of the Santonian–Lower Eocene succession in northern Negev

phorites (SPP); (3) nodular phosphorites. These three types are distinguished on the basis of their phosphate allochems, their phosphate content, and a variety of textural features. An additional, less frequent type, consists of thin beds of primary non pelletal phosphorites at places sandwiched between clay-rich horizons of the Mishash Formation (Nathan & Soudry 1982).

Coarse biodetrital phosphorites

These are mostly found within the Santonian Menuha Formation and within the lower part of the Mishash Formation (the chert member, the phosphatic carbonate and the porcellanite units of the phosphate series member; Fig. 3). They are commonly massive (apparently unstratified), coarse-grained, indurated, usually with a low phosphate content and intensively bioturbated (burrow system mainly of *Thalassinoides* assemblage) (Fig. 4a). The base of individual beds is always sharp and erosional. Imbricated platy lime pebbles and molluscan shell-lag with shells with a convex-up orientation also occur at the basal part of some layers. Other diagnostic features commonly displayed by these phosphorites such as amalgamated beds in proximal positions, scour-and-fill structures, crude graded bedding, and intraclasts of subjacent derivation,

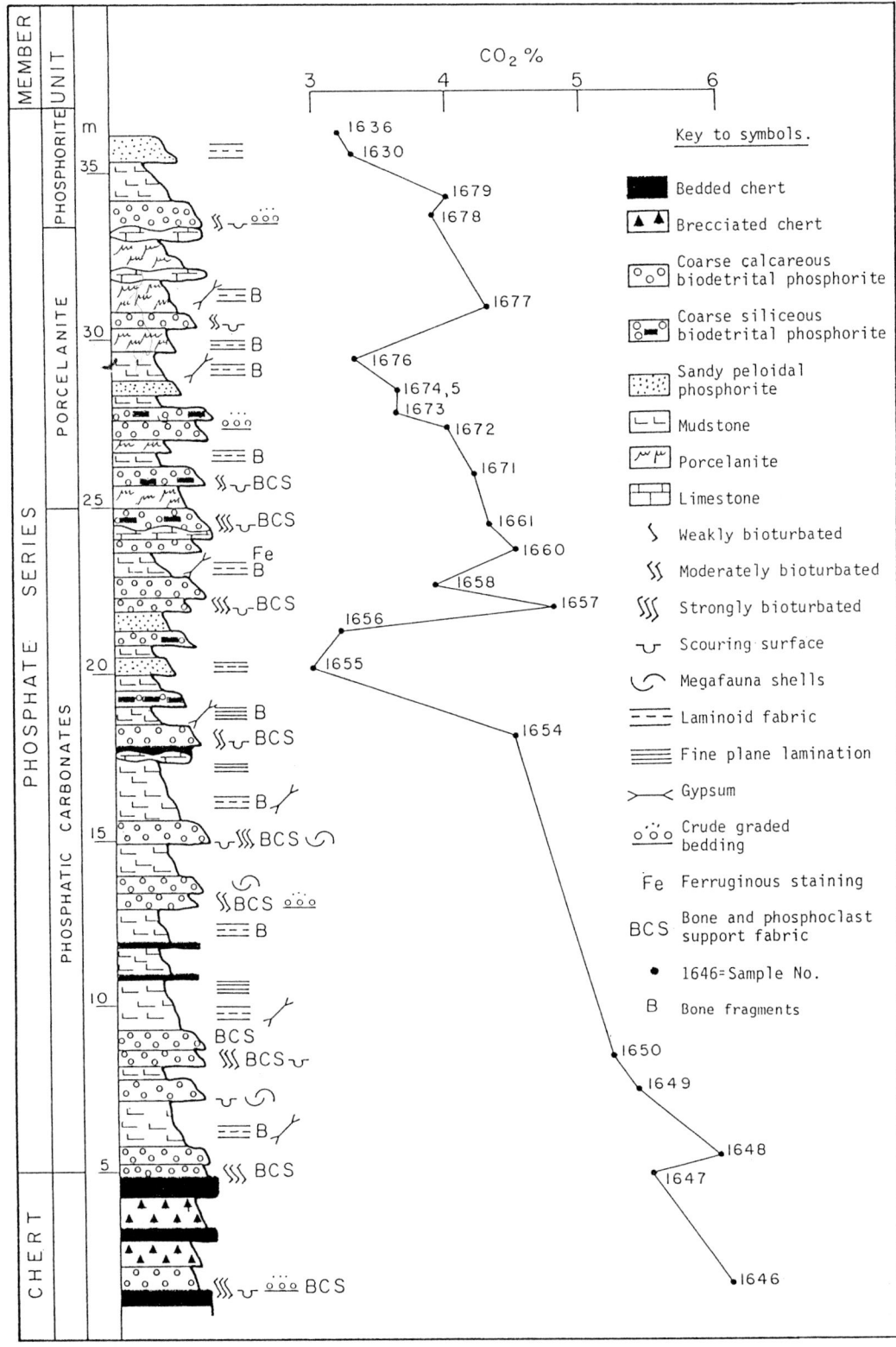

Fig. 3. CO_2 content variations of phosphate layers in the Chert and the Phosphate Series members in 'En Aqrabbim area.

Table 1. *Structural CO_2 content of apatites from various formations*

Sample	Locality[1]	Rock type[2]	CO_2%
ZL 001	Har Qanna'im	phosphatic chalk	5.2
Z1 002	Wadi Qilt. Judean Desert	phosphatic chalk	5.1
AM 001	south of Har Shahar	phosphatic chalk	5.9
ZL 004	Har Teref	nodule	5.1
BN 001	Kabri, Galilee	phosphatic chalk	6.1
Formation, member or unit: Menuha, No. of samples, 5,		Average CO_2% value	5.5
YN 082	Nahal Ya'alon	CCBP	5.2
DS 1646	'Aqrabbim	CCBP	6.1
DS 1723	'Arad (Dead Sea Road)	CCBP	4.9
GS 1136	'Arad (Rogem Valley)	CCBP	5.4
GS 1140	'Arad (Rogem Valley)	CCBP	5.6
GS 1193	NE of Oron	CCBP	5.4
Formation, member or unit: Mishash, Chert Member, No. of samples, 6,		Average CO_2% value	5.4
GS 1133	'Arad (Rogem Valley)	Altered sample	1.6
GS 1171	'Arad (Rogem Valley)	Altered sample	2.3
Formation, member or unit: Mishash, Chert Member, No. of samples, 2		Average CO_2% value	2.0
DS 784	Nahal Shilhav	CCBP	4.3
DS 787	Nahal Shilhav	CCBP	3.8
DS 794	Nahal Shilhav	CSBP	4.2
DS 801	Nahal Shilhav	SSBP	4.0
DS 813	Nahal Shilhav	CCBP	3.8
DS 824	Nahal Shilhav	CCBP	4.6
DS 1572	Nahal Shilhav	lens of primary apatite	4.6
DS 1526	'En Tamid	CSBP	5.3
DS 1647	'Aqrabbim	CCBP	5.5
DS 1648	'Aqrabbim	CCBP	6.0
DS 1649	'Aqrabbim	CCBP	5.4
DS 1650	'Aqrabbim	CCBP	5.2
DS 1654	'Aqrabbim	CCBP	4.5
DS 1655	'Aqrabbim	SPP	3.0
DS 1656	'Aqrabbim	SPP	3.2
DS 1657	'Aqrabbim	CSBP	4.8
DS 1658	'Aqrabbim	CCBP	3.9
DS 1660	'Aqrabbim	CSBP	4.5
DS 1661	'Aqrabbim	CCBP	4.3
DS 1687	Nahal Hawwa	CCBP	4.7
DS 1689	Nahal Hawwa	CCBP	4.9
DS 1691	Nahal Hawwa	SPP	3.5
DS 1701	Oron Hawwa	CCBP	4.7
DS 1707	Oron Hawwa	CCBP	3.3
DS 1712	'Arad town	CCBP	4.2
Formation, member or unit: Mishash, Phosphatic Carbonate Unit, No. of samples, 25,		Average CO_2% value	4.4
DS 1618	Oron	Solenoceras infilling	3.0
DS 1671	'Aqrabbim	CSBP	4.2
DS 1672	'Aqrabbim	CCBP	4.0
DS 1673	'Aqrabbim	SCBP	3.6
DS 1674	'Aqrabbim	SPP	3.6

Table 1. *cont.*

Sample	Locality[1]	Rock type[2]	CO_2%
DS 1675	'Aqrabbim	SPP	3.5
DS 1676	'Aqrabbim	laminated phosphatic mudstone	3.3
DS 1677	'Aqrabbim	CSBP	4.3
DS 1692	Nahal Hawwa	SCBP	4.3
DS 1693	Nahal Hawwa	SCBP	5.2
DS 1694	Nahal Hawwa	SCBP	4.6
DS 1713	'Arad town	SCBP	4.7
	Formation, member or unit: Mishash, Porcellanite Unit,		
	No. of samples, 12,	Average CO_2% value	4.0
DS 1021	Nahal Shilhav	SPP	3.9
DS 1609	Rekhes Menuha	Lybicoceras infilling	4.5
GH 334	Rekhes Menuha		4.5
GH 335	Rekhes Menuha		3.9
GH 337	Rekhes Menuha		3.6
GH 338	Rekhes Menuha		3.5
GH 339	Rekhes Menuha		3.7
GH 340	Rekhes Menuha		3.3
DS 1615	Nahal 'Ashosh	Aporrhais infilling	3.0
DS 1617	Nahal 'Ashosh	Baculites infilling	3.0
DS 1624	Nahal Shezaf	SPP	3.6
DS 1716	'Arad town	CCBP	4.4
DS 1717	'Arad town	CCBP	4.1
DS 1719	'Arad town	CCBP	4.2
ZE 21	Zefa'	BP	4.4
ZE 22	Zefa'	BP	4.8
OR 20	Oron	CCBP	3.2
OR 21	Oron	CCBP	3.0
OR 22	Oron	CCBP	3.4
DS 1731	Nahal Omer	SPP	3.5
DS 1737	Nahal Omer	SPP	3.2
DS 1695	Nahal Hawwa	CCBP	3.5
DS 1698	Nahal Hawwa	laminated SPP	3.1
DS 1639	Nahal Zin	CCBP	4.2
DS 1678	'Aqrabbim	CCBP	3.9
DS 1679	'Aqrabbim	CCBP	4.0
DS 1680	'Aqrabbim	phosphoclast	4.5
DS 1683	'Aqrabbim	bone fragment	4.0
DS 1630	'En Yorqe'am	laminated SPP	3.3
DS 1636	'En Yorqe'am	laminated SPP	3.2
	Formation, member or unit: Mishash, Phosphorite Unit.		
	No. of samples, 30,	Average CO_2% value	3.7
NI 1	Har Nishpe	baculite mould	4.0
ZL 6	Har Nishpe	baculite mould	6.1
DS 1604	Oron	Ostrea mould	4.1
DS 1605	Nahal 'Ashosh	coral mould	6.0
DS 1607	Nahal 'Ashosh	baculite mould	4.0
AR 1	'Arad industrial zone	nodule	4.9
ZL 3	Har Teref	nodule	5.2
DS 1613	Har Teref	nodule	5.2
ZL 5	Har Teref	nodule	5.9
YN 44	Har Orahot	nodule	5.6
HH 206	Har Orahot	nodule	4.6
	Formation, member or unit: Ghareb, Nodular phosphorite Unit,		
	No. of samples, 11,	Average CO_2% value	5.1

Table 1. cont.

Sample	Locality[1]	Rock type[2]	$CO_2\%$
DS 1610	Har Teref	nodule, altered sample	2.3
YN 48	Har Orahot	nodule, altered sample	0.5
HH 204	Har Orahot	nodule, altered sample	1.8
HH 205	Har Orahot	nodule, altered sample	0.7
YN 103	'En Ofarium	nodule, altered sample	1.2
Formation, member or unit: Ghareb, Nodular phosphorite Unit, No. of samples, 5,		Average $CO_2\%$ value	1.3
DS 1621	Hor Hahar	nodule (glauconitic)	6.0
DS 1622	Hor Hahar	Aturia (phosphatic mould)	5.0
YN 306	road to Oron	nodule (glauconitic)	5.3
Formation, member or unit: Mor, No. of samples, 3,		Average $CO_2\%$ value	5.4

[1] The samples come from the Negev Desert unless otherwise indicated.
[2] Abbreviations rock types:
CCBP, coarse calcareous biodetrital phosphorite; CSBP, coarse siliceous biodetrital phosphorite; SCBP, sandy calcareous biodetrital phosphorite; SSBP, sandy siliceous biodetrital phosphorite; SPP, sandy peloidal phosphorite; BP, biodetrital phosphorite.

have led Soudry (1983) to interpret these rocks as storm-generated sediments in shallow environments (Aigner 1982).

Thin sections of these phosphorites typically show a poorly sorted non compacted granular phosphate fraction dominated by bone fragments together with some amounts of laminated, (microstromatolite-like; Soudry 1987) and non laminated phosphate clasts and peloids (Fig. 4b). As a general rule, the intergranular material consists of chert (CSBP) or (? neo) spar with some lime mud (CCBP). All transitions have been observed between the two forms.

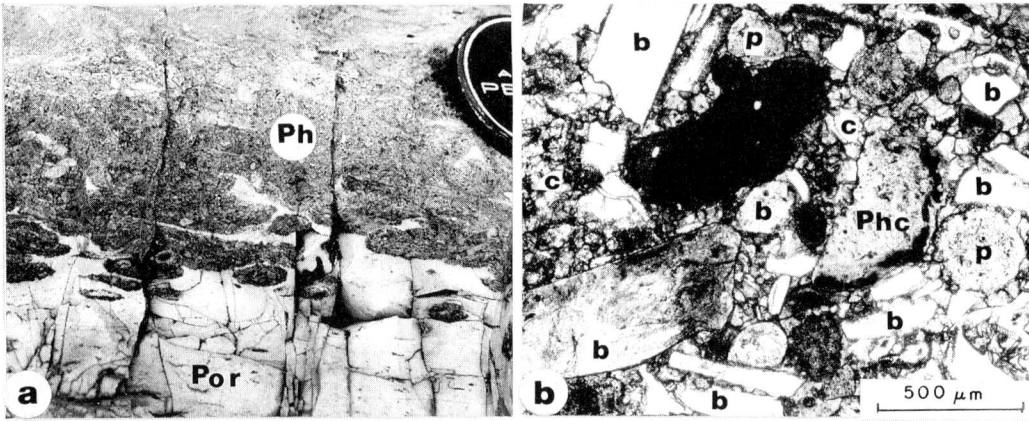

Fig. 4. Coarse biodetrital phosphorites. (a) Outcrop of a heavily burrowed coarse silicified biodetrital phosphorite (Ph) overlying a porcellanite layer (Por). Note the burrow system penetrating the top of the porcellanite layer and the resulting bio-erosional surface. (b) Photomicrograp of a coarse calcareous biodetrital phosphorite showing abundant bone fragments (b) together with phosphate clasts (Phc) and peloid (p) cemented by sparry calcite (c). The low degree of compaction of the phosphate components suggest an early lithification of this lithotype (plain light).

Sandy peloidal phosphorites

These are mostly encountered within the phosphorite unit of the Mishash Formation (Fig. 3) where they form one or more layers of economic value, interbedded with low grade phosphoritic mudstone and shales. They are typically soft and sandy and always display a more or less pronounced internal layering. Beds with a clear lamination commonly show an alternation of centimetre-sized phosphate laminae and millimetre-sized clay-carbonate laminae (Fig. 5a), in a manner reminiscent of the fabrics of biolaminated sediments (Monty 1976). In beds with more massive fabrics the layering is usually expressed by a common orientation of elongated particles parallel to the bedding.

Optical microscopy of these phosphorites generally show closely packed well-sorted phosphate peloids bound in a carbonate micritic or a phosphatic vermiform matrix (Fig. 5b), the latter is representative of rocks with a high phosphate content. The composition of the phosphate matrix is of special interest. SEM observation shows it to be formed by a variety of hollow microbial tubules displaying a range of spatial micro-organizations (Fig. 5c), whereas the phosphate peloids fixed in this filamentous meshwork are themselves composed of packed microbial remains of different types (Soudry 1987). Analogies between the fabric of the matrix and that of the corpuscles indicate that these rocks are the products of fragmentation and redeposition of recurrent, early phosphatized microbial crusts in semi-emerged to shallow submerged environments (Soudry 1987).

Nodular phosphorites

These are associated with unconformity planes at two stratigraphic positions: in the basal part of the Early Maastrichtian Ghareb Formation, and in the Lower Eocene Mor Formation (Fig. 2). They consist of two distinct components: (a) nodules (*sensu stricto*) and (b) internal megafossil moulds. These two components are differentially distributed. Nodules are prominent in proximal areas, whereas megafossil moulds (cephalopods, bivalves, gastropods, etc.) are dominant in distal environments (Soudry & Lewy 1988; Lewy 1990). Both nodules and moulds occur as hard (well-lithified) bodies with a high phosphate content (usually more than 30% P_2O_5). The nodules are commonly irregularly-shaped, 2–3 cm in diameter, and occur as individual or compound bodies.

Viewed in thin section the nodules show loosely-packed phosphate peloids or foraminiferal tests bound by a clotted phosphate matrix (Fig. 5d), the clots are arranged in clusters up to a few hundred microns in diameter. The scanning electron microscope (SEM) reveals this intergranular phosphate matter to be formed by a variety of globose structures of distinct geometries and organizations. Most of them are 2 μm sized globules wrapping the phosphate particles and binding them (Fig. 5e). These structures have been interpreted (Soudry & Lewy 1988; Lewy 1990) as the fossilized remnants of cocoid-dominated microbial populations which colonized intergranular and intragranular cavities and acted as templates for synsedimentary phosphate precipitation. Diagenetic neomorphic processes may in places transform the initial micritic coating of the globose structures into large hexagonal lamellae (Fig. 5f).

Results and discussion

The results are given in Tables 1–3. Table 1 summarizes outcrop data except for the samples from the Santonian–Campanian Sayyarim Formation which are given in Table 2. Table 3 summarizes data for the core samples. The average for each group (bed, unit, member, formation) is given in the tables. Data from the 7 samples which are either weathered or have a late epigenetic history, (Chert member and nodular phosphorite unit), were not included in the average. Table 1 shows a clear trend from relatively high values (5–6%) of CO_2 in samples which originate from the Menuha and Mor Formations to the low values of samples from the Mishash Formation (3–4.5%). Table 3 shows the same trend, within the Phosphate Series Member of the Mishash Formation, from the average value of 4.1% of beds I and V to the 3.4% average value of beds II-III and III.

A one-way analysis of variance was carried out (using Minitab) on all the data to determine the significance of differences in the means obtained. The samples were divided into five groups. Thirty-one samples from the Menuha Formation, Chert member, nodular phosphorite unit and Mor Formation which gave similar means, constituted a 'high' group, since these samples have the highest CO_2 content. The other four groups are: 24 samples from the phosphatic carbonate unit (phoscarb); 12 samples from the porcellanite unit (porce); 91 samples from the phosphorite unit (phosph); and 19 samples from the Sayyarim Formation (sayyar). The results of the analysis are given in Table 4.

Table 4 shows that the F-ratio is 41.86 which allows rejecting the null hypothesis for prac-

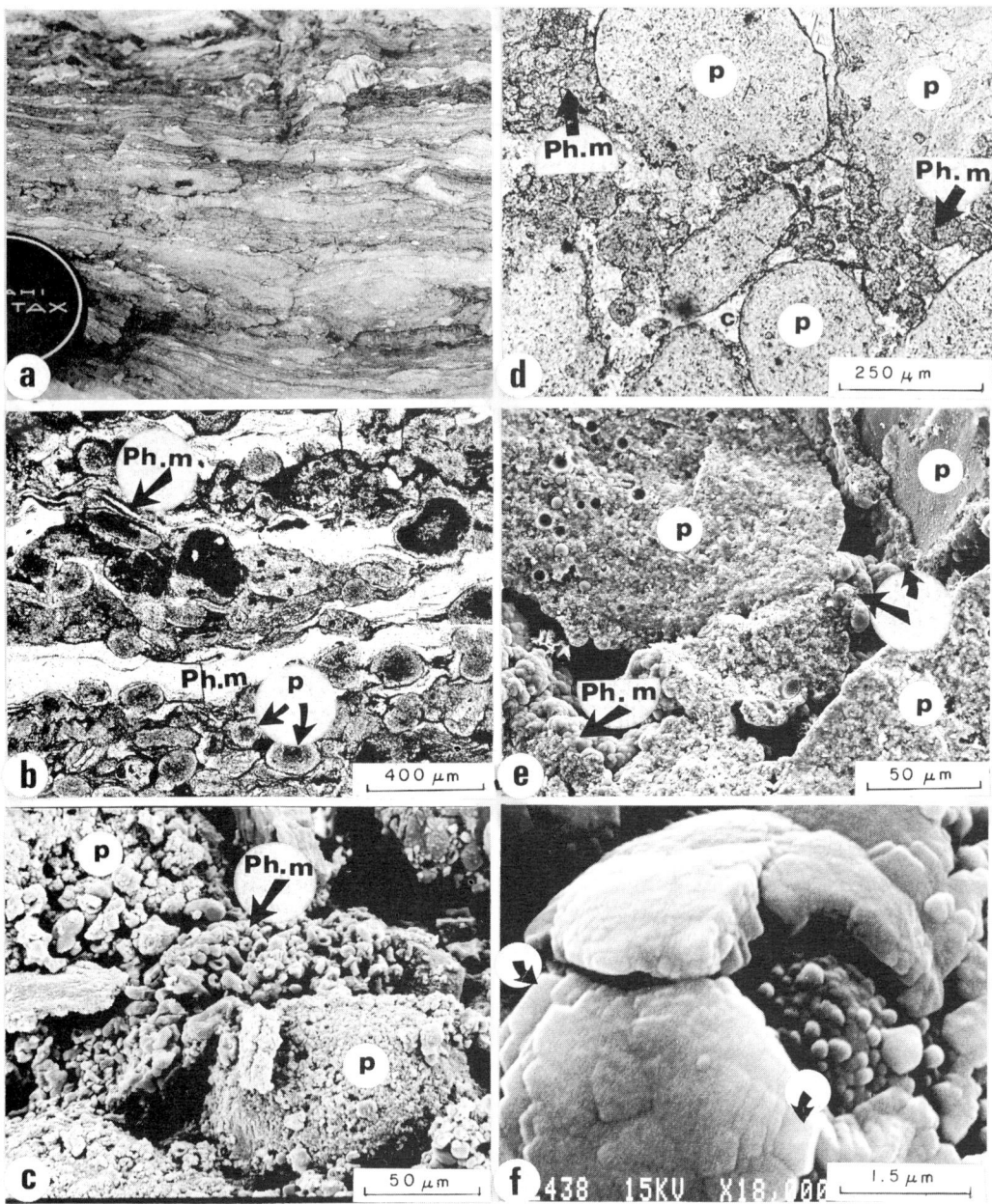

Fig. 5. Microstructure of sandy peloidal phosphorites (a–c) and nodular phosphorites (d–f).
(**a**) Outcrop of a sandy peloidal phosphorite with a laminar fabric. Lamination is due to alternation of milimetre to centimetre-sized phosphate laminae (light grey) and millimetric argillaceous, carbonate laminae (dark grey). The lamination is also emphasized by a common orientation of elongated phosphate particles (white components). The discordant white strips at the upper corner are secondary gypsum. (**b**) Photomicrograph of (a) showing closely spaced phosphate peloids (p) within a layered cloudy (vermiform) to translucent phosphate matrix (Ph. m.) (Plain light). (**c**) SEM micrograph of (b) showing the phosphate matrix (Ph. m) to consist of tightly packed hollow microbial tubules. (p) are phosphate peloids. (**d**) Thin section of a phosphate nodule showing phosphate peloids (p) bound by a clotted phosphate matrix (Ph. m.). (c) is spar-infilled remaining interparticle cavities (photomicrograph, plain light). (**e**) The clotted phosphate matrix (Ph. m.) are viewed in SEM. Clustered globose microbial structures wrapping (arrow) and binding (arrow) phosphate peloids (p). (**f**) Recrystallization processes in nodules. The original micritic phosphate coating of the globose structures is converted in tangentially-disposed hexagonal phosphate lamellae (arrows). Such an aggrading neomorphism is generally accompanied by a lowering of the CO_2 content of the apatite. (SEM micrograph).

Table 2. *Structural CO_2 content of apatites from the Sayyarim Formation, Southern Negev*

Sample	Locality	Rock type	CO_2%
GH 89	Wadi Taba	dolomitic phosphorite	4.0
GH 101	Wadi Taba	dolomitic phosphorite	4.3
Sandstone Member, No. of samples, 2			
GH 274	Nahal Shehoret		4.6
GH 285	Nahal Shehoret		4.4
Laminar Chert Member, No. of samples, 2			
GH 152	Wadi Taba	porcellanitic chalk	5.2
GH 160	Wadi Taba	silicified phosphorite	4.6
GH 162	Wadi Taba	silicified phosphorite	4.1
GH 163	Wadi Taba	silicified phosphorite	4.5
GH 165	Wadi Taba	silicified phosphorite	4.5
GH 167	Wadi Taba	silicified phosphorite	4.4
GH 169	Wadi Taba	silicified phosphorite	4.3
GH 171	Wadi Taba	silicified phosphorite	3.8
GH 173	Wadi Taba	silicified phosphorite	4.2
GH 197	Wadi Taba	nodule	4.1
GH 251	Bir-Zafra	nodule	5.5
GH 364	Nahal Rehav'am	nodule	4.4
GH 372	Nahal Rehav'am	nodule	4.3
GH 384	Nahal Rehav'am	nodule	4.0
GH 385	Nahal Rehav'am	nodule	4.3
Porcellanite Member, No. of Samples, 15			
Total No. of samples, 19		Average CO_2%	4.4

Table 3. *Structural CO_2 content of apatites from phosphorites in the Zin Phosphate Field*

Sample bed*	Locality	CO_2%
AA 6 I	Yorqe'am south	3.6
AA 7 I	Yorqe'am south	4.4
AA 21 I	Yorqe'am south	4.6
AA 60 I	Hor Hahar	4.0
AA 59 I–II	Hor Hahar	4.0
No. of samples, 5 Average CO_2% value		4.1
AA 5 II	Yorqe'am south	2.9
AA 4 II	Yorqe'am south	3.2
AA 8 II	Yorqe'am south	3.9
AA 15 II	Yorqe'am south	3.3
AA 20 II St.	Yorqe'am south	4.2
AA 19 II	Yorqe'am south	4.2
AA 29 II St.	Yorqe'am south	3.6
AA 25 II	Yorqe'am south	4.0
AA 30 II St.	Yorqe'am south	3.9
AA 31 II	Yorqe'am south	4.3
AA 42 II St.	Yorqe'am south	3.7
AA 41 II	Yorqe'am south	3.7
AA 46 II St.	Yorqe'am south	3.9
AA 47 II	Yorqe'am south	4.2
AA 51 II	Yorqe'am south	3.8
AA 57 II St.	Hor Hahar	2.9
AA 58 II	Hor Hahar	3.4
No. of samples, 17, Average CO_2% value		3.7
AA 3 II–III	Yorqe'am south	2.4
AA 9 III	Yorqe'am south	3.3
AA 14 II–III	Yorqe'am south	3.0

Table 3. cont.

Sample bed*	Locality	$CO_2\%$
AA 18 II–III	Yorqe'am south	3.1
AA 24 II–III St.	Yorqe'am south	3.3
AA 28 II–III	Yorqe'am south	3.7
AA 26 III St.	Yorqe'am south	3.6
AA 27 III	Yorqe'am south	3.4
AA 33 II–III St.	Yorqe'am south	4.0
AA 32 II–III	Yorqe'am south	4.2
AA 34 III	Yorqe'am south	3.6
AA 39 II–III St.	Yorqe'am south	3.5
AA 40 II–III	Yorqe'am south	3.3
AA 37 III St.	Yorqe'am south	3.0
AA 38 III	Yorqe'am south	3.3
AA 45 II–III	Yorqe'am south	3.6
AA 50 II–III	Yorqe'am south	3.2
AA 56 II–III St.	Yorqe'am south	3.0
AA 55 II–III	Hor Hahar	3.0
AA 54 III St.	Hor Hahar	3.7
AA 53 III	Hor Hahar	3.3
No. of samples, 21, Average $CO_2\%$ value		3.4
AA 2 IV	Yorqe'am south	3.6
AA 10 IV	Yorqe'am south	4.0
AA 13 IV	Yorqe'am south	3.7
AA 17 IV	Yorqe'am south	3.6
AA 22 IV	Yorqe'am south	3.7
AA 36 IV	Yorqe'am south	3.7
AA 44 IV	Yorqe'am south	3.6
AA 49 IV	Yorqe'am south	4.3
AA 52 IV	Hor Hahar	3.8
No. of samples, 9, Average $CO_2\%$ value		3.8
AA 1 V	Yorqe'am south	4.1
AA 11 V	Yorqe'am south	4.3
AA 12 V	Yorqe'am south	4.5
AA 16 V	Yorqe'am south	4.2
AA 23 V	Yorqe'am south	3.9
AA 35 V	Yorqe'am south	4.2
AA 43 V	Yorqe'am south	3.2
AA 48 V	Yorqe'am south	4.3
No. of samples, 8, Average $CO_\%$ value		4.1

* Economic phosphorite beds from the Phosphorite unit.

tically any level (taking into consideration the number of samples). Thus, in the northern and central Negev there are three distinct groups of phosphate rocks, distinguished according to their CO_2 content: high, phoscarb and phosph, and an intermediate group, porce, which is between phosph and phoscarb. The results for the Sayyarim Formation (sayyar) from the southernmost Negev are preliminary; there are too few samples (and most of them are from the porcellanite member) to reach definite conclusions. Nevertheless, the results are interesting and show that this formation is similar to the phosphatic carbonate unit with regard to CO_2 content.

The differences between samples in different stratigraphic horizons together with the similarity of the samples within the same group indicate that the CO_2 content is linked to a depositional and/or to a very early diagenetic origin. The results show also that rocks from horizons richer in P_2O_5 have on average a lower CO_2 content. This result is not limited to the Negev phosphorites and appears to be obtained widely elsewhere (Gulbrandsen 1970). It suggests, therefore, that the lowering of the

Table 4. *Analysis of variance on stratigraphy.*

Group	Number of samples	Mean	Standard deviation
high	31	5.229	0.729
phoscarb	24	4.396	0.765
porce	12	4.025	0.644
phosph	91	3.708	0.487
sayyar	19	4.395	0.401

Source	DF	SS	MS	F	P
Factor	4	56.644	14.161	41.86	0.000
Error	172	58.190	0.338		
Total	176	114.834			

DF is degree of freedom; SS is sum of squares. The total sum of squares is broken into two sources: factor, the variation due to differences between the groups; and error, the variation due to differences within a group. MS is mean squares = SS/DF, F is F-ratio = MS Factor/MS Error. p gives the probability that all the groups have the same mean.

CO_2 content is linked to the phosphate concentration mechanism. It should be stressed that this is not true within a stratigraphic horizon, in this case there is no correlation between P_2O_5 and CO_2. This strengthens the diagenetic hypothesis. Furthermore, it is difficult to understand how samples from different topographies could have exactly the same weathering history.

Another one way analysis of variance was made on all samples for which petrographic data were available. One group, is made up of 41 coarse-grained samples while a second group, consists of 17 fine-grained samples. Table 5 and Fig. 6 summarize the results. The results show that samples with coarser grains (CCBP or CSBP) have a significantly higher CO_2 content than the fine grained samples (SCBP or SSBP) irrespective of stratigraphic position. This again strengthens the diagenetic hypothesis. It should be emphasized that there are two ways to change the CO_2 content of an apatite.

(1) By selective dissolution: francolites in a phosphatic rock usually have a range of CO_2 values; francolites with a higher CO_2 concentration are more soluble, therefore partial dissolution of a phosphate rock leads to a lower CO_2 in the residue. (2) Dissolution and reprecipitation, which leads to a recalibration of the carbonate content in the francolite according to the ratios of the anions in solution. It should be emphasized that the changes induced by selective dissolution are limited since the range of substitution within a sample is rather narrow.

Both environment of deposition and early diagenetic processes are apparently reflected in the fabrics and grain composition of the different phosphorites. The bioturbation features commonly associated with the coarse biodetrital phosphorites, and the high energy of character of these rocks, point to deposition in aerated environments with good mixing of waters which had a chemical composition typical to sea-water.

Table 5. *Analysis of variance on petrography.*

Group	Number of samples	Mean	Standard deviation
Coarse	41	4.490	0.745
Fine	18	3.706	0.624

Source*	DF	SS	MS	F	p
Factor	1	7.702	7.702	15.24	0.000
Error	57	28.806	0.505		
Total	58	36.507			

* See Table 4 for explanation.

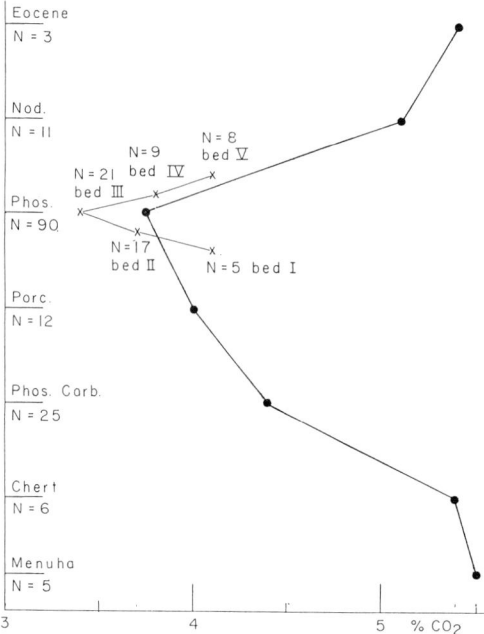

Fig. 6. Variation of CO_2 content with stratigraphy.

This is in contrast to the sandy peloidal phosphorites where the laminated fabric suggests deposition in anoxic bottoms (see also Reiss 1988) where Eh and pH changes might have somewhat changed the chemical composition. The abundance of bone fragments in coarse biodetrital is apparently related to onshore washing of vertebrate remains from phosphatizing hemipelagic environments, as suggested by Antia & Whitaker (1978) for the Upper Silurian Ludlow bone bed (see also Reif 1982). This again stands in contrast to the peloids of sandy phosphorites which are derived mainly from nearshore environments (Soudry 1987).

Another reason for the differences in CO_2 content is perhaps the difference in permeability and porosity of the rocks during early diagenesis. An early lithification of the coarse biodetrital phosphorites is suggested by the lack of compaction of the granular fraction. This might have inhibited migration of solutions in these rocks while such a migration was perhaps easier in the sandy peloidal phosphorites during a soft sediment stage and given their high intergranular and intergranular porosity (up to 40%). Early lithification may also explain the similarity in CO_2 content of the nodular phosphorites and the coarse biodetrital phosphorites although the two have very different structures. The interlocked fabric of the globose structures in the nodular phosphorites together with early phosphate cementation, has very rapidly produced close textured rigid forms. Nevertheless, epigenetic (weathering ?) neomorphic processes which occurred in some of these nodular phosphorites are accompanied by a considerable lowering of the CO_2 content.

The authors thank A. Peer, Geological Survey of Israel, for technical help. They also thank A. Notholt and an anonymous reviewer for their critical reading of the manuscript. This work was done within the framework of project 28480, Geological Survey of Israel. The authors are grateful also for partial financial support from both the Earth Science Research Administration, Ministry of Energy and the US–Israel Binational Foundation (BSF grant 3409/83).

References

AIGNER, T. 1982. Calcareous tempestites: storm-dominated stratification in Upper Muschelkalk limestones (Middle Trias, S.W. Germany). *In:* EINSELE, G. & SEILACHER, A. (eds) *Cyclic and Event Stratification*. Springer-Verlag, Berlin-Heidelberg-New York, 180–198.

ANTIA, D. D. J. & WHITAKER, J. H. McD. 1978. Scanning electron microscope study of the genesis of the Upper Silurian Ludlow bone bed. *In:* WHALLEY, W. B. (ed.) *Scanning Electron Microscopy in the Study of Sediments*. Proceedings of the Symposium, Swansea, September 1977, Geo Abstracts Publishers, Norwich, England, 119–134.

CHIEN, S. H. 1977. Thermodynamic considerations on the solubility of phosphate rock. *Soil Science*, **123**, 117–121.

GULBRANDSEN, R. A. 1970. Relation of carbon dioxide content of apatite of the Phosphoria Formation to regional facies. *US Geological Survey Professional Paper*, **700B**, B9–B13.

LEHR, J. R. & MCCLELLAN, G. H. 1972. *A revised laboratory reactivity scale for evaluating phosphate rock for direct application*. Bulletin Y-43, Tennessee Valley Authority, Muscle Shoals, Alabama.

——, ——, SMITH, J. P. & FRAZIER, W. A. 1968. Characterization of apatites in commercial phosphate rocks. *Colloque International sur les phosphates minéraux solides*, Toulouse, May 1967, 2, 29–44. Société Chimique de France, Paris, 16–20.

LEWY, Z. 1990. Pebbly phosphate and granular phosphorite (Late Cretaceous, southern Israel) and their bearing on phosphatization processes. *In:* NOTHOLT, A. J. G. & JARVIS, I. (eds) *Phosphorite Research and Development*. Geological Society, London, Special Publication, **00**, 000–000.

MCARTHUR, J. M. 1978. Systematic variations in the contents of Na, Sr, CO_3 and SO_4 in marine carbonate-fluorapatite and their relation to weathering. *Chemical Geology*, **21**, 41–52.

McClellan, G. H. & Lehr, J. R. 1969. Crystal chemical investigation of natural apatites. *American Mineralogist*, **54**, 1374–1391.

—— & Van Kauwenbergh, S. J. 1990. Mineralogy of sedimentary apatites. *In*: Notholt, A. J. G. & Jarvis, I. (eds) *Phosphorite Research and Development*. Geological Society, London, Special Publication, **51**, 23–31.

Monty, C. L. V. 1976. The origin and development of cryptalgal fabrics. *In*: Walter, M. R. (ed.) *Stromatolites*. Elsevier, Amsterdam, 193–271.

Nathan, Y. 1984. The mineralogy and geochemistry of phosphorites. *In*: Nriagu, J. O. & Moore, P. B. (eds) *Phosphate Minerals*. Springer-Verlag, Berlin-Heidelberg-New York, 275–291.

—— & Soudry, D. 1982. Authigenic silicate minerals in phosphorites of the Negev (Southern Israel), *Clay Minerals*, **17**, 249–254.

Reif W. E. 1982. Muchelkalk/Keuper bone-beds (Middle Triassic, SW-Germany) — Storm condensation in a regressive cycle. *In*: Einsele, G. & Seilacher, A. (eds) *Cyclic and Event Stratification*, Springer-Verlag, Berlin-Heidelberg-New York, 180–198.

Reiss, Z. 1988. Assemblages from a Senonian high-productivity sea, *Revue de Paléobiologie, Vol. Spec.* **2**, Benthos'86; 323–332.

Soudry, D. 1983. *Etude de la série phosphatée de la région d'Ein Yahav (Négev Central, Israel) — Logique sequentielle, pétrologie, approache de la phosphatogénèse*. Thesis, Doct. Sciences Nat., Institut National Polytechnique de Lorraine.

—— 1987. Ultra-fine structures and genesis of the Campanian Negev high-grade phosphorites (Southern Israel), *Sedimentology*, **34**, 641–660.

—— & Lewy, Z. 1988. Microbially influenced formation of phosphate nodules and megafossil moulds (Negev, Southern Israel), *Palaeogeography, Palaeoclimatology, Palaeoecology*, **64**, 15–34.

Role of microbial processes in the genesis of Jordanian Upper Cretaceous phosphorites

ABDULKADER M. ABED & KHALID FAKHOURI
Department of Geology and Mineralogy, University of Jordan, Amman, Jordan

Abstract: Samples representing granular and laminated phosphorites from the main Upper Cretaceous deposits of Jordan, have been examined by SEM for microbial structures. Globular clusters and sheaths, postulated to have formed by bacterial activity or possibly as inorganic precipitates, have been found only within cavities and in the cementing phosphatic material around phosphate grains of the granular phosphorites. These structures are interpreted to have originated in association with percolation of ground-water late in the history of the sediments, rather than as a consequence of primary grain-forming processes. Organic-rich phosphate grains are believed to have acted as a food source for the bacteria and are, therefore, the cause of the microbial structures and not the result of them. On the other hand, empty cylindrical and/or spherical sheaths, interpreted to be of primary, cyanobacterial (blue-green algae) origin, were readily observed in the laminated phosphorites. These phosphorites are interpreted to have been deposited as algal mats in shallow, subtidal to intertidal environments.

Phosphorites and phosphatic rocks have received a good deal of attention in the last ten years. Their genesis was recently reviewed by Riggs (1986). Three processes are believed to be dominant: (a) direct (authigenic) precipitation from pore water, (Birch 1980; Benmore *et al.* 1983; Loughman 1984), (b) replacement of sea-floor sediments particularly carbonates, by phosphates, (Ames 1959; Baturin 1972; Price & Calvert 1978; Birch 1980; McArthur *et al.* 1988), and (c) formation of phosphates by means of microbial activity, especially bacteria (O'Brien *et al.* 1981; Soudry & Champetier 1983; Lucas & Prevot 1985; Southgate 1986; Soudry 1987; Soudry & Lewy 1988; Soudry & Southgate 1989; Lamboy 1990; Lewy 1990). The aim of this work was to investigate the possible role of microbial processes in the genesis of some granular and laminated Upper Cretaceous phosphorites from Jordan.

Geological setting

The Jordanian phosphorites (Fig. 1) form part of Levant–North Africa Upper Cretaceous–Lower Tertiary phosphate belt (Notholt 1980). They are Lower to low Middle Maastrichtian (Table 1) in age (Abed & Ashour 1987; Hamam 1977), and constitute the upper part of the Amman Formation (B2b) of Masri (1963) or the Phosphorite Member of Bender (1974). The minable deposits of Esh-Shidiya (Fig. 1) in the southeast are 18 m thick and are made of four phosphate horizons, designated from top to bottom A0 to A3, (Fig. 2) with individual units being separated by chert, limestone and dolomite. Units A2 and A3 are the main economic horizons. Compared to other deposits in Jordan, the Esh-Shidiya phosphorites (particularly horizon A3) exhibit higher contents of siliciclastic sand reflecting their proximity to the Arabian–Nubian sandstone continent (Khalid & Abed 1982).

At Al Hasa, in central Jordan (Fig. 1), there is only one economic phosphate bed (Fig. 2) which ranges in thickness from 1 m to several metres, and is almost always underlain by massive, up to 40 m thick, oyster build-ups and overlain by marl and marly limestone. There are a further 28 isolated economic ore bodies in central Jordan.

At Ruseifa (Fig. 1) phosphorites occur at four horizons, designated from bottom to top a1 to a4, and separated by limestone, dolomite and minor chert. The phosphorites represent four rhythmic sequences. Each rhythm starts with a bioturbated omission surface overlain by a relatively thick fining-upwards granular phosphorite, which is commonly overlain in turn by a thinner laminated phosphorite. The whole sequence is terminated by a non-phosphatic horizon made of laminated early diagenetic dolomite and/or limestone with minor chert. More than one bed of granular and laminated phosphorite may be present in each rhythm, especially in unit a3 and a4 (Fig. 2).

The NW Jordan phosphorites (Fig. 1) are about 8 m thick. The lower part is composed of several beds of granular and laminated phosphorite alternating with chert, and followed by

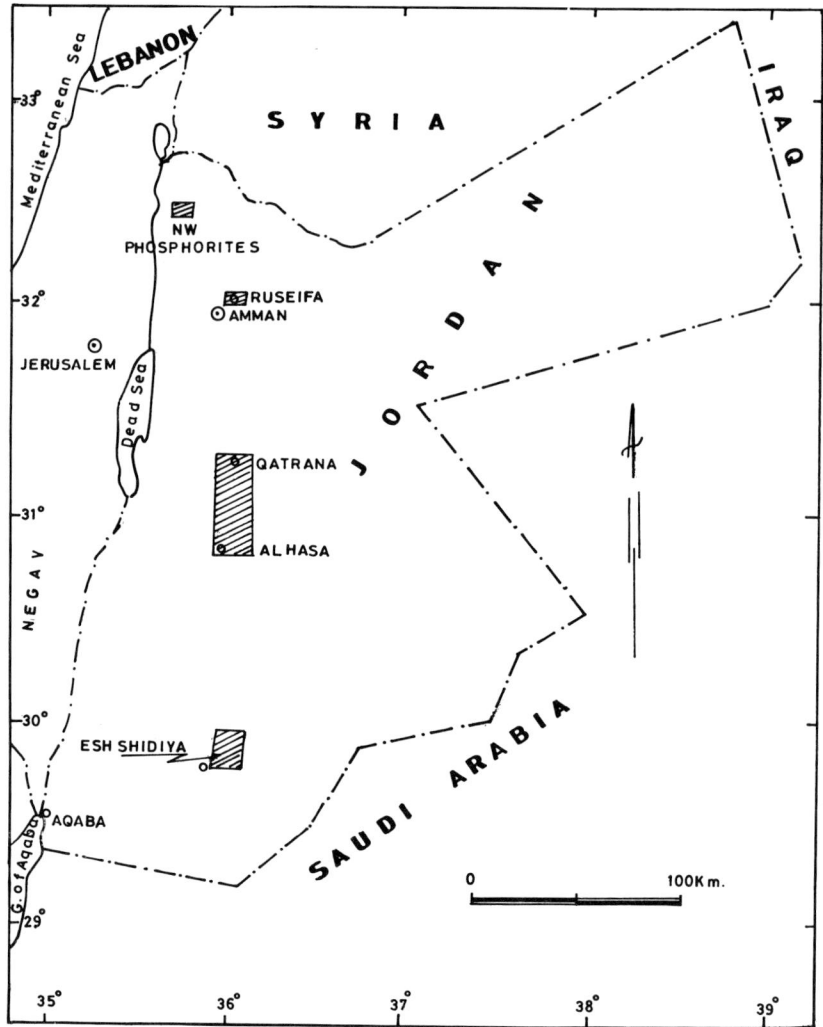

Fig. 1. Location of the Jordanian phosphate deposits. Major fields are indicated by hatched areas.

a 3 m thick bed of massive granular phosphorite (Fig. 2). The whole sequence is overlain by soft marl and is underlain by the main chert of the Amman Formation.

Sixty samples were selected from the mines to represent the main phosphate horizons, except in The Ruseifa Basin. This structure is believed to have formed in the uppermost Cretaceous due to the folding associated with the Syrian Arc System (Bowen & Jux 1987; Abed 1988), and here samples were taken from both basinal sequences (the mines) and from highs, where indurated uneconomic phosphate beds occur.

The Jordanian phosphorites consist predominantly of granular and laminated sediments. The former type dominates the main phosphate beds and occurs as phosphorite packstones or grainstones, with carbonate and minor phosphate forming the matrices and cements. The laminated type is relatively minor, and when present it occurs as the upper part of the phosphate bed in clear fining-upwards cycles (Fig. 3a). The lamination is interpreted to be mostly algal in origin (Fig. 3b). This is clearly shown in both the Ruseifa and NW phosphorites (Abed & Al Agha 1989).

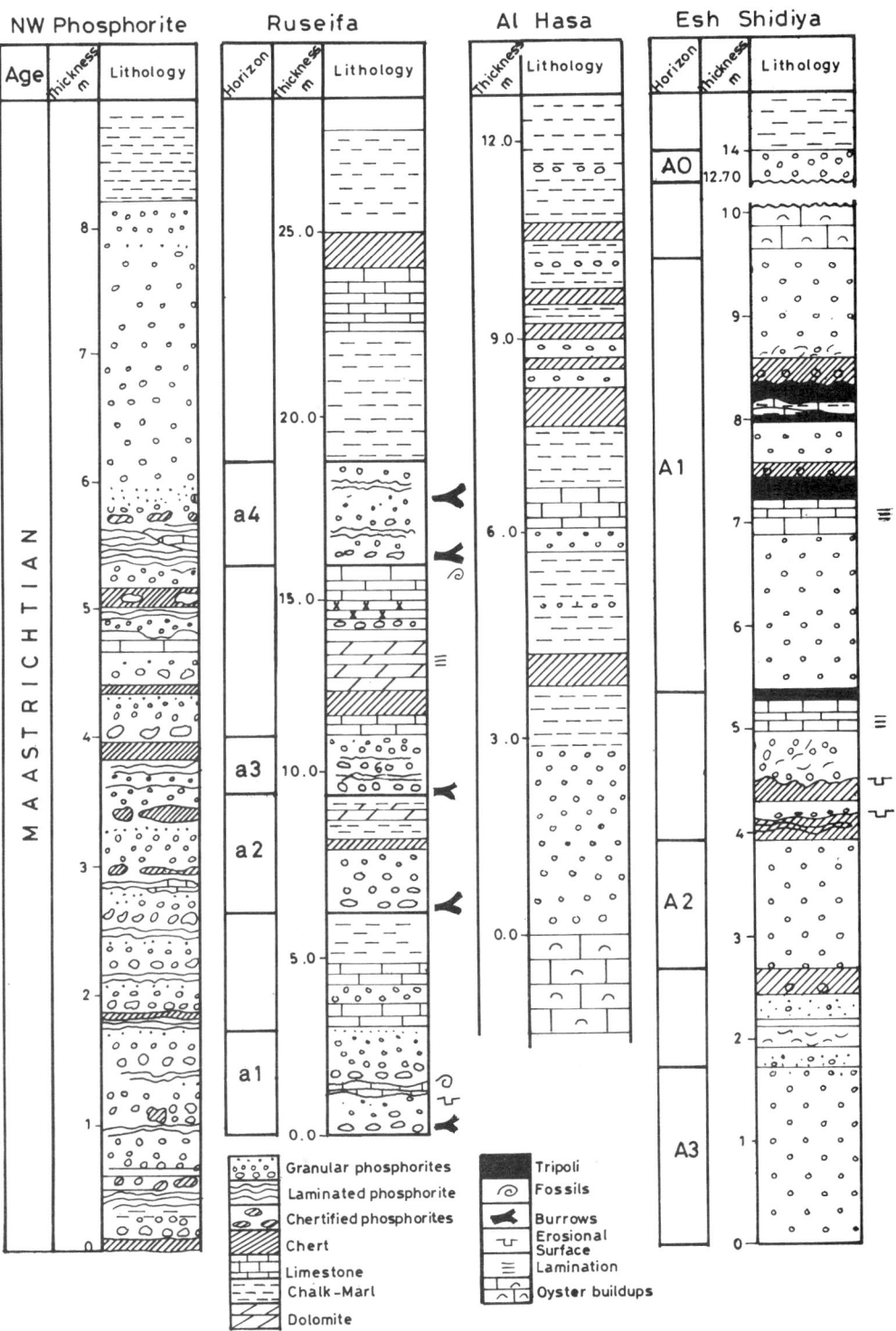

Fig. 2. Representative columnar sections for the four major phosphate areas. Note how the Esh Shidiya and Al Hasa phosphorites are dominantly granular while those of Ruseifa and the NW are granular and laminated.

Table 1. *The stratigraphical position of Jordanian phosphorites*

Series	Stage	Members (Bender 1974)	Formations (Masri 1963)	
Early Palaeogene	Palaeocene	Chalk–Marl	Muwaggar	B_3
Upper Cretaceous	Maastrichtian	Phosphorite	Amman	B_{2b}
	Campanian	Silicified Limestone		B_{2a}
	Santonian			
	Coniacian	Massive Limestone	El Ghudran	B_1
	Turonian		Wadi Sir	A_7
	Cenomanian	Echinoidal Limestone	Shueib	A_{5-6}
			Hummar	A_4
		Nodular Limestone	Fuhais	A_3
			Naur	A_{1-2}
Lower Cretaceous		Kurnub (Hathira) Sandstone		

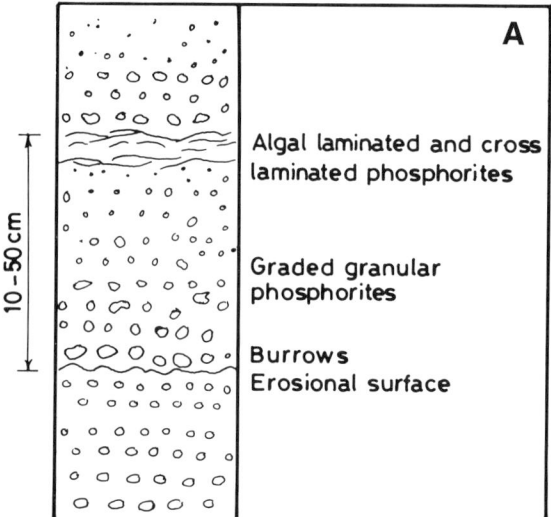
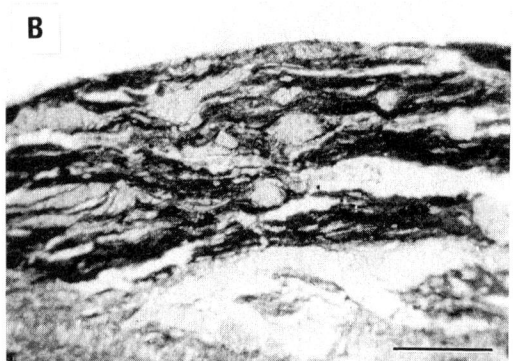

Fig. 3. (A) A typical fining upward cycle in the Ruseifa and the NW phosphate deposits showing the relationship between the granular and laminated facies. (B) Laminated phosphorite (white) alternating with micrite (dark). Plain polarized light. Scale bar is 0.4 mm.

Petrography

Jordanian granular phosphorites are made of four main types of grain: intraclasts, skeletal fragments, peloids, and coated grains. Intraclasts dominate over the other types. They are mostly structureless, but occasionally display (Fig. 4a) algal structures, normal lamination and fractures (desiccation cracks). Most peloids appear not to be of faecal origin, but are interpreted to be rounded to subangular intraclasts (Fig. 4b). This is supported by the similar chemical compositions of both types (Fakhouri 1987; Abed & Fakhouri 1989). Some well-rounded peloids are found as internal sediments in the pores and cavities of bones. Reworking of the bone material can liberate these internal moulds, and they can be easily confused with faecal pellets (Abed & Al Agha 1989). Skeletal particles are highly angular, and commonly exhibit cavities filled with phosphate mud. Coated grains are rather minor constituents of the granular phosphorites compared with the other grain types. They are composed of a nucleus, typically a bone fragment, coated by a single layer of phosphate mud. Depending on the nucleus shape, the coated grains are either rod-like or spherical (Fig. 4c). Some coated grains display a yellowish transparent isopachous, radial-fibrous phosphatic cement on their outer surfaces (Fig. 4d).

Fractured intraclasts generally show varying degrees of alterations along their fractures (Fig. 5a). Additionally, some peloids have alteration rims (Fig. 5b), which most probably have resulted from the oxidation and/or leaching by groundwater of the organic matter incorporated within the grains (cf. McArthur et al. 1987). The granular phosphorites are usually packstones in the depositional lows and grainstones on the highs. Phosphate cement and mud are rather minor but occur in some samples at all localities.

Microbial processes

Three groups of protist have been advocated as having a direct role in the formation of phosphorites. These are eubacteria or true bacteria, the cyanobacteria or blue-green algae (Riggs 1979; O'Brien et al. 1981; Soudry & Champetier 1983; Williams & Reimers 1983; Southgate 1986; Soudry 1987; Soudry & Lewy 1988; Soudry & Southgate 1989), and the fungi (Dahanayake & Krumbein 1985). The last type has not yet been found in the Jordanian phosphorites and will not be considered further in this article.

Granular phosphorites

Globular clusters, postulated by various authors (e.g.: Soudry & Champetier 1983; Soudry & Lewy 1988; Lewy 1990 in the Negev phosphorites) to be bacterial products, have been found in several samples of the granular phosphorites, especially in the Ruseifa deposits. These structures are most commonly present in the microcavities of the Jordanian samples (Fig. 6a, b), but also occur within the matrix around phosphate grains. This is in agreement with the findings of Soudry & Lewy (1988) from the geologically similar Negev phosphorites.

Scanning electron microscopy (SEM) and energy dispersive x-ray analysis (EDX) has revealed the presence of similar globular clusters composed of halite (NaCl). Halite microstructures are rather frequent and exist mostly within and around open pore-spaces (Fig. 7a, b). Some of the particles in these halite clusters are hollow with rounded edges, suggesting partial dissolution of the halite (Fig. 7b).

Another type of microbial structure present in the granular phosphorites, are empty globular to cylindrical sheaths, which are interpreted to be of blue-green algal (cyanobacterial) origin (cf. Soudry & Champetier 1983; Southgate 1986). These microstructures are found only within the phosphatic matrix of the Jordanian samples (Fig. 8a & b), and although more than 1000 phosphate grains were examined on different levels of magnifications by SEM, globular clusters and sheaths have never been identified inside phosphate grains.

Laminated phosphorites

Samples from the algally laminated phosphorites of Ruseifa and from NW Jordan, readily show that the phosphate laminae are composed of empty globular and tube-like sheaths (Fig. 9a & b), which are also interpreted to be of blue-green algal origin (see Krumbein 1983 for a review on the role of these organisms in the formation of ancient stromatolites).

Interpretation and discussion

Riggs (1979), while studying the genesis Florida phosphorites, was among the first to interpret the globular microstructures as coccoidal type bacteria. The structures were found attached to phosphate grains and in cavities. O'Brien et al. (1981) described similar microstructures in Holocene to Pleistocene phosphates from the east Australian continental shelf. They

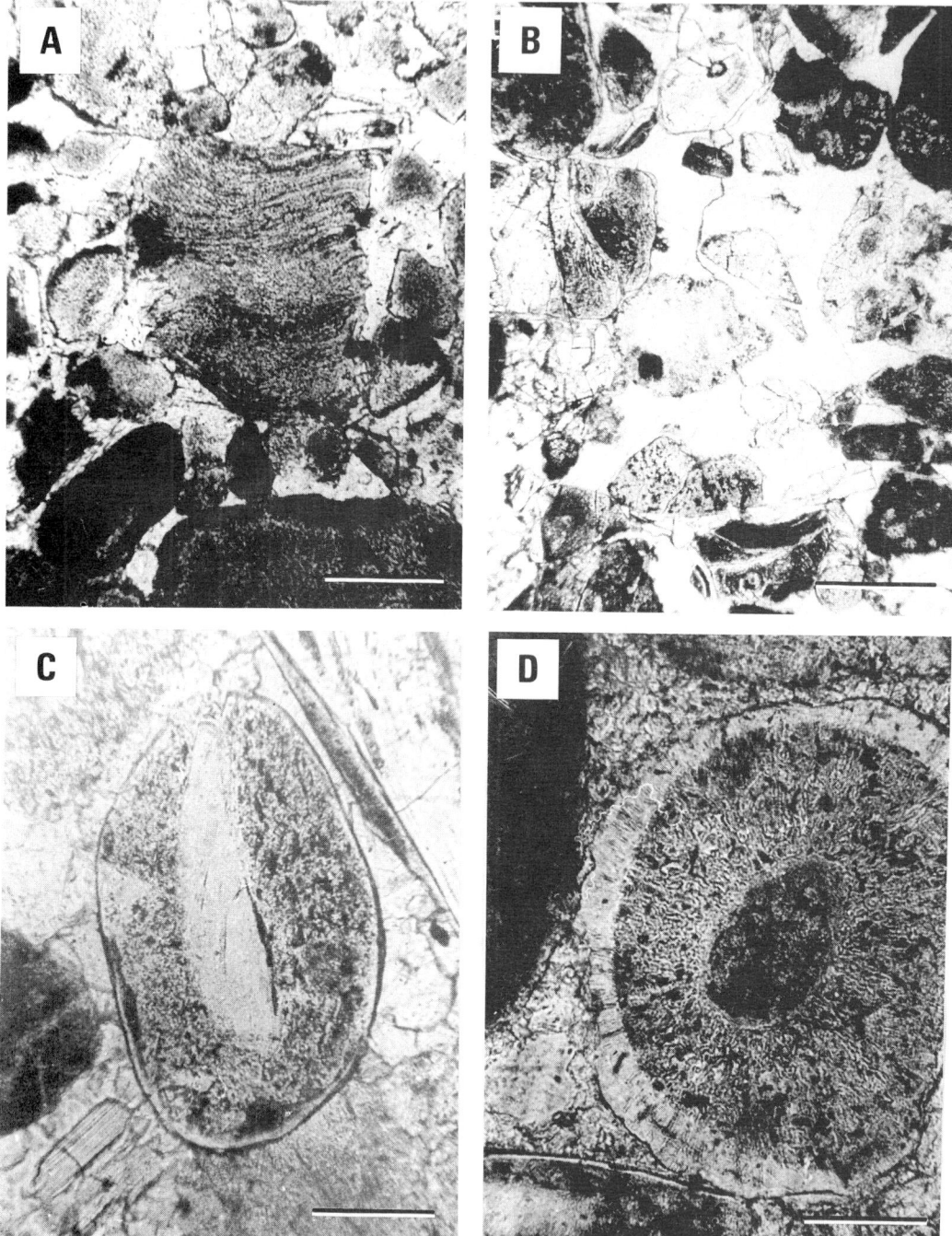

Fig. 4. (A) Grainstone phosphorite showing a laminated intraclast in the centre of the photograph. Plain polarized light. Scale bar is 0.3 mm. (B) Typical grainstone phosphorite. Grains are dominantly intraclasts. Plain polarized light. Scale bar is 0.4 mm. (C) Coated grain made of one coat of phosphate mud around a bone fragment. Plain polarized light. Scale bar is 0.13 mm. (D) Isopachous prismatic phosphate cement around a coated grain. The cement is pale yellow coloured and transparent. Plain polarized light. Scale bar is 0.07 mm.

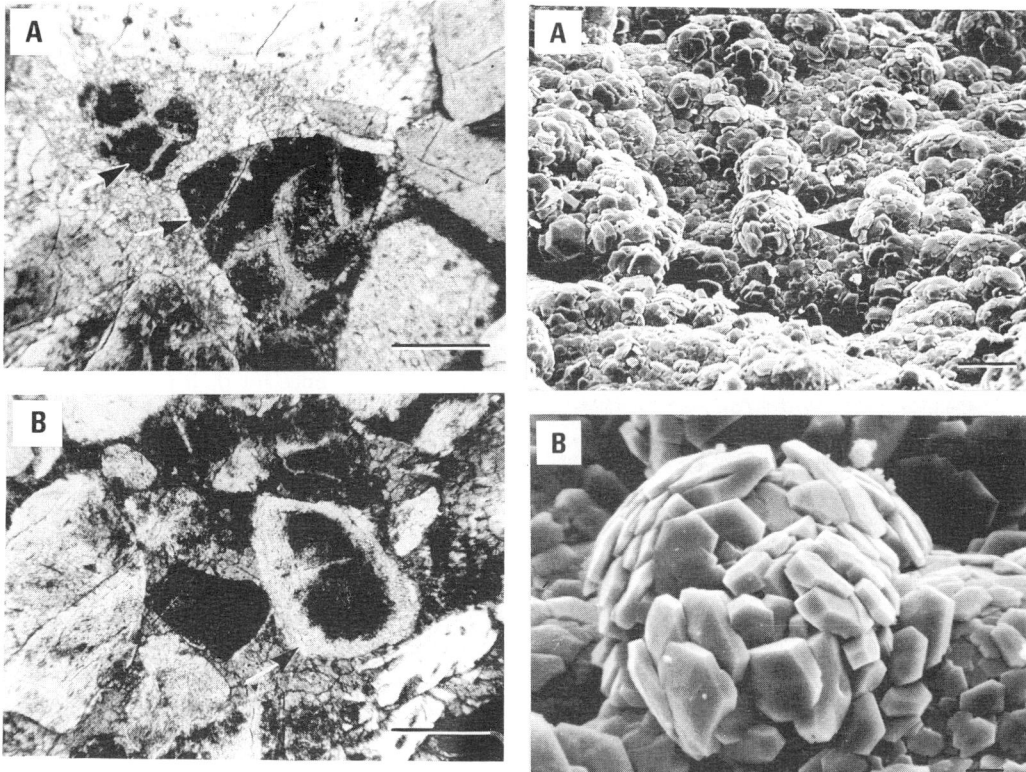

Fig. 5. (A) Dark organic-rich phosphate particles (arrowed) with strongly developed fractures. Note how alteration has proceeded along the cracks. Plain polarized light. Scale bar is 0.4 mm. (B) Alteration rim around a phosphate particle (arrowed). Note the gradational contact between the original dark interior of the grain and the altered rim. Plain polarized light. Scale bar is 0.4 mm.

Fig. 6. (A) A group of phosphate globular clusters (arrowed) found in a cavity of a granular phosphorite sample from Ruseifa. SEM photomicrograph. Scale bar is 10 μm. (B) Enlargement of one of the globular clusters in (A). SEM photomicrograph. Scale bar is 1 μm.

found bacterial structures occurring as isolated coccoidal cells scattered in the matrix or attached to phosphate grains, fossil shells, and glauconites. More recently, Soudry & Lewy (1988) proposed a microbial mat model for the genesis of certain nodular phosphate horizons in the Negev, southern Israel. These authors suggested that their phosphorites were formed through the reworking of microbial mats deposited in sabkhas of the Upper Cretaceous. However, their plates clearly show that the bacterial structures are present solely in the binding material around the phosphate grains or in cavities.

From both published examples and from our own observations of Jordanian phosphorites, it is apparent that bacterial globules and globular clusters do not generally occur inside phosphate grains, but rather are attached to them or occur in their internal cavities. This leads the present authors to believe that this type of bacterial structure was produced during weathering, in association with percolating ground water which reactivated microbial processes in the sediments. The structures were formed, therefore, very late in the history of these rocks, and where not a result of a primary mat-forming process. Such a conclusion is supported by the fact that halite (NaCl) globular clusters, somewhat like those of the bacteria, are found in similar settings. Since sodium and chlorine are amongst the most mobile elements, and since no primary evaporites are found associated with the investigated phosphorites, these halite microstructures are probably due to dissolution and precipitation processes occurring after uplift. In fact, the halite particles which make up the clusters are sometimes hollow with rounded

Fig. 7. (A) Several halite (NaCl) clusters growing in a pore space of a granular phosphorite sample from Ruseifa. Note the hollow halite crystal (arrowed) Scale bar is 10 μm. (B) Enlargement of one of the halite clusters in (A). SEM photomicrograph. Scale bar is 10 μm.

suggesting active processes of dissolution (Fig. 7b). This is not surprising because most halite in the Jordanian phosphorites has originated by washing from the arid Jordanian plateau by rain water and reprecipitation in near surface phosphate deposits (unpublished data by Abed).

There is ample evidence in the literature that the cocci-like globules and clusters may be due to contamination either in the field or in the laboratory (De Ley 1968). Bradely (1963) described bacteria from the Lower Cretaceous Nevada lake deposits, but found later (Bradely 1968) that these are inorganic globular fluorite (CaF_2) produced during the subsequent weathering of the samples. Riggs (1988, pers. comm.) has produced globules composed of fluorite compounds in his leaching experiments on the North Carolina phosphorites. Bien & Schwartz (1965) has shown that ground water percolating through the Permian Zechstein salt rocks have large numbers of bacteria that easily penetrate into these porous rocks and leave their relics around the grains.

The granular phosphorites of both Jordan and the Negev (Soudry & Lewy 1988) are essentially grainstones to packstones which are extremely porous and are situated near-surface in an arid area. The presence of both phosphate and halite globular clusters in these deposits can be explained by precipitation by bacteria in association with descending ground water. Absence of halite clusters in the samples studied by Soudry & Lewy (1988) might reflect a deeper subsurface occurrence for their samples.

It is suggested, therefore, that the Upper Cretaceous phosphate grains acted as substrates on which ground-water associated bacteria have grown and produced phosphate clusters. These bacteria could use phosphorus leached from the phosphate particles by ground water as a nutrient. An indicator of leaching of phosphate particles is the presence of relatively high fluorine contents in ground waters from the major phosphate areas of Jordan. Francolite (containing up to 4% F) is the only fluorine-bearing mineral in the whole Upper Cretaceous succession in these areas (Abed & Fakhouri 1989). In central Jordan, the F content in Al Hasa-Qatrana area (Fig. 1) ranges between 0.8 to 1.5 mg l^{-1} compared with only 0.2 mg l^{-1} away from this area (Marie 1987). The same water does not contain excess phosphorus, which may have been utilized by ground-water associated microbes, possibly bacteria, or, may have precipitated inorganically as fluorine depleted phosphate minerals within pore spaces and cavities (Marie 1987). Petrographically, leaching can be noticed along particle cracks and margins as shown in Fig. 5a and b (see above).

In addition, bacteria generally utilize organic matter as an energy source. Jordanian phosphorites underlie or are interbedded with a thick oil shale horizon having a total organic carbon (TOC) content of up to 20% (Abed & Amireh 1983), while isolated phosphate grains from Ruseifa have a TOC average of 0.47% (Reeves & Saadi 1971) and those from the NW phosphorite have an average of 2.88% (Abed & Al Agha 1989).

Clearly therefore, both unoxidized organic matter and inorganic phosphorus are both readily available for bacteria to live in the pore spaces and cavities of the Jordanian phosphorites. Consequently, it is suggested that the globular clusters were a result of the organic-rich phosphorite grains rather than their cause, as postulated by Soudry & Lewy (1988). An inorganic origin for the phosphate globular clusters cannot be totally discounted. Individual clusters (Fig. 6b) are clearly made of typical hexagonal apatite crystals, which indicate recrystallization of the original bacterial structure or possibly that they are a primary chemical precipitate. Also, an association of the phosphate clusters, postulated to be bacterial, with superficially similar the halite clusters, which are almost certainly inorganic chemical precipitates, might indicate similar origin for both types. Such an interpretation, however, necessitates an explanation for the globular shape of these phosphate microstructures, and it is difficult to envisage how they could they be produced inorganically.

In contrast to the granular phosphorites, bacterial relics are abundant within the algally laminated phosphorites (Fig. 9a & b). These globular and tubular sheaths are interpreted to be of blue-green algal origin, very much like those of the coated grains of Soudry & Champetier (1983) and the phoscrete of Southgate (1986). It is now well accepted that

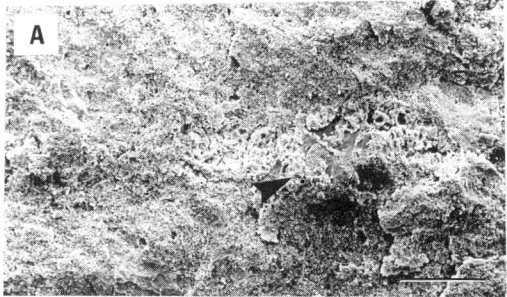

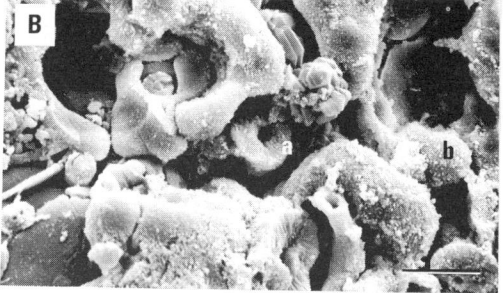

Fig. 9. (A) SEM photomicrograph of an algally laminated phosphorite. Note the lamination at the centre (arrowed). Scale bar is 100 μm. (B) Enlargement of part of the lamina of (A). Note the radial empty tubular (a) and globular (b) blue-green algal (cyanobacterial) sheaths. SEM photomicrograph. Scale bar is 10 μm.

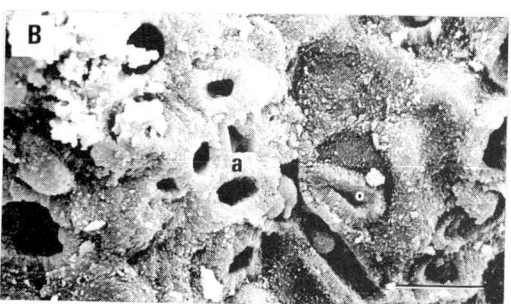

Fig. 8. (A) SEM photomicrograph of a granular phosphorite from Ruseifa showing phosphate intraclasts (I) and phosphate matrix (M) at the centre. Scale bar is 100 μm. (B) Enlargement of part of the matrix in Fig. 8a showing empty radial tubular blue-green algal (a) (cyanobacterial) sheaths. Such sheaths are not seen inside the phosphate grains. SEM photomicrograph. Scale bar is 10 μm.

stromatolites, algally laminated sediments and oncolites are generated through the activities of cyanobacteria, particularly in association with other prokaryote protists (Monty 1977; Buick *et al.* 1981; Krumbein 1983). Mineralization of these structures could explain the widely distributed occurrence of phosphatized microbial relics in laminated phosphate grains and sediments in the Jordanian phosphorites.

The authors believe that laminated Jordanian phosphorites were generated in a subtidal to intertidal environment in a low energy regime. This interpretation is based on presence of blue-green algal sheaths, algal laminations (Fig. 4a), desiccated intraclasts (Fig. 5a), and algally laminated dolomites and carbonates associated with the phosphorite horizons (cf. Banerjee *et al.* 1980; Lucas *et al.* 1980; Howard & Hough 1979; Birch 1980; Soudry & Champetier 1983; Soudry 1987). Eustatic changes of sea level would provide a mechanism for reworking the phosphatized blue-green algal mats and redepositing them as granular phosphorites (Abed & Al Agha 1989). However, the scarcity of the laminated intraclasts and coated grains relative to the structureless phosphate grains would imply another origin for the latter grain type. It is hard to believe that diagenesis would have obliterated all signs of bacterial activities inside these grains, as Soudry & Lewy (1988) postulated in their model for the Negev phosphorites, and other mechanisms of formation for such enigmatic structureless phosphate particles must be advocated (cf. Southgate 1986; Abed & Al Agha 1989).

Conclusions

(1) In granular phosphorites, bacterial microstructures are only present as globular clusters and sheaths in cavities within the grains or in the binding material around the phosphate grains. Halite clusters which are superficially similar to the phosphate clusters occur in similar settings.

(2) Phosphate clusters occurring in granular phosphorites are interpreted to be due to bacterial activities associated with ground water percolation through these organic-rich phosphorites. Such clusters are produced during the weathering of phosphorite rather than as a result of primary phosphate-forming processes.

(3) Halite clusters occurring in granular phosphorites are interpreted as inorganic precipitates from descending rain water which has leached halite from the adjacent arid Jordanian plateau. Their presence indicates that an inorganic origin for the phosphate clusters cannot be totally discounted.

(4) Cyanobacterial sheaths constitute the laminae of algally laminated phosphorites and are undoubtedly of primary origin. In this respect, these sediments are similar to carbonate stromatolites in their genesis and formation, and were deposited in shallow subtidal to intertidal environments.

An earlier version of the manuscript was improved through the critical reading and comments of I. Jarvis and the referees of the Geological Society.

References

ABED, A. M. 1988. Outlines of the paleogeography of Jordan in the phosphorite times. *Guidebook to the 11th international Phosphorite. Field Workshop and Symposium* Jordan, 71–80.

—— & AL AGHA, M. R. 1989. Petrography, geochemistry and origin of the NW Jordan phosphorites. *Journal of the Geological Society, London*, **146**, 499–506.

—— & AMIREH, B. 1983. Petrography and geochemistry of some Jordanian oil shales from north Jordan. *Journal of Petroleum Geology*, **5**, 261–174.

—— & ASHOUR, M. 1987. Petrography and age determination of the NW Jordan phosphorites. *Dirasat*, **14**, 247–63.

—— & FAKHOURI, K. H. 1989. On the chemical variability of some Jordanian phosphate particles. *Chemical Geology*, (In press).

AMES, L. L. 1959. The genesis of carbonate apatite. *Economic Geology*, **4**, 829–841.

BANERJEE, D. M., BASU, P. C. and SRIVASTAVA, R. N. 1980. Petrology, mineralogy, geochemistry and origin of the Precambrian Aravallian phosphorite deposits of Udaipur and Jhabua, India. *Economic Geology*, **75**, 1181–1199.

BATURIN, G. N. 1972. Phosphorus in interstitial water of sediments on the South West African Shelf. *Oceanology*, **12**, 849–855.

BENDER, F. 1974. *Geology of Jordan*. Gebrueder, Berlin.

BENMORE, R. A., COLEMAN, M. L. & MCARTHUR, J. M. 1983. Origin of sedimentary francolite from its sulphur and carbon isotope composition. *Nature*, **302**, 516–518.

BIEN, E. & SCHWARTZ, W. 1965. Geomicrobiologische Untersuchungen VI. Uber das Vorkommen konservieter toter und lebender Bakterienzellen in Salzgesteinen. *Zeitschrift fur Allgemeine Mikrobiologische*, **5**, 185–205.

BIRCH, G. F. 1980. A model for a penecontemporaneous phosphatization by diagenetic and authigenic mechanisms from the western margin of southern Africa. *In*: BENTOR, Y. K. (ed.) *Marine Phosphorites*. Society of Economic Paleontologists and Mineralogists Special Publications, **29**, 79–101.

BOWEN, R. & JUX, U. 1987. *Afro-Arabian Geology*. Chapman and Hall, London.

BRADELY, W. H. 1963. Unmineralized fossil bacteria. *Science*, **141**, 919–921.

—— 1968. Unmineralized fossil bacteria: a retraction. *Science*, **160**, 437.

BUICK, R., DUNLOP, I. S. R. & GROVES, D. I. 1981. Stromatolite recognition in ancient rocks: An appraisal of irregular laminated structures in an Early Archean chert — bante unit from North-Pole, Western Australia. *Alcheringa*, **5**, 161–181.

DAHANAYAKE, K. & KRUMBEIN, W. E. 1985. Ultrastructure of a microbial mat-generated phosphorite. *Mineralium Deposita* **20**, 260–265.

DE LEY, J. 1968. Molecular biology and bacterial phylogeny. *In*: DOBZHANSKY, T., HECHT, M. K. STEERE, W. C. (eds) *Evolutionary Biology*, 2. Appleto-Century Crofts, New York, 103–156.

FAKHOURI, K. 1987. *Chemical variability of francolites from Jordan and the role of microbial processes in phosphogenesis*. Thesis, Univ. Jordan.

FLUGEL, E. 1982. *Microfacies analysis of limestones*. Springer, Berlin.

HAMAM, K. A. 1977. Foraminifera from the Maastrichtian phosphate-bearing strata of Al Hasa area, Jordan. *Journal of Foraminiferal Research*, **7**, 34–43.

HOWARD, P. F. & HOUGH, M. J. 1979. On the geochemistry and origin of the D-Tree, Wonarah and Sherrin Creek phosphorite deposits of the Georgina Basin. Northern Australia. *Economic Geology*, **74**, 260–284.

KHALID, H. & ABED, A. M. 1982. Petrography and geochemistry of Esh – Shidiya phosphates. *Dirasat*, **9**, 81–101.

KRUMBEIN, W. E. 1983. Stromatolites – The challenge of a term in space and time. *Precambrian Research*, **20**, 493–531.

LAMBOY, M. 1990. Microbial mediation in phosphatogenesis: new data from the Cretaceous phosphatic chalks, northern France. *In*: NOTHOLT, A. J. G. & JARVIS, I. (eds) *Phosphorite Research and Development*. Geological Society, London, Special Publication, **52**, 157–167.

LEWY, Z. 1990. Pebbly phosphate and granular phosphorite (Late Cretaceous, southern Israel) and their bearing on phosphotization processes. *In*: NOTHOLT, A. J. G. & JARVIS, I. (eds) *Phosphorite Research and Development*. Geological Society, London, Special Publication, **52**, 169–178.

LOUGHMAN, D. L. 1984. Phosphate authigenesis in the Aramachy Formation (Lower Jurassic) of Peru. *Journal of Sedimentary Petrology*, **53**, 1147–1156.

LUCAS, J. & PREVOT, L. 1985. The synthesis of apatite by bacterial activity: Mechanism. *Sciences Géologique Memoir*, **77**, 83–92.

—— & TROMPETTE, R. 1980. Petrology, mineralogy and geochemistry of the late Precambrian phosphate deposits of the Upper Volta (W. Africa). *Journal of the Geological Society, London*, **137**, 787–782.

MARIE, A. M. 1987. *Fluoride content of the ground water resources of East Jordan*. Thesis, Univ. Jordan.

MASRI, M. 1963. *Report on the geology of the Amman-Zerga area*. Central Water Authority, Amman.

MCARTHUR, J. M., HAMILTON, P. J., GREENSMITH, J. T., BOYCE, A. J., FALLICK, A. E., BIRCH, G., WALSH, J. N., BENMORE, R. A. & COLEMAN, M. L. 1987. Phosphorite geochemistry: isotopic evidence for meteoric alteration of francolite on a local scale. *Chemical Geology*, **65**, 415–425.

—— THOMSON, J., JARVIS, I., FALLICK, A. E. & BIRCH, G. F. 1988. Eocene to Pleistocene phosphogenesis off western South Africa. *Marine Geology*, **85**, 41–63.

MONTY, C. 1977. Evolving concepts on the nature and the ecological significance of stromatolites. *In*: FLUGEL, E. (ed.) *Fossil Algae*. Springer, Berlin, 15–36.

—— 1981. *Phanerzoic Stromatolites, Case Histories*. Springer, Berlin.

NOTHOLT, A. J. 1980. Economic phosphatic deposits, mode of occurrence and stratigraphical distribution. *Journal of the Geological Society, London*, **137**, 793–805.

O'BRIEN, G., HARRIS, J., MILNES, A. & VEEH, H. 1981. Bacterial origin of East Australian continental margin phosphorites, *Nature*, **294**, 442–4.

PRICE, N. B. & CALVERT, S. E. 1978. The geochemistry of phosphorites from the Namibian shelf. *Chemical Geology*, **23**, 151–170.

REEVES, M. J. & SAADI, T. A. K. 1971. Factors controlling the deposition of some phosphate bearing strata from Jordan. *Economic Geology*, **66**, 451–465.

RIGGS, S. R. 1979. Phosphorite sedimentation in Florida — A model of phosphogenic system. *Economic Geology*, **74**, 285–314.

—— 1986. Phosphogenesis and its relationship for Proterozoic and Cambrian phosphorites. *In*: COOK, P. J. & SHERGOLD, J. H. (eds) *Phosphate Deposits of the World*. Cambridge University Press, Cambridge, 352–369.

SOUDRY, D. 1987. Ultra-fine structures and genesis of the Campanian Negev phosphorites, Southern Israel. *Sedimentology*, **34**, 641–60.

—— & CHAMPETIER, Y. 1983. Microbial processes in the Negev phosphorites, Southern Israel. *Sedimentology*, **30**, 411–423.

—— & LEWY, Z. 1988. Microbially influenced formation of phosphate nodules and megafossil moulds Negev, Southern Israel. *Palaeogeography Palaeoclimatology Palaeoecology*, **64**, 15–34.

—— & SOUTHGATE, P. N. 1989. Ultrastructure of a Middle Cambrian primary nonpelletal phosphorite and its early transformation into phosphate vadoids: Georgina Basin, Australia. *Journal of Sedimentary Petrology*, **59**, 53–64.

SOUTHGATE, P. N. 1986. Cambrian phoscrete profiles, coated grains and microbial processes in phosphogenesis, Georgina Basin, Australia. *Journal Sedimentary Petrology*, **56**, 429–441.

WILLIAMS, L. A. & REIMERS, C. 1983. Role of bacterial mats in oxygen-deficient marine basins and coastal upwelling regimes: Preliminary report. *Geology*, **11**, 267–269.

Depositional sequences of the Duwi, Sibâîya and Phosphate Formations, Egypt: phosphogenesis and glauconitization in a Late Cretaceous epeiric sea

CRAIG R. GLENN

Department of Geology and Geophysics and Hawaii Institute of Geophysics, University of Hawaii, Honolulu, HI 96822, USA

Abstract: Sedimentary facies, depositional environments and stratigraphic relationships of the Upper Cretaceous Duwi, Sibâîya and Phosphate Formations in Egypt are summarized in order to shed light on the origins and interrelationships between phosphorites and their commonly associated facies. The distribution of phosphorites, glauconitic sands, organic carbon-rich shales, cherts and porcelanites, and bioclastic and fine-grained limestones are correlated in both space and time across the central phosphorite provenance of the country. The correlations are linked by means of sequence stratigraphic analysis and are tied to global eustatic sea-level curves.

Two main depositional realms are inferred from the Campanian lithofacies assemblages: (1) a deeper-water hemipelagic environment accompanying maximum transgression dominated by deposition of phosphorites, organic carbon-rich shales and biosiliceous sediments, which, after maximum flooding, shoals upwards into (2) a progradational stage accompanying sea-level fall, during which oyster banks with brackish back-reef sediments (Eastern Desert) and deltaic sediments (upper Quseir Variegated Shale, Nile Valley) dominated eastern portions of the phosphorite belt, while greensands were reworked seaward from earlier inner shelf sediments in the west (Western Desert). These depositional realms encompass (1) transgressive to lower highstand systems tracts and (2) upper highstand to lowstand progradational wedge system tracts, respectively. The reoccurrence of lower Maastrichtian phosphatic shale, marl and associated phosphorite lithofacies at the top of the sequence indicates repetition of the sedimentary cycle.

Both phosphorites and glauconitic greensands appear to be the result of current winnowing and reworking of authigenic grains from previously deposited sediments. The resulting deposits range in scale from thin lag layers, to amalgamated beds, to giant sand waves up to 10 m thick. In the Egyptian setting, the common factors linking massive phosphorite and glauconite occurrences may have been the co-occurrence of humid climate, deep lateritic weathering in hinterlands, anoxic bottom waters on the inner shelf, and high fluvial discharge rates of phosphorus and iron. Superimposed on these factors were influences of cross-shelf currents necessary to concentrate authigenic precipitates, and a sea level change, which acted to markedly reduce the diluting effects of siliciclastic detritus during rapid sea level rise, as well as vertically stack 'authigenic facies' during sea level fall.

The Duwi, Sibâîya and Phosphate Formations occur as thin, widespread shallow-marine deposits that crop out in a generally east–west trending belt spanning the middle latitudes of Egypt (Fig. 1). These strata are phosphate-rich and of economic importance, forming part of an extensive Middle-East–North African phosphogenic province of Upper Cretaceous–Paleogene age. In total, these rocks account for the greatest accumulation of marine phosphorites known, possibly in excess of 70 billion metric tons of phosphate rock; the phosphate resources in Egypt alone are estimated to exceed 3 billion metric tons (Notholt 1985). These sediments were deposited in shallow epeiric seas which flanked the Tethyan trough to the north. In Egypt, Israel, Jordan, Saudi Arabia and Iraq the phosphorites are Upper Cretaceous (Campanian–Maastrichtian) in age. As is common to many of these occurrences (and indeed to many phosphorite sequences), the phosphatic rocks of the Duwi group, as defined below, are found to occur in association with bioclastic and fine-grained carbonates, cherts and porcelanites, organic-carbon rich shales and glauconitic sandstones.

The sedimentary facies of the main Egyptian phosphorite-bearing strata are examined below with the aim of: (1) establishing the depositional and diagenetic conditions necessary for the development of each facies, (2) outlining the relationship of each associated facies to the

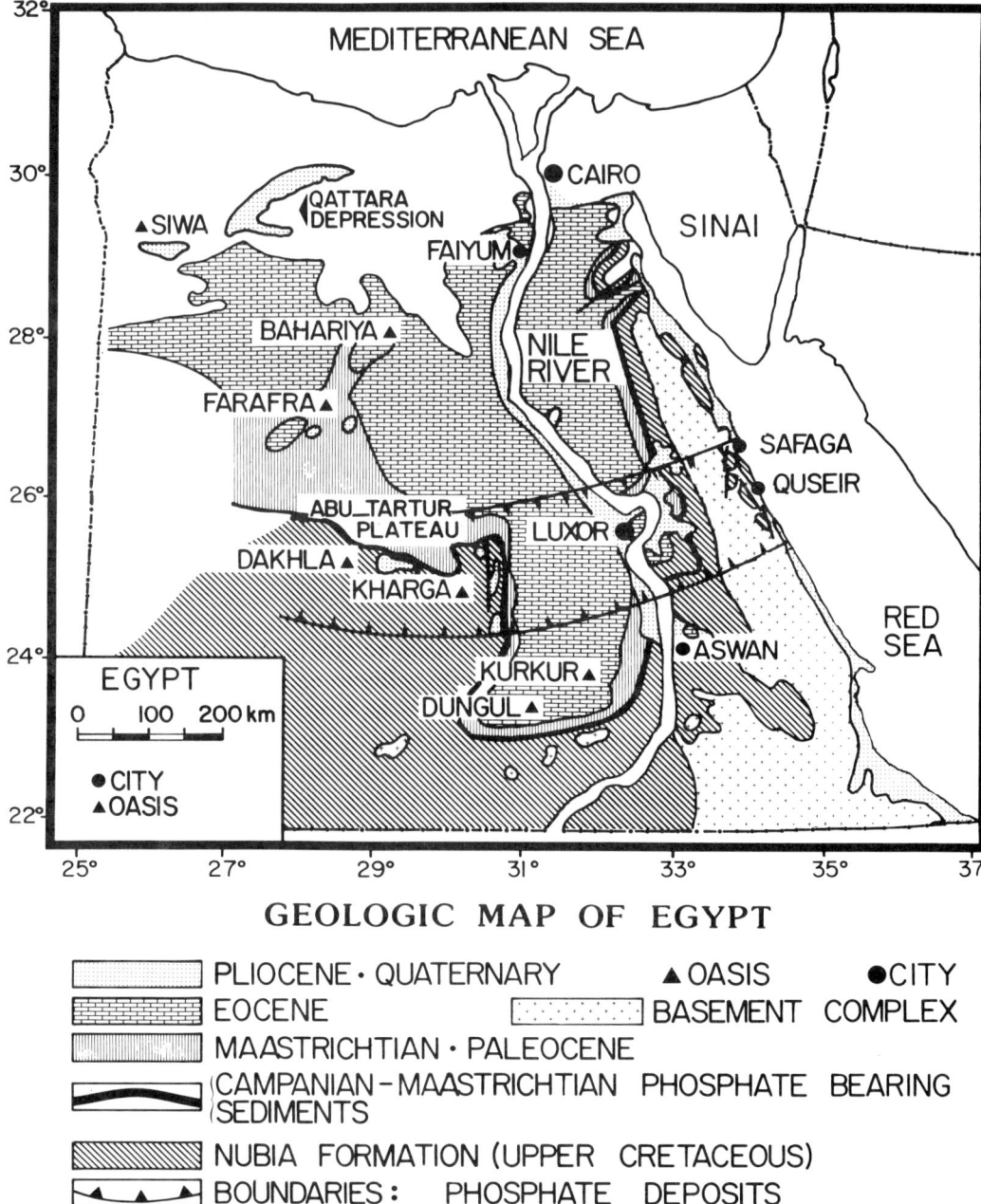

Fig. 1. Sketch map showing general geology of central Egypt and location of Campanian–Maastrichtian age sedimentary rocks. Sawtooth boundaries show location of major (economic) Late Cretaceous phosphorites discussed in this paper. Modified from Spanderashvilli & Mansour (1970).

phosphorites and (3) constructing a general depositional/diagenetic model for the development of the units that might be applied to similar sedimentary assemblages elsewhere in the world. The results presented are examined within the context of Late Cretaceous climate and major fluctuations of sea level that influenced sedimentation on the Egyptian shelf.

In the region of the Eastern Desert (Red Sea Coast) the principal phosphorite-bearing strata are known as the Duwi Formation (Youssef 1957), in the Nile Valley region as the Sibâîya Phosphate Formation (Youssef 1957; El-Nagger 1966a, b) and in the Western Desert as the Phosphate Formation (Awad & Ghobrial 1966). In this work, the mosaic of facies making up these formations are collectively referred to as the Duwi group. These rocks are Upper Campanian to lowermost Maastrichtian and cap or are intercalated with marginal marine to shallow-marine shales of the Variegated Shales Member of the upper Nubia Formation. Some of these shales are considered to be part of the Duwi group here. Overlying the phosphorite suite are deeper-water marine marls and chalks of the Maastrichtian Dakhla Formation. The Upper Cretaceous succession was deposited in generally shallow epicontinental seas that transgressed the northern margin of the Arabo-Nubian craton and deepened towards the Tethyan seaway to the north. The phosphoritic Formations are largely confined to, and best exposed in, an east–west trending belt spanning southern to central Egypt (Fig. 1). To the north of this belt, Upper Cretaceous sediments are dominated by fine-grained carbonate and sandy facies. To the south, phosphatic rocks occur interstratified with terrigenous clastics and sedimentary iron ores.

In the Eastern Desert, the Duwi group is 'conformably overlain' by strata of the Dakhla Formation that encompass a planktonic foraminiferal zone (*Globotruncana ganssseri*) as well as a calcareous nannoplankton zone (*Arkhangelskiella cymbiformis*) assigned to the Maastrichtian in other parts of the world (El-Dawoody & Barakat 1973). Faris (1984), Schrank (1984b) and Schrank & Perch-Nielsen (1985) arrived at a Maastrichtian age for the Dakhla shales and a possible late Campanian to early Maastrichtian age for the underlying upper portion of the Duwi group in the Eastern Desert on the basis of palynological studies. A similar conclusion was reached on the basis of vertebrate faunas by Richardson (1982). Dominik & Schaal (1984) compared vertebrate and invertebrate faunas of the phosphorites across Egypt and concluded that the Campanian-Maastrichtian boundary lies at the top of the 'Duwi Phosphate'. Schrank & Perch-Nielsen (1985) also report a transitional to mixed upper Campanian–lower Maastrichtian microflora from higher portions of the Duwi section southwest of the Abu Tartur mining site in the Kharga Oasis (Fig. 1). Following the biostratigraphic studies of Reiss (1962), Ghanem et al. (1970), Barakat & El-Dawoody (1973), and El-Dawoody & Barakat (1973), Glenn (1980) concluded that the Campanian/Maastrichtian stage boundary lies within the uppermost ten meters of the Duwi Formation at Gebel Duwi (Eastern Desert Type Section) and that the basal portion of the Duwi Formation is of late Campanian age. Of particular importance for correlation are upper Campanian ammonite guide fossils *Bostrychoceras polyplocum* and *Libycoceras* sp. which occur in the Eastern Desert Duwi phosphorites (Abd El Razik 1969) as well as in both the Sibâîya Formation phosphorites *and* the underlying Variegated Shale (Quseir Formation) in the Nile region (Issawi 1972). Late Campanian calcareous nannofossils have also been identified from the Variegated Shales of the Nile Valley at Abu Had (this work; K. Perch-Nielsen, pers. comm. 1988). Barthel & Herrmann-Degen (1981), Schrank (1984a, b) and Schrank & Perch-Nielsen (1985) provided evidence for a late Campanian age for the basal phosphorites of the Western Desert (Dakhla–Kharga area). The main phosphorite-bearing sequence throughout Egypt thus appears to be of late Campanian age, as do the upper Variegated Shales (Quseir Formation) in the Nile Valley region. The Campanian–Maastrichtian boundary resides near the top of the sequence in all areas considered in this study.

The correlation of the Duwi group lithofacies between the Western Desert, Nile Valley and Eastern Desert is summarized in Fig. 2. These correlations are based on the available age data and interpretations of depositional environments and relative palaeowater depths between facies using the sequence stratigraphic analysis described by Haq et al. (1987), Vail (1987) and Wagoner et al. (1987); also see examples in Wilgus et al. 1988. Seismic control was not available in this analysis and tectonic subsidence on the craton is assumed to be simple and linear. Throughout the region, phosphorites and finer-grained shales and/or porcellanites occurring at the base of the group generally give way up-section to coarse-grained bioclastic limestones and siltstones in the east (Fig. 2), and to trough- and tabular-crossbedded greensands intercalated with black shales, phosphorites and siliciclastic layers in the west. Superimposed on this pattern, however, are rapid lateral facies changes which resulted from irregularities in local bottom relief, as well as from the common occurrence of condensed intervals resulting from sediment starvation, extensive bottom winnowing and sediment redistribution. Sea-level variations have played a major role in facies distribution. These relationships are discussed further below.

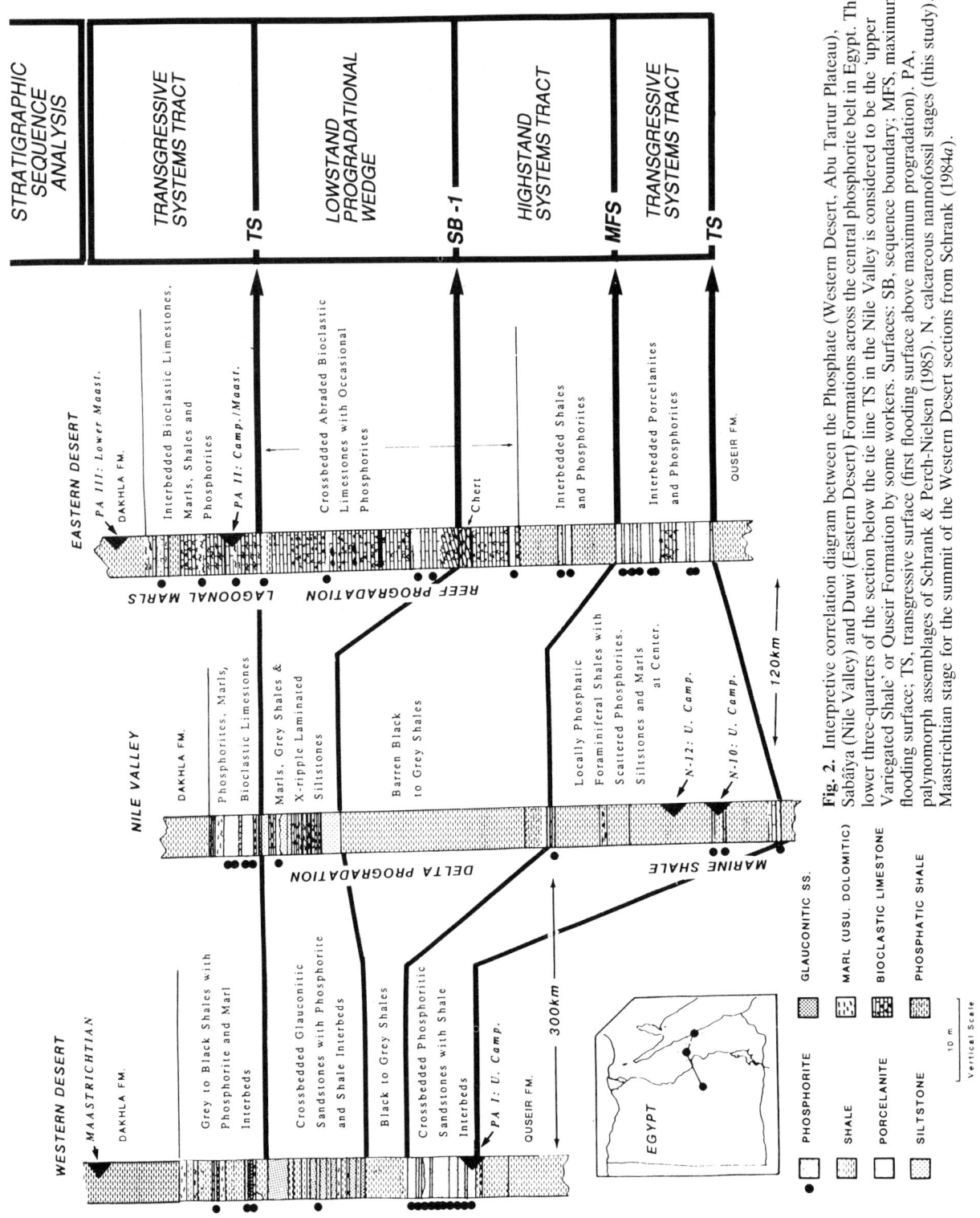

Fig. 2. Interpretive correlation diagram between the Phosphate (Western Desert, Abu Tartur Plateau), Sabâiya (Nile Valley) and Duwi (Eastern Desert) Formations across the central phosphorite belt in Egypt. The lower three-quarters of the section below the tie line TS in the Nile Valley is considered to be the 'upper Variegated Shale' or Quseir Formation by some workers. Surfaces: SB, sequence boundary; MFS, maximum flooding surface; TS, transgressive surface (first flooding surface above maximum progradation). PA, palynomorph assemblages of Schrank & Perch-Nielsen (1985). N, calcareous nannofossil stages (this study). Maastrichtian stage for the summit of the Western Desert sections from Schrank (1984a).

Phosphorite sedimentology

Individual phosphorite beds may range from a few millimetres to tens of centimetres thick, and are interbedded with all associated lithofacies described below. Commonly the phosphorites are thin (1 to 30 cm) and somewhat lenticular, and they are often associated with organic carbon-rich shales (Fig. 3). Their thickest accumulation occurs at the base of the sequence at Abu Tartur, near the top of the sequence in the Nile Valley, and both at the base and at the top of the sequence in the Eastern Desert. Coarse-grained phosphorite lags are commonest in more southerly areas and at the top of the Duwi group throughout the region. More massive units are typically composed of multiple accumulations of thinner individual beds that have developed by repeated episodes of bed amalgamation (Fig. 4), and in most instances are interbedded with either fine-grained porcellanites, shales, or occasionally marls. Extensive bioturbation of the beds is common and, as a result, internal stratification

Fig. 3. Phosphorite interbed in shales from the Nakheil mine (A-Beds), Eastern Desert. Note the random phosphate grains in the encasing phosphatic shales. Diameter of coin is 1.9 cm.

Fig. 4. Phosphorite bed amalgamation in the main phosphorite member at the base of the section at Abu Tartur (Fig. 1) in the Western Desert. The phosphorites thicken laterally at the expense of the intervening shales.

is often lacking (Fig. 5). Where not extensively bioturbated, the beds may display shallow basal scour, faint internal layering, clay intercalations or drapes, faint fining- or coarsening-upwards of phosphatic grains, or crossbedding.

The principal phosphatic components of the main Duwi phosphorite beds are diagenetic peloids ('pellets') and skeletal grains. Using Folk's classification of carbonate textures (1962), these rocks usually range from phosphatic fossiliferous pelmicrites to biosparites; that is, phosphatic and non-phosphatic allochemical constituents reside in either a micritic matrix, a sparry cement, or a poorly washed combination of both. Siliceous, dolomitic and occasionally phosphatic cements are also found to occur. The phosphatic grains are either skeletal (phosphatized or primary fish debris), peloidal or of intraclastic (compound) origin (Fig. 6). True oolitic grains have not been observed. Peloid grains usually predominate and often contain clots of organic matter, pyrite inclusions and occasional siliciclastic material. Light carbon stable isotope signatures (Glenn & Arthur 1989) of the phosphate peloids extracted from *both* the phosphorites and their associated phosphatic shales indicate that these phases precipitated in association with organic matter degradation processes in reducing pore waters.

Eastern Desert sequence: the 'Duwi Formation'

Mining geologists of the Eastern Desert generally refer to three major phosphorite 'seams' (groups of mineable beds) occurring around the Quseir–Safaga area along the Red Sea coast. These are the A-, B-, and C-beds. The C-beds occur near the base of the section in association with porcellanites or more commonly darkcoloured shales, B-beds occur near the centre of the section in association with bioclastic limestones, and A-beds occur near the top of the group in association with black shales or ochrecoloured dolomitic marls (Fig. 2). A-beds and C-beds are the most laterally extensive (as well as more weakly cemented) and remain the principal target of phosphorite mining activities in the region. Most commonly, the beds are about 10 to 35 cm thick, although occasionally they reach a metre or more. Thin millimetre to centimetre thick phosphorite intercalations may, however, be found throughout the sequence. Most of the Red Sea Coast phosphorites are rather lenticular, and appear to pinch and thin along outcrop. As with other portions of the Duwi group, the phosphorites are commonly highly bioturbated, and thus often lack internal stratification (Fig. 5). Where not completely bioturbated, the Eastern Desert phosphorites sometimes show shallow basal scour, occasional cross-stratification with claydraped bedding surfaces, or faint fining or coarsening upwards internal stratification indicative of current-related deposition. Both waning storm conditions and downslope redeposition from local highs could account for the poorly-defined fining-upwards cycles sometimes observed within these rocks (e.g. Johnson 1978; Walker 1979; Swift & Rice 1984). In some cases, phosphorite bed amalgamation is clearly indicated, and the author believes that this is how most of the thicker phosphorite beds developed.

The porcellanites (nomenclature of Bramlette 1946) at the base of the section are finelylaminated to thinly-bedded biosiliceous rocks which may contain scattered mega- or micro-

Fig. 5. Interbedded basal phosphorites and shales at Abu Tartur. Note typical massive phosphorite fabrics and well defined extensive burrow features preserved along the base of the phosphorite beds. White stringers are secondary gypsum.

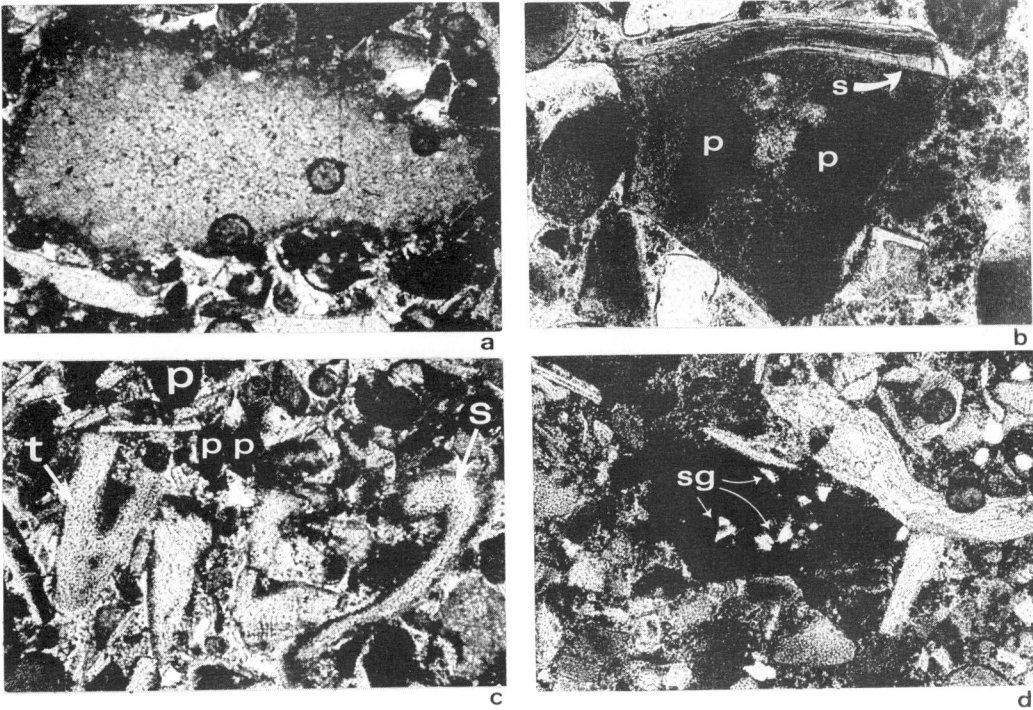

Fig. 6. Examples of phosphorite grain morphologies in thin section. (**a**) 'Internally structureless' peloid very similar to many of those found in associated shales (circles are bubbles in thin section epoxy). Plane polarized light (PPL). Horizontal field (HF) = 1.35 mm. (**b**) Phosphatic intraclast containing phosphate peloids (p), and shell fragment (s). Surrounding grains are structureless peloids (dark) and siliciclastic grains (light). PPL. HF = 1.54 mm. (**c**) Silicified phosphorite with phosphatized shell (s), sharks tooth (t), and numerous phosphate peloids (p). PPL. HF = 3.15 mm. (**d**) Angular siliciclastic grains (sg) contained in phosphatic intraclast. As with (b), this grain represents a ripped-up and redeposited phosphatic cement. PPL. HF = 1.54 mm.

scopically visible phosphate pellets or blebs (Figs 7 and 8), and may include abundant foraminifera, shell fragments, and interstitial iron oxides, the latter presumably weathered after pyrite. Intercalated phosphorite beds in these units are commonly silicified. The biogenic silica once present in these rocks (diatoms, silicoflagellates and sponge spicules; cf. Soudry et al. 1981) is diagenetically altered to quartz-chert. In southern portions of the Eastern Desert area these units grade laterally into thin intercalated glauconitic sandstones, shales and red siltstones of the uppermost Quseir Shales (Fig. 2). Analogy to similar deposits elsewhere (e.g. Ingle 1973; Calvert 1974; Pisciotto 1981a, b) suggests that these rocks are a result of relatively high rates of siliceous primary productivity in a clastic sediment-starved basin. The thin laminations, minor bioturbation and

Fig. 7. Laminated porcellanites from basal section in the Eastern Desert. Thin phosphorite intercalation below coin. Randomly disseminated phosphate peloids may be found in these porcellanites. Diameter of coin is 2.4 cm.

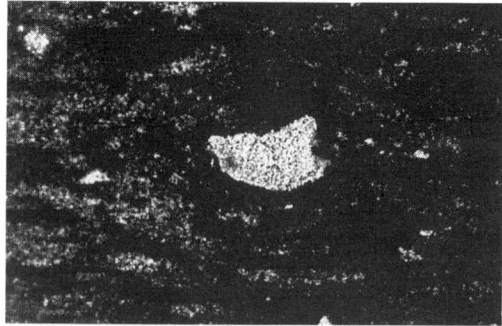

Fig. 8. Photomicrograph of early diagenetic in situ phosphatic peloid in porcellanite. Bending of laminae around pellet indicates it was hard prior to significant compaction. PPL. HF = 3.15 mm.

predominance of fine-grained clays within these rocks suggests relatively terrigenous-free, oxygen-deficient and low-energy hemipelagic depositional conditions. However, periodic aeration and reworking episodes are also indicated by occasional shell lags, isolated phosphorite burrow fills (with no connecting over-lying phosphorite), bioturbation fabrics and scoured basal contacts of intercalated phosphorites.

The porcellanites give way up-section to black or grey thinly laminated, locally phosphatic organic carbon-rich marine shales (Fig. 9). These shales often contain abundant fish and foraminiferal debris, scattered euhedral pyrite crystals, and phosphorite interbeds (Fig. 3). Organic carbon contents, ranging from a few per cent near the base of the Formation to over 20% for shales associated with the phosphorite 'A' beds near the top, suggest dysaerobic to anoxic bottom waters accompanied shale deposition. Rock pyrolysis data (Glenn 1987) indicates a mixed algal-terrigenous to marine-algal source for the organic matter of these shales. Burrow structures are usually absent, except where they occur in association with intercalated phosphorites. Glauconite may be present in small amounts. Locally, the shales may be highly phosphatic (1.0–12.0% P_2O_5; Glenn 1980), containing scattered ovules randomly disseminated in the groundmass (Fig. 10). Compactional distortion of the shale laminae surrounding these grains (cf. Fig. 8) indicates that they were indurated and thus formed prior to significant burial. These pellets are very similar to many of those found in associated phosphorites and appear to represent the in situ source grains from which, through current winnowing, the intercalated phosphorites were mechanically derived.

Relatively thick bioclastic oyster limestones cap the underlying hemipelagic porcellanites and shales (Fig. 2). Biomicrite to poorly-washed biosparite (Folk 1962) coquinas containing highly abraded thick-valved oyster debris (*Ostrea villie*) dominate this complex. Throughout the Eastern Desert these rocks are characterized by large-scale, low-angle tabular crossbedding indicative of high wave energies and off-bank shedding of abraded bioclastic debris (Fig. 11). The bulk of these rocks thus form stratigraphic, rather than ecologic reefs (cf. Dunham 1970). Confinement of in situ ecological assemblages of articulated *O. villie* to southwestern exposures, coarsening upward reef progradational sequences (Richardson 1982), and the northward thinning and eventual termination of this lithofacies (Glenn 1980) suggest that these carbonates were derived from the south. These oyster buildups are similar to

Fig. 9. Laminated shale lithofacies in the Eastern Desert.

Fig. 10. In situ peloids in black phosphatic shale. Such grains are suggested to be starting source grains from which the phosphorites are mechanically derived. Diameter of coin is 2.4 cm.

Fig. 11. Tabular crossbedding in prograding *O. villie* bioclastic limestones, Eastern Desert (Gebel Anz).

other Upper Cretaceous–Cenozoic wave-resistant structures dominating very shallow high-energy, often restricted (brackish and/or hypersaline) tropical marine environments; limited diversity and rapid community growth in these structures is often interpreted as being due to a lack of predation by stenohaline organisms, as well as to reduced competition for space and nutrients (cf. Heckel 1974). Fully open-marine assemblages, characterized by the presence of rudists, are not present in the Egyptian complex. After building to sea level, accommodation space for reefs diminished and the banks prograded toward deeper water while emergent portions of the reef serve as traps for fine-grained lime muds or marls.

Upper portions of the Duwi Formation are characterized by commonly dolomitic (44–50% $MgCO_3$, white to pinkish ochre-coloured marls. These beds interfinger with and overlie upper portions of the bioclastic limestones in the southern Eastern Desert. The faunal content of the marls (articulated large sessile clams, small bivalves, gastropods, and echinoids) indicates a quiet water, restricted environment. This faunal assemblage decreases in abundance up section and to the south and southeast, while planktonic (and subordinate benthionic) foraminifera increase in abundance within the uppermost part of this lithofacies to the north and northwest. These faunal trends are interpreted to be reflecting deeper water depositional conditions to the north of the oyster complex, with restricted shallow water, back-reef conditions predominating and intercalating with oyster buildups to the south. Sediment-infilling of aragonitic shell moulds, collapsed micritic envelops, geopetal and umbrella textures, and mouldic, fenestral, fracture, vuggy and intercrystal porosity (cf. Choquette & Pray 1970)

suggest syngenetic void development in response to periodic fresh or brackish water incursion in the back-bank marls. These facies give way to deeper water shales and marls of the Dakhla Formation above.

The Nile Valley 'Sibâiya (Phosphate) Formation' and the upper 'Variegated Shales'

In the Nile Valley Region the contact between the Sibâiya (Phosphate) Formation and the underlying Quseir Variegated Shale is difficult to define. Germann *et al.* (1985) and Soliman *et al.* (1986) placed this boundary near the top of the shaley units (approximately at the upper transgressive surface in Fig. 2) and just below the appearance of well-developed ledge-forming bioclastic limestones and phosphorites (Fig. 12). The lower portions of these shales, however, are grey at outcrop and black when freshly broken (not variegated in colour like the Quseir Formation elsewhere), contain abundant planktonic foraminifera, pyrite, rare assemblages of small (3–4 mm), possibly dwarfed molluscs, scattered phosphate pellets, and intercalated phosphorite, marl and porcelanite beds (Figs 2 and 12). These lower shales are of open-marine origin, therefore, and are very similar in both lithology and lithological association to the lower black shale lithofacies of the Duwi Formation to the east and the Phosphorite For-

Fig. 12. Stratigraphic section in the vicinity of Gebel Had, Nile Valley. Transgressive phosphorite interbeds in the shales at base of section give way up section to prograding, coarsening-upwards deltaic shales to sands (the sequence boundary in Fig. 2 is near the base of the resistant cap rock of butte), and eventually to interbedded bioclastic limestones, marls and phosphorites capping the top of the sequence. The section shown is about 68 m thick.

mation to the west. Furthermore, calcareous nannoplankton identified in these shales (this work, K. Perch-Nielsen pers. Comm., 1988) indicate an Upper Campanian age and thus equivalent to the lower phosphorites at Abu Tartur to the west (cf. Barthel & Herrman-Degen 1981; Schrank 1984; Schrank & Perch-Nielsen 1985). These lower shales of the Nile Valley sequence (upper Quseir Variegated Shales) thus appear to be relatively deeper-water hemipelagic transgressive deposits correlative with the basal hemipelagic shale-porcellanite-phosphorite sequences in the eastern and western deserts (Fig. 2). Like the Eastern Desert occurrences, the intercalated phosphorites in the shales are commonly bioturbated and represent periodic aeration and sediment reworking within an otherwise quiet-water, oxygen-deficient environment.

The lower Nile Valley shales pass subtly upwards into about 20 m of shales barren of foraminifera and phosphate pellets, and then to interbedded shales and cross ripple-laminated siltstones and fine-grained sandstones cut by erosional channels. These deposits are interpreted as upward-coarsening prodelta muds and delta front sands and silts, respectively (Soliman et al. 1986; Figs 2 and 12). Assuming simple linear subsidence in the basin, these upwards shoaling sediments mark the progradation (basinward shift) of a shallow water delta complex accompanying sea-level fall. These deltaic sediments are, in turn, overlain by about 5 to 10 m of more fully marine marls, bioclastic limestones and phosphorite beds similar to those at the top of the group in the Red Sea coast district.

The Western Desert 'Phosphorite Formation'

The thickest accumulation of minable phosphorite rocks occur at the base of the sequence at Abu Tartur where the beds locally combine to form a single seam averaging about 4 m thick (Fig. 2). Exploratory bore hole information indicates that these deposits thicken to the west and underlie much of the 1200 km^2 area occupied by the Abu Tartur Plateau (Fig. 1; Wassef 1977; Notholt 1985). These thick phosphorite accumulations appear to be offshore phosphatic sandwaves, ridges or bars (Garrison et al. 1979). To the southwest, along surface exposures and to the west in the subsurface (Wassef 1977), the phosphorite beds become thinner and increasingly intercalated with shales. While their basal contact overlying the Quseir shales is quite planar, the beds gently thicken and thin (ridge and swale) between 1.5 m to more than 10 m over distances of several hundred metres (Fig. 13). Subsurface bore-hole data (Hermina 1973) indicates that these sand bodies are up to several thousand metres long. The beds display large scale, low-angle internal crossbedding with beds dipping 5–10° to the southeast, and in some instances cross sets are overlain by thin, burrowed clay drapes. Troughs between phosphorite sandwaves are oriented to the northeast-southwest, normal to the directions indicated by the crossbedding. The indicated palaeocurrent directions are the same as those measured for the overlying glauconitic sandstones (Garrison et al. 1979; Glenn 1987) and suggest northeast–southwest trending phosphatic sandwaves (and later, glauconitic sandwaves) which may have migrated to the southeast. More work is needed to firmly establish the palaeocurrent directions.

While cross-shelf tidal currents may have played a significant role in controlling these bedforms, the appearance of burrowed clay drapes on crossbed sets and the dominantly unidirectional crossbedding directions may indicate rather punctuated development and storm reworking from the northwest (cf. Bose et al. 1988). As with the other phosphorite occurrences in Egypt, the shales intercalated and laterally associated with the phosphorites are locally phosphatic, containing randomly disseminated phosphate pellets and skeletal debris. It is probable that tidally-dominated current winnowing and erosion on the western Egyptian shelf acted to separate the phosphatic grains from such fine-grained phosphatic sediments and further concentrated and molded these accumulations into the giant sand waves.

Overlying and drowning the phosphorite sand waves are hemipelagic dark grey to black organic carbon-rich shales. These strata are, in turn, overlain by well-sorted cross-bedded sideritic glauconitic sandstones intercalated with numerous shale and phosphorite interbeds (Fig. 14). The chief mineral constituents of the greensands are dark green glauconitic–smectite peloids (c. 75%), mixed with varying proportions of siliciclastic grains (chiefly angular to subangular quartz), highly pyritized phosphate pellets and fish debris, and irregularly shaped peloidal siderite ($FeCO_3$) grains (lumps). Reworked early diagenetic ferruginous-siderite concretions are common. Phosphate nodules up to a few centimetres in diameter are also sometimes present. Relative to the underlying phosphoritic sand waves (see above), crossbedding and channel development in the greensands is generally at higher angles and of smaller scale

Fig. 13. Crossbedded phosphorite sandwaves or bars (between arrows) along the Abu Tartur Plateau in the Western Desert (Fig. 1). The basal units below the phosphorites are transitional fluvial-marine maroon to grey variegated shales and siltstones of the Quseir Formation. View is to northeast parallel to swale strike and normal to sand wave propagation; phosphorite aggregate bed thickness at right edge of photo is about 6–7 m. Shallow swale between bar flanks is infilled with black shales. Beds in sand wave to right depositionally thin and pinch out to the left. Left side of swale may be erosional.

(Fig. 14). While the development of these bedding structures indicates substantial movement and redistribution of these sands on the seafloor, the presence of thin clay drapes or seams between some crossbedding sets indicates periodic waning of otherwise relatively high-energy depositional conditions. As with the underlying phosphorites, both tidal- and storm-induced current activity may have been important in the distribution and development of these sands.

Individual glauconitic grains in thin section commonly show compactional flattening indicative of being relatively plastic during burial, and some of these have angular to subrounded outlines suggesting grain breakage prior to burial and probably during current transport. Other, rarer, glauconitic grains are somewhat tabular and are apparently a replacement product after precursory clays or micas. Obvious replacement or infilling textures after biogenic carbonates have not been observed, but this may be due to complete alteration of precursor grains (cf. Odin & Matter 1981). A high concentration of aluminium in these grains ($c.$ 10–15% Al_2O_3) is, however, suggestive of a precursory illitic phase (Glenn 1987; cf. Bornhold & Giresse 1985).

Discussion

A summary of the stratigraphic and sedimentologic interpretations presented above is shown in Fig. 2. On the basis of these interpretations, the Duwi group in Egypt may be divided into a series of system tracts representing a linkage of contemporaneous depositional systems (cf. Vail 1987). Two main depositional realms are inferred from the lithofacies assemblage discussed above: (1) a deeper-water hemipelagic environment (transgressive to highstand systems tracts) accompanying maximum transgression and dominated by deposition of phosphorites, organic carbon-rich shales and biosiliceous sedi-

Fig. 14. Upper stratigraphic section at Abu Tartur. (1) main phosphorite member currently being mined (cf. Fig. 15), (2) black to grey deeper water shales, (3) crossbedded and channelized glauconitic sandstones, (4) black shales with phosphorite and dolomitic marl interbeds. Thin arrow lies along crossbedding plane dipping to southeast. Thicker arrow points to glauconite channel scoured into shales; channel axis strikes northeast–southwest.

ments, which, after maximum flooding, shoals upwards into (2) a progradational stage accompanying sea-level fall (highstand to lowstand progradational wedge systems tracts) during which oyster banks with brackish backreef sediments (Eastern Desert) and deltaic sediments (Nile Valley) dominated eastern portions of the phosphorite belt, while greensands were reworked seaward in areas to the west. Following these progradational phases, the cycle begins to repeat itself with renewed marine transgression and the deposition of phosphatic shales and marls and their associated phosphorites in earliest Maastrichtian times. Figure 15 links these depositional interpretations, as drawn from the discussions above, to the eustatic sea level curve of Haq *et al.* (1987). The observed sequence of lithofacies appears to correlate well with the sequential stacking of contemporaneous depositional systems as predicted by the sequence stratigraphic model. Figure 16 illustrates the interpretation of the sequential development of facies preserved in the Western Desert from north to south.

While thinner, more discontinuous phosphorite beds may be found throughout the sequence, the development of more massive phosphorite deposits exposed in the Duwi group appear to be condensed intervals more tightly tied to early marine transgressive phases, i.e. near the base and the top of the sequence analyzed. The thickest deposits occur at the base of the sequence at Abu Tartur where they formed giant sandwave bedforms before drowning by shales. In the siliciclastic setting of the North American Atlantic shelf, sandwaves have developed during the Holocene rise of sea level and are the result of winnowing and reworking of relict shelf sediments which migrated shoreward over lagoonal deposits as well as the subsequent modification and maintenance of these features by storm and tidal current activity (e.g. Stubblefield *et al.* 1984; Berné *et al.* 1988; Knebel 1989). In other areas influenced by strong tidal currents, similar features were either initiated by present-day hydraulic conditions or are a partially relict feature now in dynamic equilibrium with present day tidal flow (cf. Johnson 1978). In such settings, sandwaves and ridges have not yet been drowned by modern shelf sedimentation because most terrigenous detritus (mud) is trapped in flooded estuaries and other nearshore environments or passes by to accumulate in deeper portions of the shelf or slope. At Abu Tartur, the phosphoritic sandwaves most likely developed intermittently and penecontemporaneously with deposition of offshore fine-grained phosphatic hemipelagic sediment from which they were erosionally derived. Eventually, with highstand conditions, these features were submerged below wave base and blanketed by hemipelagic muds (Figs 15 and 16).

In contrast to the phosphorites, the thick reworked glauconitic sandstones of the Western Desert appear to be progradational facies temporally correlative with reef and delta progradations to the east. They thus appear to have been reworked from more near-shore sites of precipitation, perhaps from inner shelf or estuarine sediments (Fig. 16). Glauconite is a unique authigenic mineral because it contains iron in both its reduced and oxidized states, and appears to require mildly reducing pore-waters or microenvironments in which to form. Probably most important is that iron must first be

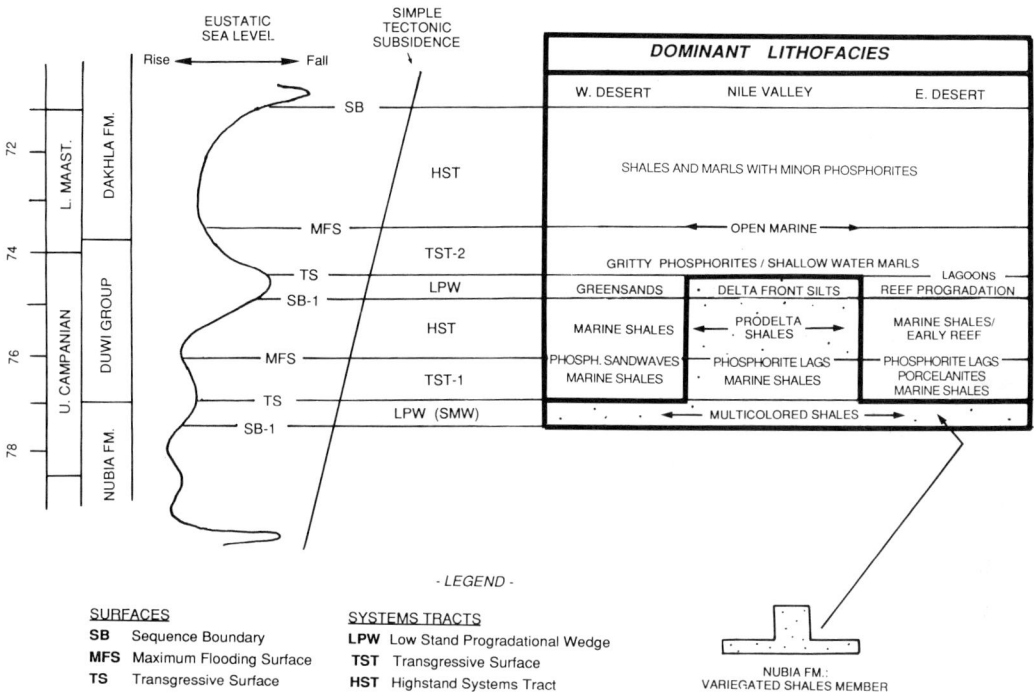

Fig. 15. Summary of the major depositional interpretations for the Duwi group in Egypt as tied through sequence stratigraphic analysis to the global eustatic sea-level curve and linear time scale of Haq *et al.* 1987. Compare with Fig. 2.

made soluble under reducing conditions, and then oxidized upon incorporation into the authigenic phase. It is thus at anoxic/oxic interfaces receiving abundant iron that extensive glauconitization might proceed across large areas of the sea floor. Assuming iron is not limited, this may occur along suboxic horizons within sediments that receive a suitable supply of sedimentary organic matter (cf. Coleman 1985; Froelich *et al.* 1988) or, in the case of oxygen-minimum zones (cf. Glenn & Arthur 1988) or stratified basins, on those portions of the shelf where oxygen-deficient bottom waters and the overlying oxygenated waters meet.

The association between phosphorites and glauconitic sands has long been recognized yet poorly understood. The abundance of glauconites and their associated iron-rich cements suggests a humid climate and deep lateritic weathering which supplied abundant iron to inner portions of the Egyptian shelf (see discussions in Glenn & Arthur 1990), and the associated organic-rich shales and the phosphorites themselves suggests fertile cratonic seas and associated bottom-water, or at least sediment pore-water, anoxia. How are these factors linked? Most studies have concluded that the ultimate driving force behind the development of important phosphatic facies is the stimulation of primary productivity by delivery of oceanic phosphate through coastal upwelling. Upwelled waters, however, are rapidly depleted of nutrients within relatively short distances of the locus of upwelling (cf. Ryther *et al.* 1971; Barber & Smith 1981) and, in the Late Cretaceous Egyptian setting, the main zone of phosphorite and black shale deposition is several thousand kilometres removed from potential areas of equatorial divergence to the east, and some 400 to 500 km removed from potential coastal upwelling zones along the open Tethyan margin to the north (cf. Sestini 1984; Scotese & Summerhayes 1986). Upwelling, however, need not be the only source of phosphorus available to such shallow-marine environments. In high-discharge settings, fluvial nutrient supplies may account for elevated primary productivity on associated shelves (e.g. Diester-Haas 1983; Fox *et al.* 1986) and, although little appreciated in most phosphorite models, such inputs may be more than adequate to account for the large tonnages of phosphate stored in major phos-

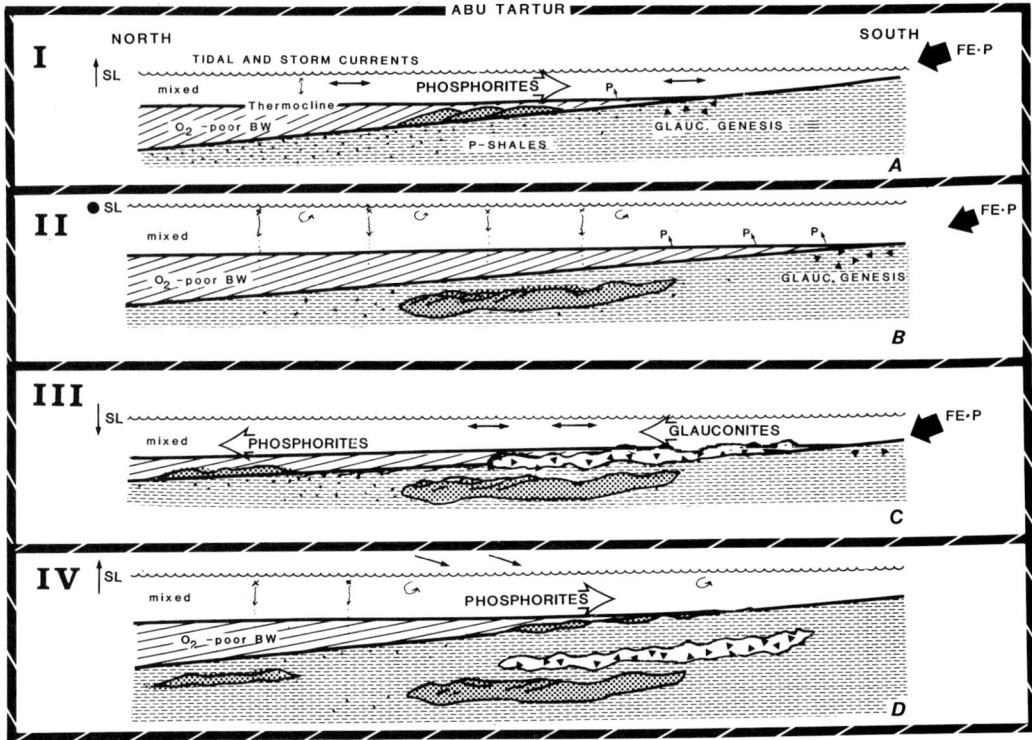

Fig. 16. Conceptual cartoon sketch of the sequential development (I–IV) of lithofacies present at Abu Tartur. The shelf in this region is relatively broad and shallow. Compare with Figs 2, 14 and 15. (**I**) Mid late Campanian marine transgression. Accompanying sea-level rise, sand waves are developed through progressive intermittent winnowing of fine-grained phosphatic muds and migration shoreward. Storms and/or tidal currents play a dominant role in maintaining these structures. Glauconite genesis is restricted to more near-shore areas along the intersection of anoxic and oxic water masses, where the delivery of detrital iron is at a maxima. Release of sorbed phosphorus from detrital particles plays a dominant role in the supplying nutrients to the basin. (**II**) Late Campanian highstand. Maximum flooding results in drowning of sand waves below wave base. Only the most intense storms rework the bottom at this phase offshore. Anoxic bottom waters impinge further inshore, perhaps into estuaries where glauconite genesis is accelerated by enhanced supply of iron. (**III**) Latest Campanian sea-level fall and low stand. Deltaic sediments in southerly areas feeding detritus sorbed nutrients further to the north. Phosphatic shale environments shift seaward with retreating oxygen-depleted bottom waters and result in formation of northern calcareous facies belt phosphorites. Glauconites reworked seaward from former sites of precipitation with marine regression. Reaching similar water depths to that of the early transgressive phosphorites in (a), the prograding greensands also become moulded into sand waves. (**IV**) Campanian-Maastrichtian marine readvance. Phosphatic shales and their winnowed phosphoritic counterparts again move shoreward to form interbedded shales and winnowed phosphorite lags, before eventual drowning beneath highly anoxic bottom waters.

phorite occurrences. Estimates of the total flux of soluble inorganic phosphorus from the modern Amazon estuary (a relatively non-disturbed drainage system), for example, suggest an average delivery of P of c 2.4×10^{11} g a^{-1} to high salinity fringes (≈ 25 ppt) on the Amazon shelf (Fox *et al.* 1986). Such a supply over a 5 million year period equates to more than a trillion metric tons of phosphorus or, assuming an average phosphorite rock content of 22% P_2O_5, more than 4000 times the estimated phosphorus (%P) reserves of Egypt, and nearly 1800 times the estimated Upper Cretaceous reserves (Notholt 1985) of Egypt, Israel, Jordan, Syria and Turkey combined. Further, the Amazon shelf may also be instructive because glauconite occurs on the shelf in association with vertically extensive iron reduction zones in muds supersaturated with respect to authigenic vivianite (iron phosphate) and siderite (iron carbonate) (Aller *et al.* 1986). Phosphorus desorption from iron colloids in vertically extended iron reduction zones has also been suggested by isotopic data from the Egyptian in situ black shale phosphate grains (Glenn & Arthur 1990). Thus, at least in the Egyptian setting, the common factors linking massive phosphorite and glauconite occurrences may be warm, humid climate with deep lateritic weathering, and abundant fluvial discharge which provides flushing of iron and phosphorus to the sea (and perhaps aiding in establishing density stratified water masses in estuaries and on the shelf). Detrital transport of phosphorus as a sorbed component on iron compounds and other particulates may have been an important step in this process (Fig. 16). Superimposed on these are the factors of shelf currents (tidal, wind-generated and cross-shelf geostrophic flow) which are necessary to concentrate authigenic precipitates during sediment reworking, and sea-level change, which acts to trap the diluting effects of siliciclastic detritus in near-shore environs during rapid rise, as well as vertically stack authigenic greensands over phosphorites during sea-level fall. The coupling of shelf-current winnowing, sediment shelf bypassing and near-shore terrigenous-sediment trapping during rapid sea level rise collectively provide a major mechanism of sequence condensation which helps to explain the apparent conflict between the low overall sedimentation rates commonly associated with phosphoritic and glauconitic sequences and the interpretations presented here for major tropical riverine input of nutrients, iron and fine-grained siliciclastic detritus.

Conclusions

The Duwi group strata (Duwi, Sibâîya, Phosphate and uppermost Quseir Variegated Shale Formations) represent a relatively sediment-starved heterogeneous collection of shallow-marine epicontinental facies which collectively provide a detailed record of organic carbon-rich shale, phosphorite, glauconite, chert/porcellanite and shallow-water reef development accompanying third order changes of sea level. On the basis of lithostratigraphy and interpretations of depositional environments, relative palaeowater depths and available age data, the Duwi group may be divided into a series of systems tracts which represent a linkage of contemporaneous depositional packages. From the base of the Duwi group upwards these are: (1) a lower transgressive systems tract (TST) composed of landward stepping phosphorites and hemipelagic sediments, (2) a highstand systems tract (HST) composed largely of seaward prograding black to grey prodelta shales in Western Desert and Nile Valley environs and the buildup of oyster reefs in the east, (3) a low stand progradational wedge (LPW) of glauconitic sandstones (west), delta-front siltstones and sandstones (Nile Valley) and progradational reefal debris (east), and (4) an uppermost Campanian–lower Maastrichtian transgressive systems tract roughly similar in lithology to that in (1) composed of marls and phosphatic shales and relatively coarse-grained phosphorites.

Phosphorites found disseminated throughout the entire upper Campanian–lower Maastrichtian sequence are viewed as the result of current winnowing and concentration of authigenic phases initially precipitated in closely associated organic carbon-rich sediments (black shales and porcellanites). In many portions of the stratigraphic section, lenticular phosphorite lags resulted from episodic sediment winnowing and redeposition. Bed amalgamation due to current reworking appears responsible for the phosphorite bed thickening. In the Western Desert, the main phosphorites appear to have formed as giant phosphorite sandwaves, produced by extensive tidal and storm reworking of penecontemporaneously deposited phosphatic muds accompanying initial stages of marine transgression in this region. In all settings, the phosphorites have been very strongly bioturbated and represent episodic periods of bottom-water oxygenation within otherwise quiet-water dysaerobic to anoxic conditions. Because the Egyptian phosphorites appear to have developed in shallow-marine settings far removed from open ocean margins, it is suggested that possible upwelling scenarios may be inadequate to account for the large tonnages of phosphorus contained in these rocks, and that a more likely principal source may have been deep lateritic weathering of the Arabo-Nubian Massif to the south. Although the application of either a single- of multi-point riverine input model has yet to be tested in other portions of the Middle East phosphogenic provenance, extrapolation of modern Amazon River data suggests river input sources large enough to account for phosphorite resources of Egypt, Israel, Jordan, Syria and Turkey combined.

Consideration of sources and interactions between iron and phosphorus in phosphorite models may also provide a key with which to unlock the mystery of giant phosphorite–greensand associations. The Western Desert sequences are condensed relative to those in the

Nile Valley and the Red Sea coast because these later areas were more strongly influenced by shoaling and progradation of deltas and reefs, respectively. The Western Desert progradational stage, as represented by crossbedded sideritic greensands, developed through the winnowing and basinward reworking of sideritic–glauconitic muds previously deposited at the time of maximum regional transgression and relative sediment starvation. At this time, glauconitization probably took place landward of the main locus of phosphorite generation where redoxclines intersected the seafloor, and close enough to shore to receive ample supplies of iron. Phosphorus desorption from reduced iron compounds in these settings may have played an important intermediate step in the transfer and regeneration of nutrients to the basin throughout the southern, landward portions of the shelf. While the initial stages of glauconitization and phosphatization appear to have been roughly penecontemporaneous and laterally segregated on the Egyptian shelf, the final distribution and vertical stacking of these facies appears to be the result of major sea-level rise and fall.

This work stems from portions of the author's graduate research under the direction of R. E. Garrison and M. A. Arthur and special thanks and appreciation are extended to them for their support and numerous stimulating ideas. Thanks go to P. R. Vail and G. P. Eberli for help with sequence stratigraphic analysis and to K. Perch-Nielsen for examination of calcareous nannofossils. I also profited from helpful discussions with S. E. A. Mansour, P. D. Snavely III, M. Richardson, E. M. El Shazly, M. E. Hilmy, A. El-Kammar, M. Hagras, E. R. Philobbos, and the many members of IGCP working group 156. Thanks go also to G. P. Eberli and R. E. Garrison for critically reviewing the manuscript. This paper is Hawaii Institute of Geophysics Contribution No. 2268.

References

ABD EL-RAZIK, T. M. 1969. Stratigraphical studies on the phosphate deposits between River Nile and Red Sea (south of Latitude 27°N). *Cairo University Faculty of Science Bulletin*, **42**, 299–324.

ALLER, R. C., MACKIN, J. E. & COX, R. T., Jr. 1986. Diagenesis of Fe and S in Amazon inner shelf muds: apparent dominance of Fe reduction and implications for the genesis of ironstones. *Continental Shelf Research*, **6**, 263–289.

AWAD, G. H. & GHOBRIAL, M. G. 1966. Zonal stratigraphy of Kharga Oasis. *Annals of the Geological Survey of Egypt*, **34**, 1–77.

BARAKAT, M. G., & EL-DAWOODY 1973. A microfacies study of the upper Cretaceous-Paleocene-lower Eocene sediments of Duwi and Gurnah sections, southern Egypt. *Hung., Foeldt. Intez., Evi Jel*, **1973**, 391–414.

BARBER, R. T. & SMITH, R. L. 1981. Coastal upwelling ecosystems. *In*: LONGHURST, A. R. (ed.). *Analysis of Marine Ecosystems*. Academic Press, New York, N.Y., 31–68.

BARTHEL, K. W. & HERRMAN-DEGEN, 1981. Late Cretaceous and Early Tertiary stratigraphy in the Great Sand Sea and its SE margins (Farafra and Dakhla oases), SW Desert, Egypt. *Mitteilungen der Bayerischen Staatssammlung für Palaeontologie und historische Geologie*, **12**, 141–182.

BERNÉ, S., AUFFRET, J. -P. & WALKER, P. 1988. Internal structure of subtital sandwaves revealed by high-resolution seismic reflection. *Sedimentology*, **35**, 5–20.

BORNHOLD, B. D. & GIRESSE, P. 1985. Gluconitic sediments on the continental shelf off Vancouver Island, British Columbia, Canada. *Journal of Sedimentary Petrology*, **55**, 653–664.

BOSE, P. K., GHOSH, G., SHOME, S. & BARDHAN, S. 1988. Evidence of superimposition of storm waves on tidal currents in rocks from the Tithonian-Neocomian Umia Member, Kutch, India. *Sedimentary Geology*, **54**, 321–329.

BRAMLETTT, M. N. 1946. *The Monterey Formation of California and the Origin of its Siliceous Rocks*. United States Geological Survey Professional Paper 212.

CALVERT, S. E. 1974. Deposition and diagenesis of silica in marine sediments. *In*: HSÜ, K. J. & JENKYNS, H. C. (eds) *Pelagic Sediments: On Land and Under the Sea*. International Association of Sedimentologists Special Publication, 273–298.

CHOQUETTE, P. W. & PRAY, L. C. 1970. Geologic nomenclature and classification of porosity in sedimentary carbonates. *Bulletin, American Association of Petroleum Geologists*, **54**, 207–250.

COLEMAN, M. L. 1985. Geochemistry of diagenetic non-silicate minerals: kinetic considerations. *Philosophical Transactions of the Royal Society of London, A*, **315**, 39–56.

DIESTER-HAAS, L. 1983. Differentiation of high oceanic fertility in marine sediments caused by coastal upwelling and/or river discharge off northwest Africa during the Late Quaternary. *In*: THIEDE, J. & SUESS E. (eds) *Coastal Upwelling-Its Sedimentary Record, Part B: Sedimentary Records of Ancient Coastal Upwelling*, Plenum Press, New York, N.Y., 399–419.

DOMINIK, W., & SCHAAL, S. 1984. Notes on the stratigraphy of the Upper Cretaceous phosphates (Campanian) of the Western Desert, Egypt. *Berliner Geowissenschaftliche Abhandlungen, Reihe A: Geologie und Paläontologie*, **50**, 153–175.

DUNHAM, R. J. 1970. Stratigraphic reefs versus ecologic reefs: *Bulletin American Association of Petroleum Geologists*, **54**, 1931–1932.

EL-DAWOODY, A. S., & BARAKAT, M. G. 1973. Nannobiostratigraphy of the Upper Cretaceous-Paleocene contact in the Duwi Range, Quseir district, Egypt. *Rivista Italiana di Paleontologia e Stratigrafia*, **79**, 103–124.

EL-NAGGER, Z. R. 1966a. Stratigraphy and planktonic foraminifera of the Upper Cretaceous Lower Tertiary succession in the Esna-Idfu region, Nile Valley, Egypt. *Bulletin of the British Museum of Natural History*, Suppl. **2**.

—— 1966b. Stratigraphy and classification of the type Esna Group in Egypt. *Bulletin American Association of Petroleum Geologists*, **50**, 1455–1477.

FARIS, M. 1984. Biostratigraphy of the Upper Cretaceous-Lower Tertiary succession of Duwi Range, Quseir District, Egypt. *Revue de Micropaléontologie*, **27**, 107–112.

FOLK, R. L. 1962. Spectural subdivision of limestone types. *In*: HAM, W. E. (ed.) *Classification of Carbonate Rocks. Memoir American Association of Petroleum Geologists*, **1**, 62–84.

FOX, L. E. SAGER, S. L. & WOFSY, S. C. 1986. The chemical control of soluble phosphorus in the Amazon estuary. *Geochimica et Cosmochimica Acta*, **50**, 783–794.

FROELICH, P. N., ARTHUR, M., BURNETT, W. C., DEAKIN, M., HENSLEY, V., JAHNKE, R., KAUL, L., KIM, K., ROE, K., SOUTAR, A. & VATHAKANON, C. 1988. Early diagenesis of organic matter in Peru continental margin sediments: phosphorite precipitation. *Marine Geology*, **80**, 309–394.

GARRISON, R. E., GLENN, C. R., SNAVELY, P. D. & MANSOUR, S. E. A. 1979. Sedimentology and origin of Upper Cretaceous phosphorite deposits at Abu Tartur, Western Desert, Egypt. *Annals of the Geological Survey of Egypt*, **9**, 261–281.

GERMANN, K., BOCK, W. D. & SCHRÖTER, T. 1985. Properties and origin of Upper Cretaceous Campanian phosphorites in Egypt. *Sciences Géologiques. Mémoir*, **77**, 23–33.

GHANEM, M. ZALATA, A. A., ABD EL RAZIK, SAID ABDEL GHANI, M., MIKHAILOV, I. A., RAZVALIAEV, A. V., & MIRTOV, Y. V. 1970. Stratigraphy of the phosphorite-bearing cretaceous and paleogene sediments of the Nile Valley between Idfu and Qena: *In*: MOHARRAM, O. (ed.) *Studies on Some Mineral Deposits of Egypt*. United Arab Republic Geological Survey, Cairo, 109–134.

GLENN, C. R. 1980. *Stratigraphy, Petrology and Sedimentology of the Duwi Formation (Late Cretaceous) Eastern Egypt*: Ms Thesis, University of California, Santa Cruz.

—— 1987 *Phosphorus Fluxes, Phosphorite Sedimentation and Associated Diagenesis in Oxygen-Deficient Basins: The Modern Black Sea and Peru Margin and the Upper Cretaceous of Egypt.* PhD. Dissertation, University of Rhode Island, University Microfilms Publication.

—— & ARTHUR, M. A. 1988. Petrology and major element geochemistry of Peru margin phosphorites and associated diagenetic minerals: Authigenesis in modern organic-rich sediments. *Marine Geology*, **80**, 231–267.

—— & —— 1990, Anatomy and origin of a Cretaceous phosphorite-greensand giant, Egypt: *Sedimentology*, **37**, 123–154.

HAQ, B. U., HARDENBOL, J. & VAIL, P. R. 1987. Chronology of fluctuating sea levels since the Triassic. *Science*, **235**, 1156–1167.

HECKEL, P. H. 1974. Carbonate buildups in the geologic record: a review. *In*: LAPORT, L. F. (ed.) *Reefs in Time and Space. Society of Economic Paleontologists and Mineralogists Special Publications*, **18**, 90–154.

HERMINA, M. H. 1973. Preliminary evaluation of Maghrabi-Liffiya phosphorites, Abu Tartur area, Western Desert, Egypt. *Annals of the Geological Survey of Egypt*, **3**, 39–74.

INGLE, J. C., Jr. 1973. Summary comments on Neogene biostratigraphy, physical stratigraphy, and paleocenography in the marginal northeastern Pacific Ocean. *In*: KULM, L. D., VON HUENE, R. L. D., DUNCAZN, J. R., INGLE, J. C., JR., KLING, S. A., MUSICH, L. F., PIPER, D. J. W., PRATT, R. M., SCHRADER, H. -J., WESER, O. E., & WISE, S. W., JR., (eds) *Initial Reports of the Deep Sea Drilling Project*, **18**, Washington, D.C., 837–855.

ISSAWI, B. (1972) Review of the Upper Cretaceous-Lower Tertiary stratigraphy in central and southern Egypt. *American Association of Petroleum Geologists, Bulletin*, **56**, 1448–1463.

JOHNSON, H. D. 1978. Shallow siliciclastic seas. *In*: READING, H. D. (ed.) *Sedimentary Environments and Facies*. Elsevier, Amsterdam, 207–258.

KNEBEL, H. J. 1989. Modern sedimentary environments in a large tidal estuary, Deleware Bay. *Marine Geology*, **86**, 119–136.

NOTHOLT, A. J. G. 1985. Phosphorite resources in the Mediterranean (Tethyan Phosphogenic Province: A progress report. *Sciences Géologiques Mémoir*, **77**, 9–21.

ODIN, G. S. & MATTER, A. 1981. De glauconiarum origine. *Sedimentology*, **28**, 611–641.

PISCIOTTO, K. A. 1981a. Diagenetic trends in the siliceous facies of the Monterey shale in the Santa Maria region, California. *Sedimentology*, **28**, 547–571.

—— 1981b. Distribution, thermal histories, isotopic compositions, and reflection characteristics of siliceous rocks recovered by the Deep Sea Drilling Project. *In*: WARME, J. E., DOUGLAS, R. G. & WINTERER, E. L. (eds) *The Deep Sea Drilling Project: A Decade of Progress*. Society of Economic Paleontologists and Mineralogists Special Publication, **32**, 129–147.

REISS, Z. 1962. Stratigraphy of phosphate deposits in Israel. *Bulletin of the Geological Survey of Israel*, **34**, 1–23.

RICHARDSON, M. 1982. *A Depositional Model for the Cretaceous Duwi (Phosphate) Formation, South of Quseir, Red Sea Coast, Egypt:*. Ms. Thesis, University of South Carolina.

RYTHER, J. H., MENZEL, D. W., HULBURT, E. M., LORENZEM, C. J. & CORWIN, N. 1971. The production and utilization of organic matter in the Peru coastal current. *Invest. Pesq.*, **35**, 43–59.

SCHRANK, E. 1984a. Organic-walled microfossils and sedimentary facies in the Abu Tartur phosphates (Late Cretaceous, Egypt). *Berliner Geowissenschaftliche Abhandlungen, Reihe A: Geologie und Paläontologie*, **50**, 177–187.

—— 1984b. Organic-geochemical and palynological studies of a Dakhla Shale profile (Late Cretaceous) in southeast Egypt: Part A: Succession of microfloras and depositional environment. *Berliner Geowissenschaftliche Abhandlungen, Reihe A: Geologie und Paläontologie*, **50**, 189–207.

—— & PERCH-NIELSEN, K. 1985. Late Cretaceous Palynostratigraphy in Egypt with comments on Maastrichtian and early Tertiary calcareous nannofossils. *Newsletters on Stratigraphy*, **15**, 81–99.

SCOTESE, C. R. & SUMMERHAYES, C. P. 1986. Computer model of paleoclimate predicts coastal upwelling in the Mesozoic and Cenozoic. *Geobyte*, Summer 1986, 28–42.

SESTINI, G. 1984. Tectonic and sedimentary history of the NE African margin (Egypt-Libya). *In*: DIXON, J. E. & ROBERTSON, A. H. F. (eds) *The Geological Evolution of the Eastern Mediterranean*. Geological Society, London Special Publication, **17**, 161–175.

SOLIMAN, M. A., HABIB, M. E. & AHMED, E. A. 1986. Sedimentologic and tectonic evolution of the Upper Cretaceous-Lower Tertiary succession at Wadi Qena, Egypt. *Sedimentary Geology*, **46**, 111–133.

SOUDRY, D., MOSHKOVITZ, S. & EHRLICH, A. 1981. Occurrence of siliceous microfossils (diatoms, silicoflagellates and sponge spicules) in Campanian Mishash Formation, southern Israel. *Eclogae Geologicae Helvetiae*, **74**, 97–107.

SPANDERSHVILLI, G. I. & MANSOUR, M. 1970. The Egyptian phosphates. *In*: MOHARRAM, O. (ed.) *Studies On Some Mineral Deposits of Egypt*. United Arab Republic Geological Survey, 89–106.

STUBBLEFIELD, W. L., MCGRAIL, D. W. & KERSEY, D. G. 1984. Recognition of transgressive and post-transgressive sand ridges on the New Jersey continental shelf: Reply. *In*: TILLMAN, R. W. & SIEMERS, C. T. (ed.) *Siliciclastic Shelf Sediments*. Society of Economic Paleontologists and Mineralogists Special Publication, **34**, 37–41.

SWIFT, D. J. P. & RICE, D. D. 1984. Sand bodies on muddy shelves: A model for sedimentation in the Western Interior Cretaceous seaway, North America. *In*: TILLMAN, R. W. & SIEMERS, C. T. (eds) *Siliciclastic Shelf Sediments*. Society of Economic Paleontologists and Mineralogists Special Publication, **34**, 43–62.

VAIL, P. R. 1987. Seismic stratigraphic interpretation procedure. *In*: BALLY, A. W. (ed.) *Atlas of Seismic Stratigraphy, Vol. 1*. American Association of Petroleum Geologists Studies in Geology, Tulsa, **27**, 1–10.

WAGONER, J. C., MITCHUM, R. M., Jr., Posamentier, H. W. & VAIL P. R. 1987. Key definitions of sequence stratigraphy. *In*: BALLY, A. W. (ed.) *Atlas of Seismic Stratigraphy, Vol. 1*. American Association of Petroleum Geologists Studies in Geology, **27**, 11–14.

WALKER, R. G. 1979. Shallow marine sands. *In*: WALKER, R. G. (ed.) *Facies Models*. Geoscience Canada Reprint Series, No. 1, Geological Association of Canada, Kitchener, Ontario, 75–90.

WASSEF, A. S. 1977. On the results of geological investigations and ore reserves calculations of Abu Tartur phosphorite deposit. *Annals of the Geological Survey of Egypt*, **7**, 1–60.

WILGUS, C. K., HASTINGS, B. S., KENDALL, C. G. St. C., POSAMENTIER, H. W., ROSS, C. A. & VAN WAGONER, J. C. (eds). 1988. *Sea-Level Changes: An Integrated Approach*. Society of Economic Paleontologists and Mineralogists Special Publication **42**.

YOUSSEF, M. I. 1957. Upper Cretaceous rocks in Kosseir area. *Bulletin de l'Institut d'Egypte*, **7**, 35–53.

Clay minerals and phosphorite genesis in the Upper Cretaceous of the northern Siberian Platform

YU. N. ZANIN, K. V. ZVEREV & E. P. SOLOTCHINA

Institute of Geology and Geophysics, Siberian Branch of the Academy of Sciences of the USSR, 630090, Novosibirsk, USSR

Abstract: Nodular phosphorites are common in the Turonian to Maastrichtian deposits (sands, silts and clays) of the Yenisei Mouth Depression in the north of the Siberian Platform. Only the silty clays of Campanian age are non-phosphatic. Study of the fine-grained fraction (<2 μm) of these deposits demonstrates that the phosphorite-bearing parts of the succession contain similar clay mineral assemblages. These consist of kaolinite and chlorite, with a small but constant admixture of gibbsite, and in some cases a small amount of montmorillonite. In the non-phosphatic rocks of the Campanian, clay minerals consist only of montmorillonite and chlorite. Kaolinite and gibbsite are completely absent. It is argued that the clay mineralogy and other data suggest that the formation of the phosphorites was linked to the removal of phosphorus from continental areas during periods of intense weathering in a humid subtropical climate.

The bulk of the Upper Cretaceous (Upper Turonian–Maastrichtian) in the Yenisei Mouth Depression of the northern Siberian Platform, is represented by epi-continental sea basinal deposits containing abundant nodular phosphorites at many stratigraphic levels (Fig. 1). General lithological and stratigraphic details of these deposits have been described by Zakharov et al. (1986). Upper Turonian deposits up to 25 m thick crop out in the eastern bank of Yenisei Bay (Fig. 1), 30 km to the north of Vorontsovo. They are represented by sands and, in the middle part, by layers of clay often enriched in chlorite, glauconite and chamosite, and containing concretions with carbonate and phosphate cements. The Coniacian is at least 20 m thick (exposure is incomplete) and crops out on the eastern side of the Yenisei Bay to the south of Vorontsovo (Fig. 1). The lower and upper parts of the Coniacian are composed of sands with interbeds of chamosite-bearing sand containing phosphorite nodules. The middle part of the Coniacian is of clay and silt composition and contains no phosphates. Deposits of Santonian, Campanian and Maastrichtian age were studied on the Tanama River (Fig. 1), a tributary of the Yenisei River. Santonian rocks are represented mainly by loose sands and silts which contain numerous phosphorite nodules. Campanian deposits are 40 m thick and consist of rather homogeneous mud and clays, without phosphorite nodules. The Maastrichtian is 35 m thick and consists of sands which are chamosite-bearing in places, with clay and silt in separate interbeds, and with frequent phosphorite-bearing concretions.

All of the stages studied from the Yenisei Mouth Depression excluding the Campanian, therefore, are phosphorite-bearing.

Lithology and mineralogy

The Upper Cretaceous rocks of the Yenisei Mouth Depression are mainly unconsolidated. The Turonian, Coniacian, Santonian and Maastrichtian deposits are very similar in composition. They often contain glauconite, chlorite and chamosite (unchanged or oxidized to goethite) which tinge the rocks green or brown, carbonized (and sometimes phosphatized) wood debris and fossil macrofauna. Chamosite is used here to describe the dominant oolitic iron-rich aluminosilicate mineral identified petrographically. This name is widely used in literature (Pettijohn 1975) to describe this type of grain, although it should probably be refered to the mineral berthierine (Bailey 1980). Unfortunately, the latter 7 Å mineral cannot be identified by XRD in the presence of chlorite. The predominant components of the rocks are quartz and plagioclase, (commonly showing signs of weathering to varying degrees), with small admixtures of potassium feldspar. Erosion surfaces are often observed.

Strongly cemented horizons are represented principally by nodules. Terrigenous grains of quartz and feldspars within the nodules are cemented by carbonate, carbonate and phosphate, or phosphate minerals. Carbonates are represented by siderite and/or calcite, although where carbonate is present in the phosphorite nodules, it generally consists only of siderite.

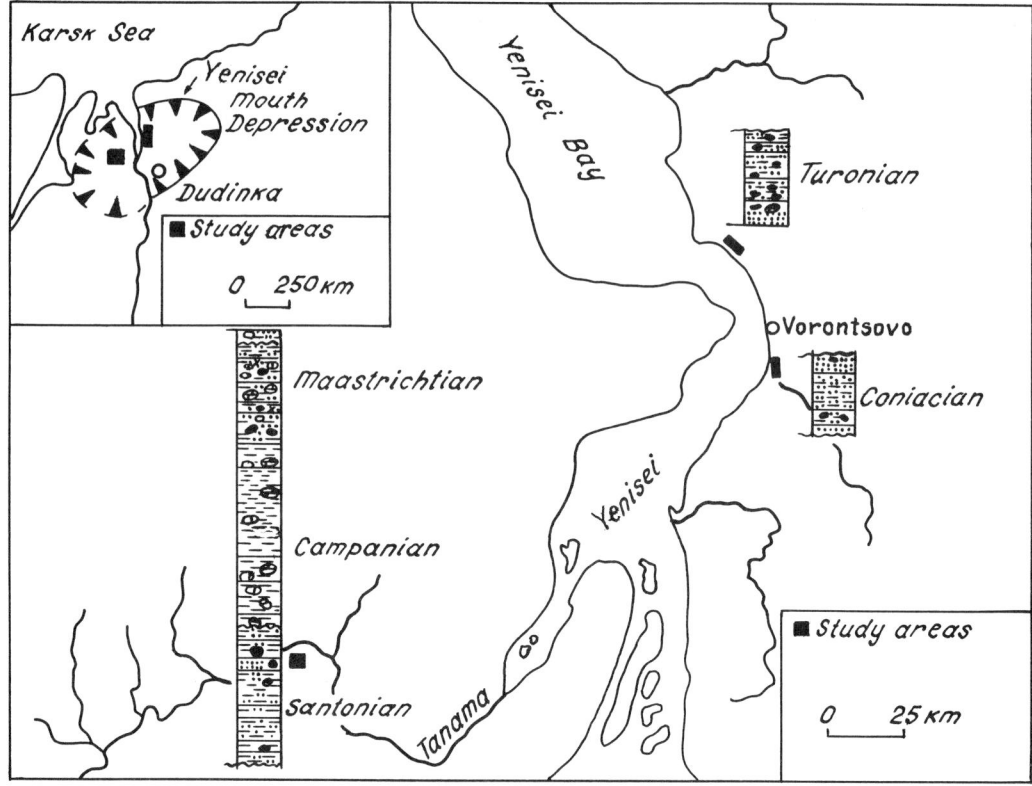

Fig. 1. Sketch map of the sections studied and the location of the Yenisei Mouth Depression. See Fig. 5 for lithological key.

Chloritic grains are often present coating diagenetically altered detrital grains of basic extrusive rocks (Ronkina 1965). Nodules occur more rarely in the mud-rocks of the Campanian; where they do occur, they consist predominantly of chert, limestone, marly lime-stone and sometimes pyrite; phosphorites are absent.

The question arises, as to whether the absence of phosphorite nodules in the Campanian really indicates a very low phosphorus content in the rocks, or whether the phosphatic material is spread more regularly throughout the sediment in fine-grained form? To answer this question representative samples from each of the stages studied were analysed by X-ray fluorescence (XRF). The data for these analysis are given in Table 1 which demonstrates that the P_2O_5 content in the Campanian samples is half that of the Turonian, Coniacian and Santonian material, and is the same as that in the Maastrichtian sediments. It should be emphasized, however, that the in Maastrichtian deposits are dominantly sands, whereas the Campanian material analysed consisted of clay-rich silts. The average phosphorus content of mudrocks is known to be several times higher than in sandstones (Turekian & Wedepohl 1961), so the similar phosphorus contents of the Campanian and Maastrichtian indicate, a relative enrichment in phosphorus in the Maastrichtian sediments of the area.

Phosphorites

The phosphorite nodules occur in sandy layers 0.1−0.3 m, sometimes up to 0.7 m thick. Phosphorite-bearing layers are separated by non-phosphatic interbeds 1−4.5 m thick. Scattered individual nodules are also observed outside the nodule beds. The nodule size ranges from 1 to 15 cm in diameter, only occasionally being significantly larger. Nodules display a range of shapes including globular, discoidal, spindle shaped, and tabular forms. The phosphate occurs as a massive crypto- and microcrystalline matrix and forms coatings on grains

Table 1. Chemical composition (weight percent) of Upper Cretaceous rocks from the Yenisei Mouth Depression.

Element	Stage									
	Turonian $n = 5$		Coniacian $n = 5$		Santonian $n = 5$		Campanian $n = 5$		Maastrichtian $n = 5$	
	$\bar{x}$	σ	$\bar{x}$	σ	$\bar{x}$	σ	$\bar{x}$	σ	$\bar{x}$	σ
SiO_2	64.35	3.40	64.91	8.63	65.77	1.68	67.20	5.02	72.44	3.40
TiO_2	0.92	0.39	0.74	0.14	0.84	0.05	0.69	0.17	0.78	0.10
Al_2O_3	9.64	2.14	9.20	2.18	12.63	1.51	12.41	2.41	11.41	0.29
Fe_2O_3*	13.33	2.36	13.23	4.37	7.75	1.02	5.18	1.00	5.34	1.52
MgO	1.21	0.32	1.22	0.43	1.32	0.02	1.80	0.09	0.71	0.34
CaO	1.23	0.34	1.08	0.34	1.30	0.15	0.73	0.26	0.95	0.10
Na_2O	1.26	0.32	1.11	0.43	1.72	0.11	1.49	0.48	1.84	0.08
K_2O	2.01	0.21	2.18	0.26	2.22	0.07	1.68	0.40	2.56	0.22
P_2O_5	0.15	0.04	0.17	0.08	0.15	0.03	0.08	0.02	0.08	0.03

Determinations by X-ray fluorescence.
* Iron total.

of detrital or authigenic minerals. The nodules are often zoned, their cores being more enriched in phosphorus than their peripheries. The phosphorus-enriched zone may occupy one-third to three-quarters of the nodule volume. Phosphatized shell fragments, burrows and other evidence of bioturbation, faecal pellets and phosphatic grains of unknown origin may be present in the nodules, and sometimes also occur in their surrounding matrix. Phosphatized bioclasts include crab tests (Fig. 2), echinoids, molluscan debris, and skeletal fragments and teeth from marine reptiles and fish, mainly sharks. Fossils commonly form the nucleus to the nodules (e.g. Figs 2, 3). Phosphatized trace fossils include worm and occasional crustacean burrows (*Ophiomorpha*). Nodules formed around burrows are elongated in shape. Phosphatized wood debris occurs only inside the nodules (Fig. 3). Phosphatized faecal pellets are represented by grains having a slightly elongated ovoid form, and a grain size of 0.5–1.2 mm. They are commonly observed within the nodules but also occur in small lens-shaped accumulations within the surrounding matrix.

Two additional types of phosphatic component are observed at microscopic and submicroscopic levels: coated grains and microbial structures. The coated grains are rare and 1.1–1.6 mm in size. Their nuclei are composed of silt and sand grains with siderite cements, and are coated by a thin layer of phosphate up to 0.11 mm thick. Microbial, coccoid-like structures occur in some of these layers. Analysis of different types of phosphorites using scanning electron microscopy (SEM) has revealed the accumulations of these coccoid-like bodies which are 2–3 μm, rarely up to 6–7 μm in diameter. The presence of phosphatized bacterial structures in phosphorites has been described repeatedly (e.g. O'Brien *et al.* 1981; Zanin *et al.* 1984). It is reasonable to hypothesize that the organic remnants which form the cores to many of the nodules acted as nutrients sources for the development of the bacterial communities which in turn concentrated phosphorus from the surrounding seawater, and ultimately led to the formation of phosphorite nodules themselves.

Fig. 2. Phosphatized test of a crab forming the nucleus to a phosphorite nodule. Scale bar, 1 cm.

Fig. 3. Fragment of phosphatized wood within a phosphorite nodule. Scale bar, 1 cm.

Table 2. *Chemical composition (weight percent) of phosphorite nodules from the Upper Cretaceous deposits of the Yenisei Mouth depression.*

Element	Sample			
	86K	223T	308	504
SiO_2	14.77	30.80	36.20	32.50
TiO_2	0.17	0.41	0.52	0.53
Al_2O_3	4.83	6.38	6.40	5.90
Fe_2O_3	6.14*	4.64*	3.22	4.02
FeO	nd	nd	2.41	16.63
CaO	36.85	27.07	22.98	11.58
MgO	1.38	1.31	0.74	1.48
MnO	nd	nd	0.16	0.43
Na_2O	0.77	1.32	1.24	1.28
K_2O	0.67	1.08	1.24	1.03
P_2O_5	26.03	17.54	14.70	7.20
H_2O^-	nd	nd	0.98	0.68
CO_2	nd	nd	2.86†	11.93†
Loss on ignition	8.02	9.25	8.17	16.87
Total	99.63	99.80	98.96	100.13

nd, not determined; *Iron total; †not included in total.

Methods: Samples 86K, 223T, X-ray fluorescence; Samples 308, 504, SiO_2, Al_2O_3, TiO_2, Fe_2O_3, spectrophotometry; Na_2O, K_2O, flame photometry; FeO, CO_2, titration; CaO, MgO, MnO, flame atomic absorption spectrometry.

The P_2O_5 content of the phosphorite nodules is considerably higher than in the surrounding sediment ranging from 5 to 20% and greater (Table 2). The only phosphate mineral identified by X-ray diffraction (XRD) (Fig. 4) and microscopic analysis was carbonate-fluorapatite (francolite). Quartz, plagioclase, potassium feldspar, kaolinite, chlorite, chamosite, muscovite, and traces of amphibole are also present in the phosphorite nodules. Siderite is abundant in some phosphorite nodules (Table 2, Sample 504) where it partly replaces phosphate. Phosphate replacement of siderite has never been observed.

Clay mineralogy

The separation of the fine-grained fraction (< 2 μm) was carried out in water using established methods (Vikulova 1957). Sediments were pretreated with 2% HCl to remove carbonates and phosphate. The composition of the fine fraction was studied using XRD, infra-red spectroscopy and chemical techniques. X-ray diffractograms were obtained using a 'DRON-3' diffractometer, and IR-spectra using a 'Specord J-75' spectrometer. Preparations for X-ray analysis of the fine fraction were made by settling from an aqueous suspension onto a glass slide. However, the production of well-oriented basal plane mounts was prevented by the presence of the detrital quartz grains. Nevertheless, sufficient grain orientation took place to increase the 001 lines of the clay minerals, enabling their identification more readily than using pressed pellet techniques.

The X-ray data for the clay fractions of Turonian, Coniacian, Santonian and Maastrichtian rocks demonstrate that they have very similar compositions (Fig. 5). Clay minerals are represented by kaolinite and chlorite, sometimes with a small admixture of montmorillonite

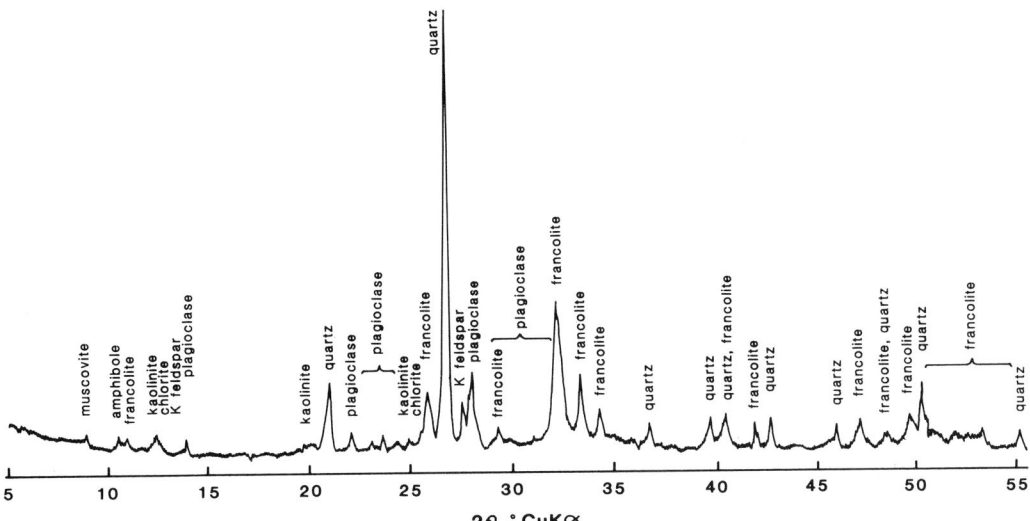

Fig. 4. X-ray diffractogram of a Coniacian phosphorite nodule (Sample 308; Cu Kα radiation, scanned at 1° min^{-1} with a chart speed of 80 cm hr^{-1}).

(see below). Other minerals present are quartz, muscovite and feldspars. Gibbsite also occurs in all samples but in minor amounts.

A representative XRD trace for the clay fraction of a typical phosphorite-bearing deposit (Sample 597) is shown in Fig. 6a. It is evident that kaolinite which (based on the diffraction lines pattern) has an average degree of structural ordering, predominates among the clay minerals. It was not possible to characterize the kaolinite further (e.g. to study defects of kaolinite crystal structure, intercolation and so on) because of the multiphase composition of the clay fraction.

Montmorillonite and chlorite also form significant portions of the clay fraction (Fig. 5). The presence of a mineral of the smectite group (montmorillonite) was corroborated by a displacement of the d_{001}-line from 14.4 Å (e.g. initial sample, Fig. 6a) to 17 Å (saturated with ethylene glycol, Fig. 6b). The presence of a small amount of chlorite is evidenced by the 4.71 Å line in the X-ray diffractogram of the initial samples. The standard technique was used for to confirm the occurrence of this mineral (heating at 600 °C for 1 hour). The $d=14.0$ Å chlorite (001) line was seen distinctly in the diffractogram of the heated samples (Fig. 6c). Gibbsite is present as a small admixture in the rocks and only one of its lines (4.85 Å) was identified on the diffractograms (Fig. 6a).

In order to better characterize the composition of the clay fractions, IR-spectra in the range of OH-valency vibrations were obtained. In addition to the absorption bands in the frequencies interval 3695–3620 cm^{-1}, which were attributed to kaolinite (Fig. 7), three bands occurred at 3580, 3450 and 3530 cm^{-1}, indicating the presence of gibbsite.

In rocks of Campanian age, montmorillonite (see below) is the main clay mineral, with chlorite occurring as a subordinate component (Fig. 5). Kaolinite and chlorite are completely absent, but a zeolite (clinoptilolite ?) is present in all Campanian samples and cristobalite occurs in most of them.

An XRD trace for the clay fraction of a typical non-phosphatic Campanian rock (Sample 554) is given in Fig. 8a. Saturation of the clay fraction with ethylene glycol leads to a displacement of the smectite (001) reflection position from $d=14.3$ Å to 16.85 Å (Fig. 8b), and the d_{060} lines is at 1.500 Å. These properties are characteristic of montmorillonite, a dioctahedral mineral of the smectite group. Heating of the samples at 600 °C for 1 hour (Fig. 8c) enabled the presence of a small amount of chlorite to be identified in all samples (Fig. 5).

Average chemical compositions of the fine-grained fractions are given in Table 3. The fine-grained fractions from the Campanian are characterized by higher SiO_2 contents and by lower Al_2O_3 contents than other samples. These differences are due to the presence of cristobalite and absence of kaolinite in the Campanian material (Fig. 5). The P_2O_5 content of the

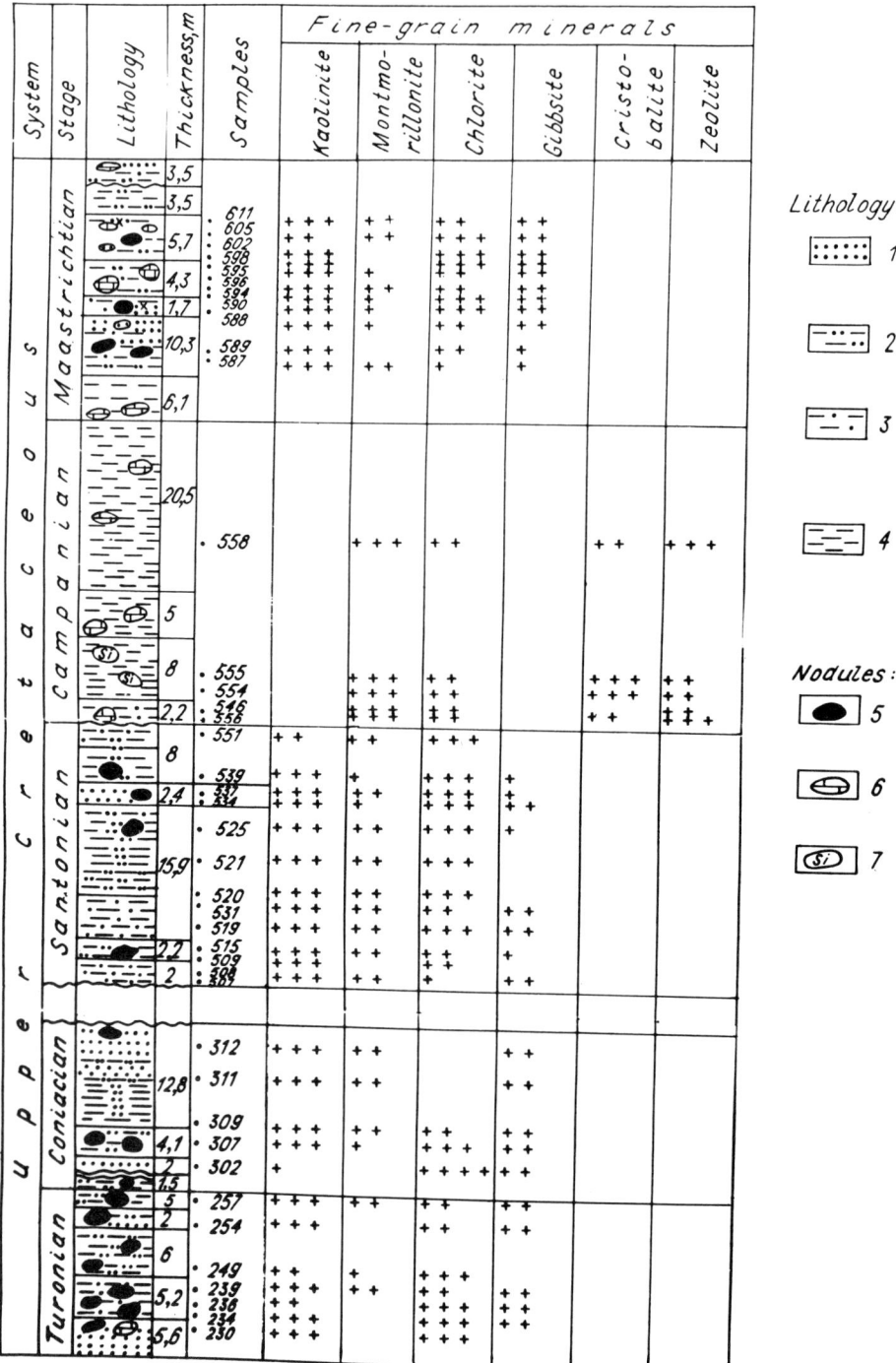

Fig. 5. Composition and distribution of the fine-grained fraction in Upper Cretaceous rocks of the Yenisei Mouth Depression. Stratigraphy after Zakharov *et al.* (1986). Lithology: 1, sand; 2, sandy silt; 3, sandy clay; 4, silty clay and clay. Nodules: 5, phosphatic; 6, carbonate; 7, siliceous.

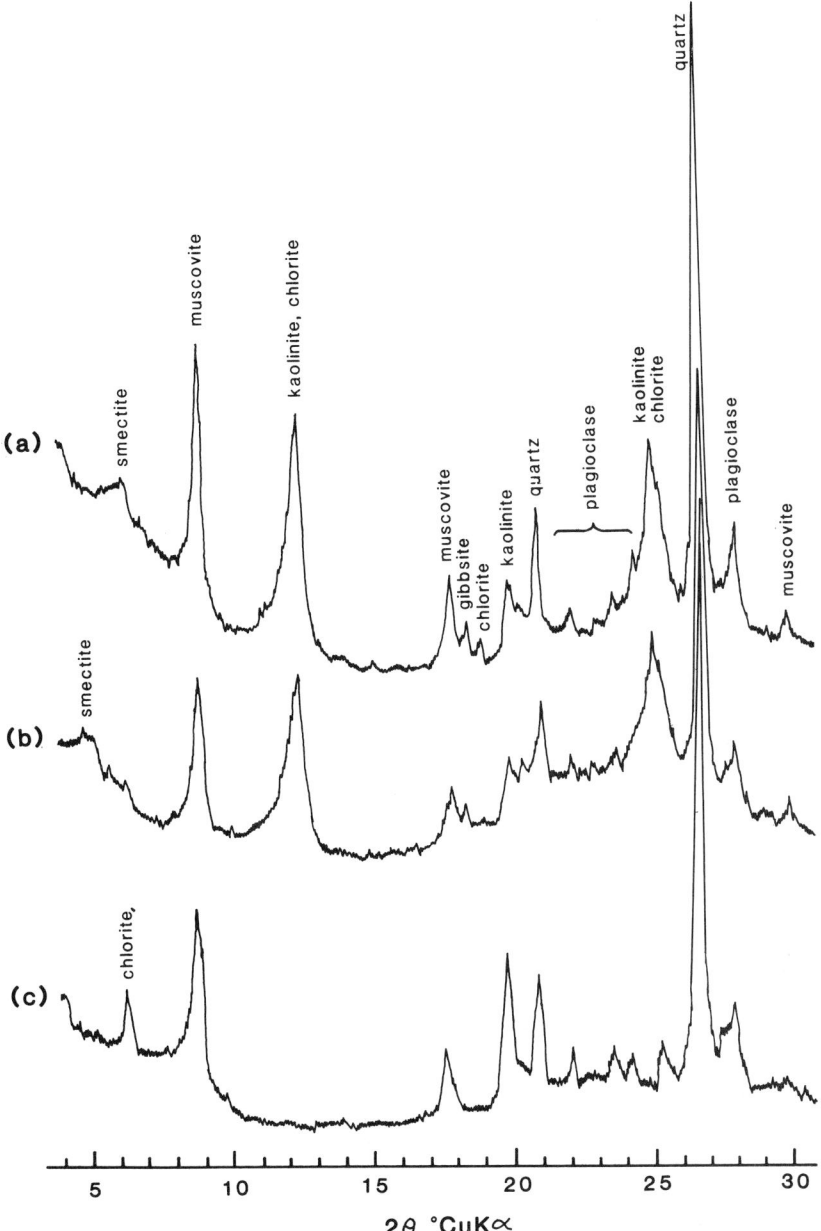

Fig. 6. Portion of the X-ray diffractogram for the fine-grained fraction (<2 μm) of a Maastrichtian sediment (Sample 597) typical of the phosphorite-bearing deposits. Analytical conditions as in Fig. 4. (a) Air dried; (b) saturated with ethylene-glycol; (c) heated to 600°C for 1 hour.

Campanian samples is two or three times lower than those from the Turonian, Coniacian, Campanian and Maastrichtian. It is difficult to conclude anything on the mineralogical controls on phosphorus in the clay fraction. In many samples, however, and particularly those from the Campanian and Maastrichtian, the CaO contents are too low for all of the phosphorus present to occur as apatite. In addition to that occurring in francolite partially preserved from dissolution by 2% HCl, therefore, part of the phosphorus must be sorbed by iron oxyhydrox-

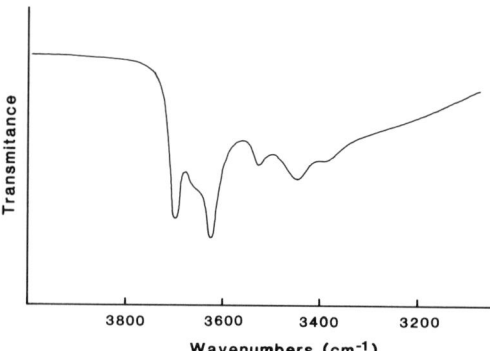

Fig. 7. Infra-red spectrum for Maastrichtian sediment Sample 597 (see Fig. 6) in the range of the OH-valency vibrations.

ides and/or clay minerals. Indeed, the P_2O_5 contents of the fine-grained material from the phosphorite-bearing stages correlate to some degree with their iron oxide contents, the ratio $P_2O_5:(Fe_2O_3 + FeO)$ being 0.020–0.030 in the Turonian–Santonian and Maastrichtian, whereas in the Campanian material this ratio is only 0.011.

Sorption of phosphorus by clay minerals, is likely to be most important in the Campanian, since here the major clay mineral is montmorillonite, which is known to have a much stronger affinity for phosphorus than kaolinite (De 1961).

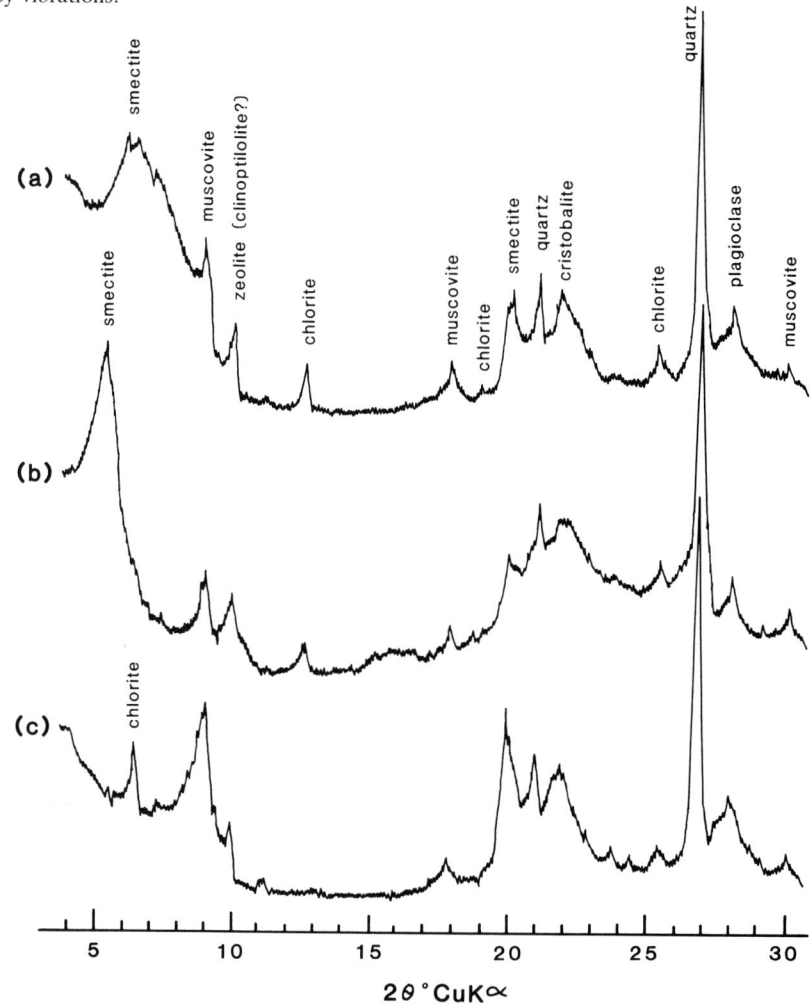

Fig. 8. Portion of the X-ray diffractogram for the fine-grained fraction of a Campanian sediment (Sample 554), typical of non-phosphatic deposits in the area. Analytical conditions as in Fig. 4. (a) Air dried (b) saturated with ethylene glycol; (c) heated at 600°C for 1 hour.

Table 3. *Chemical composition (weight percent) of the fine-grained fraction (<2 μm) separated from Upper Cretaceous rocks of the Yenisei Mouth Depression, after treatment with 2% HCl.*

Element	Turonian $n = 16$		Coniacian $n = 8$		Santonian $n = 14$		Campanian $n = 5$		Maastrichtian $n = 12$	
	$\bar{x}$	σ	$\bar{x}$	σ	$\bar{x}$	σ	$\bar{x}$	σ	$\bar{x}$	σ
SiO_2	54.80	5.44	53.56	4.85	52.38	1.34	60.30	2.50	53.22	1.65
TiO_2	1.20	0.21	1.18	0.13	1.31	0.15	0.88	0.09	1.27	0.12
Al_2O_3	18.21	6.13	19.05	4.68	21.76	1.53	15.56	1.04	22.38	2.55
Fe_2O_3	8.66	3.83	3.91	1.02	4.28	1.12	4.63	1.38	2.77	0.59
FeO	0.86	0.29	1.05	0.20	1.15	0.11	0.88	0.22	0.90	0.13
MgO	0.89	0.17	0.89	0.19	1.33	0.29	1.72	0.11	0.91	0.11
CaO	0.47	0.27	0.21	0.22	0.22	0.07	0.05	0.11	0.11	0.08
Na_2O	0.83	0.39	0.64	0.44	0.65	0.20	0.48	0.06	0.70	0.06
K_2O	1.76	0.52	1.69	0.17	2.09	0.33	1.87	0.23	2.32	0.37
P_2O_5	0.19	0.11	0.11	0.02	0.12	0.04	0.06	0.02	0.11	0.04

Methods: SiO_2, Al_2O_3, TiO_2, Fe_2O_3, P_2O_5, spectrophotometry; Na_2O, K_2O, flame photometry; FeO, CO_2 titration; CaO, MgO, MnO, flame atomic absorption spectrometry.

Discussion

The above data demonstrate a wide range of phosphorus contents in the Upper Cretaceous deposits of Yenisei Mouth Depression. The Turonian, Coniacian, Santonian and Maastrichtian rocks are characterized, in contrast to those from the Campanian, by the presence of phosphorite nodules, and higher overall phosphorus contents, both in the bulk rocks and in their fine-grained fractions. An increased phosphate content is clearly associated with a distinctive fine-grained mineral assemblage, represented mainly by kaolinite with an admixture of chlorite, plus a small, but constant, amount of gibbsite, and sometimes of montmorillonite. In contrast, fine-grained mineral assemblages from the non-phosphatic deposits of the Campanian are dominated by montmorillonite with admixtures of chlorite, zeolite (clinoptilolite ?) and cristobalite.

Differences in the clay minerals assemblages (and fine-grained minerals as a whole) from deposits of different age are dependant on such factors as: (a) the composition of material eroded from siliciclastic source areas, as controlled by both differences in its primary mineralogy and the nature of weathering processes; (b) differences in the physico-chemical conditions of sedimentation; (c) different configurations and dimensions of sedimentary basins and the temporal evolution of their sediment supply; (d) post-depositional weathering of deposits. Let us consider these factors in relation to the deposits under study.

The most important clay minerals indicative of conditions of the sedimentary rock formation are kaolinite and gibbsite. These are both generally formed by deep weathering processes. Kaolinite is occasionally observed in sandstones from the Yenisei Mouth Depression as a coarse-grained authigenic pore-fill (Ronkina 1965), but the mass generation of kaolinite and gibbsite is only possible in association with well-developed zoned weathering profiles of lateritic composition (also indicated by the presence of gibbsite). The occurrence of authigenic kaolinite and oxidized chamosite indicate that the Upper Cretaceous rocks of Yenisei Mouth Depression have been subject to some post-depositional alteration, but no in situ deep-weathering profiles have been observed. Moreover, kaolinite and gibbsite are absent n the Campanian deposits, which are both underlain and overlain by rocks containing these minerals; this would be impossible if the former minerals where being produced by modern weathering. It must be concluded, therefore, that the gibbsite and a major part of the kaolinite are detrital minerals which originated from continent areas undergoing lateritic weathering, and were deposited subsequently in adjacent Upper Cretaceous seas. The presence of chlorite (which would not be preserved under conditions of post-depositional deep weathering) in association with kaolinite and gibbsite, is consistent with a detrital origin for the latter two minerals. The occurrence of chlorite suggests further that rocks in the source area were undergoing varying degrees of weathering (corresponding, perhaps, to different zones in the weathering profiles), prior to being eroded. Alternatively, at least some of the chlorite may be of authigenic origin.

It is known (Millot 1964) that in modern sedimentary basins, kaolinite is deposited closer to the shoreline than montmorillonite. The question arises, therefore, as to whether differences in the clay minerals assemblages from the Upper Cretaceous deposits of the Yenisei Mouth Depression, are a consequence of their different depositional environments? The paleogeographic setting should in this case be different in the Campanian from that in Turonian, Coniacian, Santonian and Maastrichtian. The palaeogeography of northern Siberia, however, displays no significant changes in the shoreline positions throughout the late Cretaceous (Ronkina 1965; Golbert et al. 1968), indicating that clay mineral sorting was not a major factor.

It must be concluded, therefore, that differences in the compositions of the clay minerals assemblages studied, were due, at least primarily, to different sources and/or weathering processes in continental areas. Clay minerals from the phosphorite-bearing stages indicate lateritic weathering in a warm and humid climate. Analysis of the Upper Cretaceous climate of the area based primarily on the floral data (Golbert et al. 1977; Yasamanov 1976) also reflect dominantly subtropical conditions. The montmorillonite-dominated assemblage from the Campanian, however, suggests an arid climate and/or derivation from lower horizons of weathering (Millot 1964). Indeed, montmorillonite may form in the lower parts of weathering profiles developed on basic and ultrabasic rocks even under humid conditions, but in these cases the volume of material is small, and it is considered that the dominance of the mineral in the Campanian of the study area, could only be caused by a shift to an arid climate.

Differences in the weathering regime during the deposition of the phosphorite-bearing stages on the one hand, and the phosphate-poor Campanian on the other, are reflected in the chemical composition of the fine-grained fraction (Table 3). The higher SiO_2 and lower Al_2O_3 contents in the fine-grained material from the Campanian as compared with the other stages may partly reflect less intense weathering of the primary rocks on the continents during the former period, although SiO_2 is also present in the Campanian samples as cristobalite (Fig. 5). The presence of the latter mineral may associated with the transformation of biogenic silica derived from diatoms, the temporary occurrence of which was possibly related to a fall in temperature (Ronkina 1965).

Possible relationships between the phosphorite genesis and climate have been discussed by many previous workers. Two points of view have been expressed in literature. Many geologists have provided evidence for humid climates in areas of phosphorite formation. Shatsky (1955) discussed the origins of nodular and granular phosphorites from epicontinental seas, and concluded that the source of phosphorus was from weathering of land areas. Strakhov (1960) suggested that platform phosphorites formed both in arid and humid conditions, but that in basinal settings, they were deposited only under arid conditions. In particular, Strakhov (1960) considered that the Jurassic and Cretaceous phosphorites of the East European Platform, and the Cretaceous and Palaeogene phosphorites of the West Siberian Platform were deposited in humid environments. Gerasimov et al. (1966) noted the high phosphate contents of oolitic ironstones and their connection, in some cases, with bauxites, and suggested that weathering crusts were the source of phosphorus. Kazarinov (1966) decided that weathering processes in general controlled marine phosphorite formation, and Bushinskii (1969) suggested that the main source of phosphorus in marine phosphate accumulations was via dissolved phosphorus in the rivers from the continent. This idea of phosphorus having been derived from the weathering of continental areas has been supported by a number of later workers (Sinitsin 1967, Kholodov 1970; Sagunov 1971; Howard & Hough 1979; Zanin 1981, 1982; Arthur & Jenkyns 1981; Glenn 1990).

Other authors have proposed that the climate of areas undergoing phosphorite formation is generally arid. Cook & McElhinny (1979) discussed phosphorite genesis on continental shelves in relation to upwelling and noted climate aridity in adjacent areas of the continents, while Hite (1978) and Yanshin & Zharkov (1986) indicated a possible connection between phosphorite and evaporite genesis. In the succession studied here, however, arid conditions (Campanian) appear to have promoted the deposition of phosphate-poor, rather than phosphate-rich sediments. From the above discussion, therefore, it is concluded that deposition of the phosphorite-bearing strata of Yenisei Mouth Depression occurred during periods of increased chemical weathering associated with warm, humid phases, and that phosphorite genesis may be related to the influx of phosphorus leached from developing weathering profiles on adjacent land areas.

The assemblage of authigenic minerals in the phosphorite-bearing deposits which (in addition to francolite) includes glauconite, chamosite, siderite and calcite, is indicative of weakly-alkaline, mildly reduced conditions during the formation of the phosphorites (Krumbein & Garrels 1952). The assemblage of authigenic minerals in the Campanian sediments includes pyrite as well as calcite, zeolite and cristobalite, which also indicate reducing conditions during early diagenesis.

The association of authigenic silica (in this case cristobalite) with zeolites is widespread (e.g. Suprychev 1979; Sheppard & Cude 1973). Zeolite genesis is often related to the diagenetic transformation of acid volcanic rocks (Deffeyes 1973; Kossovskaia & Muraviev 1975), but not all geologists share this opinion. Some authors have suggested that zeolites form diagenetically in alkaline environments (Perosio et al. 1961; Brown et al. 1969; Melnik 1972; Peres 1976). According to Melnik (1972) and Suprychev (1979), the high solubility of biogenic silica in alkaline medium may lead to clinoptilolite formation as a result of the interaction of dissociated silica with K- and Na-aluminosilicates; sources of alumina and alkaline elements can include feldspars and micas (Brown et al. 1969). The latter mechanism seems to be the more likely source of zeolites in the Campanian deposits of Yenisei Mouth Depression, given the absence of acid volcanic rock debris. According to Kaplan & Chirva (1977), similar authigenic clinoptilolites occur in the Upper Cretaceous deposits of the Khatanga Depression, an area adjacent to the East Yenisei Mouth Depression.

Finally, it is interesting to note that the products of lateritic weathering developed in northern Siberia, which include aluminophosphate rocks (Zanin et al. 1972) and bauxites (Drenov et al. 1975; Zabirov et al. 1978), may also be of late Cretaceous age.

Conclusions

(1) Nodular phosphorites from the Upper Cretaceous of the Yenisei Mouth Depression, northern Siberian Platform, occur in sandy mudstones of Turonian, Coniacian, Santonian and Maastrichtian age. The main clay minerals in these deposits are kaolinite and chlorite, with small admixtures of gibbsite and occasionally montmorillonite. In contrast, the main clay mineral in the non-phosphatic Campanian clays is montmorillonite.

(2) The composition and distribution of clay minerals indicates that the phosphorite-bearing deposits of the Yenisei Mouth Depression were formed during times of intensified continental weathering, which was caused by climatic warming and increasing humidity. The phosphorus which was ultimately incorporated into the phosphorite nodules by bacterial processes may have been derived from the continents during these periods of the enhanced deep weathering.

The authors wish to thank V. A. Zakharov and A. S. Mikhailov for discussion on some of the problems discussed in this paper, I. Jarvis and two anonymous referees for their written comments and suggestions, and A. D. Kireev and I. M. Fominych for providing the geochemical analyses.

References

ARTHUR, M. A. & JENKYNS, H. C. 1981. Phosphorites and paleoceanography. In: 26 Congrés Géologique International, Paris, 1980. Oceanologia Acta, Colloque C4. Géologie des Océans, Paris, 83–96.

BAILEY, S. W. 1980. Structures of layer silicates. In: BRINDLEY & BROWN (eds) Crystal structures of clay minerals and their X-ray identification. Mineralogical Society, London, 1–124.

BROWN, G. CATT, J. A. & WEIR, A. H. 1969. Zeolites of clinoptilolite-heulandite type in sediments of south-east England. Mineralogical Magazine, 37, 480–488.

BUSHINSKII, G. I. 1969. Old phosphorites of Asia and their genesis. [Translation from Russian]. Israel Program for Scientific Translations.

COOK, P. J. & MCELHINNY, M. W. 1979. A reevaluation of the spatial and temporal distribution of sedimentary phosphate deposits in the light of plate tectonics. Economic Geology, 74, 315–330.

DE, S. K. 1961. Intake of phosphate by Indian montmorillonite and kaolinite. Neues Jahrbuch für Mineralogic Monatshefte, 10, 234–237.

DEFFEYES, K. S. 1959. Zeolites in sedimentary rocks. Journal of Sedimentary Petrology, 29, 602–609.

DRENOV, N. A., ISAEVA, L. L., MITAEV, A. G. BRYZGALOVA, M. M. & BIDZHIEV, R. A. 1975. New data on the presence of bauxites in the Siberian Platform. Doklady of the Academy of Sciences of the USSR, 220, 5, 1176–1179. (In Russian).

GERASIMOV, E. K., RODIN, R. S. & SHATKO, A. V. 1966. On the question of phosphorite genesis. In: Conditions of the localization of the phosphate-bearing sediments in Siberia. Nedra Publishers, Leningrad, 40–49. (In Russian).

GLENN, C. R. 1990. Depositional sequences of the Duwi, Sibaiya and Phosphate Formations, Egypt: phosphogenesis and glauconitization in a Late Cretaceous epeiric sea. In: NOTHOLT, A. J. G. & JARVIS, I. (eds) Phosphorite Research and Development. Geological Society, London, Special Publication, 52, 205–222.

GOLBERT, A. V., GRIGORIEVA, K. N., ILIENOK, L. L., MARKOVA, L. G. & TESLENKO YU. K. 1977.

Palaeoclimates of Siberia in the Cretaceous and Palaeogene periods. *In*: SACHS (ed.) Moscow, Nedra Publishers. (In Russian).
——, MARKOVA, L. G., POLYAKOVA, I. D., SACHS, V. N. & TESLENKO YU. V. 1968. *Palaeolandscapes of West Siberia in the Jurassic, Cretaceous and Palaeogene*. Moscow, Nauka Publishing House. (In Russian).
HITE, R. I. 1978. Possible genetic relationship between evaporites, phosphorites and iron-rich sediments. *The Mountain Geologist*, **14**, 97–107.
HOWARD, P. F. & HOUGH, M. J. 1979. On the geochemistry and origin D Tree, Wonarah, and Sherrin Creek phosphorite deposits of the Georgina Basin, Northern Australia. *Economic Geology*, **74**, 260–264.
KAPLAN, M. E. & CHIRVA, S. A. 1977. Zeolites in the Cretaceous sediments of the Khatangskii depression (Northern part of Eastern Siberia). *Lithology and Mineral Resources*, **6**, 120–124. (In Russian).
KAZARINOV, V. P. 1966. The problem of the exploration of the large deposits of the rich phosphorites in Siberia. *In*: *Conditions of the localization of the phosphate-bearing sediments in Siberia*. Nedra Publishers, Leningrad, 3–14. (In Russian).
KHOLODOV, V. N. 1970. On Vendian and Cambrian metallogeny of Eurasia. 2. Vanadium, iron, manganese sedimentary ores and the conditions of their formation. *Lithology and Mineral Resources*, 29–44 (In Russian).
KOSSOVSKAIA, A. G. & MURAVIEV, V. A. 1975. Identity of ocean and platform zeolite-cristobalite rocks. *Doklady of the Academy of Sciences of the USSR*, **223**, 431–433.
KRUMBEIN, W. C. & GARRELS, R. M. 1952. Origin and classification of chemical sediments in terms of pH and oxidation–reduction potentials. *Journal of Geology*, **60**, 1–33.
MELNIK, J. M. 1972. On the formation of frame silicates under hypergene conditions. *Mineralogical Sbornic of the Lvov University*, **26**, 89–92. (In Russian).
MILLOT, G. 1964. *Géologie des argiles*. Masson, Paris.
O'BRIEN, G. W., HARRIS, J. R., MILNES, A. R. & VEEH, H. H. 1981. Bacterial origin of East Australian continental margin phosphorites. *Nature*, **294**, 442–444.
PERES, F. S. 1976. New data on zeolites of the Moldavian SSR. *Doklady of the Academy of Sciences of the USSR*, **231**, 977–980 (In Russian).
PEROSIO, G. N., PROSOROVICH, G. A. & SOROKINA, E. G. 1961. Heulandite from Mesozoic and Cenozoic deposits of West Siberian Lowland. *In*: *Material on Geology, Hydrogeology and Mineral Resources of Western Siberia*. Gostoptechizdat, Leningrad, 128–131. (In Russian).
PETTIJOHN, F. J. 1975. *Sedimentary rocks*, 3rd edition. Harper & Row Publishers, New York.
RONKINA, Z. Z. 1965. *Composition and formation conditions of Jurassic and Cretaceous deposits in the North of Central Siberia*. Nedra Publishers, Leningrad. (In Russian).

SAGUNOV, V. G. 1971. *Geology of agronomical ores of Kazakhstan*. Nauka Publishing House, Alma-Ata. (In Russian).
SHATSKY, N. S. 1955. Phosphorite-bearing formations and the classification of phosphorite deposits. *In*: *Conference on sedimentary rocks, Part 2*. Academy of Sciences of the USSR Publishing House, Moscow, 7–100. (In Russian).
SHEPPARD, R. A. & CUDE, A. J. 1973. Zeolites and associated authigenic silicate minerals in tuffaceous rocks of Big Sandy Formation, Mohave County, Arizona. *Professional Paper of the US Geological Survey* **830**, 1–36.
SINITSIN, V. M. 1967. *Introduction to palaeoclimatology*. Nedra Publishers, Leningrad. (In Russian).
STRAKHOV, M. M. 1960. Climate and phosphorus accumulation. *Geology of Ore Deposits*, **1**, 3–15. (In Russian).
SUPRYCHEV, V. A. 1979. Diagenetic zeolites of Mesozoic and Cenozoic chert-carbonate formation. *Proceedings of the USSR Academy of Sciences. Geological Series.* **5**, 83–93. (In Russian).
TUREKIAN, K. K. & WEDEPOHL, K. H. 1961. Distribution of elements in some major units of the Earth's crust. *Geological Society of America Bulletin*, **72**, 175–192.
VIKULOVA, (ed.) 1957. *Methods in the petrographic and mineralogic research of clays*. Gosgeoltechizclat, Moscow. (In Russian).
YANSHIN, A. L. & ZHARKOV, M. A. 1986. *Phosphorus and potassium in nature*. Nauka Publishing House, Novosibirsk. (In Russian).
YASAMANOV, N. A. 1976. *Climates and landscapes of Mesozoic and Cenozoic of West and Middle Siberia*. Nedra Publishers, Moscow. (In Russian).
ZABIROV, YU. A. & CHEKHA, V. P. 1978. First bauxite manifestation and perspective of the North-Siberian bauxite-bearing province. *Lithology and Mineral Resources.* **2**, 31–39. (In Russian).
ZAKHAROV, V. A., ZANIN, YU. N., ZVEREV, K. V., LEBEDEVA, N. K., KHLONOVA, A. F., KHOMENTOVSKII, O. V., BEIZEL, A. L. & ENDELMAN, L. G. 1986. *Stratigraphy of the Upper Cretaceous deposits of Northern Siberia*. Institute of the Geology and Geophysics of the Siberian Branch of the USSR Academy of Sciences, Novosibirsk. (In Russian).
ZANIN, YU. N. 1981. Climatic aspects of the evolution of phosphate accumulation in the Phanerozoic. *In*: *Problems of the Evolution of the Geological Processes*. Nauka Publishing House, Novosibirsk, 122–133. (In Russian).
ZANIN YU. N. 1982. Epicontinental phosphate accumulation and some problems of humid phosphogenesis. *In*: *Geology of Phosphorite Deposits and Problems of Phosphorite genesis*. Institute of Geology and Geophysics of the Siberian Branch of the USSR Academy of Sciences, Novosibirsk, 52–64. (In Russian).
——, ZHIROVA, L. T. & SERDYUKOVA, P. A. 1972. Phosphatic zones of weathering in the Essey Massif (North of Siberian Platform). *Geology*

and Geophysics, **3**, 112–114. (In Russian).
——, LETOV, S. V., KRASIL'NIKOVA, N. A. & MIRTOV, YU. V. 1984. Phosphatized bacteria from Cretaceous phosphorites of the East-European Platform and Palaeocene phosphorites of Morocco. *In*: *Phosphorites*, Sciences Géologiques, Mémoire, **77**, 79–81.

Condensation and phosphogenesis: example of the Helvetic mid-Cretaceous (northern Tethyan margin)

KARL B. FÖLLMI

Earth Sciences Board, University of California, Santa Cruz, CA 95064, USA
Present address: Geological Institute, ETH-Zentrum, CH 8092 Zürich, Switzerland

Abstract: The Aptian to Lower Cenomanian condensed phosphatic beds, deposited in the Helvetic Shelf along the northern Tethyan margin, consist of thin strata (generally <50 cm) of densely packed phosphatized particles and crudely laminated crusts embedded in glauconitic sands, marls, and pelagic micrites. The beds record very low net sediment-accumulation rates (typically 2–20 cm Ma^{-1}). The presences of complex internal and laterally rapidly changing microstratigraphies, multiple phosphate generations, intimate mixtures of fossils from different biostratigraphical zones and ecological habitats, a proximal transition into bioturbated glauconitic sands, and a distal transition into allochthonous sediments deposited in channel and fan systems, indicate that the condensed phosphatic beds formed along the axis of a stable westward-flowing current system contouring the northern Tethyan margin. Repetitive cycles of deposition, phosphogenesis, re-exposure and reworking, and deposition (Baturin Cycles), caused by current-induced lateral migration of sand bodies, shaped the internal stratification and composition of the condensed phosphatic beds. Catastrophic burial of entire benthic communities was crucial to phosphogenesis. The sudden presence of large amounts of organic matter buried in siliciclasts created a reactive environment favourable for the concentration and precipitation of phosphates, probably with the help of physicochemical cycles of the $Fe^{3+}-Fe^{2+}$ and $Mn^{4+}-Mn^{2+}$ redox pairs.

Condensed phosphatic sediments form in marine environments, characterized both by lowered net sediment-accumulation rates and episodic sediment remobilization, and by (bio-)chemical conditions suitable for phosphogenesis. This exceptional combination of physical and chemical oceanographic circumstances limits the occurrence of condensed phosphatic beds to specific settings in space and time, and renders these sediments an unique value as palaeoceanographic marker beds.

Although extensively studied over the last fifty years, condensed phosphatic beds remain enigmatic in many aspects (e.g., Heim & Seitz 1934; Schaub 1936; Baturin 1971; Kennedy & Garrison 1975; Krajewski 1984). For instance, the cause of very rapid lateral changes in the internal composition of such beds is largely unknown. An apparent paradox lies also in the good to excellent preservation of phosphatized fossils within condensed beds, taking into account the very low sediment-accumulation rates under which these beds formed. Moreover, the process of phosphatization in such environments remains an open question: the physical and (bio-)chemical paths, along which dissolved phosphates concentrate, precipitate, and accumulate, are still poorly understood (cf. Bentor 1980; Burnett & Froelich 1988).

The mid-Cretaceous, strongly condensed phosphatic beds of the northern Alpine Helvetic Unit embody an archetypal example of condensed phosphatic beds studied for more than a century, and recently subjected to a suite of detailed analyses (Gebhard 1983, 1985; Föllmi 1986, 1989 *a*, *b*; Föllmi & Ouwehand 1987; Delamette 1985, 1988*a*, *b*; Ouwehand 1987). The exceptional degree of condensation (net sediment-accumulation rates of 2–20 cm Ma^{-1} for several million years), and their persistance along the northern Tethyan margin from the Iberian peninsula to the western Carpathians, underscore the importance of these beds for the reconstruction of palaeoceanographic conditions along the mid-Cretaceous northern Tethyan margin (Föllmi 1986, 1989*b*, Ouwehand 1987; Delamette 1988*a*, *b*). The well-known distribution of the condensed phosphatic beds in space and time and the unraveled relationships with adjacent non-condensed beds, allow for a detailed reconstruction of their genesis, thereby touching on the open questions mentioned above.

Condensed phosphatic sediments of the mid-Cretaceous Helvetic Shelf

General setting

Helvetic Shelf sediments of the mid-Cretaceous are reflective of a profound deepening-upward trend. Barremian to Lower Aptian shallow-water platform carbonates (Schrattenkalk Formation) of proximal areas, attributed to an inner-shelf facies, are overlain by a thin sequence of Aptian to Upper Albian or Lower Cenomanian glauconitic sands and marls intercalated with condensed phosphatic beds (Garschella Formation), which, in turn, are blanketed by Upper Albian to Upper Santonian pelagic limestones (Seewen Formation).

Barremian to Lower Aptian rhythmically bedded limestones and marls (Drusberg Formation) of distal areas, considered herein as representing outer-shelf facies, are overlain by Upper Aptian to Upper Albian, commonly laminated, dark marls and muds (Garschella Formation), which are overlain in turn by the pelagic Seewen Formation (Fig. 1).

The transitional area between the inner and outer shelf changed from a gentle homocline in the Barremian, early, and early late Aptian, into a distally steepened margin from the latest Aptian onward. Steepening of the margin is indicated by the development of deep-cutting (maximum 25 m) stable channels and by its function as a distinct facies boundary, and is contributed to differential subsidence caused by differential compaction of the rigid inner shelf carbonate platform, and the muddy outer shelf, as well as by a phase of enhanced tectonic activity close to the Aptian–Albian boundary (Föllmi 1989b).

Trends in condensation and phosphogenesis

Condensed phosphatic beds of the Garschella Formation typically consist of thin (<50 cm) and well-individualized strata characterized by abundant (10–90% of total volume) particulate phosphatized fossil fragments, lithoclasts, and/or phosphatized crusts. The latter display several phosphate generations and a complex internal microstratigraphy due to the superposition and lateral replacement of different types and generations of nonphosphatized sediments (glauconitic sands, marls, and micrites; Fig. 2). Included ammonoids, inoceramid bivalves, and globotruncanid foraminifera indicate that the process of phosphatization and condensation continued throughout the time interval from earliest Aptian (*turkmenicum* Zone; Gebhard 1983, 1985) to early Cenomanian (*brotzeni* or *reicheli* Zone; Ouwehand 1987). During this interval, four phases of sand replenishment interrupted or geographically limited the formation of phosphatic beds. A first phase is observed in the middle late Aptian where the large imput of siliciclasts and their distribution over the entire inner shelf overwhelmed the processes of phosphatization and condensation for approximately 1 million years (Brisi and Gams Beds; Fig. 1). Three independent phases of replenishment occurred in the middle early Albian (middle *tardefurcata* Zone; Niederi Beds), late early Albian (late *mammillatum* Zone; Sellamatt Beds; Fig. 1), and middle late Albian (late *inflatum* Zone; Aubrig Beds; Fig. 1). The Albian phases were less extensive and powerful in comparison to the middle late Aptian phase; they did not entirely overpower the processes of condensation and phosphogenesis, but limited those to distal regions of the inner shelf. This resulted in a distally directed bundling of Albian phosphatic beds into one single bed along the inner-shelf margin (Fig. 1).

Space-time patterns are recorded by the above mentioned guide fossils within the phosphatic beds. They indicate the following major trends.

(1) Onset of the condensation–phosphogenesis sedimentary system in the Aptian was diachronous and trended from east to west, as documented by the Luitere Bed (Fig. 1; earliest Aptian in southeastern Germany; middle early Aptian in western Austria; early late Aptian in eastern and central Switzerland).

(2) Termination of the condensation–phosphogenesis sedimentary regime in the Albian and early Cenomanian was diachronous as well, and typified by a distal-proximal trend (south–north; termination in distal inner shelf areas: late Albian, Plattenwald Bed; in proximal inner-shelf areas: early Cenomanian, Kamm Bed; Fig. 1).

(3) Although condensation persisted at least until the late Albian, phosphogenesis ceased earlier in distal inner-shelf areas. The termination of phosphogenesis trended also in proximal directions through time (cessation along the inner-shelf margin: middle *tardefurcata* Zone; in distal areas north of the inner-shelf margin: late *mammillatum*

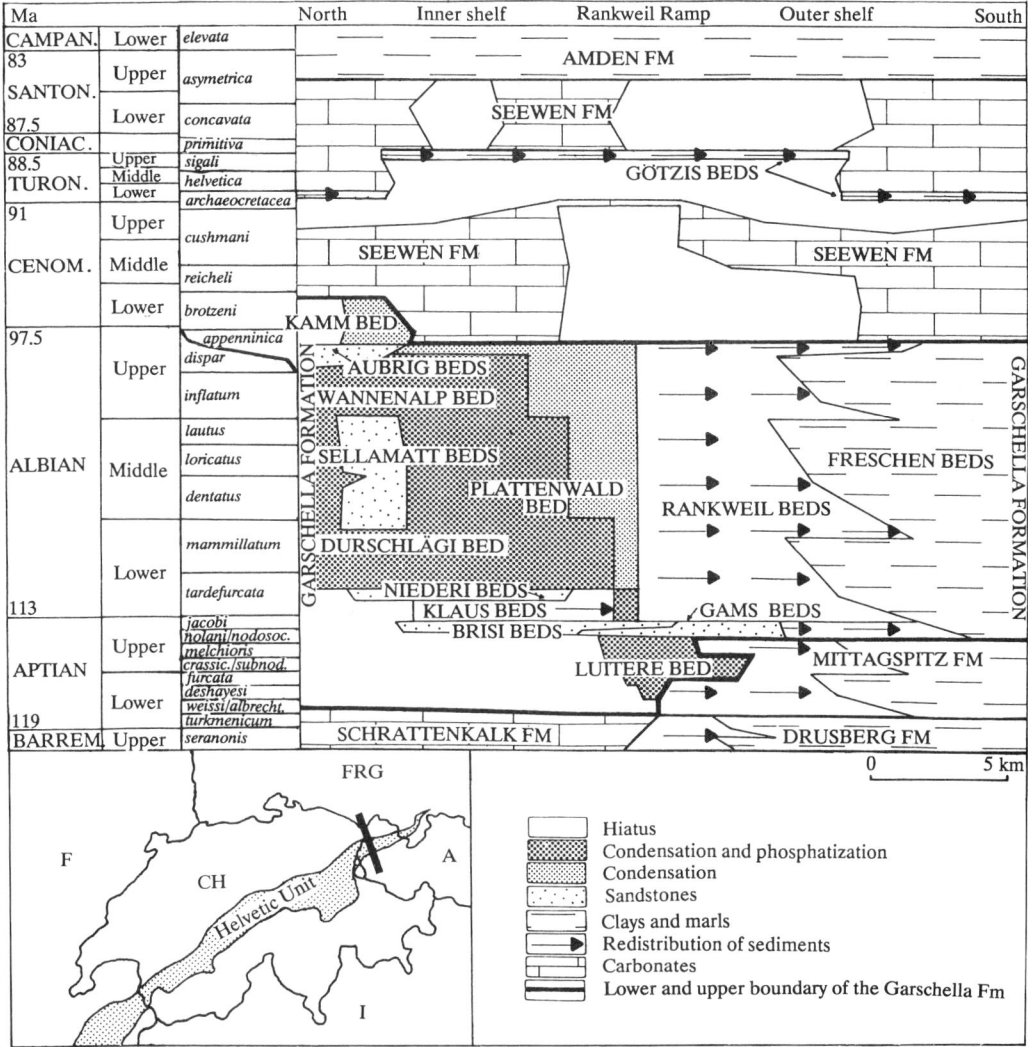

Fig. 1. Time-space distribution of Barremian to Campanian sediments of the western Austrian Helvetic Unit. Location of the proximal-distal transect is indicated on the map. Ammonite biozonation (*seranonis* to *dispar* Zones) is based on the work of Mikhailova (1979), Casey (1961), and Owen (1971). Foraminifera biozonation (*appenninica* to *elevata* Zones) is after Robaszynksi & Caron (1979). The absolute time scale is adopted from Harland *et al.* (1982). The lithostratigraphic terminology is after Föllmi & Ouwehand (1987).

Zone; in intermediate areas: *loricatus* Zone; in more proximal areas: late *inflatum* and locally *dispar* Zones; in northernmost areas: *brotzeni* or *reicheli* Zone).

In general, the following sedimentary trends are observed.

(1) The complexity of internal nonphosphatized sediments tends to decrease and micritic sediments tend to dominate toward distal areas (Fig. 2).

(2) The degree of condensation tends to increase toward distal inner-shelf areas resulting in the bundling of phosphatic beds into one. This tendency is also reflected by the phosphates, which are dominated by particles and nodules in proximal areas and by complex crusts in distal areas (Figs 2, 3A and B).

Types of phosphatic sediments

Temporal and spatial trends in the development of the phosphatic beds are mirrored by the types of phosphate deposited (Föllmi et al. in prep.).

Pristine phosphates (i.e., those excluded from any sedimentary interaction after phosphatization; not re-exposed and reworked) are present in the form of particles scattered in small amounts throughout noncondensed marly or sandy sequences (e.g., Sellamatt Beds), and typically consist of one phosphate generation. They appear also in form of a single-generation crust or rim; for instance, as a coating on rugged carbonate surfaces, or as a thin layer consisting of phosphatized microbial mats capping condensed phosphatic beds stromatolites; Figs 3C and D).

Condensed phosphates have been re-exposed to hydrodynamics and/or biodynamics after phosphatization. Commonly, they enter renewed phases of phosphogenesis after re-exposure, which are documented by the presence of different phosphate generations. The different generations may include different types and amounts of detritus, and the generation interfaces may be coated by iron oxyhydroxides or overgrown by encrusting sessile organisms (e.g., foraminifera and serpulids). The presences of intimate mixtures of fossils of different ages and different ecological habitats, of a varying amount of phosphate generations, and of complex internal microstratigraphic relationships, suggest that the condensed phosphates went through repetitive cycles of burial, phosphogenesis, erosion and re-exposure to hydrodynamics and biodynamics, burial, phosphogenesis, etc. According to Mullins & Rasch (1985), such cycles may be termed Baturin Cycles (cf. Baturin 1971; Kennedy & Garrison 1975; Krajewksi 1984; Brandt 1985; Föllmi 1989b). During Baturin cycling, particulate phosphates are concentrated into multi-event condensed and particulate phosphatic beds, and phosphatized laminae are stacked up to multi-event condensed phosphatic crusts characterized by complex internal boundaries between the laminae (presence of low-angle truncations and scouring marks).

Allochthonous phosphates appear within gravity-flow deposits, thereby displaying the characteristic features of redeposited sediments such as sharp and unconformable bases, internal grading, and the mixture of phosphatic diaclasts with nonphosphatic lithoclasts.

The three types described above represent end-members, and mixtures are usually present. Condensed phosphates can be covered with a final layer of pristine phosphates, and condensed phosphatic beds can include a fine-grained allochthonous fraction due to winnowing and redeposition. Nevertheless, the different types are attributable to varying geographic settings on the shelf or to different phases in the sedimentary process of condensation and phosphogenesis. Pristine particulate phosphates are commonly found outside of the main area of condensation and phosphogenesis, either in a proximal (inner shelf) or distal direction (outer shelf). The absence of Baturin Cycles prevented a transformation into condensed phosphates. The termination of Baturin Cycles within the area of condensation and phosphogenesis resulted in an ultimate pristine phosphate layer on top of the condensed phosphatic beds (e.g., Fig. 3C). Gravity flows including allochthonous phosphatic particles are abundant along the boundary region between the inner and outer shelf (Figs 4 and 5).

Genesis of the Helvetic condensed phosphatic beds

Current-induced condensation and the presence of an oxygen-minimum zone

The area of condensation and phosphogenesis is part of a conspicuous sediment zonation aligned approximately parallel to the inner shelf margin and traceable along the mid-Cretaceous northern Tethyan margin.

(1) The proximal inner shelf is characterized by the accumulation of glauconitic, commonly bioturbated, sands and marls. Sediment-accumulation rates were moderate ($10-30$ m Ma^{-1}) and bottom waters moderately to well-oxygenated. This belt embodied the main site of glauconite formation (zone A in Figs 4 and 5).

(2) The distal inner shelf hosted the erosion-winnowing-condensation-phosphogenesis sedimentary regime typified by ultra-low sediment-accumulation rates ($2-20$ cm Ma^{-1}), the presence of Baturin Cycles, and thriving biotic communities dominated by microbial colonies, suspension feeders (hexactinellid sponges; solitary corals; terebratulid and rhynchonellid brachiopods; shallow endobiontic bivalves; crinoids), as well as gastropods and ammonoids (Figs 3E and F; zone B in Figs 4 and 5).

(3) The inner-shelf margin predominantly received and accumulated eroded and transported sediments derived from the two

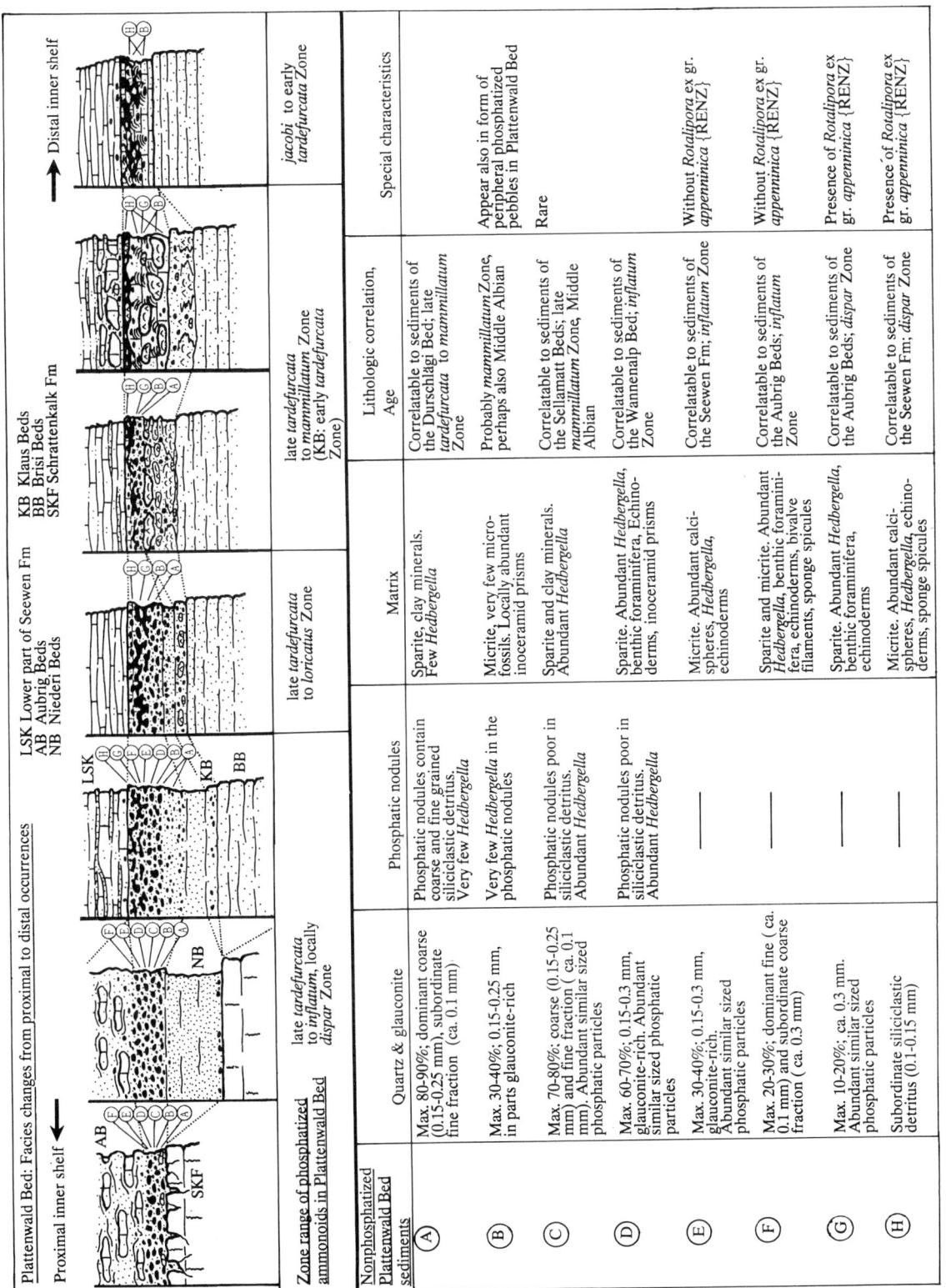

Fig. 2. Proximal-distal facies changes within the Albian Plattenwald Bed. Overview of the facies, distribution, and inferred age of included nonphosphatized sediments, and timing of phosphogenesis. Key to lithological symbols is provided in Fig. 4 (from Föllmi 1989*b*). See Appendix for locality details.

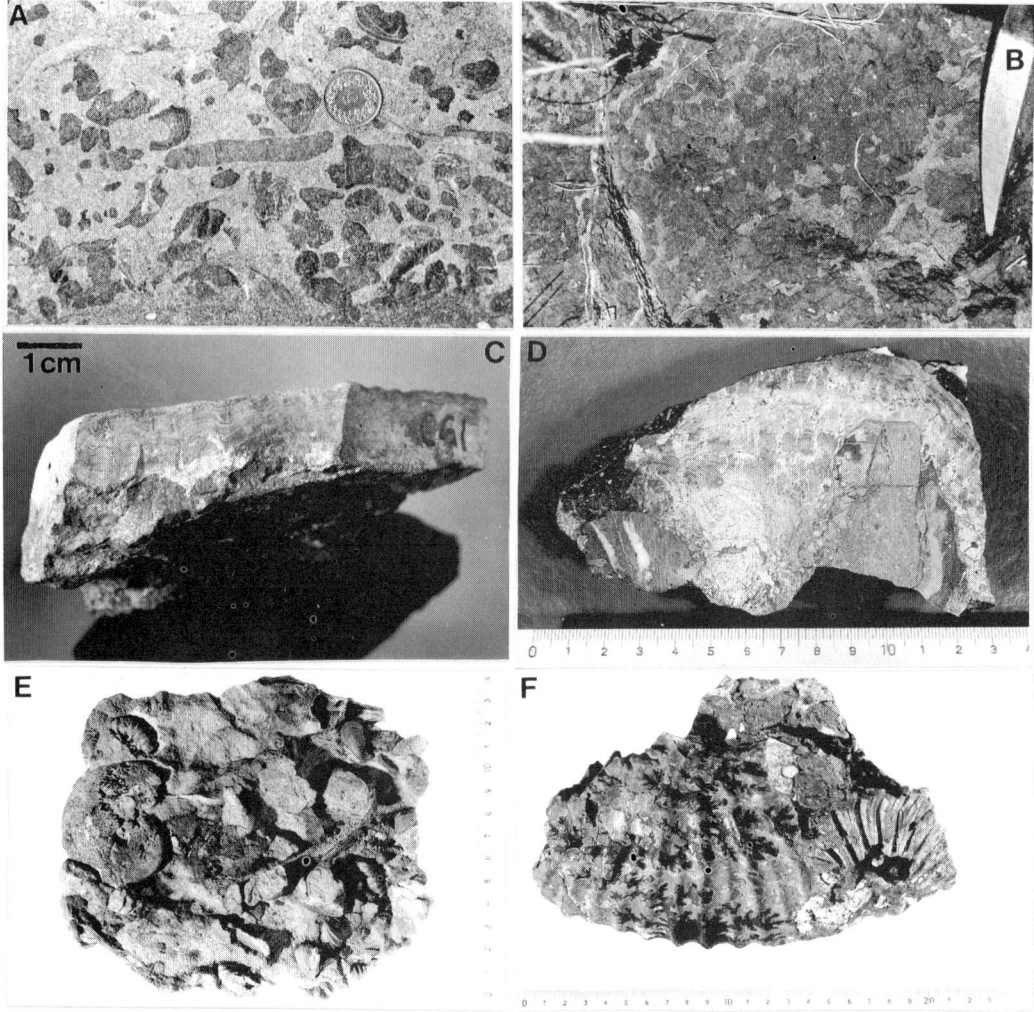

Fig. 3. (A) Phosphatic fossil debris of early to late Albian age in a multi-event winnowed and condensed bed (Plattenwald Bed). (B) Phosphatic particles embedded in phosphatized glauconitic sands cemented by different phosphate generations, thereby forming a complex phosphatic crust. Depressions are infilled with younger nonphosphatized micrites (Plattenwald Bed). (C) Pristine phosphatized columnar microbial mat on top of condensed phosphates (Plattenwald Bed). (D) Phosphatized columnar microbial mats colonizing Schrattenkalk pebbles. Base of the mat on the left side is tilted (Plattenwald Bed). (E) Densely packed fossils on weathered surface of a multi-event winnowed and condensed bed (Wannenalp Bed). (F) Phosphatized internal mould of *Hypacanthoplites* sp. An external mould of the same genus is present on the right side (all photographs from Föllmi 1989*b*) See Appendix for locality details.

inner-shelf facies belts. These sediments were redeposited in channel and fan systems (zone C in Figs 4 and 5).

(4) The outer shelf was blanketed by fine-grained, dark and commonly laminated, hemipelagic sediments. Accumulation rates were moderate (maximum 10 m Ma^{-1}) and the bottom waters poorly oxygenated, indicative of the presence of an oxygen-minimum zone during Aptian and Albian (zone D in Figs 4 and 5).

This zonation of four different sedimentary regimes is interpreted as the record of a westward-flowing current system contouring along the northern Tethyan margin, the zone of condensation and phosphogenesis being located within the erosive radius of the current axis (Delamette 1985, 1988*a*, *b*; Föllmi 1986, 1989*b*;

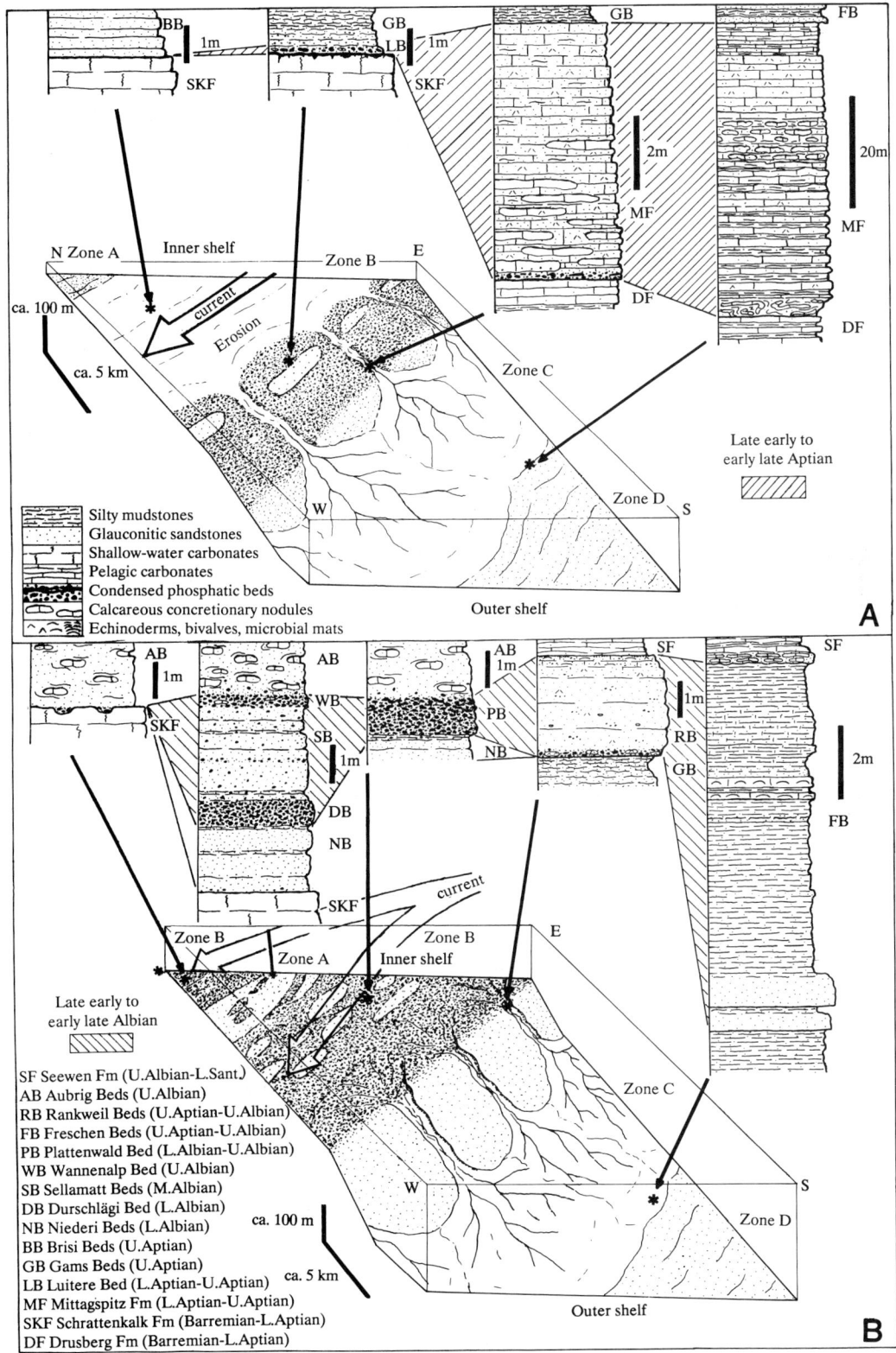

Fig. 4. Distribution of inferred coeval sections of (**A**) late early to early Aptian, and (**B**) late early to early late Albian ages. The palaeogeographic positions of sections is indicated on simplified block diagrams of the distal inner and proximal outer shelf. The vertical scale is exaggerated. See Appendix for locality details.

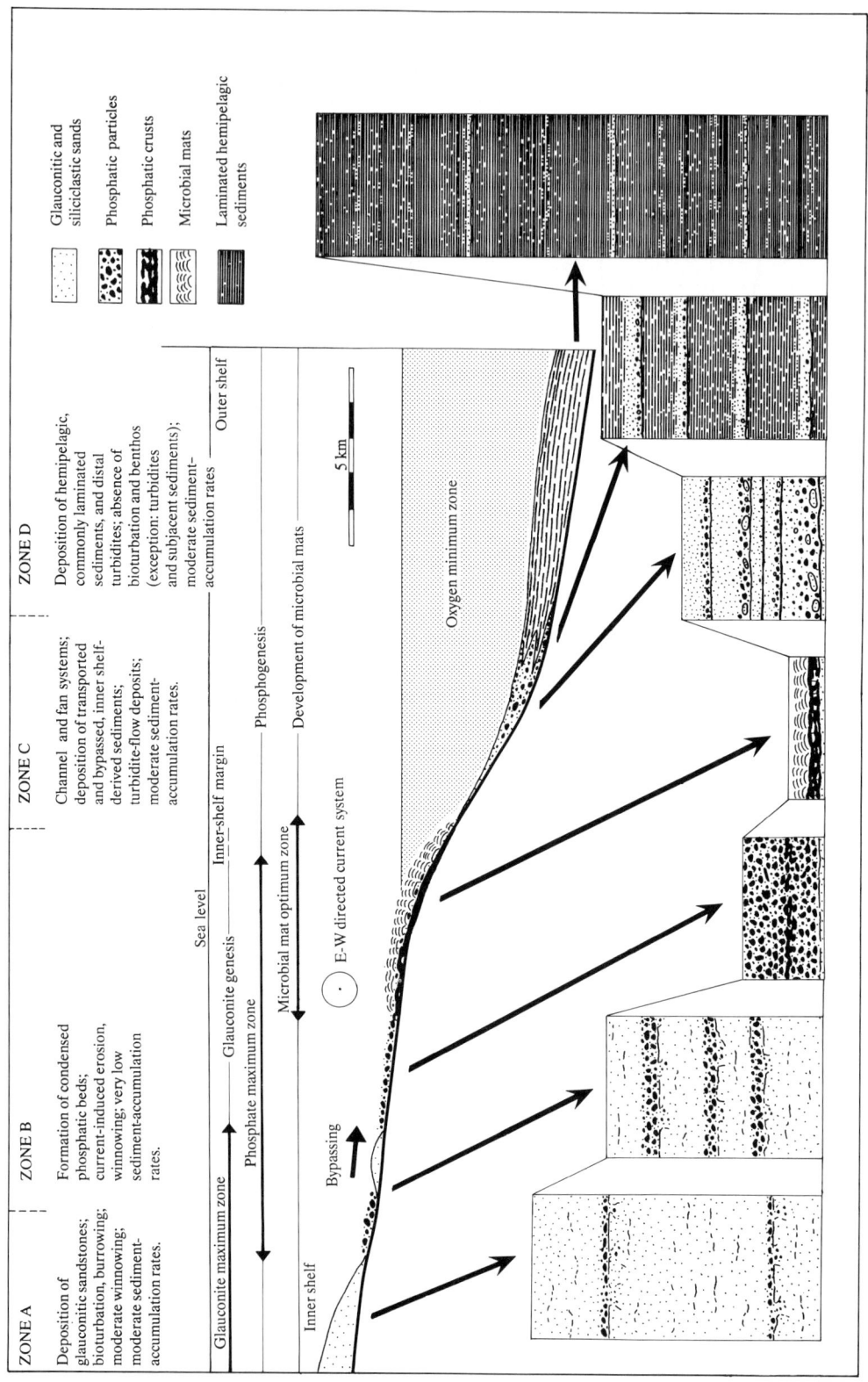

Fig. 5. Schematic transect perpendicular to the Helvetic Shelf. Note the position of the different sedimentary systems in a setting where the current axis is located on the distal inner shelf (e.g., Lower Albian Plattenwald Bed). The bottom sections are representative and approximately time-equivalent (from Föllmi 1989b).

Ouwehand 1987). The Baturin Cycles, the ultra low accumulation rates, and the resulting complex microstratigraphies inherent to the condensed phosphatic beds are thought to be the result of small-scale lateral shifts in the current axis and to variations in the current velocity (cf. Gross et al. 1988; Culvin et al. 1988).

Large proximal-distal shifts in the sediment zonation mirror significant lateral shifts in the current system and are thought to be related to relative sea-level changes (cf. Pinet & Popenoe 1985). Such migrations are observed in the middle late Aptian where the belt of glauconitic sediments expanded onto the entire inner shelf and the area of condensation and phosphogenesis was eliminated; in the latest Aptian where the belt of condensation and phosphogenesis migrated onto the proximal inner shelf and the belt of redeposited sediments expanded onto the distal inner shelf; and in the late Albian where the belt of condensation and phosphogenesis retrograded in a proximal inner-shelf direction, and the belt of hemipelagic sedimentation gradually spread over the inner shelf, thereby eliminating the zone of redeposition.

The establishment of two discrete belts of glauconitic sediments and condensed phosphatic beds on the inner shelf during the middle early through middle Albian is probably related to bifurcation of the current system evoked by topographic irregularities in the eastern Helvetic inner shelf, and by a broadening of the inner shelf toward the west (succession of zones A-B-A-B-C-D in the distal direction; Fig. 4B; Föllmi 1986, 1989b; Ouwehand 1987).

The current-system apparently limited the upper extension of the oxygen-minimum zone, as is indicated by the degree of bioturbation in the belt of glauconitic sediments (Fig. 5; cf. Vercoutere et al. 1987).

The belt of phosphogenesis and condensation is considered as a valuable palaeoceanographic marker belt, indicative of the palaeoposition of the current axis and of the upper boundary of the oxygen-minimum zone. Lateral migrations in the area of condensed phosphatic beds are commensurable to shifts in these important oceanographic systems.

Biological evidence of sudden burial and early diagenetic phosphogenesis

Phosphatized microbial colonies are ubiquitous in the mid-Cretaceous condensed phosphatic beds. They are preserved as columnar stromatolites (Figs 3C and D), thin phosphatic coatings around fossils and lithoclasts, corpuscular phosphatized particles within voids of ammonoids and other organisms, laminations within phosphatic crusts, or microstromatolites in phosphatized siliciclasts sediments (cf. Krajewski 1984; Delamette 1988a; Föllmi 1989b; Lamboy 1990. The optimum in preservation of microbial mats is observed in distal areas of the belt of condensation and phosphogenesis, and was probably related to the proximity of the oxygen-minimum zone (Fig. 5).

The degree of preservation of the microbial assemblages is remarkable. The excellent textural fossilization indicates that the concentration and precipitation of PO_4^{3-} occurred prior to and independently from the decay of the microbial organic matter, during a very early stage of diagenesis.

The phosphatized fossils within the condensed phosphatic beds are well-preserved, a feature generally known from condensed phosphatic beds (Figs 3E and F; e.g., Kennedy & Garrison 1975; Baturin 1982; Soudry & Lewy 1988; Lewy 1990). Signs of abrasion, erosion and biologic activity (encrustings; borings) on phosphatized fossils are commonly due to reworking and re-exposure to hydro- and biodynamic process after phosphogenesis.

The lack of pre-phosphatization epizoans encrusting phosphatized shells and other signs of biological activity (e.g., borings and traces of biological abrasion), and the good preservation suggest that the organisms were buried rapidly (cf. Brett 1988; Speyer & Brett 1988). Prolonged exposure at the sea floor would have destroyed the organisms, either through disintegration by scavenging organisms, chemical dissolution, or physical disarticulation (cf. Cummins et al. 1986).

These observations suggest that phosphogenesis was an early diagenetic process which took place after the rapid burial of the to be phosphatized organisms, but prior to the total decay of buried organic matter of microbial colonies and probably of other organisms as well. This obviously conflicts with the observed ultralow net sediment-accumulation rates within the area of condensation and phosphogenesis, unless the effect of Baturin Cycles is taken into consideration (see next section).

Sedimentological evidence of sudden burial and early diagenetic phosphogenesis

A sedimentary feedback on phosphogenesis was exerted by the presence of siliciclasts. This feedback was negative in the middle late Aptian where the first phase of siliciclastic replenishment overwhelmed the condensation–phosphogenesis system. A positive feedback is observed in the Albian where the presence of siliciclasts in the belt of condensation and phosphogenesis went hand-in-hand with phosphogenesis. As soon as the siliciclasts were removed by the prevailing current activity, phosphogenesis ceased. This is indicated by the oldest age of micritic sediments incorporated within the condensed phosphatic beds, which is younger than the youngest phosphatized guide fossils (Fig. 2). The southward extension of the three Albian phases of sand replenishment was retrogressive during the Albian and the subsequent, current-induced removal of sand trended, therefore, sequentially in a proximal direction, both probably due to a persistent sea-level rise (Haq et al. 1987). The cessation of phosphogenesis, trending in a proximal direction through the Albian, is positively correlatable with the sequential removal of siliciclasts on the inner shelf, which followed each phase of sand replenishment with a delay of approximately 1.5 million years (Föllmi 1989b).

(1) The inner-shelf margin was only influenced by the middle late Aptian phase of sand replenishment. This sand was removed in the early Albian and phosphogenesis expired at the early-late *tardefurcata* Zone boundary (phosphatized fossils cover the *jacobi* and early *tardefurcata* Zones). Condensation, however, persisted throughout the Albian, and generally lasted until early Turonian. This resulted in a condensed phosphatic bed of 20–30 cm thickness, which represents the time interval between late Aptian (*jacobi* Zone) and early Turonian (*archaeocretacea* Zone) (approximately 25 Ma, after Harland et al. 1982; Fig. 1).

(2) The outermost inner-shelf belt, adjacent to the inner-shelf margin, was influenced by the first Albian phase of sand replenishment (at the early/late *tardefurcata* Zone boundary) and underwent total sand depletion at the end of the *mammillatum* Zone. Phosphatized fossils indicate the time span between late *tardefurcata* and *mammillatum* Zones. Condensation persisted until the *dispar*, locally until the *brotzeni* Zone.

(3) In a small inner-shelf belt proximal to the above, the influence of the first two Albian phases of sand replenishment (second phase in the late *mammillatum* Zone) is detected. The removal of sand was completed in the *loricatus* Zone. Phosphogenesis lasted from the late *tardefurcata* to the *loricatus* Zone, whereas condensation processes terminated in the *dispar* Zone.

(4) In a more proximal inner-shelf belt, the impact of all three phases of replenishment (the third occurred in the late *inflatum* Zone) is observed, and the termination of sand removal was approximately coeval with that of condensation (*dispar* Zone). Phosphogenesis and condensation occurred from the late *tardefurcata* to the late *inflatum*, locally *dispar* Zone, persistent in distal areas of this belt and interrupted in proximal parts, because of a temporary overpowering of the condensation-phosphogenesis system by the phases of sand replenishment (in a complex pattern due to late early to late middle Albian bifurcation of the current system; Ouwehand 1987; Föllmi 1989b).

(5) The innermost inner-shelf belt, observable in the Helvetic Unit, experienced sand depletion in the *brotzeni* or *reicheli* Zone and phosphogenesis stopped at that time.

Although larger and larger areas of the distal inner shelf became sand-depleted during the Albian, the inner-shelf margin continued receiving siliciclasts throughout the Albian, as is documented in the sand-rich sediments of the uppermost Aptian to uppermost Albian Rankweil Beds, which have been derived from the inner shelf and redeposited in channel and fan systems along the inner shelf margin (zone C in Figs 4 and 5). This documents that sand deposition bypassed the sand-depleted distal inner-shelf zone of condensation and phosphogenesis, probably via bed-load transport. However, before certain distal inner shelf areas became sand-depleted and while phosphogenesis was still an ongoing process, this sand may have been removed episodically in the form of distally migrating sand bodies. In analogy to modern, current-dominated, siliciclastic shelf areas (cf. Hamilton et al. 1980; Flemming 1988; Smith 1988; Stride 1988), the following genetic sequence is inferred for the mid-Cretaceous Helvetic inner shelf: episodically, the inner shelf experienced the influx of large amounts of siliclastic sands (four phases during Aptian and Albian). Being subjected to the westward-flowing current system, the sand became remobilized in the form of moving sand-bodies until depletion of the supply which was first reached in the distal portions of the sand sheet. Subsequently, sand derived from more proximal parts was dynamically bypassed over these regions of sand-depletion.

Also in analogy to modern counterparts, the remobilization of sand bodies presumably occurred episodically, in response to changes in current velocity, or lateral changes in the position of the current axis. In this stop-and-go rhythm of the shifting sand bodies, whole benthic communities might have been buried in catastrophic way, subsequently phosphatized, and re-exposed again. This scenario would explain the good preservation of the fossils, their early diagenetic phosphatization, even before the total decay of organic matter, and the positive temporal correlation between the termination of phosphogenesis and the point of total sand removal. It explains as well how Baturin Cycles functioned in the distal inner-shelf zone of condensation and phosphogenesis. Sudden burial by migrating sand bodies is also indicated by the preferential preservation of benthic and benthos-oriented nektonic organisms, and the virtual absence of preserved active nekton in the condensed phosphatic beds.

Physicochemical cycles of phosphate, sedimentary cycles of phosphogenesis

Employing the above scenario of sudden burial of benthic communities by sand bodies, episodically migrating in the rhythm of changes in current intensity or position, and subsequent phosphatization of these communities, the question of how phosphogenesis took place remains to be answered. The decay of organic matter and the subsequent release of dissolved PO_4^{3-} appears to have been of minor influence, since phosphogenesis took place prior to the entire putrefaction of the microbial colonies and probably of the other organisms as well.

An alternative possibility of PO_4^{3-} concentration in the interstitial waters around the buried organisms is given with the physico-chemical cycling of the redox-pair $Fe^{3+}-Fe^{2+}$ (and eventually $Mn^{4+}-Mn^{2+}$) around the redox boundary. This mechanism is known to mobilize PO_4^{3-} in lake sediments (cf. Stumm & Leckie 1970) and is postulated to concentrate PO_4^{3-} near marine redox boundaries, either within the water column (Shaffer 1986), or in sediments (O'Brien & Heggie 1988; Froelich et al. 1988; Heggie et al. 1990). In this model, particulate iron (and manganese) oxyhydroxides settles through the water column and interstitial waters, thereby scavenging PO_4^{3-} from the bottom and interstitial waters. Underneath the redox boundary, the hydroxides are dissolved and the metal anions reduced, thereby releasing PO_4^{3-}. The anions diffuse back into the bottom waters and become oxidized. Particulate oxyhydroxides form and the cycle can repeat.

A important precondition of this physico-chemical 'pump and shuttle' (Shaffer 1986) is the presence of burrowing endofauna, which facilitates the irrigation of the siliciclastic sediments (Heggie, pers. comm. 1988).

A second important mechanism of PO_4^{3-} concentration was probably the transfer of dissolved bottom-water PO_4^{3-} into interstitial waters by current-induced convection-diffusion processes (Föllmi & Garrison in prep.).

If it is assumed that this mechanism was responsible for the concentration of PO_4^{3-} in the mid-Cretaceous condensed phosphatic beds, the formation of these beds may have occurred as following (Fig. 6).

(1) The westward-flowing Tethyan boundary current lowered sediment-accumulation rates to virtually zero along the distal inner shelf (proximal inner shelf in latest Aptian and near the Albian/Cenomanian boundary). Bottom waters were probably nutrient-rich due to the proximity of the oxygen-minimum zone (cf. Mullins & Rasch 1985). A diverse benthic community was present, consisting of microbial colonies stabilizing sediments at the sediment/water interface, suspension feeders, and their predators. Bioturbation was minimized, probably because of a shallow oxic/anoxic interface within the sediments.

(2) Changes in the current velocity and position induced perturbations in this steady-state system. Sediments were reworked and sand bodies remobilized, thereby burying benthic communities in a catastrophic way.

(3) The large amounts of buried organic matter attracted burrowers and scavenging microbes, the latter creating local anoxia around the buried organisms. Burrowing and sediment mixing by the endofauna led to irrigation of the sand body, thereby opening the system to the phosphate pump and shuttle.

(4) The PO_4^{3-} pump and shuttle led to high concentrations of PO_4^{3-} near the anoxic areas around the buried organisms. Precipitation of apatite began, eventually using microbial phosphate-rich granules as nucleation germs (Reimers et al. 1990).

(5) Current-induced convection-diffusion processes transferring bottom-water PO_4^{3-} across the sediment-water interface contributed further to the accretion of solid phosphates.

(6) Variation in current intensity and position caused the selective removal of the siliciclastic matrix (winnowing), until a gravel of phosphatized clasts remained (start of

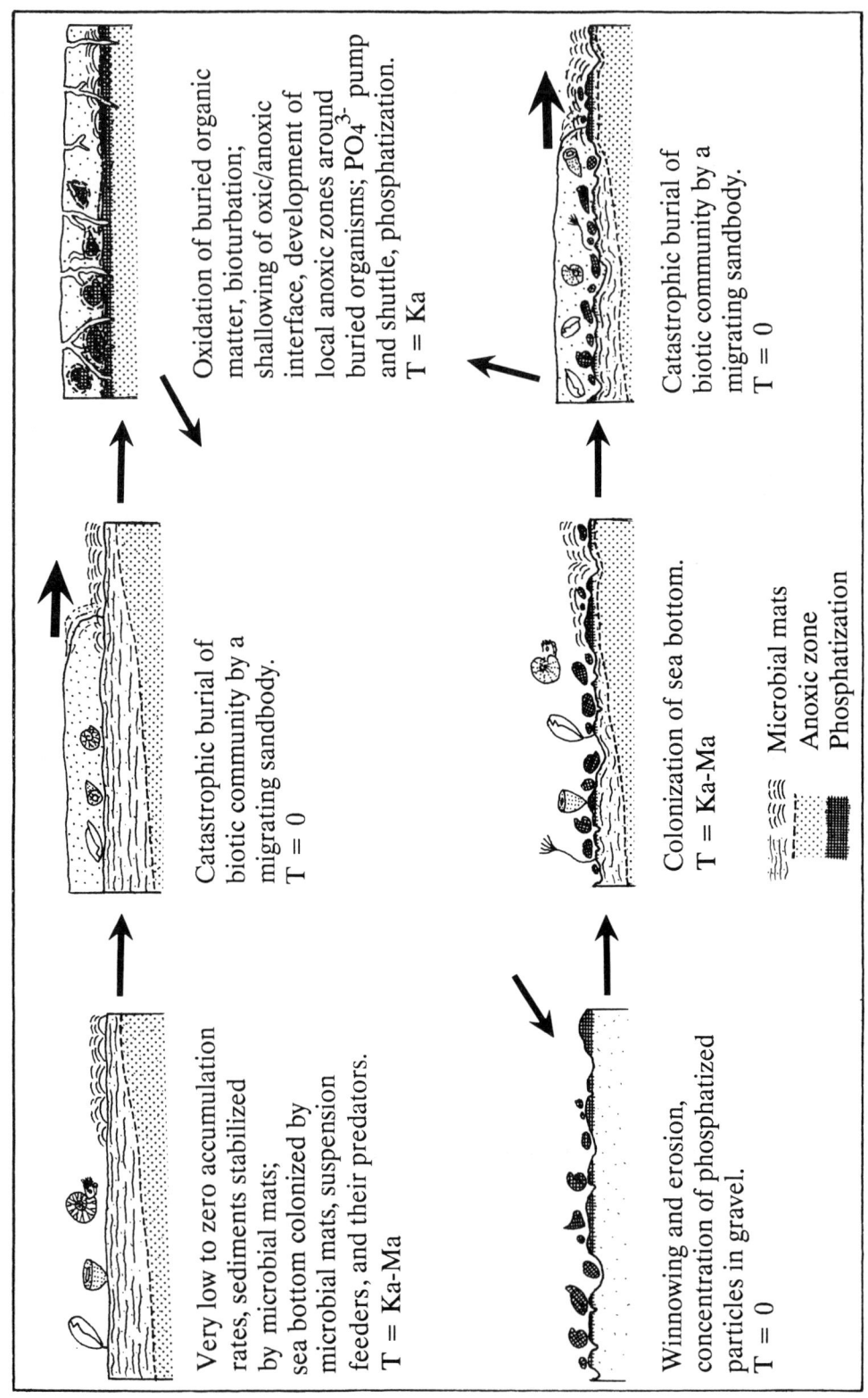

Fig. 6. Diagrammatic scenario of condensation and phosphogenesis on the current-dominated mid-Cretaceous Helvetic Shelf. The path of Baturin cycling is indicated by the thinner arrows. T is an estimate of the time span during which an event or phase occurred (from Föllmi 1989b).

Baturin cycling). Stronger current pulses transported some of the phosphatized clasts as well, leading to their inclusion in allochthonous beds, particularly on the inner-shelf margin (zone C in Figs 4 and 5).

Repeated cycles (1–6) of rapid burial, phosphatization, and re-exposure (Baturin Cycles) were responsible for the characteristic multiple generation composition and the complex internal microstratigraphies of the mid-Cretaceous Helvetic phosphatic beds.

Although the belt of condensation and phosphogenesis represented the main area of phosphate formation, minor amounts of phosphate were also generated outside this belt; for instance in the inner-shelf area of glauconitic sand accumulation, or in the outer-shelf area of hemipelagic deposition. These phosphates may have been also the result of sudden burial of organisms (e.g., by gravity flows), but the higher sediment-accumulation rates and the absence of current-induced Baturin Cycles prevented their accumulation into phosphatic beds. They represent pristine phosphates.

Concluding remarks on condensation

As a result of his detailed observations on the mid-Cretaceous, condensed phosphatic beds in central and eastern Switzerland, western Austria, and southern F.R. Germany, Arnold Heim (1934) implicitly defined condensation as the result of strongly reduced sedimentation rates, generally caused by current activity. He considered the concentration of fossils of different ages as a prime characteristic of condensed beds; however, he excluded sedimentary reworking as a possible agent in the formation of such faunal mixtures.

Based on work in coeval beds of the Helvetic Unit in western Switzerland, Schaub (1936, 1948) clearly recognized the essential function of sediment reworking and transport in the formation of these condensed phosphatic beds. He was one of the first sedimentologists to describe Baturin Cycles. This crucial divergence in the interpretation of the mid-Cretaceous. Helvetic condensed phosphatic beds, and therefore in the definition of condensation, is reflected in the ongoing debate of the role of sediment reworking in the process of condensation (e.g., Rod 1946; Mensink 1960; Wendt 1970; Jenkyns 1971; Fürsich 1971, 1978; Kennedy & Garrison 1975; Krajewski 1984; Haq et al. 1987; Kidwell 1989; Glenn 1990).

The present reexamination of the mid-Cretaceous Helvetic beds suggests that current-induced Baturin cycling in the form of episodic migrating of sand bodies along the area of condensation was inherent to the process of condensation. Sediment reworking, therefore, represented an important agent in the concentration of the phosphatized particles, at least in this type area of condensation.

Transitions are observed to nonphosphatized sequences (e.g., glauconitic sandstones, laminated marls; Fig. 5), which document low sediment-accumulation rates as well (10–30 m Ma^{-1}). These beds do show signs of episodic erosion events such as hiatuses and intercalations of turbiditic beds, but not to the extend that the primary bedding is disturbed on a large scale. Such sequences are sediment-starved, probably due to current activity, but not condensed *sensu* Heim and Schaub, because of the lack of faunal mixtures of different time zones and Baturin cycling. Therefore, it appears inappropriate to term beds 'condensed', which are sediment-starved, but still preserve their primary bedding, and lack the features of condensation, described below. The process of condensation requires active sediment reworking to an extend that primary bedding is distorted and complex microstratigraphies take their place. Transitions between sediment-starved and condensed sequences are possible (Fig. 5).

The mid-Cretaceous condensed beds are characterized by: (1) complex and laterally rapidly changing, internal microstratigraphies; (2) lateral transitions into phosphatic crusts and hardgrounds, as well as hiatuses; (3) faunal mixtures of different biostratigraphic zones and of different ecological habitats, (4) extremely low sediment-accumulation rates of 2–20 cm Ma^{-1} over several million years, (5) multiple generations of early diagenetic mineralization by glauconite and/or phosphate.

Within the context of the adjacent sediments, they represent unique palaeoceanographic marker beds in terms of detectors of intense hydrodynamics such as current-activity, sediment dynamics such as episodically migrating sand bodies, as well as geochemical dynamics embodied in the rapid, early diagenetic precipitation of phosphate.

This work is a 'reworked and condensed' excerpt from my PhD thesis. I am grateful to my advisors R. Trümpy and H. Rieber (both Zürich), and R. Oberhauser (Vienna). Amongst my Swiss colleagues, I especially thank P. Ouwehand (Solothurn) and M. Delamette (Fribourg) for sharing our experiences in the Helvetic mid-Cretaceous. I acknowledge with many thanks the constructive comments on the manuscript of R. Garrison and J. Dolan (both Santa Cruz), L. Pray (Madison), of the reviewers of the Geological

Appendix: section localities

Coordinates and locality names used in the tables correspond to the topographic map 'Hoher Freschen', 1:50 000, no. 228, Swiss Federal Topographic Survey, CH-3084 Wabern. Descriptions of sections and localities are found in Föllmi (1986).

Figure 2. (From left to right)

Breiterberg	772.800/249.550/890
Rappenloch Bridge	776.820/250.820/610
Örfla Gulch	767.850/244.400/525
Hohe Lug, Emmabach	769.370/244.600/830
Klaus	766.950/242.470/500
Ache	777.300/247.900/980
Stein	764.300/233.400/520

Figure 3.

(A)	Steinriesler Bach	782.100/252.800/590
(B)	Osanken	768.020/243.070/640
(C)	V. Schaner Alp	776.470/247.320/980
(D)	Unterklien	772.050/250.300/460
(E)	Schwarzenberg	773.050/248.950/1040
(F)	Simonsbach	783.750/248.400/780

Figure 4.

(A) (from left to right)	Moos, Emmabach	770.470/244.970/1050
	U. Wäldle Alp	774.900/245.180/1070
	Hinterwang	773.030/240.250/1330
	Mittagspitz	785.000/242.700/1900
(B) (from left to right)	Fallbach	774/500/250.250/890
	Schwarzenberg	773.050/248.950/1040
	Müselbach	777.750/250.650/730
	Rankweil	767.950/238.440/500
	S Hoher Freschen	777.600/241.620/1900

Society (M. Pedley and J. Kennedy), and especially of I. Jarvis. The manuscript was written during a postdoctoral tenure at the University of California, Santa Cruz, supported by the Swiss National Science Foundation and the ETH Zürich.

References

BATURIN, G. N. 1971. Stages of Phosphorite Formation on the Ocean Floor *Nature*, **232**, 61–62.
—— 1982. *Phosphorites on the Sea Floor*. Elsevier, Amsterdam.
BENTOR, Y. K. 1980. Phosphorites – the Unsolved Problems. *In*: BENTOR, Y. K. (ed.) *Marine Phosphorites – Geochemistry, Occurrence, Genesis*. Special Publication of the Society of Economic Paleontologists and Mineralogists, **29**, 3–18.
BRANDT, K. 1985. Sea Level Changes in the Upper Sinemurian and Pliensbachian of Southern Germany *In*: BAYER, U. & SEILACHER, A. *Sedimentary and Evolutionary Cycles*, Lecture Notes in Earth Sciences No. 1. Springer, 113–126.
BRETT, C. E. 1988. Comparative Taphonomy and Ecology of Fossil "Mother Lodes". *Paleobiology*, **14**, 214–220.
BURNETT W. C. & FROELICH, P. N. 1988. Preface *In*: BURNETT, W. C. & FROELICH, P. N. (eds) *The Origin of Marine Phosphate. The Results of the R.V. Robert D. Conrad Cruise 23–06 to the Peru Shelf*. Special Issue of Marine Geology **80**, III–VI.
CASEY, R. 1961. The Stratigraphical Palaeontology of the Lower Greensand. *Palaeontology*, **3**, 487–621.
CULVIN, S. J., BRUNNER, C. A. & NITTROUER, C. A. 1988. Observations of a Fast Burst of the Deep Western Boundary Undercurrent and Sediment Transport in South Wilmington Canyon from DSRV Alvin. *Geo-Marine Letters*, **8**, 159–165.
CUMMINS, H., POWELL, E. N., STANTON JR., R. J. & STAFF, G. 1986. The rate of taphonomic loss in modern benthic habitats: how much of the potentially preservable community is preserved. *Palaeogeography, Palaeoclimatology, Palaeoecology*, **52**, 291–320.
DELAMETTE, M. 1985. Phosphorites et Paléocéanographie: l'Exemple des Phosphorites du Crétacé

moyen Delphino-Helvétique. *Comptes Rendus des Seances de l'Académie des Sciences, Série II*, **300**, 1025–1028.

—— 1988a. L'Evolution du Domaine Helvétiques (entre Bauges et Morcles) de l'Aptien supérieur au Turonien: Séries condensées, phosphorites et circulations océanique. *Publications du Département de Géologie et Paléontologie de l'Université de Genève*, **5**, 1–316.

—— 1988b. Relation between the Condensed Albian Deposits of the Helvetic Domain and the Oceanic Current-influenced Continental Margin of the Northern Tethys. *Bulletin du Société Géologique de France*. **8**, (IV, 5), 739–745.

FLEMMING, B. W. 1988. Pseudo-tidal Sedimentation in a Non-Tidal Shelf Environment (Southeast African Continental Margin). *In*: DE BOER, P. L., VAN GELDER, A. & NIO S. D. (eds) *Tide-Influenced Sedimentary Environments and Facies*, Reidel, Dordrecht, 167–180.

FÖLLMI, K. B. 1986. Die Garschella und Seewer Kalk Formation (Aptian-Santonian) im Vorarlberger Helvetikum und Ultrahelvetikum. *Mitteilungen des Geologischen Instituts der ETH und Universität Zürich*, **262**, 1–392.

—— 1989a. Beschreibung Neugefundener Ammonoidea aus der Vorarlberger Garschella-Formation (Aptian-Albian). *Jahrbuch der Geologischen Bundesanstalt, Vienna*, **132**(1), 105–189.

—— 1989b. Evolution of the mid-Cretaceous Triad: Platform Carbonates, Phosphatic Sediments, and Pelagic Carbonates Along the Northern Tethys Margin. *Lecture Notes in Earth Sciences*, **23**, 1–157.

—— & OUWEHAND, P. J. 1987. Garschella-Formation und Götzis-Schichten (Aptian-Coniacian): Neue Stratigraphische Daten aus dem Helvetikum der Ostschweiz und des Vorarlbergs. *Eclogae Geologicae Helveticae*, **80**, 141–191.

FROELICH, P. A., ARTHUR, M. A., BURNETT, W. C. DEAKIN, M., HENSLEY, V., JAHNKE, R., KAUL, L., KIM, K.H., ROE, K., SOUTAR, A. & VATHAKANON, C. 1988. Early Diagenesis of Organic Matter in Peru Continental Margin Sediments; phosphorite precipitation. *In*: BURNETT, W. C. & FROELICH, P. N. (eds) *The Origin of Marine Phosphate. The Results of the R.V. Robert D. Conrad Cruise 23–06 to the Peru Shelf*. Special Issue of Marine Geology. **80**, 309–343.

FÜRSICH, F. 1971. Hartgründe und Kondensation im Dogger von Calvados. *Neues Jahrbuck für Geologie und Paläontologie, Abhandlungen*, **138**, 313–342.

—— 1978. The Influence of Faunal Condensation and Mixing on the Preservation of Fossil Benthic Communities. *Lethaia*, **11**, 243–250.

GEBHARD, G. 1983. *Stratigraphische Kondensation am Beispiel Mittelkretazischer Vorkommen im Perialpinen Raum*. PhD Thesis University of Tübingen.

—— 1985. Kondensiertes Apt und Alb im Helvetikum (Allgäu und Vorarlberg), Biostratigraphie und Fauneninhalt. *In*: KOLLMANN, H. A. (ed.) *Beiträge zur Stratigraphie und Paläogeographie der Mittleren Kreide Zentral-Europas. Schriftenreihe der Erdwissenschaftlichen Kommission der Österreichische Akademie der Wissenschaften* **7**. Springer, 271–284.

GLENN, C. R. 1990. Depositional sequences of the Duwi, Sibaiya and Phosphate Formations, Egypt phosphogenesis and glauconitization in Late Cretaceous epeiric sea. *In*: NOTHOLT, A. J. G. & JARVIS, I. (eds) *Phosphorite Research and Development*. Geological Society, London, Special Publication, **52**, 205–222.

GROSS, T. F. WILLIAMS A. J. & NOWELL, A. R. M. 1988. A Deep-Sea Sediment Transport Storm. *Nature*, **331**, 518–521.

HAMILTON, D., SOMMERVILLE, J. H. & STANFORD, P. M. 1980. Bottom Currents and Shelf Sediments, Southwest of Britain. *Journal of Sedimentary Petrology*, **26**, 115–138.

HAQ, U. B., HARDENBOL, J. & VAIL, P. R. 1987. Chronology of Fluctuating Sea Levels since the Triassic. *Science*, **235**, 1156–1167.

HARLAND, W. B., COX, A. V., LLEWELLYN, P. G., PICKTON, C. A. G., SMITH, A. G. & WALTERS, R. 1982. *A Geological Time Scale*. Cambridge University Press, Cambridge.

HEGGIE, D. T., SKYRING, G. W., O'BRIEN, G. W., REIMERS, C., HERCZEG, A., MORIARTY, D. J. W., BURNETT, W. & MILNES, A. R. 1990. Organic carbon cycling and modern phosphorite formation on the East Australian continental margin: an overview. *In*: NOTHOLT, A. J. G. & JARVIS, I. (eds) *Phosphorite Research and Development*. Geological Society, London, Special Publication, **52**, 87–117.

HEIM, A. A. 1934. Stratigraphische Kondensation. *Eclogae Geologicae Helveticae*, **27**, 372–383.

—— & SEITZ, O. 1934. Die Mittlere Kreide in den Helvetischen Alpen von Rheintal und Vorarlberg und das Problem der Kondensation. *Denkschriften der Schweizerischen Naturforschenden Gesellschaft, Bern*, **69**(2), 185–310.

JENKYNS, H. C. 1971. The Genesis of Condensed Sequences in the Tethyan Jurassic, *Lethaia*, **4**, 327–352.

KENNEDY, W. J. & GARRISON, R. E. 1975. Morphology and Genesis of Nodular Phosphates in the Cenomanian Glauconite Marl of South East England. *Lethaia*, **8**, 339–360.

KIDWELL, S. M. 1989. Stratigraphic Condensation of Marine Transgressive Records: Origin of Major Shell Deposits in the Miocene of Maryland. *Journal of Geology*, **97**, 1–24.

KRAJEWSKI, K. P. 1984. Early Diagenetic Phosphate Cements in the Albian Condensed Glauconitic Limestone of the Tatra Mountains, Western Carpathians. *Sedimentology*, **31**, 443–470.

LAMBOY, M. 1990. Microbial mediation in phosphatogenesis: new data from the Cretaceous phosphatic chalks, northern France. *In*: NOTHOLT, A. J. G. & JARVIS, I. (eds) *Phosphorite Research and Development*. Geological Society, London, Special Publication, **52**, 157–167.

Lewy, Z. 1990. Pebbly phosphate and granular phosphorite (Late Cretaceous, southern Israel) and their bearing on phosphatization processes. *In*: Notholt, A. J. G. & Jarvis, I. (eds) *Phosphorite Research and Development*. Geological Society, London, Special Publication, **52**, 169–178.

Mensink, H. 1960. Beispiel für die Stratigraphische Kondensation, Schichtlücke und den Leitwert von Ammoniten aus dem Jura Spaniens. *Geologische Rundschau* **49**, 70–82.

Mikhailova, I. A. 1979. The Evolution of Aptian Ammonoids. *Paleontological Journal*, **13**, 267–274.

Mullins, H. T. & Rasch R. F. 1985. Sea-Floor Phosphorites along the Central California Continental Margin. *Economic Geology*, **80**, 696–715.

O'Brien, G. W. & Heggie, D. 1988. East Australian Continental Margin Phosphorites. *EOS, Transactions of the American Geophysical Society*, **69**, 2.

Ouwehand, P. J. 1987. *Die Garschella Formation ("Helvetischer Gault", Aptian-Cenomanian) der Churfirsten-Alvier Region (Ostschweiz). Sedimentologie, Phosphoritgenese, Stratigraphie*. PhD. Thesis ETH Zürich.

Owen, H. G. 1971. Middle Albian Stratigraphy in the Anglo-Paris Basin. *Bulletin of the British Museum of Natural History, Geological Supplement*, **8**, 1–164.

Pinet, P. R. & Popenoe, P. 1985. A Scenario of Mesozoic-Cenozoic Ocean Circulation over the Blake-Plateau and its Environs. *Geological Society of America Bulletin* **96**, 618–626.

Reimers, C. E., Kastner, M. & Garrison, R. E. 1990. The Role of Bacterial Mats in Phosphate Mineralization with Particular Reference to the Monterey Formation *In*: Burnett, W. C. & Riggs, S. R. (eds) *Genesis of Neogene to Modern Phosphorites* Cambridge University Press, Cambridge, (in press).

Robaszynski, F. & Caron, M. 1979. *Atlas de Foraminifères planctoniques du Crétacé moyen*, **142**.

Rod, E. 1946. Über ein Fossillager im Oberen Malm der Melchtaleralpen. *Eclogae Geologicae Helveticae*, **39**, 177–198.

Schaub, H. P. 1936. Geologie des Rawilgebietes. *Eclogae Geologicae Helveticae*, **29**, 89–94.

—— 1948. Über Aufarbeitung und Kondensation. *Eclogae Geologicae Helveticae* **41**, 89–94.

Shaffer, G. 1986. Phosphate Pumps and Shuttles in the Black Sea. *Nature*, **321**, 515–517.

Smith, D. B. 1988. Bypassing of Sand over Sand Waves and through a Sand Wave field in the Central Region of the Southern North Sea. *In*: De Boer, P. L., Van Gelder, A. & Nio, S. D. (eds) *Tide-Influenced Sedimentary Environments and Facies*. Reidel, Dordrecht, 39–50.

Soudry, D. & Lewy, Z. 1988. Microbially influenced Formation of Phosphate Nodules and Megafossil Moulds (Negev, Southern Israel). *Palaeogeography, Palaeoclimatology, Palaeoecology*, **64**, 15–34.

Speyer, S. E. & Brett, C. E. 1988. Taphonomic Models for Epeiric Sea Environments: Middle Paleozoic Examples. *Palaeogeography, Palaeoclimatology, Palaeoecology*, **63**, 225–262.

Stride, A. H. 1988. Preservation of Marine Sand Wave Structures. *In*: De Boer, P. L., Van Gelder, A. & Nio, S. D. (eds) *Tide-Influenced Sedimentary Environments and Facies*. Reidel, Dordrecht, 13–22.

Stumm, W. & Leckie, I. O. 1970. Phosphate exchange with sediments; its role in the productivity of surface waters. *Advances in Water Pollution Research*, **2**(III–26), 1–16.

Vercoutere, T. L., Mullins, H. T., McDougall, K. & Thompson, J. B. 1987. Sedimentation across the Central California Oxygen Minimum Zone: an Alternative Coastal Upwelling Sequence. *Journal of Sedimentary Petrology*, **57**, 709–722.

Wendt, J. 1970. Stratigraphische Kondensation in triadischen und jurassischen Cephalopodenkalke der Tethys. *Neues Jahrbuch für Geologie und Paläontologie, Monatshefte*, 433–448.

Early Ordovician shelly phosphorites of the Baltic Phosphate Basin

A. V. ILYIN[1] & H. N. HEINSALU[2]

[1] Institute of the Lithosphere, USSR Academy of Sciences, Moscow 109180, USSR
[2] Geological Institute, Estonian Academy of Sciences, Tallinn 200101, Estonia

Abstract: The Baltic Phosphate Basin contains a condensed sequence of late Upper Cambrian–early Ordovician sediments. These consist mainly of mature quartz sandstones with debris of phosphatic shells of inarticulate brachiopods, notably *Schmidtites* and *Ungula*. The sequence also contains *Dictyonema* shales and glauconitic sandstones. A large-scale exploration project completed recently has delineated several new promising areas, including the Rakvere deposit in northeastern Estonia where resources amount to as much as 700 Mt of P_2O_5 contained in easily beneficiated ores grading 10% P_2O_5. Palaeogeographically, the Baltic phosphate basin opened westwards into a large euxinic basin that occupied Sweden, southern and central Norway, and Denmark. Alum shales accumulated in the latter basin from early Middle Cambrian times. Further west, the basin joined the Iapetus Ocean.

Shelly phosphorites have been known in Estonia since around the mid-19th century (see Veiderma & Puura 1988). For example, Schmidt (1861) studied the chemistry of these phosphatic sediments, at the same time advocating their possible use as a source of mineral fertilizer. The phosphorites are exposed along the entire length of the steep erosional limestone escarpment ('glint', Estonian 'clint'), that stretches along the Baltic coast of Estonia from near Tallinn eastwards to the central part of the Leningrad area, a total distance of about 500 km (Fig. 1).

Mining of Estonian shelly phosphorites began in 1919 and most of the production has come from Maardu, about 16 km east of Tallinn, where underground mining commenced in 1940. An openpit mine came into operation in 1965. Phosphorite ore is extracted at the rate of about 500 000 t per annum and is beneficiated by washing, conditioning and flotation. The concentrate is then finely ground for direct application to the soil as fertilizer or used in the manufacture of superphosphate fertilizer. Some of the limestone removed as overburden is used in road construction. A comparable tonnage of

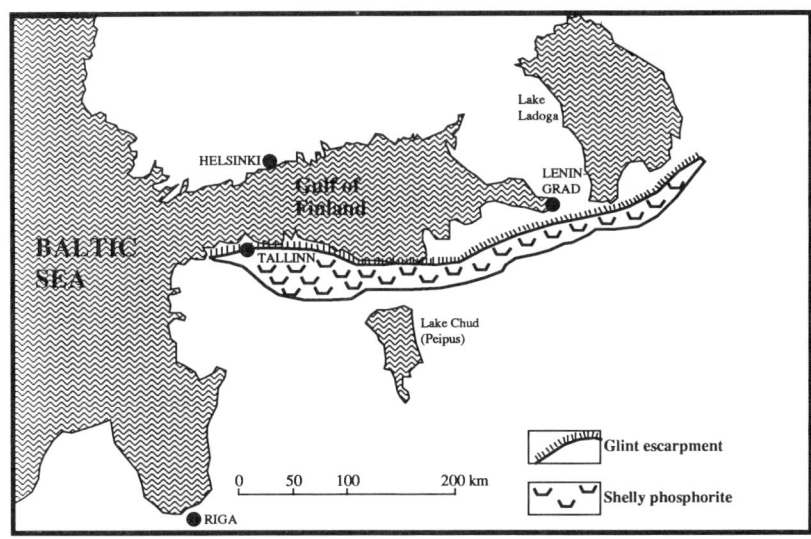

Fig. 1. Location of the Baltic Phosphate Basin.

marketable phosphate rock containing about 28% P_2O_5 is produced at the Kingisepp combine in the westernmost part of the neighbouring Leningrad district. The Kingisepp ore contains 6–7% P_2O_5.

In the late 1970s–early 1980s, a major phosphate exploration project was undertaken in the Baltic Basin which resulted in the discovery of large deposits south of the glint line where phosphate-bearing sediments become thicker and richer in shell debris. The largest of these deposits is in the Rakvere area, where about 700 Mt of P_2O_5 (Raudsep 1982; Mustjõgi, et al. 1988) has been explored by drilling with overburden thicknesses ranging from 40 m to 200 m (Mustjõgi, et al. 1988; Puura 1987; Raudsep & Khazanovich 1980). In spite of the very large resources that exist, mining in the Rakvere area, if undertaken, would have an adverse environmental impact, especially on agriculture, and on an area already relatively densely populated and economically prosperous. Consequently, the commercial development of the phosphate deposits in the Rakvere area has been postponed for the immediate future.

Structure

The major part of Estonia lies on the southern slope of the Baltic Shield, the large Precambrian structure consolidated during the Archaean. During early Palaeozoic times, the slope subsided at a very slow rate. The monoclinal disposition of the Lower Palaeozoic rocks is broken by the generally east–west trending Paide fault zone east of Tallinn, by means of which both a West Estonian monocline and East Estonian monocline can be distinguished. To the south occur the South Baltic Syneclise and the Latvian Saddle.

Stratigraphy

The Estonian Lower Palaeozoic (Fig. 2) may be divided into two parts: a Lower Vendian, Cambrian, and Lowermost Ordovician (Tremadoc) section made up mainly of terrigenous sediments and a mainly Ordovician and Silurian section that is composed predominantly of carbonate strata with fine-grained terrigenous sequences. The latest Upper Cambrian–Lower Ordovician (Tremadoc, Arenig, and Llanvirn) section is no more than 40–50 m thick and is a typical condensed sequence, lying on Lower Cambrian sandstones and siltstones with considerable hiatus (Anon. 1960; Rõõmosoks 1983); The sequence consists of four principal units (Krasilnikova & Ilyin 1989), from the base upwards as follows:

(1) a productive phosphate series up to 15 m thick, composed of quartz sands with debris of phosphatic shells and thin impersistent lense-like bodies made up of unbroken brachiopod shells;
(2) *Dictyonema* shale, an organic dark grey to black graptolitic argillite (locally, 'shale-like siltstone'), 4 m in thickness, rich in organic matter and self igniting when removed in the course of mining;
(3) green, glauconitic sandstone, 3 m thick;

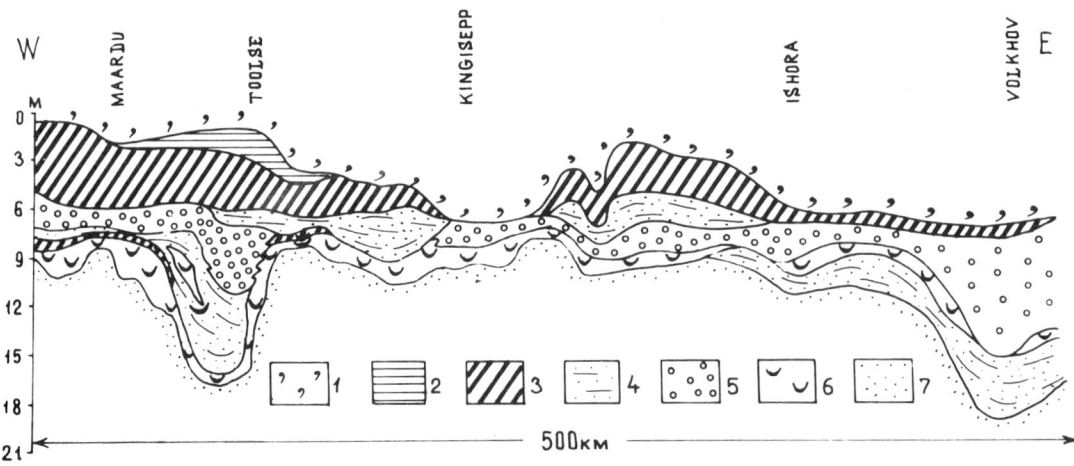

Fig. 2. A W–E cross section of Latest Cambrian–Early Ordovician sediments along the Baltic glint. 1, glauconite sands; 2, clay; 3, graptolitic (black) argillite; 4, cross-bedded quartz sandstone; 5, detrital phosphatic sandstone; 6, shelly phosphorite, composed of debris and complete shells; 7, Lower Cambrian siltstone.

(4) a light coloured, platy limestone and dolomite sequence of Lower–Middle Ordovician age, up to 20 m thick, used extensively as a building material.

The Cambrian–Ordovician biostratigraphic boundary defined by conodont and graptolite zonation is located within the productive phosphate series, the basal part of which is latest Upper Cambrian in age. However, the main part of the series is of Early Tremadoc age, comprising the Pakerort Stage, an almost horizontal sedimentary sequence (Kaljo *et al.* 1984, 1986, 1988; Arsenjev 1979; Loog & Kivimägi 1968, 1970) which has a regional dip of only 15′ to the south.

The fine-grained, mineable phosphorite bed at *Maardu* (59°30′N, 25°E), the so-called *Obolus conglomerate*, contains 9%–12% P_2O_5 and is 0.7–1.6 m, averaging, about 1.1 m, in thickness (Fig. 3). The bed lies close to the surface between similar, weakly cemented, quartz sandstones containing less than 5% P_2O_5, thin (2–3 cm) interbeds of black *Dictyonema* shale occurring only 3–4 m above the shelly phosphorites, as well as clay beds completely devoid of phosphate detritus. The phosphorite bed was first worked underground in 1940 and by openpit methods since 1965; remaining reserves are limited, perhaps sufficient for only 1–2 years' exploitation. The principal deposits in the large Rakvere field contain mineable thicknesses of 5–6 m at depths of up to nearly 200 m (Petrosyants *et al.* 1981, see Table 1).

The phosphorite deposits at *Kingisepp* (59°10′N, 29°E), approximately 112 km southwest of Leningrad, are essentially a continuation

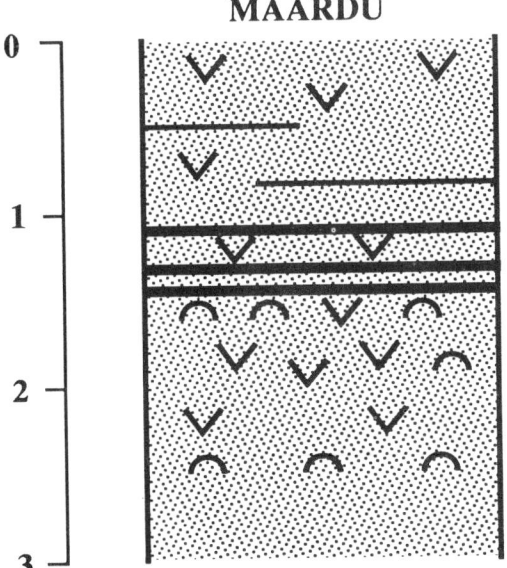

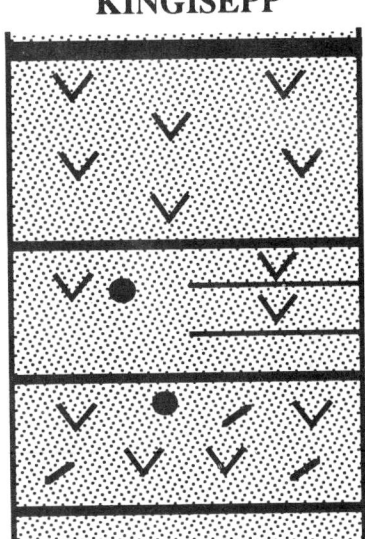

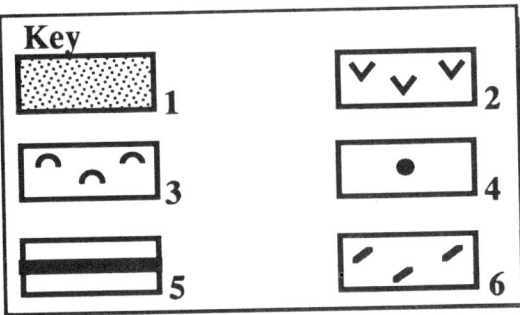

Fig. 3. Generalized sections of the productive series at Maardu and Kingisepp. 1, quartz sandstone; 2, phosphatic shell detritus; 3, unbroken shells; 4, phosphorite concretions; 5, *Dictyonema* shale; 6, secondary dolomitization.

Table 1. *Geological parameters of the Rakvere deposits, Estonia.*

Deposit	1	2	3	4
Assamala (Assamalla)	5.0	9.6	2.72	110–195
Ryagavere (Rägavere)	5.7	9.4	2.46	50–133
Cabala (Kabala)	5.2	13.4	4.02	42–107

1 Average thickness (m) of mineable phosphorite bed.
2 Average grade (% P_2O_5).
3 Yield of productive unit (t P_2O_5 m^{-2}).
4 Thickness of overburden (m).

of those occurring extensively along the Estonian coast. The productive bed is from 1.5 m to 3.0 m in thickness, averaging 7% P_2O_5 (Fig. 3). The deposits lie in a marshy area and the productive phosphate series is consequently saturated with water. Secondary dolomite, arising from surface waters percolating through the overlying Upper–Middle Ordovician dolomites is common and has, unfortunately, produced a hard dolomite which has to be removed as waste by selective mining.

Petrography

The phosphorites of the Baltic Phosphate Basin are essentially light-coloured, quartz sandstones with a detritus of phosphatic shells of inarticulate brachiopods, notably *Schmidtites* and *Ungula* (Heinsalu *et al.* 1987). The sandstones are usually weakly cemented, although sometimes rather hard where cemented by secondary dolomite. They are probably beach sediments, with crossbedding reflected in colour and grain-size differences and also in shell fragment distribution. Small-scale, randomly orientated crossbedding, with individual sets 20–30 cm thick, are distinctly visible. Frequently, there occur impersistent lens-like masses 0.1–0.3 m thick and traceable for tens of metres that consists of debris and unbroken shells 0.7–1.0 cm in size and only a few millimetres thick. Such thin-walled shells are para-autochthonous (Raudsep & Khazanovich 1980) and are usually found in the lower part of the productive phosphate series. Upwards in the sedimentary sequence, the occurrence of disintegrated detritus of thick-walled phosphatic shells reflects a transition from low-energy to high-energy depositional environments.

The quartz grains are mature, clean and devoid of any clay admixture, with some 50% of the grains being 0.25–0.5 mm in size. Most are brightly polished, and some grains have perhaps been wind-blown and redeposited, reflecting periodic elevation of the original sand bars above sealevel. Accessory minerals include ilmenite, zircon, rutile, tourmaline, and garnet (Heinsalu & Vijding 1978). In this section, the phosphatic shells are shown to be composed of thin, uniformly orientated plates that consist of a mixture of (1) microcrystalline, light coloured or colourless apatite, and (2) a highly dispersed, dark-coloured variety of apatite. Some samples show a deficiency in F and generally there is a relatively low CO_2 content (Table 2). Minor elements found in the phosphatic shells correspond to the Clarke value, in contrast to the *Dictyonema* shales which are notable for their higher content of Cu, Ni, Ag, Ga, and V. Individual crystals, concretions and thinly dispersed inclusions of pyrite are confined to oil shale interlayers in the upper part of the productive phosphate series. Phosphate also occurs filling fine fractures in brachiopod shells, as a thin coating around quartz, and as a phosphatic cement. These varieties are most probably of secondary origin resulting from dissolution of phosphatic shells. Small amounts of dolomite, calcite, gypsum, feldspar, glauconite, limonite and finely scattered pyrites also occur, the oxidation of the last named mineral imparting a black colour to the organic debris. The mode of occurrence of carbonate minerals in Estonian phosphorites has been discussed by Oja & Pirrus (1986).

Palaeogeography

The palaeogeography of the Baltic Phosphate Basin has been established with considerable

Table 2. *Chemical composition of phosphatic shells, Pakerort Stage, Maardu and Kingisepp areas (Bliskovsky 1983).*

	Maardu	Kingisepp
	wt %	
P_2O_5	36.8	35.7
CaO	49.8	48.9
F	3.3	3.2
CO_2	3.3	2.6
MgO	0.4	0.7
Fe_2O_3	1.5	2.4
Al_2O_3	0.4	0.3
K_2O+Na_2O	0.3	0.2
SO_3	0.5	0.8
SiO_2	0.6	0.8
Total	96.9	95.6

certainty. It was a shallow-water, marine bay which opened broadly to the west, with the shoreline extending in a general east–west direction (Heinsalu 1986, see Fig. 4). Facies change from north to south from mostly coarse clastics to sandy-clayey sediments. Brachiopods thrived along the northern side of the basin, i.e. on the southern slope of the Baltic Shield which subsided very slowly, as is evident from the condensed nature of deposition and the biochemical nature of the Early Palaeozoic sediments. Individual beds or units of strata are persistent, such as a bed of glauconite sand 2–3 m thick, which can be traced without lithological variation for several hundred kilometres. In longitudinal section, several sedimentational features can be discerned. Both *Dictyonema* shale (organic dark grey to black graptolitic argillite) and the phosphorite beds gradually thin eastwards and wedge out, and in the southeastern part of the Leningrad area, the phosphorites are not accompanied by these black shales.

To the south, in the central part of the Baltic region and eastern Poland, the basin became deeper and quartz sands accumulated there with only rare phosphate detritus. However, a thin horizon of phosphatic 'Obolus conglomerate' is always present at the base of the Late Cambrian–Early Ordovician sequence. A landmass was sited further south marking the southern shore of the Baltic Phosphate Basin, and a narrow, flat, latitudinal strip of land probably existed immediately south of the Rakvere area because of the absence of Pakerort sediments in the area. Eastwards, the Baltic Basin joined the Moscow Basin where much thicker terrigenous sediments accumulated. To the west, the basin opened to the Alum Shale euxinic basin that covered parts of Sweden, Norway and Denmark and, in turn, was connected to the Iapetus Ocean to the west. Alum Shale started to accumulate throughout Scandinavia in early Middle Cambrian time. Phosphogenesis started in later Upper Cambrian–early Ordovician time in the eastern part of Sweden (Öland, Lake Siljan area) and was most pronounced in the Baltic-Ladoga glint area where sedimentation of black shales during the Cambrian did not occur. Yazmir (1989) has suggested that at a period of progressively increasing biomass and accumulation of black shale phosphorus was a deficient element, with extensive phosphogenesis, taking place only

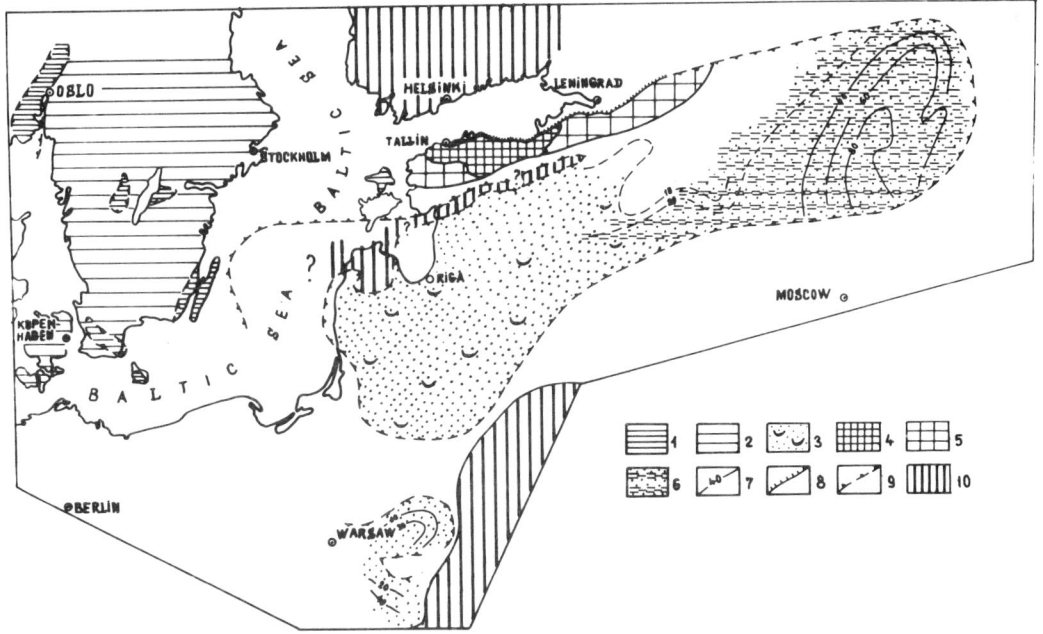

Fig. 4. Palaeogeography of the Pakerort Stage (Early Ordovician). 1, black shale (exposed); 2, black shale (buried or eroded); 3, sands with scattered phosphatic shells; 4, quartz sand rich in phosphatic shells; 5, quartz sand with some phosphatic shells; 6, sandstone and siltstone; 7, isopach (m); 8, glint line; 9, contour of probable emergent land; 10, land.

at a time of declining black shale formation. According to Karpova *et al.* (1986), brachiopods colonized a submerged plateau located close to, but isolated from, the shore by a lagoon in which black shales were forming. The latitudinally elongated phosphorite basin was about 500 km long and 100 km wide. Brachiopod banks scattered on the plateau surface were eroded during regressions and debris accumulated in depressions. Second-order transgressions caused deepening of the sea and the formation of black (*Dictyonema*) shales in areas where phosphorites had been forming previously. Several transgressive-regressive cycles are represented in the upper part of the productive phosphate series (Karpova *et al.* 1986). The Upper Palaeozoic transgression appears to have developed from west to east across the northwestern part of the East European craton. It came to the Baltic Phosphate Basin at the end of Upper Cambrian–Early Ordovician times, while a euxinic basin existed in the west from the beginning of the Middle Cambrian.

Conclusions

The peculiar lithological assemblage contained in the Baltic Phosphate Basin and the absence of carbonate sediments points to a temperate climate and the location of the basin outside low latitudes. The surrounding land areas were most probably peneplaned and supplied only a small amount of terrigenous material.

The Baltic Phosphate Basin is the best developed example of Ordovician phosphogenesis, notably in terms of the very large resources of phosphorite that it contains. Ordovician phosphogenesis is of global extent, however, Shelly phosphorites of Middle Ordovician age occur in several areas along the western margin of the Siberian craton, and Ordovician phosphorites are developed on southeastern Yugoslavia. Early Ordovician glauconitic–phosphatic shale occurs above alum shale in southern Sweden, and there are Upper Ordovician nodule beds in Wales (United Kingdom). In northwestern Argentina, phosphatic sandstones with debris of obolid and lingulid shells are found in the Lower and Upper Tremadoc stages. Thus, the development of Ordovician biogenic phosphogenesis, although global in extent, is not everywhere synchronous, a feature also of Vendian–Early Cambrian phosphogenesis. Compared with Vendian–Early Cambrian phosphorite sequences, there is a marked absence of carbonate sediments, however, attesting to the higher latitudinal position of Ordovician shelly phosphorites.

References

ANON. 1960. *Geology of the USSR.* Vol. 28, *Estonian SSR*. Geological Review and Mineral Resources. 'Nedra', Moscow. (In Russian.)

ARSENJEV, A. A. 1979. *Ordovician Phosphorite-bearing Sediments of the Baltic Sea Area.* 'Nedra', Moscow. (In Russian.)

BLISKOVSKY, V. Z. 1983. *Composition and Beneficiation of Phosphate Ores.* 'Nedra', Moscow. (In Russian.)

HEINSALU, H. N. 1986. Lithofacial zonality of Early Tremadocian sedimentation on the East European platform. *Eesti NSV Teaduste Akadeemia Toimetised, Geoloogia*, **35**, 3, 115–121. (In Russian.)

—— & VIJDING, H. A. 1978. Mineral composition of the Lower Ordovician sediments in Northern Estonia. *Eesti NSV Teaduste Akadeemia Toimetised, Geoloogia*, **27**, 2, 46–52. (In Russian.)

——, VIIRA, V., MENS, K., OJA, T. & PUURA, I. 1987. The section of the Cambrian–Ordovician boundary beds in Ulgase, northern Estonia. *Eesti NSV Teaduste Akadeemia Toimetised, Geoloogia*, **36**, 4, 154–165. (In Russian.)

KALJO, D., MUSTJÕGI, E. & ZEKCER, I. (eds) 1984. *Guidebook for Excursions 027 and 028.* International Geological Congress, Moscow. Tallinn, Academy of Sciences of the Estonian SSR.

——, BOROVKO, N., HEINSALU, H., KHAZANOVICH, K., MENS, K., POPOV, L., SERGEJEVA, S., SOBOLEVSKAYA, R. & VIIRA, V. 1986. The Cambrian–Ordovician boundary in the Baltic–Ladoga clint area (North Estonia and Leningrad region, USSR). *Eesti NSV Teaduste Akadeemia Toimetised, Geoloogia*, **35**, 3, 97–108.

——, HEINSALU, H., MENS, K., PUURA, I. & VIIRA, V. 1988. Cambrian–Ordovician boundary beds at Tonismagi, Tallinn, North Estonia. *Geological Magazine*, **125**, 4, 457–463.

KARPOVA, M. I., MIKHAILOV, A. S., MUSTJÕGI, E. & RAUDSEP, R. V. 1986. Lithological-facies model of shelly phosphorite deposits. *Razvedka i Okhrana Nedr*, **9**, 10–15. (In Russian.)

KRASILNIKOVA, N. A. & ILYIN, A. V. 1989. The Ordovician Baltic phosphorite basin, USSR. *In*: NOTHOLT, A. J. G., SHELDON, R. P. & DAVIDSON, D. F. (eds) *Phosphate Deposits of the World, Vol. 2, Phosphate Rock Resources.* Cambridge University Press, Cambridge, 494–496.

LOOG, A. & KIVIMÄGI. E. 1968. On the lithostratigraphy of the Pakerort Stage in Estonia. *Eesti NSV Teaduste Akadeemia Toimetised, Keemia, Geoloogia*, **17**, 4, 433–435. (In Russian.)

—— & ——, E. 1970. On the distribution of Estonian obolid phosphorites. *Eesti NSV Teaduste Akadeemia Toimetised, Keemia, Geoloogia*, **19**, 1, 57–61 (In Russian.)

MUSTJÕGI, E. A., DYACHENKO, V. J. & ZAGURAYEV, V. G. 1988. Guide to the Geological Excursion to Phosphorite Deposits. *In: Problems of Phosphorite Geology.* Abstracts and Guidebook of VI All-Union Conference, Tallinn, 167–175. (In Russian.)

OJA, T. & PIRRUS, E. 1986. Carbonate mineral occurrences in the phosphorites deposits of Estonia. *Eesti NSV Teaduste Akadeemia, Geoloogia*, **35**, 122–130. (In Russian.)

PETROSYANTS, E., ZAGURAYEV, V. & MUSTJÕGI, E. 1981. The state and prospects of industrial exploitation of the Rakvere phosphorite field. *Eesti NSV Teaduste Akadeemia, Keemia, Geoloogia*, **30**, 2, 79–82. (In Russian.)

PUURA, V. (ed.) 1987. *Geology and Mineral Resources of the Rakvere Phosphorite-bearing Area*. Academy of Sciences of the Estonian SSR, Tallinn. (In Russian.)

RAUDSEP, R. 1982. Estonian phosphorite and its recently discovered deposits. *Eesti Loodus*, **8**, 517–523. (In Estonian.)

—— & KHAZANOVICH, K. 1980. Lithological classification of the shelly phosphorites of the Baltic Basin. *Eesti NSV Teaduste Akadeemia, Geoloogia*, **29**, 4, 147–153. (In Russian.)

RÕÕMUSOKS, A. 1983. *Geology of Estonian Bedrock*. Academy of Sciences of the Estonian SSR, Tallinn. (In Estonian.)

SCHMIDT, F. B. 1861. Agrikultur-chemische Untersuchungen. *Livländiche Jahrbücher der Landwirtschaft*, 169.

VEIDERMA, M. A. & PUURA, V. A. 1988. *Phosphorite Deposits*. VINITI Publishers, Moscow (In Russian.)

YAZMIR, M. M. 1989. *Mechanism of global cyclicity of phosphogenesis*. Obzory VIEMS Mingeo SSSR, Moscow. (In Russian.)

The distribution of phosphatic facies in the Georgina, Wiso and Daly River Basins, Northern Australia

PETER F. HOWARD

School of Earth Sciences, Macquarie University, North Ryde, NSW 2109, Australia

Abstract: The lithological logging and chemical testing of cuttings from some 600 water bores and Bureau of Mineral Resources stratigraphic holes, together with the modelling of aeromagnetic, gravity and elevation data, have defined a carbonate−siltstone−chert phosphatic lithofacies of Middle Cambrian age. This in turn has led to the recognition of two new phosphorite deposits, the deposition of which was related to the basement configuration and its depth. The phosphatic lithofacies, which have been followed entirely in the subsurface, occur as belts peripheral to and within the Georgina, Wiso and Daly River Basins. These phosphatic belts have an average width of 32 km, a thickness of 10 to 190 m, and have been traced over a distance of 2,100 km. Of the newly discovered phosphorite deposits, the Lady Judith in the Wiso Basin rests on volcanic rocks and interdigitates with carbonates of the Montejinni Limestone, while the Ammaroo belt in the southwestern portion of the Georgina Basin is contained within a depression bounded by limestones of the Arthur Creek Formation. The phosphatic sediments are believed to have been deposited primarily as an Ordian Middle Cambrian event in the west with a 'younging' transition through Ordian and/or Early Templetonian to Late Templetonian in the southeast. The basins are extensional, exhibiting a series of broad downwarps crossed by peripheral aulacogens, grabens, half grabens formed in the Late Proterozoic and modified subsequently by the development of plateaux, narrow horst blocks and adjacent deeps during the Middle Cambrian along basin-dividing arches. The basement to the shallow-water phosphatic lithofacies consists of Proterozoic sediment or Early Cambrian volcanic plateaux or peripheral sloping platforms which in the Brunette Sub-basin have present elevations of 0−300 m ASL.

The early Middle Cambrian phosphorite deposits of the Georgina Basin discovered in 1966−68 comprise 18 named deposits occurring over a distance of approximately 1000 km along the palaeo-periphery and palaeo-insular portions of the basin. Exploration ceased because reserves were more than adequate for the long term. Reserves have been placed at 3400 M tonnes (Howard 1986) or 2800 M tonnes using more conservative specifications (Notholt & Sheldon 1986).

These discoveries were the stimulus for a broad range of studies of both the deposits and the adjacent terrains by universities, the Bureau of Mineral Resources, the Northern Territory Geological Survey, and exploration companies. Most of the studies related to the eastern edge of the basin. The only attempt at tracing the phosphate facies on a regional scale (Howard & Hough 1979; Howard 1986) involved mapping a belt from Sherrin Creek to Wonarah and into the Brunette Sub-basin. A detailed study of a portion of the belt on the eastern side of the basin has been undertaken by Southgate (1988). There were reasons for the lack of additional phosphate exploration and research further afield in the Georgina and the contiguous Wiso and Daly River Basins. These were (1) the extent of the basins, aggregating 600 000 km^2; (2) poor outcrop, often lateritized or silicified, in a largely flat soil or sand-covered arid terrain except along the uplifted basin edges; and (3) the paucity of drill holes and, as such, the lack of subsurface data. For example, only about 50 holes (excluding those on the fringes of the basin) reach basement, a hole density of one per 1200 km^2.

With the availability of new drill hole data, it was possible in this study to extend the mapping of the phosphatic lithofacies in the subsurface through a large portion of the three basins.

Method of study

Prior to 1979 few cores or cuttinngs from water bore or exploration programmes were retained by the Northern Territory Administration. Mandatory retention of cuttings from all holes drilled on cattle stations in the 1970s provided a cuttings library of several hundred holes and these, together with the cuttings from exploration holes drilled by IMC Development

Corporation in the northeastern and central portions of the Georgina Basin, provided a data base of some 600 holes for this study. The water bore data were supplemented by cores and cuttings of some 45 Bureau of Mineral Resources and Northern Territory Geological Survey stratigraphic holes and oil exploration wells, and while these holes contributed additional data on phosphatic sequences, only about 14 were effective for stratigraphic control.

Lithological logging of the cuttings was rapid and designed to record the principal lithological types of Ordian and Templetonian sediments. The cuttings were simultaneously tested with ammonium molybdate reagent to determine their phosphatic or non-phosphatic nature. Samples yielding positive reactions for phosphate were analysed for P_2O_5 content. The phosphate sediments mapped in this study have above-average geochemical abundance and, in general, fall in the range of $0.2-5\%$ P_2O_5. The basic lithological information from these logs was collated as simple columnar logs and plotted on a series of 1:250 000 data sheets.

To assist in the interpretation of lithofacies distribution, Bureau of Mineral Resources aeromagnetic, gravity and elevation data were modelled to assist in contouring the Cambrian basement configuration and gain an understanding of the location and strike of the subsurface fault systems.

Tectonic setting

The Early Palaeozoic basins in Northern Australia developed near the close of the Proterozoic, and their tectonic style can be directly related to their basements (Plumb et al. 1981). Those which lie over or within the Northern Australian Craton, such as the Georgina, Wiso and Daly River Basins, have much thinner fill and are relatively undeformed, exhibiting essentially flat lying sedimentary sequences. Palaeogeography and the nature of the contained sediments suggest that the three basins were formed by mild Cambrian structural downwarps following the Late Proterozoic Adelaidean depressions along the southeastern margin (Fig. 1) (Plumb et al. 1981). Second order pre-Templetonian (Middle Cambrian) upwarping produced structural ridges which formed the division between the Georgina, Wiso and Daly River Basins. Three distinct depositional regimes formed and exhibit divergent sedimentary history and phosphorites of markedly different character (Howard 1986). Firstly, the Smokey Anticline structural rise (Fig. 1) separates the southern open-marine

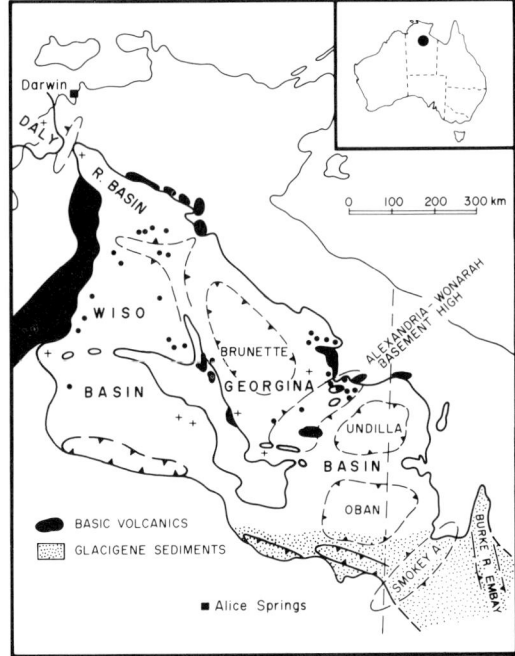

Fig. 1. General outline of the Georgina, Wiso and Daly River Basins showing the zones of upwarping and the related sub-basins, aulacogens along the southern edge, Adelaidean glacigene sediments (hatched), and Early Cambrian flood basalts. Drill hole intersections of basalt shown by bold dots and Precambrian by crosses.

shelf region and its component Burke River Embayment from the central epicontinental Undilla and Oban sub-basins and, secondly, the Alexandria–Wonarah Basement High further isolated the Brunette Sub-basin which is dominated by an extensive central area of sediments of both continental and marine character similar to that of the Wiso Basin.

This basement ridge was tectonically active and rising during Middle Cambrian sedimentation, as evidenced by drape structures and concomitant sediment thinning. Uplifted blocks are bounded by reactivated Early Proterozoic faults which have been periodically reactivated from pre-Templetonian times to Recent, as evidenced by structural lineaments in Mesozoic and Tertiary sediments and in present day by soil-type variations and surface drainage patterns (Howard 1989a).

However, several features suggest extensional events were important in the formation of the basins apart from simple downwarping, namely, structural data, distribution of flood basalts and tectonic-subsidence curves. Diastrophism which occurred in the mid-Adelaidean coincided with the development of narrow deep northwesterly-trending graben or aulacogens which incised into the southern edge of the Georgina Basin and became the sites for Adelaidean cover consisting of glacigene sediments which uncomfortably underlie marine Cambrian sediments (Plumb et al. 1981; Shergold & Druce 1980). Modelling of aeromagnetic data in the Brunette Sub-basin and the northwest Undilla Sub-basin (discussed later) and the distribution of phosphatic facies related to basement platforms and peripheral half graben in that region suggest that the sloping surfaces of half graben are important but largely unrecognized extensional structures flanking the basin edges which controlled water depth and phosphogenesis. Additional evidence of such extensional events is the extensive subsurface distribution of extrusive basic rocks at the base of the Cambrian sediments in the Wiso Basin and Brunette Sub-basin, the extent of which was not previously known (Fig. 1). These flood basalts include the Antrim Plateau, Helen Springs and Nutwood Volcanics in the Wiso Basin, and the Helen Springs, Peaker Piker and Colless Volcanics in the Georgina Basin. The holes shown on Fig. 1 represent all holes which reached basement and it is evident that the majority bottom in volcanics. The incidence of the volcanic rocks does not appear to extend far beyond the Brunette Sub-basin.

The study of tectonic-subsidence curves indicates that two major extensional events occurred, one in the Late Proterozoic and the other in the Early Palaeozoic. Dating shows that the volcanism and formation of the Wiso Basin and Brunette Sub-basin are related to a 600 Ma event, whereas the development of the southeastern portion of the Georgina Basin dated back to a 900 Ma event. Palaeomagnetic data support the interpretation that the two events are simple extensions with angular integrity between terrains, implying that in the second event structures formed in the first were reactivated (Lindsay et al. 1987).

The southeastern limit of overlap of the younger system is approximately along the axis of the Alexandria–Wonarah Basement High, an area of structural complexity where Middle Cambrian uplift has isolated a deeper portion of the Georgina Basin stage 1 (900 Ma) into asymmetrically deeper portions of the Brunette and Undilla sub-basins which abut the basement high (C and D, Fig. 2).

Stratigraphy

The stratigraphic succession of the Georgina Basin is characterized by complex facies variations which have resulted in the naming of 70 Upper Proterozoic and Cambro-Ordovician stratigraphic units. Shergold & Druce (1980) summarized their inter-relationships, thicknesses and temporal–spatial relationships. The Cambrian and Ordovician sediments are thicker in the structures along the southern margin of the basin and in the Burke River Embayment where they are 3000 and 2000 m thick, respectively. Elsewhere in the basin the sediments are mostly Middle Cambrian and less than 450 m thick (Smith 1972). Modelling of magnetic data, however, suggests that deeps either side of the Alexandria–Wonarah Basement High (C and D, Fig. 2) could contain 2500 to 3000 m of sediments (modelling of magnetic data, this study).

In the Wiso and Daly River Basins where there is less complexity and a correspondingly lesser number of named stratigraphic units, the stratigraphy has been summarized by Kennewell & Huleatt (1980), Brown (1968) and Kruse (1990). The sediments in general are in the order of less than 100 m to a maximum of a few hundred metres, except in the trough on the southern margin of the Wiso Basin where a seismic survey indicates at least 1000 m.

Palaeogeographic maps of the Georgina Basin were first prepared by Smith (1972). Subsequently, Howard (1972, 1976), Howard & Hough (1979) and Cook (1976) produced palaeogeographic maps which included facies relationships of the sediments in those portions of the basin containing known phosphate deposits or phosphatic sediments. The most comprehensive maps are those of Shergold & Druce (1980) which are arranged by time series and faunal zones from the Vendian to Early Ordovician and show the distribution of formations by their dominant facies. Most recently, Cook (1982, 1988) has constructed Cambrian palaeogeographic maps of Australia on the basis of six sedimentological time intervals which portray the sedimentation and depositional environments.

Regional distribution of phosphatic sediments

The 600 holes used in this study (Fig. 3) define a belt of phosphatic sediments which are peripheral to the Georgina, Wiso and Daly River Basins. For the most part, the belt lies entirely

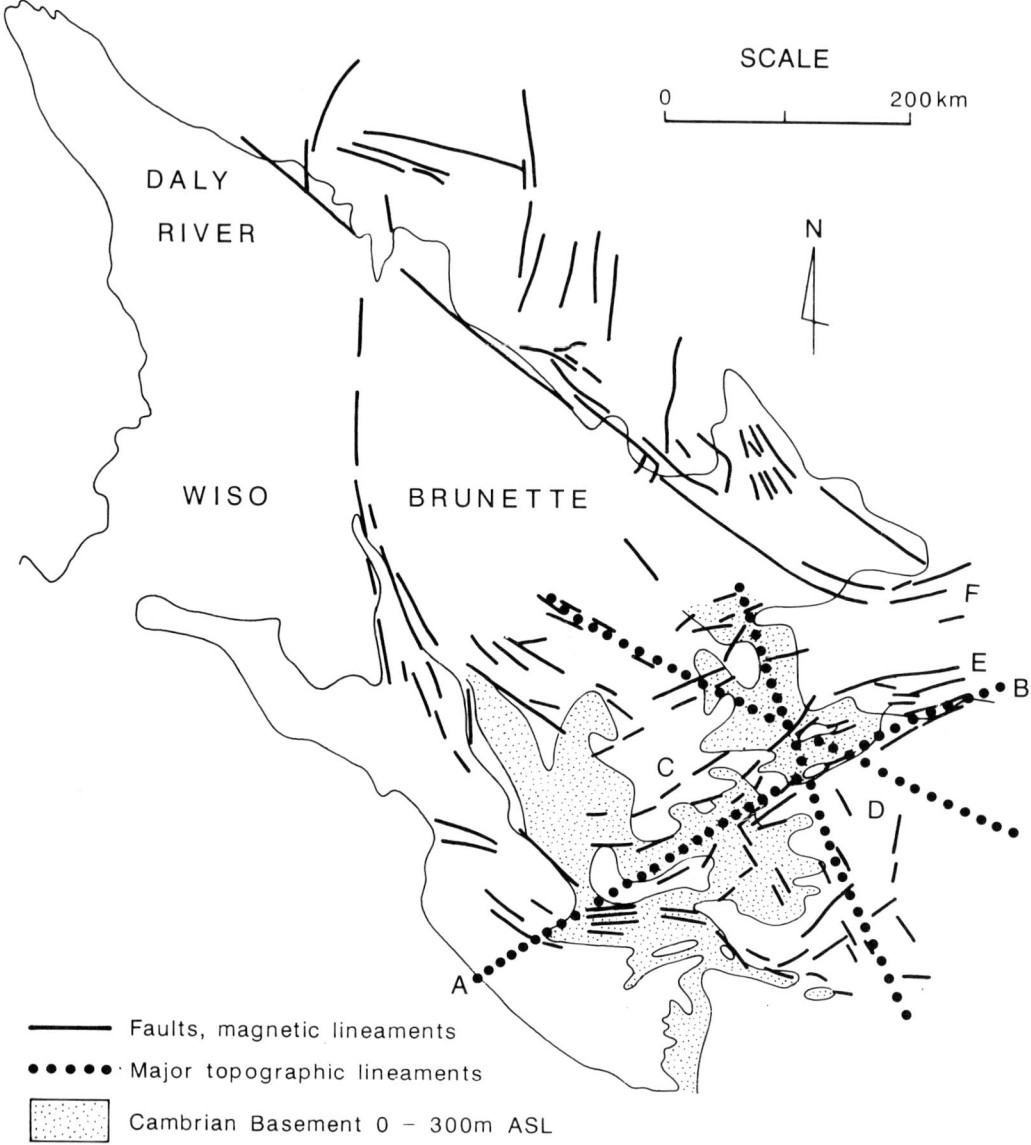

Fig. 2. Structures within the Brunette Sub-basin showing the Cambrian basement configuration in the range of 0–300 m above sea level. AB approximates to the Alexandria–Wonarah Basement High, E to the Mitchiebo Fault and related faults, and F to the Murphy Tectonic Ridge. C and D are east–west trending deep portions of the Brunette and Undilla Sub-basins.

within the basin outlines except where truncated by steep bounding faults of post-sediment age: such faulting is a reactivation of earlier Proterozoic systems. The defined length of the belt is approximately 2100 km and probably exists in the soil-covered intervening gaps in the Wiso and Daly River Basins. The southeastern portion of the Georgina Basin has not been studied by the author but the belt would certainly extend from Sherrin Creek through the Quita Creek–Ardmore area to Duchess, a further 300 km. The phosphatic belt varies from 20 to 50 km in width, averaging 32 km, and the phosphatic facies has a range of thickness of 10–190 m and averages 50 m. Everywhere the facies is essentially flat lying and undeformed.

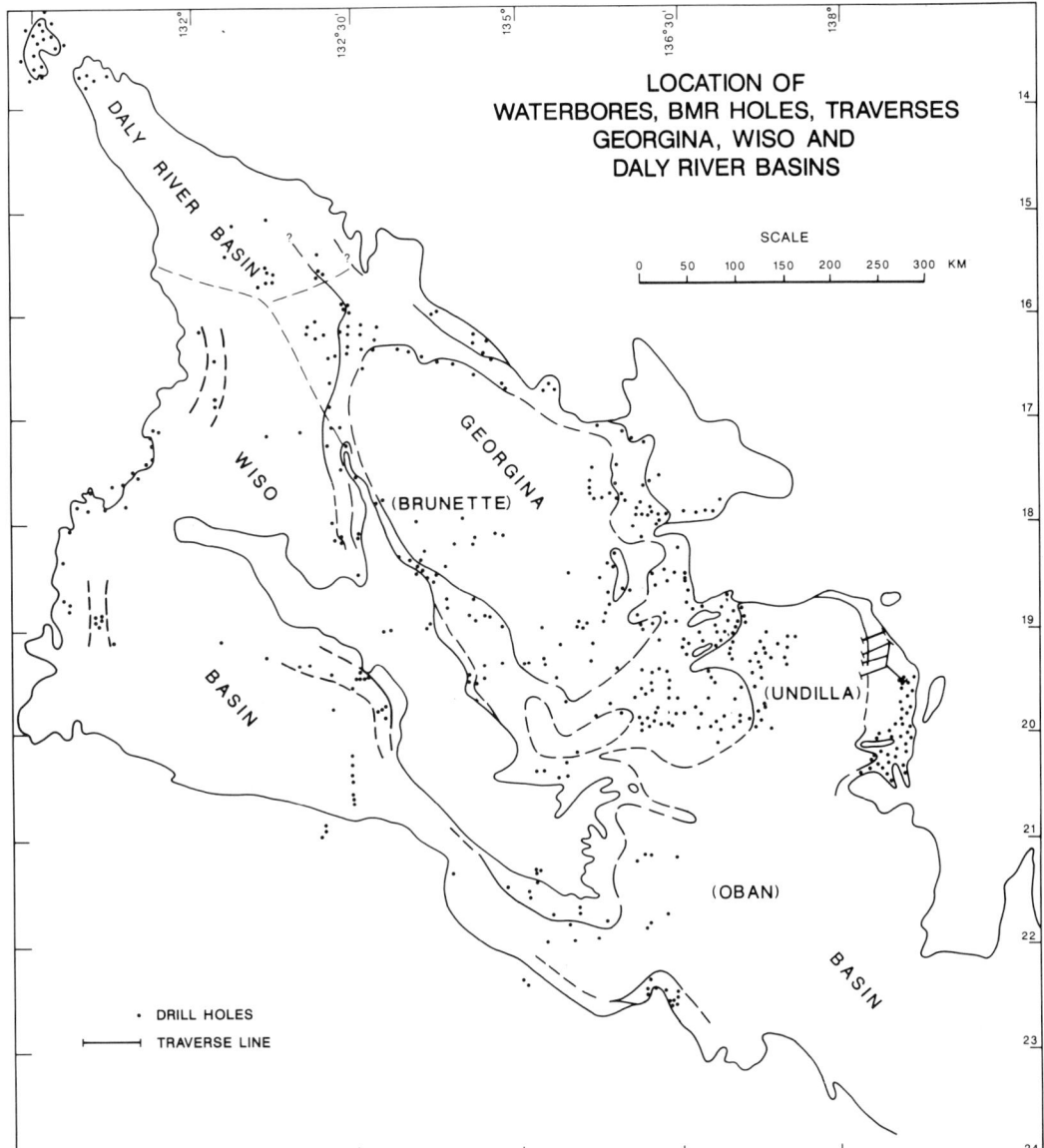

Fig. 3. The location of drill holes within the Georgina, Wiso and Daly River Basins logged in this study.

The nature of the sediments and degree of phosphogenesis

The phosphatic sediments include the dominantly siltstone–chert, mudstone, phosphorite lithofacies of six formations in the Georgina Basin (Beetle Creek Formation, Border Water Hole Formation, Gum Ridge Formation, Burton Beds, Wonarah Beds, and Arthur Creek Formation), one in the Wiso Basin (Montejinni Limestone) and one in the Daly River Basin (Tindall Limestone). The phosphatic sediments include also the laterally equivalent limestone–dolomite–chert lithofacies of the above formations and the Thorntonia Limestone, the Top Springs Limestone and the Chabalowe Formation.

The concentration of phosphate is greatest in the sediments of the Burke River Embayment, the Undilla and Oban Sub-basins and across the Alexandria–Wonarah Basement High. Phosphogenesis remains strong along the western margin of the Brunette Sub-basin but is very weak along its northern flank in the Daly River Basin and the northern part of the Wiso Basin. Too little information is available in the southern part of the Wiso Basin to judge the phosphate distribution. The high-grade section present at the Lady Judith deposit is associated with only weakly phosphatic siltstone, and the belt on the eastern side is either very weak or is very ferruginous and/or lateritized. Corresponding with the weakening of phosphogenesis along a trend from the southeast of the Georgina Basin along the northern flank of the Brunette Sub-basin and into the northern Wiso and Daly River Basins, the composition of phosphatic sediments changes to a dominantly carbonate lithofacies with patchy phosphatic zones. This does not necessarily imply a decreased likelihood of phosphate deposit occurrence as evidenced by the discovery of the high-grade Lady Judith deposit.

Towards the palaeoshore, the phosphatic sediments grade to non-phosphatic equivalents often without noticeable lithological change, though sandstones do occur. Basinward, no lithological change is apparent except in the depositional centre of the basins where phosphatic sediments interdigitate with and are overlain by Camooweal Dolomite, the Anthony Lagoon Beds, and the Hooker Creek Formation-Lothari Hill Sandstone (Fig. 4).

Basinal deposition model

Kennewell & Huleatt (1980) proposed a model to explain the distribution of the continental regressive lithofacies in relation to their marine counterparts in the Wiso Basin. In this interpretation (Fig. 5), time lines represent depositional surfaces at points in time, and regressions illustrated by a series of sea levels. This model fits the data acquired in this study, not only in respect of the Wiso Basin but also in the Brunette and Undilla Sub-basins where the continental facies are confined to the depositional centres with only partial overlap of the marine and phosphatic facies.

Dating and correlation

Dating of the phosphatic sediments and their contained phosphorite deposits in the eastern Georgina Basin has proved difficult in detail. The initial study of the regional distribution of phosphatic and laterally equivalent non-phosphatic lithofacies of the eastern and central portions of the Georgina Basin by Howard & Hough (1979) and Howard (1986) was based on the dating of A.A. Öpik in the 1950s (Öpik & Pritchard 1960) which placed the occurrences as a Templetonian Middle Cambrian event. However, recent studies by Shergold & Southgate (1988) assign an Ordian and/or earliest Templetonian age. Higher but non ore-bearing zones exist in the Undilla Sub-basin (Southgate 1986) and at Duchess above the Ordian/Templetonian, but are not considered in this study. The uppermost portion of the column at the Wonarah Deposit (Howard 1989a) yields a faunal assemblage from the weakly phosphatic siltstone and sandstone above the phosphorites which are of Early Templetonian age (Öpik 1975) and therefore a similar range of Ordian Early Templetonian is inferred for D Tree and Wonarah. On the western flank of the Brunette Sub-basin near Helen Springs, low-grade phosphorites are contained in the siltstone-chert facies of the Gum Ridge Formation of Ordian age (Shergold in Shergold & Palmer 1985), while on the south-western portion of the Georgina Basin the Ammaroo phosphorite belt which occurs in the organic-rich lithofacies of the Arthur Creek Formation (Morris 1986) is dated as Ordian/Undillan (Shergold in Shergold & Palmer 1985). In the latter case there has been little formal resolution as yet, and stratigraphic breaks can be predicted (Shergold, pers. comm. 1989).

In the Wiso Basin, phosphogenesis is confined to the Montejinni Limestone dated as Ordian age (Shergold & Palmer 1985). The Lady Judith phosphorite occurs in the middle clastic Unit 2 as defined by Kennewell & Huleatt (1980). The correlation with sediments in the Daly River Basin is afforded by the Bureau of Mineral Resources holes Larrimah 1 and 2 where the Montejinni Limestone Unit 1 in Larrimah 2 has lithologies and faunal content identical to the Tindall Limestone in Larrimah 1. Thus, Ordian phosphorites at Lady Judith (Unit 2) correspond to the siltstone in the mid-portion of the Daly River Basin (Fig. 6a–d).

The 600 lithological columns representing these time intervals show striking similarity basin-wide, as illustrated by the 14 selected sections of Fig. 7. In the Georgina Basin, there are four distinct lithological variations (Fig. 6e–h) which include sections that are wholly carbonate or wholly siltstone–chert. Generally, however, siltstone–chert overlies carbonate (Fig. 6g) and where phosphorite deposits occur, limestone and dolomitic limestone interdigitate with the siltstone–chert–phosphorite facies (Fig. 6h).

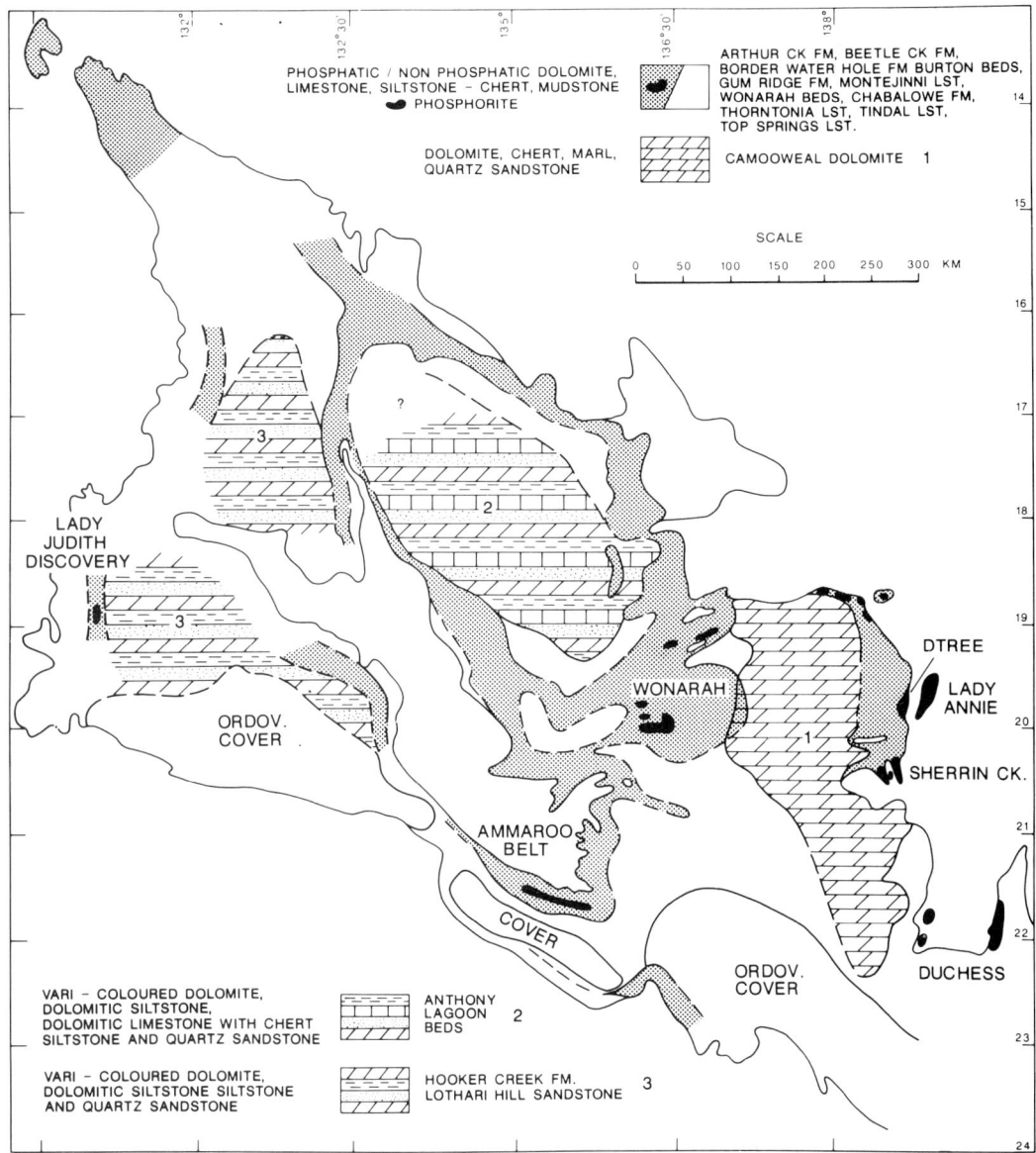

Fig. 4. Distribution of Ordian to Templetonian phosphatic and non-phosphatic lithofacies in the Georgina, Wiso and Daly River Basins.

Detailed description of the key columns are contained in Howard (1989b), Southgate (1988) and Shergold & Southgate (1986) in respect of D Tree; Howard (1989a) for Wonarah; Morris (1986) for Ammaroo; Kennewell & Huleatt (1980) for the Montejinni Limestone; Brown (1986) and Kruse (in press) for the Daly River Basin.

Structure and phosphogenesis

Structural studies and the interpretation of Cambrian basement from the modelling of aeromagnetic and gravity data in a portion of the Brunette Sub-basin give an insight into a control on phosphate deposition.

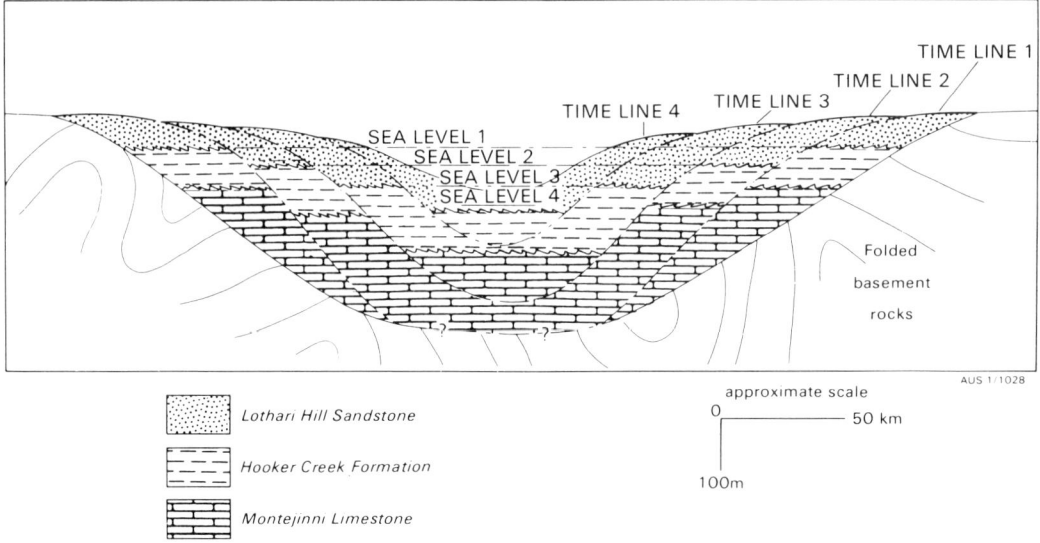

Fig. 5. A depositional model for the continental regressive facies in the Wiso Basin (Kennewell & Huleatt 1980). Time lines represent depositional surfaces in time, and a series of lowering sea levels illustrate the lithofacies relationships following regression.

The northern margin of the Brunette Sub-basin is defined by a prominent Landsat lineament which is colinear with the Murphy Tectonic Ridge at its eastern end, while the western and northwestern margins are defined by a Nimbus 7 lineament which is colinear with a portion of the phosphatic sediment belt (Figs 2 & 4). A syn-sedimentary secondary phosphate occurrence along this lineament suggests fault reactivation and uplift, at least in part, as a post-Templetonian event. To the south the lineament bifurcates and defines the eastern and western sides of the Tennant Creek Inlier.

The modelling of aeromagnetic data utilizing volcanic and Precambrian basement control obtained from company drilling programmes provides a definition of the basement surfaces in respect of present-day exposures. Regional trend analysis using magnetic, gravity and elevation data defines geological trends and faulting directions in the basement. Figure 2 portrays the basement configuration in the depth range of 100–300 m ASL (surface approximates 300 m ASL) and depicts a plateau surface extending across the Alexandria–Wonarah Basement High and planar sloping surfaces extending along the periphery of the basin, which are comparable with the interpretation of half-graben surfaces.

The elevated plateau has steep fault-controlled slopes, a series of horsts and intervening depressions or grabens, some of which are now expressed as elongate inliers (Figs 1, 2 and 4), representing submarine ridges and flanking deeps. Many of the controlling faults can be recognized in the subsurface as magnetic and gravity linear anomalies, gradients and discontinuities which demonstrate linear continuity with Early Proterozoic fault systems in the Precambrian hinterland (Fig. 2). The basement colinear features are aligned in complimentary directions between 30 to 60° parallel to the Alexandria–Wonarah Basement High, and at 120 to 135°. A number of linear magnetc features trend approximately east-west and, finally, magnetic and gravity discontinuities reflect major faults striking 150° which have played a significant part in modifying the basement topography. The major sets, namely, 30–60°, 120–135°, and 150°, are also expressed as topographic divides delineated by enhanced displays of elevation data, and these lineaments attest to present-day uplift.

Comparison of the distribution of the phosphatic facies (Fig. 4) and basement configuration in the range of 100–300 m ASL (Fig. 2) shows a close relationship. Where previous workers have deduced the depositional environments in which the phosphorite deposits formed, namely, nearshore environments ranging through lagoonal, estuarine, littoral, intertidal and supratidal (Howard 1986; Cook & Elgueta 1986;

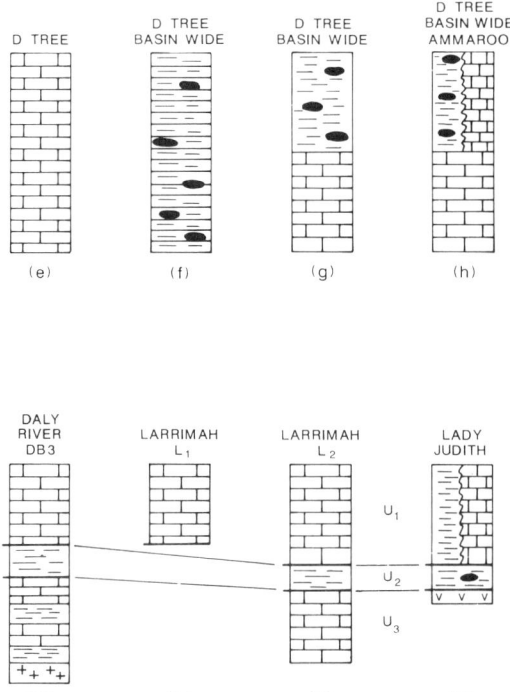

Fig. 6. Phosphatic lithologic columns in the Daly River and Wiso Basins (a, b, c, d) and in the Georgina Basin (e, f, g, h) illustrating the variation of Ordian/Early Templetonian lithologies which may be wholly carbonate (e) or wholly siltstone with or without chert (f). Sections are schematic and not to scale.

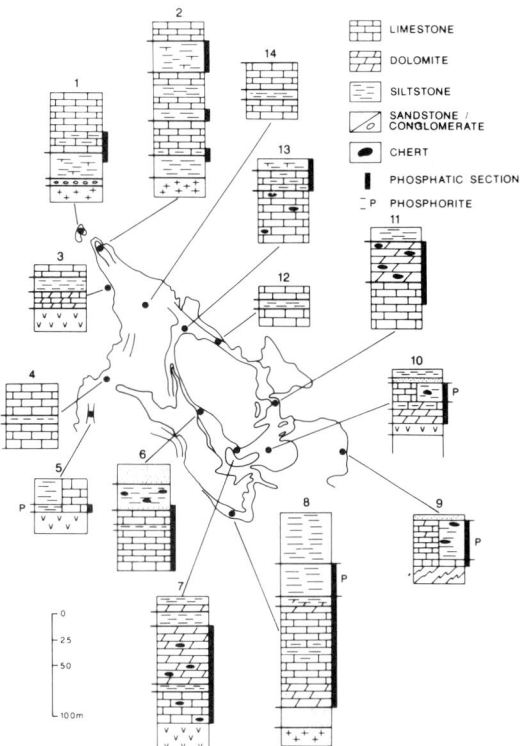

Fig. 7. Selected Ordian/Early Templetonian phosphatic sections in the Georgina, Wiso and Daly River Basins: (1) Reynolds River, N.T.G.S. 82/42; (2) Daly River DB3; (3) Top Springs; (4) Wave Hill WE; (5) Lady Judith Winnecke Creek Water Bore 43–46 and B.M.R. WC2 composite; (6) Helen Springs Water Bore 126; (7) C.R.A. AL1; (8) Ammaroo No. 2; (9) D Tree deposit; (10) Wonarah deposit composite; (11) Brunette Water Bore 220; (12) and (13) Tanumbirini Water Bores 10 and 45; and (14) Larrimah Water Bore 57. Basement shown as granite (crosses, volcanics (vees) and Precambrian (slanted).

Southgate 1986, 1988; Howard 1989a, b; Southgate & Shergold 1988), this study has mapped the aerial distribution of the phosphatic belts in which those depositional environments occur. The data further suggest that their palaeogeography and the water depth of those environments were controlled by plateaux and the sloping surfaces or platforms of half grabens peripheral to the basin.

Characteristic features of the deposits

Many publications describe the geometry and facies relationships of the Georgina Basin phosphorites (Thomson & Russell 1971; Howard 1972; Howard & Cooney 1976; Howard & Perrino 1976; Howard & Hough 1979; Rogers & Keevers 1976), in addition to those papers listed above.

Within the Undilla Sub-basin, the phosphorites at D Tree, Lady Annie, Lily Creek and Sherrin Creek accumulated in lagoons, estuaries or embayments in which the shallow-water environments were isolated from more open-marine environments by carbonate banks. Such traps are readily discernible along the periphery of the basin. In the case of the Alexandria–Wonarah Basement High, deposits formed peripheral to islands or over shallow basement highs but in both instances these structures were ringed by carbonate banks, and the phosphorite loci were therefore similar to the nearshore deposits in the Undilla Sub-basin. A third type of trap is illustrated by the Alroy deposit which lies on the western margin of the Alexandria–Wonarah Basement High and appears to occupy a submarine erosion channel or canyon in carbonate near the edge of the slope into the Brunette Sub-basin, the strike of which parallels the direction of the Mitchiebo fault system on the north side of the basin.

The phosphorites are mostly underlain or partly underlain by carbonates, though shoreward they may rest on volcanic rocks or Precambrian basement. Basinward the phosphorites interdigitate with carbonate bank facies.

The thickness of sediment between the base of the phosphorites and the top of the fringing carbonate bank range from 20 to 60 m and average 25 m. The deposits found to date vary from 10 to 25 km in length and from 1 to 7 km in width and are discrete, changing in, say, 100 m from ore grade phosphorite into phosphatic sediments grading 0.2 to 5% P_2O_5 but with a strong bias to 0.2%.

Phosphorite discoveries of this study

In the course of this study, significant phosphorite intersections were recorded at Lady Judith in the Wiso Basin and at Ammaroo in southwestern Georgina Basin. The Lady Judith deposit discovery is based on a six metre intersection of 32.1% P_2O_5 within siltstone–chert of the Montejinni Limestone. It is underlain by basalt and overlaid by red, brown and purple calcareous siltstones of the Hooker Creek Formation. The phosphorite at Ammaroo was found in 1968 during exploration activities (Morrison 1968). This study suggests that the phosphorite is extensive within an elongate trough in the carbonate–chert facies of the Arthur Creek Formation. The trap occurs on a well defined platform 12 km wide over which the lithofacies change from nearshore sandstone to siltstone, siltstone–chert, siltstone–chert–phosphorite and carbonate–chert into carbonaceous shale–carbonate at the outer edge. Along the platform, Arthur Creek Formation interdigitates with the Chabalowe Formation consisting of laminated and domal algal dolostone, sabkha evaporites and supratidal and lagoonal mixed siliciclastic rocks and algal dolostones (Morris 1986).

Conclusions

(1) Phosphatic sediments extend in a belt more than 2000 km long, 20 to 50 km wide, and 10 to 190 m thick around the periphery of the Georgina, Wiso and Daly River Basins.

(2) The belt contains 20 known deposits, most of which are concentrated in clusters over about 500 km, and based on the types of phosphorite depositional traps which are recognizable, targets for further resource assessment within the belt are now clear.

(3) The age of the phosphorites and associated sediments vary from Ordian to Late Templetonian Middle Cambrian with younging extending from the west (Wiso and southwestern Georgina Basins, western Brunette Sub-basin, and possibly the Daly River Basin) through the Undilla Sub-basin to the southeast (Burke River Embayment).

(4) The basins show extensional and warping features with younging of their structural development from southeast to the northwest. Based on the dating of the volcanic rocks, comparison with other related intracratonic basins and tectonic-subsidence curves, the Georgina Basin formed at 900 Ma and the Wiso and Daly River Basins at 600 Ma, the overlap of the two systems occurring in the region of the Brunette Sub-basin and, in particular, the structurally complex Alexandria–Wonarah Basement High. Subsurface interpretations show that half grabens, horst blocks and a fault bounded plateau are prominent structural features.

(5) The phosphatic sediments and phosphorites were deposited in nearshore environments on either downwarped surfaces or on the platform surfaces of half grabens which are extensional structures along the basin edges, at least in the Brunette Sub-basin.

(6) There are enormous gaps in our knowledge of the Palaeozoic sediments in the Wiso, Daly River and Georgina Basins for reasons enumerated earlier in the paper, but especially due to the lack of subsurface data. Few of the stratigraphic holes in the basins reach Cambrian basement even when this is predictably shallow in the range of a few hundred metres or less. In terms of national objectives, much can be done to overcome this shortcoming at low cost by modelling the existing complete coverage of aeromagnetic, gravity and elevation data in the three basins complimented by a Landsat, Nimbus 7 and photo interpretation of major structural features along the lines followed in this study in a portion of the Brunette Sub-basin (Fig. 2a).

The collection of data in this paper was made possible by Macquarie University Research Grants and C.R.A. Exploration Pty. Ltd. who supported the programme in the period 1981-84 and undertook the modelling of the aeromagnetic, gravity and elevation data which gave new insight into the understanding of a complex portion of the basins. Thanks go to C.R.A. for their permission to publish and also to International Minerals and Chemical Corporation for the years of phosphate exploration and discovery in their employ in the 1960s and as a consultant in the early 1970s, which provided the background, experience and much of the data to pursue the project. Thanks also go to P.J. Cook and J.H. Shergold, of the Bureau of Mineral Resources, Geology and Geophysics, for many useful discussions.

References

BROWN, M. C. 1968. *Middle and Upper Cambrian sedimentary rocks in the northern part of the Northern Territory*. Records of the Bureau of Mineral Resources, Geology and Geophysics, Australia, 1968/115.

COOK, P. J. 1972. Petrology and geochemistry of the phosphate deposits of northwest Queensland, Australia. *Economic Geology*, **67**, 1193-1213.

—— 1976. Georgina Basin phosphatic province, Queensland and Northern Territory — regional geology. *In*: KNIGHT, C. L. (ed.) *Economic Geology of Australia and Papua New Guinea*, **4**. Monograph 5, Australasian Institute of Mining and Metallurgy, Melbourne, 245-250.

—— 1982. The Cambrian palaeogeography of Australia and opportunities for petroleum exploration. *Australian Petroleum Exploration Association Journal*, **22**, 42-64.

—— 1988. *Palaeogeographic Atlas of Australia*, **1**, *Cambrian*. Department of Primary Industries and Energy, Bureau of Mineral Resources, Geology and Geophysics. Australian Government Publishing Service, Canberra.

—— & ELGUETA, S. A. 1986. Proterozoic and Cambrian phosphorite-deposits: Lady Annie, Queensland, Australia. *In*: COOK, P. J. & SHERGOLD, J. H. (eds) *Phosphate Deposits of the World*, **1**, Proterozoic and Cambrian Phosphorites. Cambridge University Press, Cambridge, 132-148.

KRUSE, P. 1990. Cambrian Palaeontology of the Daly River Basin. *Report of the Northern Territory Geological Survey*, **7**, (in press).

HOWARD, P. F. 1972. Exploration for phosphorite in Australia — a case history. *Economic Geology*, **67**, 1180-1192.

—— 1986. Proterozoic and Cambrian phosphorites — regional review: Australia. *In*: COOK, P. J. & SHERGOLD, J. H. (eds) *Phosphate Deposits of the World*, **1**, Proterozoic and Cambrian Phosphorites. Cambridge University Press, Cambridge, 20-41.

—— 1989a. The Wonarah phosphate deposit, Georgina Basin, Australia. *In*: NOTHOLT, A. J. G., SHELDON, R. P. & DAVIDSON, D. F. (eds) *Phosphate Deposits of the World*, **2**, World Phosphate Resources. Cambridge University Press, Cambridge, 545-550.

HOWARD, P. F. 1989b. The D Tree phosphate deposit, Georgina Basin, Australia. *In*: NOTHOLT, A. J. G., SHELDON, R. P. & DAVIDSON, D. F. (eds) *Phosphate Deposits of the World*, **2**, World Phosphate Resources. Cambridge University Press, Cambridge, 551-557.

—— & COONEY, A. 1976. D Tree phosphate deposit, Georgina Basin, Queensland. *In*: KNIGHT, C. L. (ed.) *Economic Geology of Australia and Papua New Guinea*, **4**, Industrial Minerals and Rocks. Monograph 8, Australasian Institute of Mining and Metallurgy, Melbourne, 265-273.

—— & HOUGH, M. J. 1979. On the geochemistry and origin of the D Tree, Wonarah and Sherrin Creek phosphorite deposits of the Georgina Basin, northern Australia. *Economic Geology*, **74**, 260-84.

—— & PERRINO, F. A. 1976. Wonarah phosphate deposits, Georgina Basin, Northern Territory. *In*: KNIGHT, C. L. (ed.) *Economic Geology of Australia and Papua New Guinea*, **4**, Industrial Minerals and Rocks. Monograph 8, Australasian Institute of Mining and Metallurgy, Melbourne, 273-277.

KENNEWELL, P. J. & HULEATT, M. B. 1980. Geology of the Wiso Basin, Northern Territory. *Bulletin of the Bureau of Mineral Resources, Geology and Geophysics, Australia*, **205**.

LINDSAY, J. P., KORSCH, R. J. & WILFORD, J. R. 1987. Timing the break-up of a Proterozoic supercontinent: Evidence from Australian intracratonic basins. *Geology*, **15**, 1061-64.

MORRIS, D. G. 1986. *Stratigraphy, potential source rock and petroleum geochemistry of western Georgina Basin, Northern Territory*. Report of the Northern Territory Geological Survey, Australia, GS 86/5.

MORRISON, M. E. 1968. *Progress report on phosphate exploration to October 31, 1968, Ammaroo, N. T.* Open File Report **CR 68/51**, Department of Mines and Energy, Northern Territory Geological Survey, Australia.

NOTHOLT, A. J. G. & SHELDON, R. P. 1986. Proterozoic and Cambrian phosphorites — regional review: World Resources. *In*: COOK, P. J. & SHERGOLD, J. H. (eds) *Phosphate Deposits of the World*, **1**, Proterozoic and Cambrian Phosphorites. Cambridge University Press, Cambridge, 9-19.

ÖPIK, A. A. 1975. Templetonian and Ordian Xystridura trilobites of Australia. *Bulletin of the Bureau of Mineral Resources, Geology and Geophysics, Australia*, **121**.

—— & PRITCHARD, P. W. 1960. Cambrian and Ordovician. *In*: HILL, D. & DENMEAD, A. K. (eds) *The Geology of Queensland. Journal of the Geological Society of Australia*, **7**, 89-114.

PLUMB, K. A., DERRICK, G. M., NEEDHAM, R. S. & SHAW, R. D. 1981. The Proterozoic of Northern Australia. *In*: HUNTER, D. R. (ed.) *Precambrian of the Southern Hemisphere*. Elsevier, Amsterdam, 205-307.

ROGERS, J. K. & KEEVERS, R. E. 1976. Lady Annie-Lady Jane phosphate deposits, Georgina Basin, Queensland. *In*: KNIGHT, C. L. (ed.) *Economic*

Geology of Australia and Papua New Guinea, **4**, *Industrial Minerals and Rocks*. Monograph 8, Australasian Institute of Mining and Metallurgy, Melbourne, 251–265.

SHERGOLD, J. H. & BRASIER, M. D. 1986. Proterozoic and Cambrian phosphorites — specialist studies: biochronology of Proterozoic and Cambrian phosphorites. *In*: COOK, P. J. & SHERGOLD, J. H. (eds) *Phosphate Deposits of the World*, **1**, *Proterozoic and Cambrian Phosphorites*. Cambridge University Press, Cambridge, 293–326.

—— & DRUCE, E. C. 1980. Upper Proterozoic and Lower Palaeozoic rocks of the Georgina Basin. *In*: HENDERSON, R. A. & STEPHENSON, P. J. (eds) *The Geology and Geophysics of northeastern Australia*. Geological Society of Australia, Queensland Division, 149–174.

—— & PALMER, A. R. (eds) 1985. *Cambrian system in Australia, Antarctica and New Zealand*. International Union of Geological Science Publication, **19**.

—— & SOUTHGATE, P. N. 1986. Middle Cambrian phosphatic and calcareous lithofacies along the eastern margin of the Georgina Basin, western Queensland. *In*: BELPERIO, A. P. & HARVEY, N. (eds) *Australasian Sedimentologists Group Field Guide Series* No. 2, Geological Society of Australia, 1–61.

—— & —— 1988. Timing and Distribution of Middle Cambrian phosphogenetic events in the Georgina Basin (abstract). *In*: *A Decade of Phosphorite Research and Developmet*, 11th International Field Workshop and Symposium, Project 156 Phosphorites, International Geological Correlation Programme, Oxford, 27–28.

SMITH, K. G. 1972. Stratigraphy of the Georgina Basin. *Bulletin of the Bureau of Mineral Resources, Geology and Geophysics*, Australia, **111**.

SOUTHGATE, P. N. 1986. Proterozoic and Cambrian phosorites — specialist studies: Middle Cambrian phosphatic backgrounds, phoscrete profiles and stromatolites and their implications for phosphogenesis. *In*: COOK, P. J. & SHERGOLD, J. H. (eds), *Phosphate Deposits of the World*, **1**, *Proterozoic and Cambrian Phosphorites*. Cambridge University Press, Cambridge, 327–351.

—— 1988. A model for the development of phosphatic and calcareous lithofacies in the Middle Cambrian Thorntonia Limestone, northeast Georgina Basin, Australia. *Australian Journal of Earth Sciences*, **35**, 111–130.

—— & SHERGOLD, J. H. 1988. Sequence stratigraphy and repeated facies mosaics in Middle Cambrian sediments of the Georgina Basin, Australia (abstract). *In*: *A Decade of Phosphorite Research and Development*, 11th International Field Workshop and Symposium, Project 156 Phosphorites, International Geological Correlation Programme, Oxford, 31–32.

THOMSON, L. D. & RUSSELL, R. T. 1971. Discovery, exploration and investigations of phosphate deposits in Queensland. *Proceedings of the Australasian Institute of Mining and Metallurgy, Melbourne*, **240**, 1–14.

Events leading to global phosphogenesis around the Proterozoic/Cambrian boundary

T. H. DONNELLY[1], J. H. SHERGOLD[2], P. N. SOUTHGATE[2] & C. J. BARNES[1]

[1] *CSIRO Division of Water Resources, GPO Box 1666, Canberra ACT, 2601, Australia*
[2] *BMR Division of Continental Geology, GPO Box 378, Canberra ACT, 2601, Australia*

Abstract: In the Late Proterozoic the world's oceans changed from being relatively oxic and well-mixed, into to a less mixed and more stagnant system. This resulted in the accumulation of massive amounts of organic matter and pyrite in anoxic sediments on the sea floor, and the enrichment of P in the anoxic deep ocean waters. Except for relatively short periods of increased turnover and better ventilation during times of Late Proterozoic glaciation, the stagnant ocean system appeared to be the 'normal' condition. A Cambrian seawater $^{87}Sr/^{86}Sr$ curve is presented, and it is concluded that continental rather than magmatic inputs were the major influence on the Sr isotopic composition of the latest Proterozoic and Cambrian oceans. Significant rises in atmospheric O_2 levels must have accompanied the periods of greatly enhanced organic matter burial. A return to a stable more oxic ocean system occurred around the Proterozoic/Cambrian boundary, at which time rifting of the supercontinent(s) created a large number of epicontinental seas at low latitudes, enabling deep P-rich ocean waters to be moved into shallow-water environments. While increasing pO_2 levels during the latest Proterozoic may have been largely responsible for the Cambrian 'radiation' event, the increase in pCO_2 levels at the beginning of the Cambrian, as the oceans became more oxic, may have been responsible for the acquisition of mineralized skeletal structures by soft-bodied organisms.

Major evolutionary, tectonic and geochemical changes occurred in the Late Proterozoic and Early Cambrian. Brasier (1979, 1982, 1990) described the evolutionary explosion (Cambrian radiation event) which occurred at the close of the Proterozoic and Conway Morris (1987) discussed how hard skeletal parts abruptly appeared in the fossil record at that time, after some 90% of Earth's history had elapsed. The break-up of the supercontinent(s) (cf. Smith *et al.* 1981) and the creation of many shallow epicontinental lobes off major seaways at low palaeolatitudes (Cook & McElhinny 1979) also occurred around the Proterozoic/Cambrian boundary, and a major episode of global phosphogenesis commenced about this time and extended into the Middle Cambrian (Cook & Shergold 1984, 1986).

The common characteristics of phosphate deposits which occur during this interval have been discussed by Cook & Shergold (1984, 1986), who have suggested a model for phosphogenesis which involves a complex array of physicochemical processes interacting on a variety of temporal and spatial scales. Among them, the development of an 'oceanic anoxic event' (OAE) is a significant factor. OAEs are periods of unusually high organic matter preservation (organic C being ^{12}C-enriched) to the extent that it isotopically alters the inorganic carbon pool and leads to ^{13}C-enrichment in ocean bicarbonate. Positive $\delta^{13}C$ excursions (i.e. ^{13}C-enrichment) in marine carbonate rocks have been related to coeval periods of unusually high organic matter burial rates (Jenkyns 1980; Scholle & Arthur 1980; Schlanger *et al.* 1987). Cook & Shergold (1984) suggested that P-rich bottom water created by an OAE could be upwelled onto continental margins to form phosphate deposits. A similar genetic proposal was made for major phosphogenic episodes in the Mesozoic and Cenozoic (Arthur & Jenkyns 1981), but these authors noted that OAEs do not directly coincide with phosphogenic episodes. Nevertheless, Cook & Shergold (1984) advanced the argument that as OAEs appeared to preceed most of the major phosphogenic episodes, there could be a genetic link between the two.

Here we examine some of the geochemical evidence critical to understanding phosphogenesis at the Proterozoic/Cambrian boundary. This has accrued mainly as a result of: (1) recent C-isotope studies of Late Proterozoic marine carbonates which indicate that a number of global ^{13}C excursions occurred over the period from *c.*800 Ma to around the Proterozoic/Cambrian boundary (Tucker 1986; Magaritz *et al.* 1986; Knoll *et al.* 1986; Aharon *et al.* 1987; Lambert *et al.* 1987), and (2) the establishment of a link between Middle Cambrian phosphogenesis in the Georgina Basin and OAEs under conditions of ocean ventilation similar to those of the present-day (Donnelly et al. 1988a, b). In this paper a Cambrian seawater $^{87}Sr/^{86}Sr$ curve is presented, and the global fluxes of Sr, C and S, from around 800 Ma through to the Middle Cambrian, are examined to further our understanding of the world's environment at this time. Finally, a model is proposed to explain the global phosphogenesis which occurred between the Late Proterozoic and Middle Cambrian.

Geochemistry

Ocean $^{87}Sr/^{86}Sr$ values

Strontium has a long residence time in the sea (c.4 Ma) compared to present ocean mixing times (100–1000 years), and consequently $^{87}Sr/^{86}Sr$ values are homogenous in the modern ocean (cf. Burke et al. 1982). Peterman et al. (1970) and Veizer & Compston (1974) have shown that under certain conditions, the $^{87}Sr/^{86}Sr$ values of calcareous fossils and marine carbonate sediments correspond to the average coeval seawater values, and that ancient oceans showed marked $^{87}Sr/^{86}Sr$ changes through time. A Sr isotope curve has been constructed for the last 1000 Ma of Earth's history (Burke et al. 1982; Veizer et al. 1983), but until now Cambrian data has been both limited and unreliable (see later dicussion). However, during the Cambrian, seawater $^{87}Sr/^{86}Sr$ values do appear to have reached the highest values recorded throughout the Earth's history.

In the modern ocean the $^{87}Sr/^{86}Sr$ value (0.7092; Hess et al. 1986) is predominantly the result of mixing two major Sr fluxes: (1) continental weathering (average river values c.0.712), and (2) the interaction of seawater with basalt (0.703) at mid-ocean ridges (see discussion, Holland 1984, pp. 227–241). Therefore, over geological time, seawater $^{87}Sr/^{86}Sr$ values recorded past tectonic, climatic and oceanographic events.

Veizer & Compston (1974) reported the earliest Cambrian $^{87}Sr/^{86}Sr$ values, but stated that their data were uncertain and in need of confirmation. Burke et al. (1982) constructed a seawater Sr-isotope curve from 786 points, but their data ended in the Late Cambrian with a scatter of $^{87}Sr/^{86}Sr$ values from around 0.7090 to 0.7096. A Cambrian seawater $^{87}Sr/^{86}Sr$ curve was constructed by Keto & Jacobsen (1987) using data from Burke et al. (1982), Shaw & Wasserburg (1985) and from their own work.

This curve shows a wide scatter of values in the Late Cambrian (essentially data from Burke et al. 1982), a dotted band with limited data points c.0.7091–0.7093 in the Middle Cambrian, and a sudden rise in the Early Cambrian. The Early Cambrian values are from two samples of phosphorite, both assumed to have an age of 610 Ma (Shaw & Wasserburg 1985). One of these samples, from the Hazara phosphorite deposit of Pakistan, is now known to have an Early Cambrian age (Hasan 1986), and the other from Guizhou Province of south-central China, may be either of latest Proterozoic (Sinian) or Early Cambrian age (Li 1986), depending on which of the phosphorite-bearing formations it came from. A demonstrable bias is obtained if such curves are constructed on the basis of individual samples not assessed for their retention of primary geochemical signals, and from uncontrolled stratigraphy.

Using recently obtained Middle Cambrian $^{87}Sr/^{86}Sr$ data from the Georgina Basin of northern Australia, combined with all other available data for the Cambrian (Table 1), a revised curve has been constructed (Fig. 1a). The number of samples available for the Middle Cambrian (42) provides a good data base (Table 1) to determine the average coeval seawater Sr isotope value. The width of the band over this epoch (Fig. 1a), from 0.7094–0.7102, is one standard deviation. Our results clearly show that early diagenetic carbonate concretions (Donnelly et al. 1988b) have slightly lower $^{87}Sr/^{86}Sr$ values compared to associated finely laminated carbonate mudstones (Table 1). It is suggested that when dilute HCl is used to extract Sr from these carbonates some Sr is also extracted from clays present in the associated finely laminated sediment (see also Banner et al. 1988). Therefore, using only the Sr isotope data from concretions (Donnelly et al. 1988b; and this work) and the essentially unaltered biophosphate samples reported by Keto & Jacobsen (1987), the preferred Middle Cambrian $^{87}Sr/^{86}Sr$ value is calculated at 0.7096 (Fig. 1a). Three Late Cambrian phosphatic brachiopod samples (Keto & Jacobson 1987) have $^{87}Sr/^{86}Sr$ values between 0.7092 and 0.7094, and although marginally above the band shown by Burke et al. (1982), appear to be part of a slowly rising trend that begins in the Early Cambrian (Fig. 1a). The 20 Early Cambrian samples are the least controlled, in terms of understanding what post depositional alteration they may have undergone, and show the largest range of $^{87}Sr/^{86}Sr$ values (0.7084–0.7105, Table 1). The band shown in Fig. 1a for the Early Cambrian is also one standard deviation in

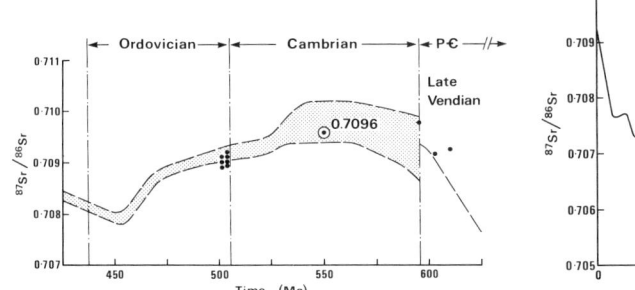 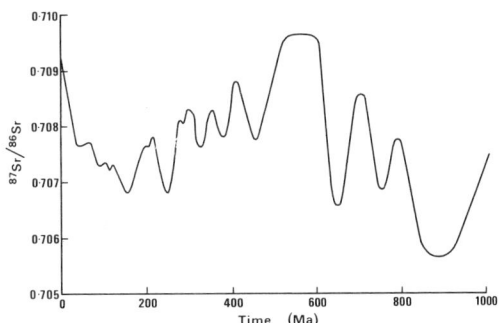

Fig. 1. (a) Cambrian seawater ^{87}Sr/^{86}Sr curve using data from Table 1. The width of the stippled band is one standard deviation, and the circled data point is the preferred value for the Middle Cambrian (0.7096). Some recent values for the Early Ordovician and Late Vendian (Aharon et al. 1987; Kovach & Miller 1988a, b) have been added to the diagram to show the trends just outside of the Cambrian boundaries. (b) Combined seawater ^{87}Sr/^{86}Sr curve for 1000 Ma of Earth history. Data for the Phanerozoic from Burke et al. (1982), for the Precambrian from Veizer et al. (1983) and Derry & Jacobsen (1988), and for the Cambrian from (a).

width. Obviously better controlled samples are needed to show if there are significant Sr isotope changes during the Cambrian. For ^{87}Sr/^{86}Sr values to reach 0.7096 in the Cambrian, continental input of Sr to the world's oceans must have been dominant towards the end of the Late Proterozoic. Such continental domination has subsequently only been matched during the Late Cenozoic (0.7092, Fig. 1b).

To model the Sr fluxes a simplified Holland (1984) approach was adopted. Following Holland's mathematical procedure the ocean is represented as a system with input of strontium fluxes F_R from continental rivers, and F_S out in the form of oceanic sediments. In addition, there is a flux F_B through reaction of seawater with basalt at mid-ocean ridges (MOR). The mass balance for the strontium isotopes may be written:

$$F_R Q_R + F_B (Q_B - Q_O) = F_S Q_O. \quad (1)$$

where Q represents the mole fraction of the particular isotope (either ^{86}Sr or ^{87}Sr), and the subscripts B, O and R refer to MOR basalts, the oceans and continental rivers, respectively. At steady state, $F_R = F_S$, and

$$F_R (Q_R - Q_S) + F_B (Q_B - Q_O) = 0 \quad (2)$$

By defining, $r = Q\,(^{87}\text{Sr})/Q(^{86}\text{Sr})$, and assuming that $Q\,(^{86}\text{Sr})$ is constant in each system and $Q_B = Q_O$, the ratio of the two fluxes F_B and F_R is given by:

$$F_R/F_B = \frac{r_O - r_B}{r_R - r_O}. \quad (3)$$

Some values are fairly well defined, for example, r_B for MOR activity has been accepted as 0.703 (see Holland 1984), and this value should remain constant through time. The isotopic composition of strontium entering the oceans via rivers (r_R) has been calculated from present-day inputs as 0.713 (Goldstein & Jacobsen 1988). This value is a function of the distribution of rock types in the major drainage basins of the world, and it seems likely that continents in the Late Vendian and Cambrian had a similar distribution of rock types and therefore a similar r_R value. At the present time the flux of strontium from seawater/basalt reactions ($c.2 \times 10^{10}$ Mole a^{-1}; Holland 1984) is dominated by the continental strontium flux, which is controlling the relatively high homogenous modern ocean value of 0.7092 (Fig. 1b).

Veizer et al. (1980) identified a major mantle event in the Late Proterozoic at around 900 Ma. Following this event the ^{87}Sr/^{86}Sr curve, with some exceptions, rose steadily into the Cambrian, where the highest ocean strontium values recorded over geological time (0.7096, Fig. 1b) have been found. In Fig. 2 the δ^{13}C values of Late Proterozoic marine carbonates are plotted with the ^{87}Sr/^{86}Sr curve for the same time. The occurrence of two major Late Proterozoic glaciations, the latest being around 640 Ma (Hambrey & Harland 1981), appear to be recorded in the sharp negative excursion of the δ^{13}C curve at this time (see later discussion), followed by a similar response from the ^{87}Sr/^{86}Sr curve. The lack of synchroneity with the other older negative excursion (Fig. 2) is not unexpected as worldwide synchroneity of Late Proterozoic glaciation is unlikely (see Crawford & Daily 1971; Hambrey & Harland 1981), and unless marine carbonates from the same area

Table 1. Cambrian $^{87}Sr/^{86}Sr$ values for marine carbonate rocks, skeletal carbonate, phosphorites and some sulphate minerals

Age	Locality	Formation	Lithology	$^{87}Sr/^{86}Sr$	Source
Early Cambrian	Officer Basin, SA.	Ouldburra Fm.	Sulphate	0.70886	1
Early Cambrian	Officer Basin, SA.	Ouldburra Fm.	Limestone/Dolostone	0.70886	1
Early Cambrian	Officer Basin, SA.	Ouldburra Fm.	Limestone/Dolostone	0.70960	1
Early Cambrian	Officer Basin, SA.	Ouldburra Fm.	Sulphate	0.70960	1
Early Cambrian	Officer Basin, SA.	Ouldburra Fm.	Sulphate	0.70917	1
Early Cambrian	Officer Basin, SA.	Ouldburra Fm.	Sulphate	0.70864	1
Early Cambrian	Officer Basin, SA.	Ouldburra Fm.	Limestone/Dolostone	0.70840	1
Early Cambrian	Officer Basin, SA.	Ouldburra Fm.	Limestone/Dolostone	0.70980	1
Early Cambrian	Warragee Bore, SA.	Wilkawillina Lst.	Limestone/Dolostone	0.71017	2
Early Cambrian	Warragee Bore, SA.	Wilkawillina Lst.	Limestone/Dolostone	0.70974	2
Early Cambrian	Brachina Gorge, SA.	Wilkawillina Lst.	Limestone/Dolostone	0.70956	2
Early Cambrian	Brachina Gorge, SA.	Wilkawillina Lst.	Limestone/Dolostone	0.71047	2
Early Cambrian	Old Wirrealpa Springs, SA.	Pararra Lst.	Limestone/Dolostone	0.70965	2
Early Cambrian	Old Wirrealpa Springs, SA.	Pararra Lst.	Limestone/Dolostone	0.71005	2
Early Cambrian	Old Wirrealpa Springs, SA.	Pararra Lst.	Limestone/Dolostone	0.70940	2
Early Cambrian	Hazara, Pakistan	Unknown Fm.	Biophosphate	0.71026	3
Early Cambrian	Guizhou, China	Unknown Fm.	Biophosphate	0.70944	4
Middle Cambrian	Georgina Basin, NT, Aust.	Upper Hay River Fm.	Carbonate Concretion	0.70990	5
Middle Cambrian	Georgina Basin, NT, Aust.	Upper Hay River Fm.	Carbonate Concretion	0.70999	5
Middle Cambrian	Georgina Basin, NT, Aust.	Lower Hay River Fm.	Carbonate Concretion	0.70923	5
Middle Cambrian	Georgina Basin, Qld. Aust.	Inca Fm.	Carbonate Concretion	0.70967	5
Middle Cambrian	Georgina Basin, Qld. Aust.	Inca Fm.	Carbonate Concretion	0.70923	6
Middle Cambrian	Georgina Basin, Qld. Aust.	Devoncourt Lst.	Carbonate Concretion	0.70989	6
Middle Cambrian	Georgina Basin, Qld. Aust.	Thorntonia Lst.	Carbonate Concretion	0.70963	6
Middle Cambrian	Georgina Basin, NT, Aust.	Upper Hay River Fm.	Limestone/Dolostone	0.71031	5
Middle Cambrian	Georgina Basin, NT, Aust.	Upper Hay River Fm.	Limestone/Dolostone	0.71010	5
Middle Cambrian	Georgina Basin, NT, Aust.	Lower Hay River Fm.	Limestone/Dolostone	0.71028	5
Middle Cambrian	Georgina Basin, Qld., Aust.	Inca Fm.	Limestone/Dolostone	0.70994	5
Middle Cambrian	Georgina Basin, Qld., Aust.	Inca Fm.	Limestone/Dolostone	0.70963	6
Middle Cambrian	Georgina Basin, Qld., Aust.	Devoncourt Lst.	Limestone/Dolostone	0.71066	6
Middle Cambrian	Georgina Basin, Qld., Aust.	Beetle Creek Fm.	Phosphorite	0.70966	6
Middle Cambrian	Georgina Basin, Qld., Aust.	Beetle Creek Fm.	Phosphorite	0.70976	6
Middle Cambrian	Georgina Basin, Qld., Aust.	Beetle Creek Fm.	Phosphorite	0.70959	6
Middle Cambrian	French Central Massif	Unknown Fm.	Limestone/Dolostone	0.70948	7
Middle Cambrian	(Montagne Noire)	Unknown Fm.	Limestone/Dolostone	0.71003	7
Middle Cambrian	(Montagne Noire)	Unknown Fm.	Limestone/Dolostone	0.70901	7
Middle Cambrian	(Montagne Noire)	Unknown Fm.	Limestone/Dolostone	0.70876	7
Middle Cambrian	Wirrealpa Creek, SA.	Wirrealpa Lst.	Limestone/Dolostone	0.70980	2

Age	Location	Formation	Lithology	$^{87}Sr/^{86}Sr$	Source
Middle Cambrian	Wirrealpa Creek, SA.	Wirrealpa Lst.	Limestone/Dolostone	0.71015	2
Middle Cambrian	Wirrealpa Creek, SA.	Wirrealpa Lst.	Limestone/Dolostone	0.79045	2
Middle Cambrian	Brachina Gorge, SA.	Wirrealpa Lst.	Limestone/Dolostone	0.70899	2
Middle Cambrian	Wave Hill NT.	Montejinni Lst.	Limestone/Dolostone	0.70929	2
Middle Cambrian	Victoria River Downs NT.	Montejinni Lst.	Limestone/Dolostone	0.70940	2
Middle Cambrian	Georgina Basin, Qld. Aust.	Thorntonia Lst.	Limestone/Dolostone	0.71015	2
Middle Cambrian	Georgina Basin, Qld. Aust.	Thorntonia Lst.	Limestone/Dolostone	0.71004	2
Middle Cambrian	Georgina Basin, Qld. Aust.	Currant Bush Lst.	Limestone/Dolostone	0.71039	2
Middle Cambrian	Georgina Basin, Qld. Aust.	Currant Bush Lst.	Limestone/Dolostone	0.70986	2
Middle Cambrian	Georgina Basin, Qld. Aust.	Quita Fm.	Limestone/Dolostone	0.70998	2
Middle Cambrian	Georgina Basin, Qld. Aust.	Quita Fm.	Limestone/Dolostone	0.70924	2
Middle Cambrian	Amadeus Basin, NT. Aust.	Hugh River Shale	Limestone/Dolostone	0.71022	2
Middle Cambrian	Amadeus Basin, NT. Aust.	Hugh River Shale	Limestone/Dolostone	0.71053	2
Middle Cambrian	Amadeus Basin, NT. Aust.	Jay Creek Lst.	Limestone/Dolostone	0.71033	2
Middle Cambrian	Amadeus Basin, NT. Aust.	Jay Creek Lst.	Limestone/Dolostone	0.71046	2
Middle Cambrian	Amadeus Basin, NT. Aust.	Jay Creek Lst.	Limestone/Dolostone	0.70998	2
Middle Cambrian	Amadeus Basin, NT. Aust.	Jay Creek Lst.	Limestone/Dolostone	0.71015	2
Middle Cambrian	Eau Claire, Wisconsin, USA.	Unknown Fm.	Biophosphate	0.70963	8
Middle Cambrian	Pavlovsk, USSR.	Unknown Fm.	Biophosphate	0.70933	8
Late Cambrian	Ellery Creek, NT, Aust.	Goyder Lst.	Limestone/Dolostone	0.70914	2
Late Cambrian	Ellery Creek, NT, Aust.	Goyder Lst.	Limestone/Dolostone	0.70961	2
Late Cambrian	Ellery Creek, NT, Aust.	Goyder Lst.	Limestone/Dolostone	0.70994	2
Late Cambrian	Near Saint John, NB Canada	St John Group	Biophosphate	0.70921	8
Late Cambrian	St Croix, Wisconsin, USA.	Potsdam Sandstone	Biophosphate	0.70943	8
Late Cambrian	Lion Mt. Burnet, Texas USA.	Riley Fm., Cap Mtn. Member	Biophosphate	0.70939	8

Data sources: 1, I. B. Lambert (Officer Basin results, pers. comm.); 2, Veizer & Compston (1974); 3, Hassan (1986); 4, Li (1986); 5, Donnelly et al. (1988b); 6, this study; 7, Gebauer & Grünenfelder (1974); 8, Keto & Jacobsen (1987).

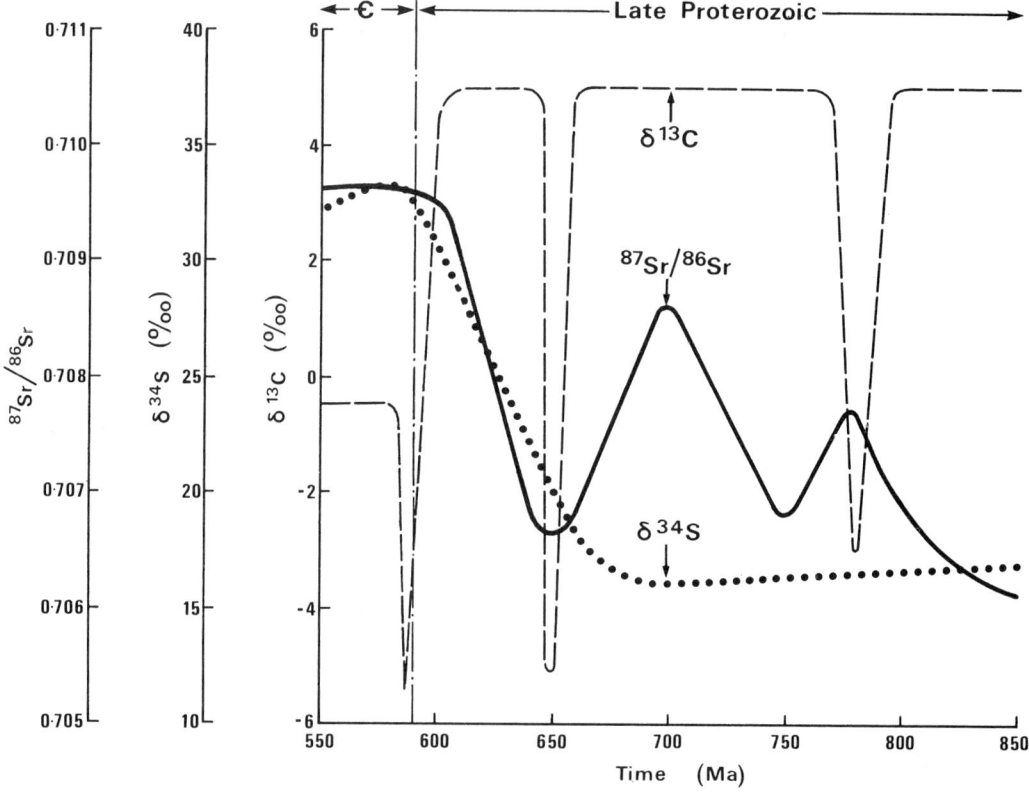

Fig. 2. Combined secular isotope curves of marine carbonates, sulphates and strontium showing the changes which occurred in their isotopic compositions during the Late Proterozoic through to the Middle Cambrian (see text for sources of data).

are analysed for both $^{87}Sr/^{86}Sr$ and $\delta^{13}C$, some error in age dating may occur.

The $^{87}Sr/^{86}Sr$ curve for the Late Proterozoic through Cambrian appears to have been driven mainly by high erosion rates (dissolved Sr fluxes are directly related to rates of continental weathering, Raymo et al. 1988). Such enhanced erosion rates may be related to: (1) the effects of two major glaciations and rising atmospheric O_2 levels during the late Proterozoic leading to increased rates of mineral oxidation, and (2) rifting at low palaeolatitudes at the end of the Proterozoic and rising atmospheric CO_2 levels around the Proterozoic/Cambrian boundary (see later sections). Certainly the evidence of Late Proterozoic tectonics suggests that the flux of strontium from seawater/basalt reactions at this time must have been reasonably significant. We assume in this study that it was about the same as during the Cenozoic, when it appears that the $^{87}Sr/^{86}Sr$ curve was driven upwards mainly by high continental erosional rates.

Therefore, using equation (3) and present-day strontium values, the ratio of continental to oceanic fluxes (F_R/F_B) is 1.0034, while using Cambrian strontium values the ratio is 1.0046. Veizer et al. (1983) presented $^{87}Sr/^{86}Sr$ data for Late Proterozoic marine carbonates and suggested the low values (c.0.7055) at around 900 Ma indicate the occurrence of a major mantle event. These authors argued that the river flux could not have been markedly less radiogenic than the present day, and that manifestations of modern-type plate tectonics were responsible for the very large basalt/seawater reactions. Certainly rifting at the close of the Late Proterozoic suggests strong basalt/seawater reactions may have occurred even at a time when the seawater $^{87}Sr/^{86}Sr$ curve was climbing toward the high values of the Cambrian (Fig. 1b).

It is suggested that the seawater $^{87}Sr/^{86}Sr$ curve, from 900 Ma to the Cambrian, is in broad terms a record of, (1) initially a return to an oceanic strontium balance (basaltic versus river input) after the domination of MOR activity at around 900 Ma; (2) continuously increasing river input of dissolved strontium both before and after the major glaciations; and (3) a decrease in continental strontium input to the oceans during the two major Late Proterozoic glaciations when processes of mechanical erosion dominated. The flux of mechanically-weathered rock material of glacial origin presently being deposited in the oceans, while significant, accounts for only some 10% of total riverine inputs. In rivers the dissolved load of calcium, and presumably also Sr, accounts for some 50% of the calcium flux. This contrasts with the glacially-derived flux of calcium which, (1) constitutes only some 5–6% of the total riverine input and (2) is present in particulate mineral form (Garrels & McKenzie 1971). Consequently, the calcium and Sr in detritus of glacial origin, once deposited in the oceans is likely to be preserved as unaltered sediment. However, the situation following a glaciation may be reversed. During periods of glacial retreat, sea-level rise and global warming, the quantity of fresh bedrock and fine particulate detritus of glacial origin available for chemical wethering would have markedly increased the total amount of dissolved solids transported to the oceans. An increase in the availability of chemically unweathered material with the increase in atmospheric O_2 levels in the late Proterozoic and CO_2 levels around the Proterozoic/Cambrian boundary are likely to have combined to produce accelerated rates of chemical weathering. Thus, whereas glacial periods probably resulted in a decline in riverine inputs of dissolved Sr to the oceans, post-glacial times may have produced accelerated rates of chemical weathering and Sr flux to the oceans.

An estimated *prehuman* (Carboniferous–Recent) erosion rate 110×10^{14} g a^{-1} was proposed by Gregor (1970) and this agrees well with calculations of present-day erosion rates (225×10^{14} g a^{-1}, Garrels & Mackenzie 1971), particularly when the degree to which man's activities have increased erosion (some 47%, Bondarev 1974; Gorshkov 1980) is taken into account. The very steep rise to high seawater $^{87}Sr/^{86}Sr$ values in the Vendian (latest Proterozoic; Fig. 1b) indicates a dominance of the strontium river flux over MOR activity. If the flux of strontium from basalt/seawater reactions during the Vendian was at least as great as present day values, then our calculations indicate that erosion rates in the Vendian were at least of the order of 110×10^{14} g a^{-1}.

Ocean carbonate $\delta^{13}C$ values

In the modern ocean temperature gradients between the equator and the polar regions are responsible for intense water circulation. Cold O_2-charged water masses ensure that sediments are predominantly deposited under oxic bottom water conditions and organic matter is largely oxidized rather than accumulated. Under such conditions the oceans are rich in CO_2 from oxidized organic matter, Ca^{2+} contents are high and P, a nutrient crucial to life in the oceans, is removed as an insoluble compound (Froelich et al. 1982; Holland et al. 1986).

In contrast, OAEs are suggested to be times of widespread deep- and intermediate-water O_2 deficits which occur as a result of a number of factors, many related to low climatic gradients between the equator and polar regions (Holland 1984; Arthur et al. 1987). Under such conditions deep waters tend to have higher temperatures and lower pO_2 contents; the metabolic rates of organisms increase and anoxia is produced in parts of the ocean floor normally oxygenated. These conditions lead to enhanced organic matter preservation. Bacterial decomposition of the organic matter on the ocean floor liberates soluble P (and N) compounds (Froelich et al. 1982) and these nutrients, through diffusion or local upwelling, can reach the zone above the redoxcline and substantially increase primary productivity. The enhanced preservation of organic matter can lock away large quatities of very ^{12}C-enriched reduced C. These processes can occur on a global or local scale and result in an isotopic change of the inorganic and organic C sources. In practice investigators of OAEs have mostly monitored the isotopic changes in marine carbonates by linking periods of positive $\delta^{13}C$ excursions to OAEs. However, to demonstrate that a positive $\delta^{13}C$ excursion is the result of an OAE, a global distribution of carbonate $\delta^{13}C$ values, or widespread accumulations of black shales should be proven.

Local positive $\delta^{13}C$ excursions may be caused by, (1) methanogenic diagenesis (Irwin et al. 1977), which may be detected by $\delta^{13}C$ measurement of coexisting inorganic and organic C phases (Knoll et al. 1986), or (2) basin-wide persistent evaporative conditions. This second possibility must be examined in terms of the sedimentological evidence, or demonstration of the global nature of the excursion. Some examples of OAEs which appear to be global events occur in the Late Cretaceous (Scholle & Arthur 1980; Arthur et al. 1987), the Early

Jurassic (see Jenkyns 1980), the Late Permian (see Magaritz et al. 1983), and, the period of interest to this study, the Late Proterozoic (Veizer & Hoefs 1976; Veizer et al. 1980; Tucker 1986; Magaritz et al. 1986; Knoll et al. 1986; Lambert et al. 1987; Aharon et al. 1987).

Although the secular isotope curve for carbon in unaltered Archaean to Recent marine carbonates (Veizer & Hoefs 1976; Fig. 3a) indicated a positive ^{13}C excursion in the Late Proterozoic (+1.5±0.36‰, 100 samples; Veizer et al. 1980), it was not until the late eighties that this excursion was examined with respect to global environments. Knoll et al. (1986) presented $\delta^{13}C$ values for Upper Proterozoic marine carbonates from Svalbard and East Greenland (Fig. 3b). Their data, covering the period from around 900–540 Ma, indicated generally sustained positive $\delta^{13}C$ excursions (Fig. 3b) with two geologically short episodes of strongly negative excursions, and a return to values of near 0‰ around the beginning of the Cambrian. However, the latest Proterozoic (Vendian) sedimentary sequence is incomplete in Svalbard and East Greenland (Knoll et al. 1986). A positive $\delta^{13}C$ excursion (up to around +5‰) occurs in late Vendian marine carbonates (Fig. 3c) of the Himalaya (Aharon et al. 1987). Examination of Late Vendian marine carbonates from other parts of the world, including Siberia (Magaritz et al. 1986), China (Lambert et al. 1987), and Morocco (Tucker 1986), confirms the global nature of the latest Proterozoic $\delta^{13}C$ excursion. However, there is some doubt as to whether it was a synchronous event (Conway Morris 1987; see also Fig. 3c).

Figure 3d has been drawn by combining Figs 3b and c to show diagrammatically a secular $\delta^{13}C$ curve for marine carbonates of Late Proterozoic and Cambrian age. This curve (Fig. 3d) suggests that from around 800 Ma to the Proterozoic/Cambrian boundary, vast amounts of organic matter accumulated and were preserved in anoxic parts of the ocean floor (see Knoll et al. 1986). This period of prolonged sluggish circulation was arrested during the Late Proterozoic glaciations. Enhanced temperature gradients in the oceans resulted in vigorous oceanic circulation and the movement of cold O_2-charged waters into the deep ocean. Such a change in circulation resulted in the movement of vast amounts of isotopically light C from the reduced to the oxidised part of the C cycle (sharp negative $\delta^{13}C$ excursions shown by carbonates, Fig. 3d). However, around the Proterozoic/Cambrian boundary the C cycle became similar to the present day. Although there is another marked negative $\delta^{13}C$ excursion around the Proterozoic/Cambrian boundary (Fig. 3d), this excursion appears to have been of short duration, since marine carbonates of Cambrian age have $\delta^{13}C$ values which average near zero per mil (-0.57±0.17‰ based on 75 samples, Veizer et al. 1980; see also James & Klappa 1983; Donnelly et al. 1988b).

The $\delta^{13}C$ values from Late Proterozoic carbonate rocks (Fig. 3d), indicate that poorly oxygenated marine basins remained globally stable for geologically significant periods. Vast amounts of organic C must have been removed from the C cycle and deposited in anoxic basins. Using the approach of Holland (1984) rough estimates can be made of organic C burial rates during an OAE and the corresponding net gain of atmosphere O_2. To carry out these calculations a number of parameters must be known or estimated. For example, the carbon isotopic difference between carbonate and organic matter found by Knoll et al. (1986), from their examination of Late Proterozoic carbonates (28.5‰), was used. It is assumed also that the isotopic composition of all carbon deposited from the oceans remained constant, at least over the period from the Late Proterozoic to the Cambrian. This assumption is supported by Schidlowski (1987) who indicated that the isotope fractionation between carbonate and organic matter has been constant over the last 2000 Ma of Earth's history. Similarly, Knoll et al. (1986) and Donnelly et al. (1988b) have shown that the isotopic compositions of carbonate and kerogen from Late Proterozoic and Middle Cambrian sedimentary rocks are inversely correlated with fractionation factors close to the standard value. The $\delta^{13}C$ values of Late Proterozoic marine carbonates prior to the

Fig. 3. (a) Secular curve of $\delta^{13}C$ changes in unaltered marine carbonates from Archaean to Ordovician (data from Veizer & Hoefs 1976; Veizer et al. 1980; for the Cambrian see also Donnelly et al. 1988b and James & Klappa 1983). The discontinuous curved line defines a positive $\delta^{13}C$ excursion of approximately +1.5‰ in the Late Proterozoic. The $\delta^{13}C$ values shown are means and standard deviations (1σ); the number of samples are shown in brackets. (b) Secular $\delta^{13}C$ curve of unaltered Late Proterozoic marine carbonates from Svalbard and East Greenland (Knoll et al. 1986). The curve is shown only for carbonate/kerogen pairs. The wavy lines indicate the positions of major hiatuses in the successions. Note that the latest Vendian (670–700 Ma) section is missing, and that $\delta^{13}C$ values for Cambrian samples are around 0‰. (c) Secular $\delta^{13}C$ curve of Late Vendian (590–650 Ma) marine carbonates from the Lesser Himalaya (Aharon et al. 1987). Similar curves have been

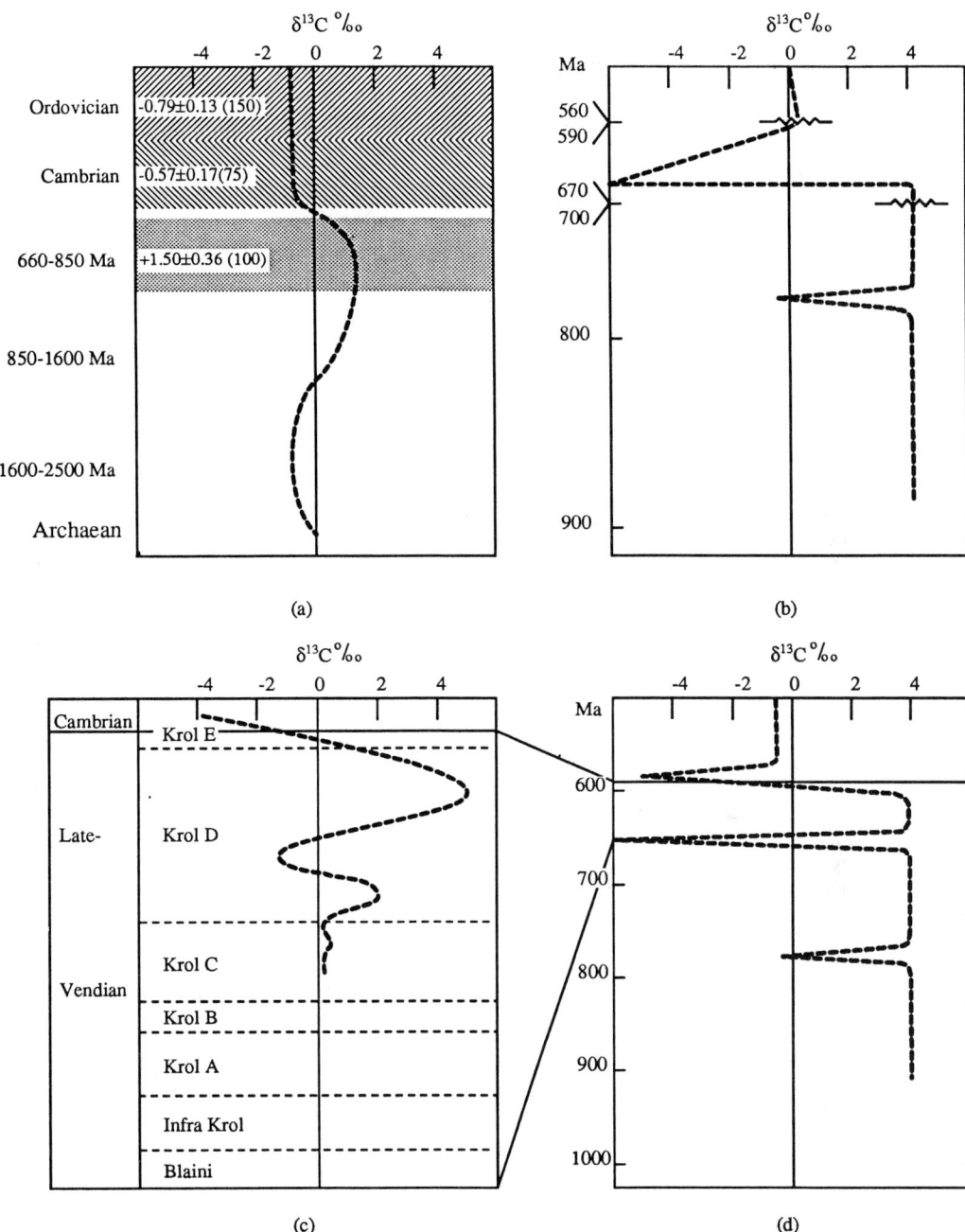

published for the Siberian Platform (Magaritz *et al.* 1986) and the Anti-Atlas (Tucker 1986). Note that the isotope curve is not complete, and that there is a sharp negative $\delta^{13}C$ excursion around the Proterozoic/Cambrian boundary. (**d**) Combined secular curves from (b) and (c). Note that except for sharp negative $\delta^{13}C$ excursions at around 650 and 770 Ma, which could be the result of the two major Late Proterozoic glaciations, positive $\delta^{13}C$ values were the norm in the Late Proterozoic until around the Proterozoic/Cambrian boundary. Some evidence of older OAEs comes from Zempolich *et al.* (1988) who has suggested that bicarbonate in Proterozoic seawater, of age 1200–1400 Ma, was ^{13}C-rich with an estimated $\delta^{13}C$ value of +5.3 ±0.5‰. In the Cambrian the $\delta^{13}C$ values of c.0‰ suggests the C cycle was similar to that of the present day (i.e. a generally oxic ocean system).

OAEs, were similar to Holocene values (c.0‰, Veizer & Hoefs 1976), while Cambrian values were slightly enriched in ^{12}C (-0.57 ± 0.17‰, Veizer et al. 1980; see also James & Klappa 1983; Donnelly et al. 1988b). At the peaks of the Late Proterozoic OAEs, δ^{13}C excursions were around $+5$‰ (Knoll et al. 1986; Aharon et al. 1987). The fraction (f) of carbon leaving the oceans as a constituent of organic matter can be expressed as:

$$f_1 = \frac{-\delta^{13}\text{C (all carbon)} + \delta^{13}\text{C (carbonate prior to OAEs)}}{\delta^{13}\text{C (carbonate–kerogen)}}$$

$$f_2 = \frac{-\delta^{13}\text{C (all carbon)} + \delta^{13}\text{C (carbonate at peak of OAEs)}}{\delta^{13}\text{C (carbonate–kerogen)}}$$

The difference between the two fractions ($\Delta f = f_2 - f_1$) for the Late Proterozoic is $5/28.5 = 0.18$. In other words, some 18% more C was being deposited as organic C at the peak of the Late Proterozoic OAE compared to the present-day. Using the present-day estimated burial rate of marine carbon of 36×10^{13} g a^{-1} as a guide this increase (ΔC) is: $0.18 \times 36 \times 10^{13} = 6.5 \times 10^{13}$ g a^{-1}. As a result the net gain in atmospheric oxygen (ΔO_2) would be: $6.5 \times 10^{13} \times 32/12 = 17.3 \times 10^{13}$ g a^{-1}. Thus if the OAE(s) persisted for some 7 Ma the present inventory of atmospheric O_2 (1.22×10^{21}g; Holland 1984) would be doubled.

This oxidizing potential could be lessened by significant global increases in sulphate formation, and the isotopic covariance shown by Phanerozoic marine carbonates and sulphates (see Veizer et al. 1980) indeed suggests that episodes of oxidation in the Phanerozoic were largely accommodated by shifts in the redox balance of sulphur. Iron, another redox sensitive element, does not appear to have played any role in buffering atmospheric O_2 levels during the Phanerozoic. The isotopic covariance indicates that times of heavy δ^{13}C values were times of light δ^{34}S values. As covariance is lost in the Late Proterozoic during times of OAEs and where ^{87}Sr/^{86}Sr values indicate that continental input to the oceans (rather than MOR activity) dominated ocean chemistry, the positive δ^{13}C excursion(s) must have increased atmospheric O_2 levels. A sustained period of OAEs like that occurring in the Late Proterozoic, must also have resulted in a time of important atmospheric O_2 buildup (see Knoll et al. 1986). Despite some periods of Late Proterozoic glaciation, which restarted oceanic circulation and oxidized vast quantities of organic matter (global δ^{13}C excursions became negative, Fig. 3d), the oceans at this time were generally stagnant. Vast amounts of organic matter were preserved and large quantities of H_2S produced. At the same time the bacterial decomposition of some of this organic matter in anaerobic environments released P to the bottom waters enabling soluble phosphorus compounds to form.

Ocean sulphate δ^{34}S values

The broad outlines of the secular curve of ocean sulphate δ^{34}S values (from gypsum and anhydrite deposits) are known (see Claypool et al. 1980). Considerable reliance can be placed on the Phanerozoic section of the curve, but data from the Proterozoic are limited and at times unsatisfactory because of the broad spread of δ^{34}S values recorded from individual areas. It would appear, however, that at some time in the Late Proterozoic seawater sulphate δ^{34}S values began to rise from around $+15$‰, reaching the highest values known ($c. +35$‰) during the Cambrian.

The activities of sulphate-reducing bacteria in anoxic sediments result in the evolution of H_2S, which is strongly enriched in ^{32}S. The δ^{34}S values of ocean sulphate, therefore, would become increasingly more ^{34}S-enriched if the scale of bacterial sulphate reduction was large and the evolved H_2S was trapped as pyrite. In the Late Proterozoic the flux of riverine sulphate, from crustal sulphate and sulphide, would have provided relatively low δ^{34}S values. The fact that ocean sulphate δ^{34}S values continued to rise towards the Proterozoic/Cambrian boundary suggests that riverine inputs must have been small compared to oceanic processes. Thus the major shift to Late Proterozoic ocean sulphate δ^{34}S values must have been due to a marked change in the flux balance of ocean sulphide and sulphate. The lack of ocean sulphate δ^{34}S data for the period around 800–600 Ma (Claypool et al. 1980) means that only the broad outlines of the secular curve of significant ^{34}S-enrichment are known for this time.

The isotopic compositions of Phanerozoic marine carbonates (δ^{13}C) and sulphates (δ^{34}S) show a negative correlation (Veizer et al. 1980). The fact that the calculated regression line lies close to that determined from mass-balance considerations, assuming neither net gain, nor loss, of O_2 in the ocean-atmosphere system as a result of interaction of the C and S cycles (Holland 1984), indicates that the operation of these cycles controls atmospheric O_2 levels.

These data demonstrate that the Fe redox cycle is not important in controlling atmospheric O_2 levels in the Phanerozoic. While the correlation suggests atmospheric O_2 levels varied little during the Phanerozoic, the scatter of data points (Veizer et al. 1980) may represent times of O_2 imbalances. Holland (1984) has shown that an imbalance of 5% can increase oxygen pressure by 50% over 40 Ma, and such an imbalance would be within the degree of scatter on the correlation diagram. In addition, consideration of the isotopic data available for marine carbonate and ocean sulphate for the period prior to the Proterozoic/Cambrian boundary (Claypool et al. 1980; Knoll et al. 1986; Aharon et al. 1987) suggests that at times over this period, considerable imbalances in the C and S cycles may have occurred. For example, at c.800 Ma, $\delta^{13}C$ (carbonate) was around +5‰ and $\delta^{34}S$ (sulphate) around +15‰; at c.700 Ma, $\delta^{13}C$ values were around +5‰ while $\delta^{34}S$ was around +25‰; and at c.600 Ma, $\delta^{13}C$ was around +5‰ and $\delta^{34}S$ around +35‰. Even allowing for some inaccuracy in these data, they suggest that during the Late Proterozoic significant imbalances occurred, which must have resulted in net gains in atmospheric O_2 levels. A similar suggestion was made by Holland (1984) for the marked isotopic imbalance (see Veizer 1985) at the Cretaceous Albian–Aptian boundary.

Holser (1977) thought that the apparent sudden rise in ocean sulphate $\delta^{34}S$ values from around +15‰ in the Late Proterozoic to +35‰ in the Cambrian (Claypool et al. 1980), was too abrupt to be the result of natural processes. Using data from Nielsen (1965) and Holland (1973), he calculated that the rate of $\delta^{34}S$ change (as a result of the actions of the sulphate-reducing bacteria) was far too slow to account for abrupt rises in the secular $\delta^{34}S$ curve. Three such rises are documented in the Earth's history, the first being the rise in the Late Proterozoic. However, compilation of many more data points for the Late Proterozoic section of the secular $\delta^{34}S$ curve (H. Strauss, pers. comm.) suggests that in fact there was a very gradual rise in ocean sulphate $\delta^{34}S$ values which reached a maximum during the Cambrian. The rise in $\delta^{34}S$ values in the Late Proterozoic ocean sulphate (c.20‰), assuming a fractionation of 1.045, would have meant the conversion of some 40% of the total ocean sulphate flux to H_2S (Chambers & Trudinger 1979), which would have been locked away in the deep ocean as ^{32}S-enriched pyrite. From the secular S-isotope curve of Claypool et al. (1980) it appears that significant time elapsed before the bulk of this sulphide was returned to the oceanic S cycle. It is significant that by the Early Cambrian carbonate $\delta^{13}C$ and sulphate $\delta^{34}S$ values again lie on the regression line, and as indicated by Zharkov (1981), this was also a time of large scale evaporitic sulphate deposition. While the Cambrian ocean was generally oxic, as demonstrated by a return to carbonate $\delta^{13}C$ values similar to those of the present day (Veizer et al. 1980; James & Klappa 1983; Donnelly et al. 1988b), there is evidence that even as late as the mid-Cambrian, P-rich deep ocean water (sulphate depleted and with ^{13}C-enriched HCO_3^-) occasionally moved up into restricted shallow-water environments (Donnelly et al. 1988b).

It is interesting to note that similar global positive $\delta^{34}S$ excursions have occurred subsequently, but none of these younger events have matched the degree of ^{34}S-enrichment shown by early Cambrian ocean sulphate (Holser 1977). The formation of pyrite in anoxic marine sediments is usually only limited by the amount of organic matter available to the sulphate-reducing bacteria (Berner 1984). Iron is normally not a limiting factor, because, while ocean waters may only contain a few µg ml^{-1} Fe, terrigenous sediments can average around 5%, much of which is available for pyrite formation. When bottom waters become anoxic, pyrite can form both in the sediments and in the anoxic part of the water column (Leventhal 1983) and because anoxic bottom waters increase the solubility of Fe, more Fe becomes available for pyrite formation. Even under conditions of the highest rates of bacterial sulphate reduction in marine terrigenous sediments Fe should not be a limiting factor (Skyring 1987). It is suggested, therefore, that the reason why the Late Proterozoic ocean sulphate $\delta^{34}S$ values rose to such high values, was mainly the result of the extended period of OAEs (Fig. 2) which occurred sporadically over some 300 Ma. This would be well within Nielsen's (1965) calculated rate of isotopic change, due to the sulphate-reducing bacteria, which he suggested would not exceed 0.3‰ Ma^{-1}.

Late Proterozoic events and global phosphogenesis

Large accumulations of phosphatic sedimentary rocks (phosphorites) have formed at relatively well-defined times throughout Earth history. At these times phosphate-rich sediments occur on a global scale (i.e. they represent major phosphogenetic events) and the areas of P-enrichment are the sites of many economic phosphate deposits. One such phosphogenic event commenced shortly after the latest Proterozoic glaciation c.640 Ma (Hambrey & Harland 1981), reached a peak during the Early Cambrian between c.590–540 Ma, and then rapidly declined and finally terminated in the late mid-Cambrian at c.540 Ma (Cook & Shergold 1984, 1986).

While it is now recognized that phosphogenesis is a complex multistage process (Cook & Shergold 1986), the scale and duration of phosphogenic episodes, such as those which occurred around the Proterozoic/Cambrian boundary, indicate that they were initiated by global events. It has been postulated that this global event is related to a period of continental extension and rifting during the Late Proterozoic and Early Cambrian (Smith et al 1981; Bond et al. 1984; Lindsay et al. 1987). Extension caused continental breakup and the creation of many shallow epicontinental seaways at low palaeolatitudes (Cook & McElhinny 1979) in which phosphate could be deposited. There is increasing evidence that some phosphate deposition occurred in peritidal environments, occasionally associated with evaporites (Southgate 1986, 1988) and bacterial mats (Soudry & Southgate 1989).

From this point of view phosphogenesis was only one expression of an event starting around 800 Ma ago which caused greatly enhanced organic matter (and pyrite) burial (see Knoll et al. 1986) with accompanying atmospheric O_2 buildup. Such dramatic increases in atmospheric O_2 levels may have been a major factor in the evolution of macroscopic metazoans (Cloud 1968) and the resulting Cambrian metazoan radiation event (Brasier 1979, 1982, 1985). A change to a more oxic ocean with the return of vast amounts of organic matter back into the C cycle, to the extent that $\delta^{13}C$ values of marine carbonates changed from sustained positive values of up to +5‰, to sustained negative values of around −0.5‰, would also have resulted in significantly increased pCO_2 levels. Conway Morris (1987) summarized the change from predominantly aragonitic carbonates in rocks older than the Vendian to calcitic in younger carbonate rocks and stressed that of the controlling factors for this change, increasing pCO_2 levels would be the most important. In addition, the change from soft-bodied organisms to those with mineralized skeletons (mainly calcitic) occurred during this time interval and may also, to some degree, be related to rising pCO_2 levels. The increase in pCO_2 levels was suggested by Conway Morris to be linked to processes connected to plate tectonics. However, such an increase would be the natural result of oxidation of large amounts of organic matter, as occurred at the beginning of the Cambrian. In addition to this source of CO_2, basaltic volcanism associated with extension and rifting (Bultitude 1976) may have provided a volcanic source of ^{12}C rich CO_2 and the supply of increased quantities of CO_2 to the atmosphere.

During the latest Proterozoic oceanic conditions changed from predominantly stagnant to relatively well ventilated in the Cambrian. A swing to excessively light carbonate $\delta^{13}C$ values followed the Late Vendian OAE (e.g. Aharon et al. 1987). Knoll et al. (1986) noted a similar excessive light swing around the time of the last major Proterozoic glaciation. The change to a more oxic ocean (either on a temporary or permanent basis) appears to result in an initial very large oxidizing event before the C cycle returns to equilibrium. It is significant that the timing of phosphate availability for deposition closely follows changing oceanic geochemistry (Cook & Shergold 1984). Continental rifting enhanced the possibility of phosphate deposition by providing abundant restricted shallow-water environments at low-latitude locations (Cook & McElhinny 1979). The movement of P-rich deep ocean water into these restricted environments lead to the formation of major phosphorite deposits in selected settings (Southgate 1988). Elsewhere some shallow-water environments resulted in the widespread accumulation of evaporative sulphate deposits in the Early Cambrian (Zharkov 1981). The stability of some anoxic parts of the oceans during the Early Cambrian meant that P-rich deep ocean waters could still be tapped to yield P-rich sediments as late as the mid-Cambrian (Donnelly et al. 1988b).

Conclusions

There is now strong evidence that the positive $\delta^{13}C$ excursions shown by Late Proterozoic marine carbonates from various parts of the world, were the result of a series of Oceanic Anoxic Events (OAEs) which occurred between 800–600 Ma ago. These OAEs resulted in part from low climatic gradients between the equator and polar regions. In addition to causing the deposition and preservation of vast amounts of organic C, they lead to phases of an expanded redoxcline and the concentration of P in the anoxic bottom waters. The OAEs also caused significant increases in oxygen production which was not taken up by either the S or the Fe cycles thus leading to a significant rise in atmospheric O_2 levels. During the Late Proterozoic glaciations vigorous oceanic circulation took place causing the oxidation of vast amounts of organic C and a decrease in continental Sr flux to the oceans. These processes resulted in a marked negative $\delta^{13}C$ excursion and a lowering of ocean $^{87}Sr/^{86}Sr$ values. Immediately prior to the beginning of the Cambrian lithospheric plate activity caused renewed vigorous oceanic circulation, promoting oceanic overturn through upwelling and the creation of shallow-water environments on broad epicontinental shelves which were suitable sites for phosphate deposition.

The authors would like to thank P. J. Cook, I. B. Lambert and S.-S. Sun for their helpful reviews of an early draft of this paper, and M. D. Brasier, H. C. Jenkyns (referees) and I. Jarvis (scientific editor) for their very useful and constructive comments. J. H. S. and P. N. S. publish with the permission of the Director of the Bureau of Mineral Resources. The paper is a contribution to the International Geological Correlation Programme, Project 156, Phosphorites and the Bureau of Mineral Resources phosphate research programme PHOSREP.

References

AHARON, P., SCHIDLOWSKI, M. & SINGH, I. B. 1987. Chronostratigraphic markers in the end-Precambrian carbon isotope record of the Lesser Himalaya. *Nature*, 327, 699–702.

ARTHUR, M. A. & JENKYNS, H. C. 1981. Phosphorites and paleoceanography. *Oceanologica Acta*, 45, 83–96.

——, SCHLANGER, S. O. & JENKYNS, H. C. 1987. The Cenomanian-Turonian Oceanic Anoxic Event, II. Palaeoceanographic controls on organic-matter production and preservation. *In*: BROOKS, J. & FLEET, A. J. (eds), *Marine Petroleum Source Rocks*. Geological Society, London, Special Publication, 26, 401–420.

BANNER, J. L., HANSEN, G. N. & MEYERS, W. J. 1988. Determination of initial Sr isotopic compositions of dolostones from the Burlington-Keokuk Formation (Mississippian): Constraints from cathodoluminescence, glauconite paragenesis and analytical methods. *Journal of Sedimentary Petrology*, 58, 673–687.

BERNER, R. A. 1984. Sedimentary pyrite formation: an update. *Geochimica et Cosmochimica Acta*, 48, 605–615.

BOND, G. C., NICKESON, P. A. & KOMINZ, M. A. 1984. Breakup of a supercontinent between 625 Ma and 555 Ma: new evidence and implications for continental histories. *Earth and Planetary Science Letters*, 70, 325–345.

BONDAREV, L. G. 1974. *Perpetual Motion. Man and preliminary transfer of matter*. Mysl, Moscow (in Russian).

BRASIER, M. D. 1979. The Cambrian radiation event. *In*: HOUSE, M. R. (ed.) *The Origin of Major Invertebrate Groups*, Systematics Association Special Volume, 12, Academic Press, London and New York, 103–159.

—— 1982. Sea-level changes, facies changes and the Late Precambrian–Early Cambrian evolutionary explosion. *Precambrian Research*, 17, 105–123.

—— 1985. Evolution and geological events across the Precambrian–Cambrian boundary. *Geology Today*, Sept.–Oct. 1985, 141–146.

—— 1990. Phosphogenic events and skeletal preservation across the Precambrian/Cambrian boundary interval. *In*: NOTHOLT, A. J. G. & JARVIS, I. (eds) *Phosphorite Research and Development*. Geological Society, London, Special Publication, 52, 289–303.

BULTITUDE, R. J. 1976. Flood basalts of probable Early Cambrian age in northern Australia. *In*: JOHNSON, R. W. (ed.) *Volcanism in Australia*. Elsevier Scientific Publishing Co., Amsterdam, 1–20.

BURKE, W. M., DENISON, R. E., HETHERINGTON, E. A. KOEPNICK, R. B., NELSON, M. F. & OMO, J. B. 1982. Variations of seawater $^{87}Sr/^{86}Sr$ throughout Phanerozoic time. *Geology*, 10, 516–519.

CHAMBERS, L. A. & TRUDINGER, P. A. 1979. Microbiological fractionation of stable isotopes: a review and critique. *Geomicrobiological Journal*, 1, 249–293.

CLAYPOOL, G. E., HOLSER, W. T., KAPLAN, I. R., SAKAI, H. & ZAK, I. 1980. The age curves of sulfur and oxygen isotopes in marine sulfate and their mutual interpretations. *Chemical Geology*, 28, 199–260.

CLOUD, P. E. 1968. Atmospheric and hydrospheric evolution on the primitive earth. *Science*, 160, 729–736.

CONWAY MORRIS, S. 1987. The search for the Pre–Cambrian boundary. *American Scientist*, 75, 157–167.

COOK, P. J. & MC ELHINNY, M. W. 1979. A re-evaluation of the spatial and temporal distribution of sedimentary phosphate deposits in the light of plate tectonics. *Economic Geology*, 74, 315–330.

—— & SHERGOLD, J. H. 1984. Phosphorus, phosphorites and skeletal evolution at the Precambrian–Cambrian boundary. *Nature*, 308, 231–236.

—— & —— 1986. Proterozoic and Cambrian phosphorites-nature and origins. *In*: COOK, P. J. & SHERGOLD, J. H. (eds) *Phosphate Deposits of the World, vol. 1. Proterozoic and Cambrian Phosphorites*, Cambridge University Press, Cambridge, 369–386.

CRAWFORD, A. R. & DAILY, B. 1971. Probable non-synchroneity of Late Proterozoic glaciations. *Nature*, 230, 111–113.

DERRY, L. A. & JACOBSEN, S. B. 1988. The Nd and Sr isotopic evolution of Proterozoic seawater. *Geophysical Research Letters*, 15, 397–400.

DONNELLY, T. H., SHERGOLD, J. H. & SOUTHGATE, P. N. 1988*a*. Pyrite and organic matter in normal marine sediments of Middle Cambrian age, southern Georgina Basin, Australia. *Geochimica* *et Cosmochimica Acta*, **52**, 259–263.

——, —— & —— 1988*b*. Anomalous geochemical signals from phosphatic Middle Cambrian rocks in the southern Georgina Basin, Australia. *Sedimentology*, **35**, 549–570.

FROELICH, P. N., BENDER, M. L., LUEDTKE, N. A., HEATH, G. R. & DE VRIES, T. 1982. The marine phosphorus cycle. *American Journal of Science*, **282**, 474–511.

GARRELS, R. M. & MACKENZIE, F. T. 1971. *Evolution of sedimentary rocks*. W. W. Norton and Co. Inc., New York.

GEBAUER, D. & GRUNENFELDER, M. 1974. Rb-Sr whole-rock dating of late diagenetic to anchimetamorphic, Palaeozoic sediments in southern France (Montagne Noire). *Contributions to Mineralogy and Petrology*, **47**, 113–130.

GOLDSTEIN, S. J. & JACOBSEN, S. B. 1988. Nd and Sr systematics of river water suspended material: implications for crustal evolution. *Earth and Planetary Science Letters*, **87**, 249–265.

GORSHKOV, S. P. 1980. The cycle of products of land denudation. *In*: RYABCHIKOV, A. M. (ed.) *Cycle of Matter in Nature and its Variation under the Impact of Man's Economic activity*. Izd. MGU, Moscow, (in Russian), 34–55.

GREGOR, C. B. 1970. Denudation of the continents. *Nature*, **228**, 273–275.

HAMBREY, M. J. & HARLAND, W. B. 1981. *Earth's pre-Pleistocene Glacial Record*. Cambridge University Press, Cambridge.

HASAN, M. T. 1986. Proterozoic and Cambrian phosphorites-deposits: Hazara, Pakistan. *In*: COOK, P. J. & SHERGOLD, J. H. (eds.) *Phosphate Deposits of the World, Vol. 1. Proterozoic and Cambrian Phosphorites*. Cambridge University Press, Cambridge, 190–201.

HESS, J., BENDER, M. L. & SCHILLING, J. G. 1986. Evolution of the ratio of strontium-87 to strontium-86 in seawater from Cretaceous to Present. *Science*, **231**, 979–984.

HOLLAND, H. D. 1973. Systematics of the isotopic composition of sulfur in the oceans during the Phanerozoic and its implications for atmospheric oxygen. *Geochimica et Cosmochimica Acta*, **37**, 2605–2616.

—— 1984. *The chemical evolution of the atmosphere and oceans*. Princeton University Press, Princeton, N.J.

——, LAZAR, B. & MCCAFFREY, M. 1986. Evolution of the atmosphere and oceans. *Nature*, **320**, 27–33.

HOLSER, W. T. 1977. Catastrophic chemical events in the history of the ocean. *Nature*, **267**, 403–408.

IRWIN, H., CURTIS, C. D. & COLEMAN, M. L. 1977. Isotopic evidence for source of diagenetic carbonates formed during burial of organic-rich sediment. *Nature*, **269**, 209–213.

JAMES, N. P. & KLAPPA, C. P. 1983. Petrogenesis of Early Cambrian reef limestones, Labrador, *Canadian Journal of Sedimentary Petrology*, **53**, 1051–1096.

JENKYNS, H. C. 1980. Cretaceous anoxic events: from continents to oceans. *Journal of the Geological Society, London*, **137**, 171–188.

KETO, L. S. & JACOBSEN, S. B. 1987. Nd and Sr isotopic variations of Early Paleozoic oceans. *Earth and Planetary Science Letters*, **84**, 27–41.

KNOLL, A. H., HAYES, J. M., KAUFMAN, A. J., SWETT, K. & LAMBERT, I. B. 1986. Secular variation in carbon isotope ratios from Upper Proterozoic successions of Svalbard and East Greenland. *Nature*, **321**, 832–838.

KOVACH, J. & MILLER, J. F. 1988*a*. Strontium analysis of biogenic apatite near the Cambrian–Ordovician boundary. *Cambrian–Ordovician Boundary Working Group (IUGS) Newsletter*, **24**, June 1988, 16–18.

—— & —— 1988*b*. Eustatic sea-level changes near the Cambrian–Ordovician boundary: data from Sr isotope analysis of biogenic apatites. *Geological Society of America, Abstracts with Programs*, **20**, A393.

LAMBERT, I. B., WALTER, M. R., ZANG WENLONG, L. S. & GUOGAN, M. A. 1987. Palaeoenvironment and carbon isotope stratigraphy of Upper Proterozoic carbonates of the Yangtze Platform. *Nature*, **325**, 140–142.

LEVENTHAL, J. S. 1983. An interpretation of carbon and sulfur relationships in Black Sea sediments as indicators of environments of deposition. *Geochimica et Cosmochimica Acta*, **47**, 133–137.

LI, Y. 1986. Proterozoic and Cambrian phosphorites — regional review: China. *In*: COOK, P. J. & SHERGOLD, J. H. (eds) *Phosphate Deposits of the World, Vol. 1. Proterozoic and Cambrian Phosphorites*. Cambridge University Press, Cambridge, 42–62.

LINDSAY, J. F., KORSCH, R. J. & WILFORD, J. R. 1987. Timing the breakup of a Proterozoic supercontinent: evidence from Australian intracratonic basins. *Geology*, **15**, 1061–1064.

MAGARITZ, M., ANDERSON, R. Y., HOLSER, W. T., SALTZMAN, E. S. & GARBER, J. 1983. Isotopes shifts in the Late Permian of the Delaware Basin, Texas, precisely timed by varved sediments. *Earth and Planetary Science Letters*, **66**, 111–124.

——, HOLSER, W. T. & KIRCHVINK, J. L 1986. Carbon-isotope events across the Precambrian/Cambrian boundary of the Siberian Platform. *Nature*, **320**, 258–259.

NIELSEN, H. (1965) Schwefelisotope im marinen Kreislauf und das $\delta^{34}S$ der fruheren Meere. *Geologische Rundschau*, **55**, 160–172.

PETERMAN, Z. E., HEDGE, C. E. & TOURELOT, H. A. 1970. Isotopic composition of Sr in seawater throughout Phanerozoic time. *Geochimica et Cosmochimica Acta*, **34**, 105–120.

RAYMO, M. E., RUDDIMAN, W. F. & FROELICH, P. N. 1988. Influence of Late Cenozoic mountain building on ocean geochemical cycles. *Geology*, **16**, 649–653.

SCHIDLOWSKI, M. 1987. Application of stable carbon isotopes to early biochemical evolution on earth. *Annual Review of Earth and Planetary Science*, **15**, 47–72.

SCHLANGER, S. O., ARTHUR, M. A., JENKYNS, H. C. & SCHOLLE, P. A. 1987. The Cenomanian–

Turonian Oceanic Anoxic Event, 1. Stratigraphy and distribution of organic-rich beds and the marine $\delta^{13}C$ excursion. *In*: BROOKS, J. & FLEET, A. J. (eds) *Marine Petroleum Source Rocks*. Geological Society London, Special Publication, **26**, 371–399.

SCHOLLE, P. A. & ARTHUR, M. A. 1980. Carbon isotopic fluctuations in Cretaceous pelagic limestones: potential stratigraphic and petroleum exploration tool. *Bulletin of the American Association of Petroleum Geologists*, **64**, 67–87.

SHAW, H. F. & WASSERBURG, G. J. 1985. Sm-Nd in marine carbonates and phosphates: Implications for Nd isotopes in seawater and crustal ages. *Geochimica et Cosmochimica Acta*, **49**, 503–518.

SKYRING, G. W. 1987. Sulfate reduction in coastal ecosystems. *Geomicrobiological Journal*, **5**, 295–374.

SMITH, A. G., HURLEY, A. M. & BRIDEN, J. C. (1981) *Phanerozoic Paleocontinental World Maps*, Cambridge Earth Science Series, Cambridge University Press, Cambridge.

SOUDRY, D. & SOUTHGATE, P. N. 1989. Ultrastructure of a Middle Cambrian primary nonpelletal phosphorite and its early transformation into phosphate vadoids: Georgina Basin, Australia. *Journal of Sedimentary Petrology*, **59**, 53–64.

SOUTHGATE, P. N. 1986. Cambrian phoscrete profiles coated grains, and microbial processes in phosphogenesis: Georgina Basin, Australia. *Journal of Sedimentary Petrology*, **56**, 429–441.

—— 1988. A model for the development of phosphatic and calcareous lithofacies in the Middle Cambrian Thorntonia Limestone, northeast Georgina Basin, Australia. *Australian Journal of Earth Sciences*, **35**, 111–130.

TUCKER, M. E. 1986. Carbon isotope excursions in Precambrian/Cambrian boundary beds, Morocco. *Nature*, **319**, 48–50.

VEIZER, J. & COMPSTON, W. 1974. $^{87}Sr/^{86}Sr$ composition of seawater during the Phanerozoic. *Geochimica et Cosmochimica Acta*, **38**, 1461–1484.

—— & HOEFS, J. 1976. The nature of $^{18}O/^{16}O$ and $^{13}C/^{12}C$ secular trends in sedimentary carbonate rocks. *Geochimica et Cosmochimica Acta*, **40**, 1387–1395.

——, HOLSER, W. T. & WILGUS, C. K. 1980. Correlations of $^{13}C/^{12}C$ and $^{34}S/^{32}S$ secular variations. *Geochimica et Cosmochimica Acta*, **44**, 579–587.

——, COMPSTON, W., CLAUER, N. & SCHIDLOWSKI. M. 1983. $^{87}Sr/^{86}Sr$ in Late Proterozoic carbonates: Evidence for a 'mantle event' at 900 Ma ago. *Geochimica et Cosmochimica Acta*, **47**, 295–302.

——, ——, —— & —— 1985. Carbonates and ancient oceans: Isotopic and chemical record on time scales of 10^7-10^9 years. *In*: SUNDQUIST, E. T. & BROECKER, W. S. (eds) *The Carbon Cycle and Atmospheric CO_2: Natural Variations Archaean to Present*. American Geophysical Union Geophysical Monograph Series, **32**, 595–601.

ZEMPOLICH, W. G., WILKINSON, B. H. & LOHMANN, K. C. 1988. Diagenesis of Late Proterozoic carbonates: the Beck Spring Dolomite of eastern California. *Journal of Sedimentary Petrology*, **58**(4), 656–672.

ZHARKOV, M. A. 1981. *History of Paleozoic Salt Accumulation*. Translated by R. E. SSORKINA, R. V. FURSENKO & T. I. VASILIEVA. Springer-Verlag, Berlin.

Phosphogenic events and skeletal preservation across the Precambrian−Cambrian boundary interval

M. D. BRASIER

Department of Earth Sciences, The University, Parks Road, Oxford OX1 3PR, UK

Abstract: Refined biostratigraphic data from India and new data from Iran confirm that several major phosphogenic events took place within the Precambrian−Cambrian boundary interval across southern and central Asia. The palaeoceanographic setting of these event is briefly reviewed with reference to changes in the biosphere, where it is shown to have coincided with a widespread radiation of skeletons of diverse mineralogy (siliceous, apatitic, aragonitic, calcitic). Although a connection between palaeoceanographic changes and biomineralization cannot yet be demonstrated, the possibility that phosphogenesis enhanced the fossil record at this point is tested.

Preliminary criteria are put forward and illustrated for the recognition of primary phosphatic skeletons, secondarily phosphatized organic remains and secondarily phosphatized carbonate skeletons. This latter category comprises a major proportion of phosphatic small shelly fossils in the Chinese and Iranian sequences. Three kinds of phosphate taphofacies are then distinguished for early skeletal biotas: residual phosphatic, phosphatized carbonate and phosphatized organic facies (i.e. lägerstatten conditions). In several phosphogenic regions, the skeletal fossil record begins with residual phosphatic assemblages, followed by phosphatized carbonate taphofacies. Such diagenesis played an important role in enhancing the fossil record at this time because the first skeletons were thin (5−40 μm) and vulnerable to neomorphism in dolomitic or clastic facies. Phosphatized organic remains appeared locally during the major phosphogenic events, perhaps associated with more anoxic seafloor conditions that supressed the shelly calcareous benthos.

The explosion of animal life across the Precambrian−Cambrian boundary marks a turning point in the history of the biosphere (e.g. Brasier 1979; Conway Morris 1987). It appears to have coincided with the rifting of cratons (Cook & McElhinny 1979; Thorpe *et al.* 1984) and palaeoceanographic changes that involved sea level rise (Brasier 1980, 1981), the widespread formation of phosphatic deposits and phosphorites (Brasier 1980; Cook & Shergold 1984, 1986a) and distinctive shifts in the stable isotopes of sulphur (Holser 1977) and carbon (Tucker 1986; Magaritz *et al.* 1986; Aharon *et al.* 1987; Brasier & Magaritz 1989).

Biostratigraphic, palaeontological and geological evidence for ocean−atmosphere changes has been reviewed elsewhere (Conway Morris 1987; Brasier & Cowie 1989; Donnelly *et al.* 1990; Brasier 1990) and will be outlined further below. Of all the events contained within the Precambrian−Cambrian boundary interval, it could be argued that the sequence of shifts from heavy to light stable isotopes of carbon in carbonates, and the contemporaneous widespread appearance of phosphorites, represents the major turning point from Precambrian to Cambrian palaeoceanography.

Phosphatic skeletons are known to have been widespread amongst the earliest skeletal assemblages, but were they predominant as some have implied (e.g. Lowenstam & Margulis 1980) or has secondary phosphatization blurred the picture? And what connection could there have been between the phosphogenic events and the contemporaneous Cambrian radiations of invertebrates? Several kinds of interaction require investigation: the influence of phosphorus on the microplankton, invertebrate trophic type, biomass, and trophic stability (e.g. Brasier 1980; Cook & Shergold 1984; cf. Tyson 1987); on excretion and secretion of the earliest phosphatic and calcareous shells (Rhodes & Bloxam 1969; Cook & Shergold 1984); and on preservation of the earliest carbonate skeletons (Brasier 1979; Brasier & Hewitt 1979; Runnegar 1985).

Of the foregoing points, the question of preservation has paramount implications for stratigraphy and interpretation of the fossil record. This is because selection of a marker point and stratotype for the Precambrian−Cambrian boundary will be guided by biostratigraphic principles, to be placed as close as practicable to the lowest known appearance of diverse shelly fossils with a good potential for correlation (Cowie 1989). Both the Siberian and South Chinese candidate stratotype sections have boundary points marked by the appearance of distinctive early skeletal assemblages but these are largely known from phosphatized, glauconitized or phosphatic skeletal remains retrieved from acid residues of carbonate sediments (Matthews & Missarzhevsky 1975; Rozanov & Sokolov 1984; Luo et al. 1984). The Newfoundland candidate section, however, has a boundary point marked by changes in trace fossils (Narbonne et al. 1987). Appearance of the first calcareous skeletons there depended upon locally reducing conditions, leading to pyritic infilling of the shells (Landing pers. comm. 1987).

It follows that if early skeletal assemblages are to be used for stratigraphic or evolutionary interpretations, it is important to learn more about taphonomic aspects of preservation, particularly in relation to phosphate. This paper examines the evidence for widespread phosphogenic events and briefly reviews their temporal and palaeoceanographic setting. The possibility that phosphatic taphofacies enhanced the fossil record across the Precambrian–Cambrian boundary is also explored. The Precambrian–Cambrian boundary is taken to lie near the top of the Zhongyicun Member phosphorite, at Marker B of the Meishucun section in China; this level and stratotype awaits ratification and its correlation with Siberian, Avalonian and other sections is under review (see Brasier 1989a).

Phosphogenic events near the boundary

The stratigraphy of Precambrian and Cambrian phosphorites has been reviewed by authors in Cook & Shergold (1986b), with overviews of the biostratigraphic framework by Shergold & Brasier (1986) and Notholt & Brasier (1986). These authors suggested that phosphogenesis occurred episodically through the late Precambrian to Middle Cambrian, associated with sea level cycles that may have been eustatic.

Although a major phosphogenic episode has long been known from the Zhongyicun Member, lower Meishucunian Stage (basal Cambrian) of South China, the correlation of comparable deposits elsewhere in Asia has been rather poorly constrained (e.g. Cook & Shergold 1984, 1986a; Shergold & Brasier 1986; Chaudhuri 1990) and even the relative age of the Chinese deposits has been in considerable doubt until recently (e.g. Cowie 1985). Three factors have latterly improved the potential for correlation of Asiatic boundary phosphorites. First, further work on the taxonomy of Chinese taxa (e.g. Yu Wen 1987; Qian & Bengtson 1989; Conway Morris Chen Meng'e 1990), Indian taxa (Brasier & Singh 1987) and Iranian taxa (Hamdi et al. 1989) has somewhat clarified understanding of both the systematics and stratigraphic associations within and beyond Chinese assemblages. Second, work has also progressed on the distribution and first appearance datum (FAD) of Chinese taxa, leading to the suggested correlation of the Zhongyicun Member phosphorites with the Lower Tal phosphorites in India, and with phosphorites at similar levels in Pakistan and Iran (Brasier & Singh 1987; Jiang 1987; Hasan 1986; Jiang et al. 1988; Hamdi et al. 1989). Finally, stable isotopic work on this boundary phosphorite has shown distinctive stable carbon isotope excursions (Banerjee 1986, Lambert et al. 1987, Aharon et al. 1987) with potential for correlation if calibrated against regional and global bioevents and sequence stratigraphy (e.g. Brasier 1987, 1989a, Brasier & Magaritz 1989; Brasier et al. 1990).

One possible correlation scheme put forward by Brasier & Singh (1987) and Brasier (1989a, fig. 3.12) and developed here (Fig. 1) is based on the FADs of selected taxa, adjusted to the assumption of a double, phosphogenic event (and light carbon isotope excursions) close to the Precambrian–Cambrian boundary. The age of some phosphorites, however, is not yet closely resolved: e.g. those with skeletal fossils below the Lower Tal phosphorite in the Lesser Himalaya (Azmi & Pancholi 1983; Banerjee 1986), and unfossiliferous phosphorites (and stone coal = ?tasmanite) up to 16 m thick at the base of the Badaowan Member of the Qiongzhusi Formation, above Zone II and below Zone III in South China. The latter lies at about the second level of very light carbon isotopes, reported by Xu Daoyi et al. (1985) and Hsü et al. (1985). Hence, there were at least two phosphogenic events in this interval, of which the Zhongyicun event was arguably the most widespread.

How do these events correlate with the fabled Siberian sections? The writer infers, from combined stable isotopic and palaeontological evidence (Brasier et al. 1990), that Zones I and II in South China are of pre-Tommotian age, correlating with the *Anabarites trisulcatus* and *Purella antiqua* Zones of the Nemakit–Daldynian as defined by Khomentovsky (1986). The basal Tommotian (as defined at Bed 8, Ulukhan Sulugur) may correlate with upper-most Zone

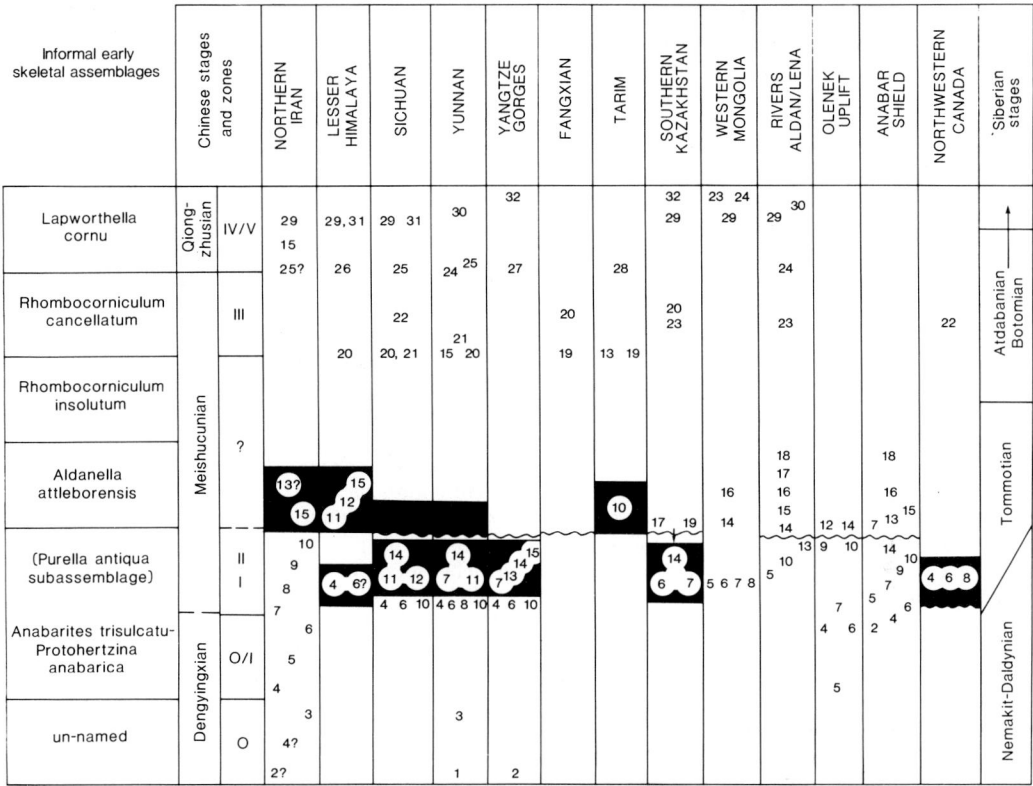

Fig. 1. A possible correlation of major phosphorite horizons, southern and central Asia, based on occurrences of late Precambrian *Chuaria* group megaflora (1), Ediacara fauna (2), *Tyrasotaenia* and *Vendotaenia* group megaflora (3) and first appearance datum of early skeletal fossils (4 to 32): 4, *Protohertzina anabarica* group protoconodonts; 5, *Cambrotubulus decurvatus* tubes; 6, *Anabarites trisulcatus* tubes; 7, *Tiksitheca licis* tubes; 8, *Maikhanella multa* group, spinose, cap-shaped molluscs; 9, *Purella antiqua* group cap shaped molluscs; 10, Circothecidae hyoliths; 11, *Arthrochites* (=*Barbitositheca*) *ansatus* group ?conulariids; 12, *Maldeotaia bandalica* group protoconodonts; 13, *Aldanella attleborensis* group pelagiellid snails; 14, *Latouchella korobkovi* group monoplacophorans; 15, Allathecidae hyoliths; 16, *Torellella lentiformis* tubes; 17, *Torellella biconvexa* tubes; 18, *Mobergella radiolata* discs; 19, *Cambroclavus* spp. sclerites; 20, *Allonnia tripodophora* group sclerites; 21, *Tannuolina zhangwentangi* group sclerites; 22, *Lapworthella filigrana* group sclerites; 23, *Rhombocorniculum cancellatum* pseudoconondonts; 24, *Lapworthella cornu* group sclerites; 25, *Eoredlichia* sp. trilobites; 26, redlichiid trilobite impression; 27, *Tsunyidiscus* sp. trilobites; 28, *Schizudiscus* sp. trilobites; 29, *Pelagiella lorenzi* group pelagiellid snails; 30, *Botsfordia caelata* brachiopods; 31, *Diandongia pista* brachiopods; 32, *Hupeidiscus orientalis* trilobites. Major phosphatic levels are shown in black. From data in Brasier (1989*a*).

II and the basal Qiongzhusi Formation, or perhaps with the profound hiatus between them.

Palaeoceanography and evolution near the boundary

Questions of environmental and evolutionary change have made the establishment of a stratigraphic framework for the boundary interval a matter of some urgency. Donnelly *et al.* (1990) have used evidence from strontium, sulphur, and carbon isotopes to reconstruct the sequence of palaeoceanographic events through the late Proterozoic to Cambrian, building on earlier models (e.g. Brasier 1980; Cook & Shergold 1984, 1986*a*; Knoll *et al.* 1986; Aharon *et al.* 1987). The writer has also reviewed isotopic and other geochemical, petrographic and palaeontological criteria to constrain models relating to changes in atmospheric oxygen, carbon dioxide and the evolutionary radiation of trace fossils and skeletons across the Precambrian–

Cambrian boundary interval (Brasier 1990).

Poorly oxygenated water masses appear to have been widespread during the late Precambrian (i.e. up to and including parts of the Rovnian, Nemakit–Daldynian and Meishucunian Stages). This can be inferred from the phosphogenic events themselves (Brasier 1980; Cook & Shergold 1984), from isotopic evidence for the sequestration of reduced sulphide and carbon (Holser 1977; Knoll et al. 1986; Lambert et al. 1987; Aharon et al. 1987) and from the widespread preservation of carbonaceous algal and invertebrate remains (Sokolov & Fedonkin 1986). This putative anoxia may, in turn, imply a release of oxygen into the atmosphere, steepening the oxygen gradient in the water column. Disappearance of soft-bodied preservation seen in the Ediacara fauna during the Kotlinian to Nemakit–Daldynian Stages, and the ensuing radiation of deeper burrowing traces, need to be investigated in the light of possible changes in oxygen levels (Brasier 1990).

One or more episodes of improved circulation and upwelling may account for the massive deposition of commercial grade phosphorites in the early Meishucunian and Tommotian Stages, as discussed above (Fig. 2). Skeletal invertebrates and skeletal algae appeared abundantly (Luo et al. 1984; Riding & Voronova 1984). Reported transitions from aragonitic to calcitic skeletal preservation and ooid precipitation (e.g. Riding 1985; Kazmierczak et al. 1985; Tucker 1987) require further investigation, not least in the context of presumed release of large amounts of dissolved carbon dioxide to the atmosphere at times of upwelling (Donnelly et al. 1990). Widespread appearance of diverse primary phosphate skeletons also coincided with wide phosphorus availability over this interval (Brasier 1980; Cook & Shergold 1984, 1986a). Even so, a chemical control over biomineralization in invertebrates cannot explain the great variety of materials that appear at this time (organic, agglutinated, siliceous, apatitic, aragonitic, calcitic; Brasier 1979). Such biomineralization may have been a response to skeletalization pressures, driven largely by predation, grazing and ecological escalation (e.g. Hutchinson 1961; Stanley 1976; Brasier 1979) but such pressures conceivably increased with increasing phosphorus, calcium and nutrient availability associated with upwelling.

The early skeletal fossil record

Before considering the potential impact of phosphogenesis on the early skeletal fossil record, it may be helpful to outline the broad sequence of biological events. Three main phases can be distinguished, broadly equivalent to the Nemakit–Daldynian, Tommotian and Atdabanian Stages of the Siberian Platform.

The first phase consists of simple tubes, mainly of phosphatic composition (*Hyolithellus*, *Rugatotheca*), or calcareous composition (*Cambrotubulus*, *Anabarites*); plus small phosphatic jaw-like protoconodonts of *Protohertzina*, and organic tubes of *Sabellidites*. These are found in the lower parts of the Nemakit–Daldynian Stage (*sensu* Khomentovsky 1986 = Manykayan Stage of Missarzhevsky 1982, 1983). Other Soviet authors largely restrict the term Nemakit–Daldynian to this kind of assemblage (e.g. Rozanov 1984) and the term Rovnian has been used for the East European Platform (Sokolov & Fedonkin 1984). The writer has used the term *Anabarites trisulcatus*–*Protohertzina anabarica* Assemblage for these faunas (Brasier 1989b).

The succeeding phase bears assemblages in which tubes and protoconodonts are joined by cap-shaped molluscs (e.g. *Maikhanella*, *Canopoconus*, *Purella*), coiled molluscs (e.g. *Latouchella*, *Archaeospira*, *Aldanella*), hyolith tubes (e.g. *Ladatheca*, *Conotheca*, *Turcutheca*), and by an array of phosphatic sclerites (e.g. *Tommotia*, *Sunnaginia*, *Lapworthella*), calcareous sclerites (e.g. *Chancelloria*, *Halkieria*) and many other problematica. This kind of assemblage is found from the upper part of the Nemakit–Daldynian (*sensu* Khomentovsky 1986) or lower part of an extended Tommotian Stage (*sensu* Rozanov 1984). It occurs sparingly in Meishucunian Zone I but is characteristic of Zone II of China (Luo et al. 1982, 1984; Brasier 1989a). This phase ranges from higher parts of the *Anabarites trisulcatus*–*Protohertzina anabarica* Assemblage through to the *Rhombocorniculum insolutum* Assemblage of Brasier (1989b).

The third phase is marked by the appearance of trilobite exoskeletons. This phenomenon of delayed calcification appears to have been diachronous, appearing first in the early Atdabanian strata in Siberia and Baltica, in mid- to late Atdabanian strata in Avalonia and in Qiongzhusian strata of about this age in China.

A three-phase succession such as this could be interpreted as ecologically biased (e.g. Brasier 1979), reflecting a shift from predominant inshore to offshore environments during transgression. But the widespread association between assemblages of phase 2 type (above) and shallow marine facies in China, Siberia and Newfoundland, indicates that evol-

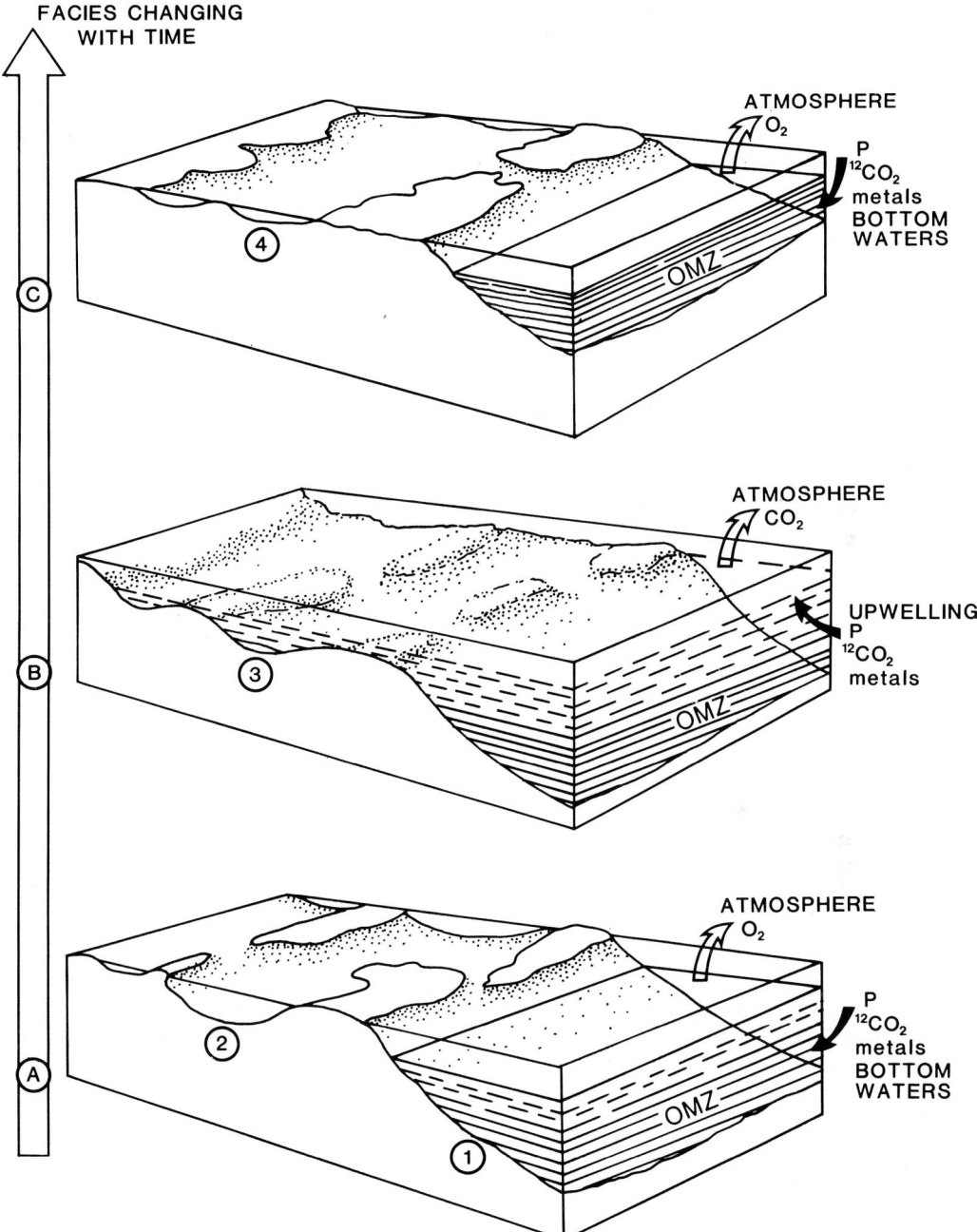

Fig. 2. A palaeoceanographic model for phosphogenesis and isotopic changes across the Precambrian–Cambrian boundary. Stage A, an oxygen deficient water mass (OMZ) lies offshore, storing phosphorus, carbon dioxide (enriched in ^{12}C), metals. Extensive deposition of reduced carbon and pyrite (1) cause heavy carbon and sulphur isotope excursions in shallow water carbonates (2), e.g. Baiyanshao to Xiaowaitoushan dolomites of Meishucun, or the Krol E unit of Lesser Himalaya. Stage B, transgression and/or upwelling cause the oxygen deficient water masses to spill over the shelf, depositing phosphorite and isotopically light carbonates (3), e.g. Zhongyicun phosphorites of Meishucun, Lower Tal phosphorites of Lesser Himalaya. Stage C returns to the conditions in A, e.g. Dahai Member of Meishucun.

utionary, migratory and taphonomic factors were perhaps more important. The question of taphonomy is considered further below.

Primary and secondary phosphatic skeletons

If the earliest skeletons were all primarily phosphatic in composition, as has often been implied in the literature (e.g. Lowenstam & Margulis 1980; Jiang Zhiwen 1984), then evolutionary and biostratigraphic interpretation of the fossil record would seem straight forward. Unfortunately, taphonomic analysis indicates that many early skeletal remains were of carbonate or organic matter that was phosphatized secondarily.

Below are set out some criteria which have assisted distinction between these various kinds of phosphatic remains.

Primary phosphate skeletons

A primary apatite composition for the skeleton for a given taxon may be inferred from the following criteria.
(i) Consistently apatitic preservation in the absence of evidence for phosphatic diagenesis (e.g. overgrowths, neoformed internal casts, phosphatic pellets).
(ii) Consistently apatitic composition when associated with other taxa preserved in primary carbonate or organic matter.
(iii) Similar thickness of apatite within specimens of similar size from the same horizon.
(iv) Evidence for consistent, finely preserved ultrastructure, such as microgranular apatitic fibres (Fig. 3b). Complex ultrastructures, such as mural pores (Fig. 3a) may be well-preserved.
(v) General absence of microborings in skeleton.
(vi) Relatively consistent chemical composition of skeleton within and between horizons (e.g. Müller 1979). This suggestion requires further study.

The following taxa meet these criteria: *Hyolithellus*, *Torellella* and other Hyolithelminthes, *Sunnaginia*, *Camenella*, *Tommotia*, *Lapworthella*, *Eccentrotheca*, *Tannuolina* and other tommotiids; *Siphogonuchites* and other siphogonuchitids; *Protohertzina*, *Hertzina*, *Amphigeisina*, *Gapparodus* and other protoconodonts; *Rhombocorniculum* and other pseudoconodonts, Paterinida, Lingulida and early Cambrian Acrotretida (Brachiopoda), Phosphatocopina (Ostracoda), *Mobergella*, *Hadimopanella*, *Microdictyon* and *Cowiella* (Problematica).

Secondarily phosphatized carbonate skeletons

Criteria which suggest that a primary carbonate skeleton was subsequently phosphatized are as follows:
(i) Phosphatized skeleton preserved elsewhere as $CaCO_3$, giving indications of calcitic or aragonitic composition in thin section. Phosphatic rims may be seen on inner or outer walls of the carbonate skeleton and even replace the skeleton; phosphate often infills cavities or metasomatically replaces sediment infills.
(ii) Carbonate skeleton dissolves away on treatment with dilute acetic or formic acid, to leave an empty mould and/or a phosphatic cast/steinkern.
(iii) Phosphatic casting of the inner skeleton surface may mould carbonate ultrastructure, such as the inner faces of prisms, nacre, foliated calcite, or crossed lamellar structures (Fig. 4a; see also Runnegar 1985).
(iv) Phosphatic replacement of the skeleton may replicate carbonate shell fabrics (see Runnegar 1985).
(v) Phosphatized carbonate skeletons may contain microborings of algae or fungi, or endolithic casts may occur on the phosphatic steinkern (Fig. 4b; see also Runnegar 1985). The margins of such casts may replicate carbonate shell ultrastructure.
(vi) Variable chemical composition of phosphatic skeleton between horizons, and variable preservation, may result from variable degrees of phosphate metasomatism. This aspect requires further study.

The following had carbonate skeletons according to these criteria: tubes *Anabarites*, *Tiksitheca*, *Cambrotubulus*, *Coleoloides*, *Coleolus*, *Spinulitheca*, *Ladatheca*; all Hyolitha; all monoplacophoran, paragastropod, rostroconch and bivalve molluscs; chancelloriid and halkieriid scelerites; archaeocyathan sponges, obolellid brachiopods and trilobite arthropods. This list includes the bulk of taxa and individuals in Precambrian–Cambrian boundary assemblages.

Secondarily phosphatized organic remains

Phosphatization of primarily organic-walled or weakly biomineralized organisms is reasonably easy to recognize with regard to arthropod ap-

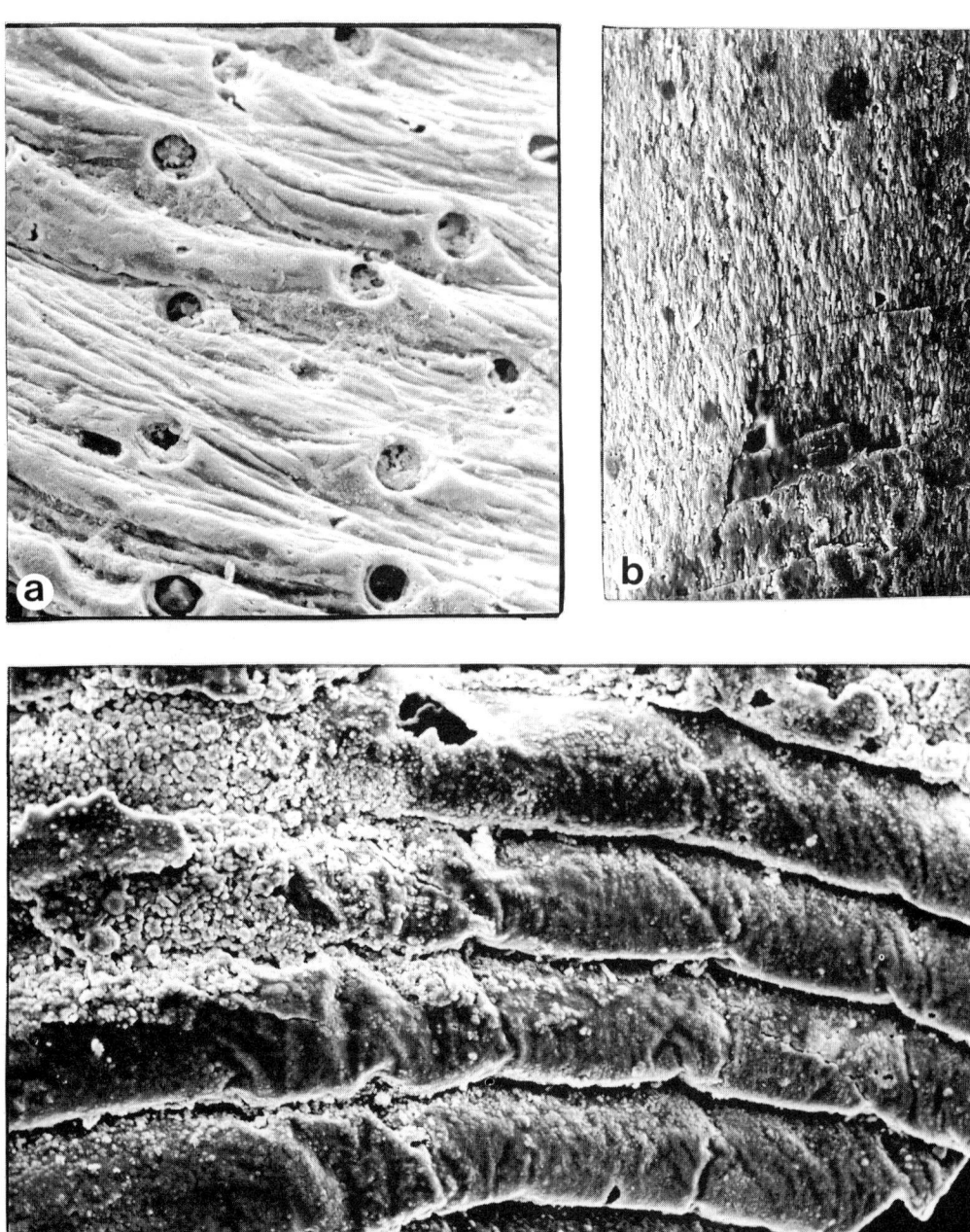

Fig. 3. Examples of primary phosphatic (a, b) and phosphatized organic (c) skeletal ultrastructure.
(**a**) External view of primary phosphatic skeleton of *Tannuolina* sp. showing growth ridges of microcrystalline apatite traversed by small pores (× 300), Qiongzhusi Formation, Meishucun, China. (**b**) External view of primary phosphatic protoconodont element of *Maldeotaia* sp., showing small apatitic fibres arranged parallel to the element axis (× 300), Lower Tal Formation, Maldeota, India. (**c**) Internal view of phosphate-impregnated flexible organic skeleton of conulariid-like *Hexangulaconularia*, showing wrinkled annulaae weakly joined together, with localized formation of secondary apatite microspherules at top left (× 300), Lower Tal Formation, Maldeota, India.

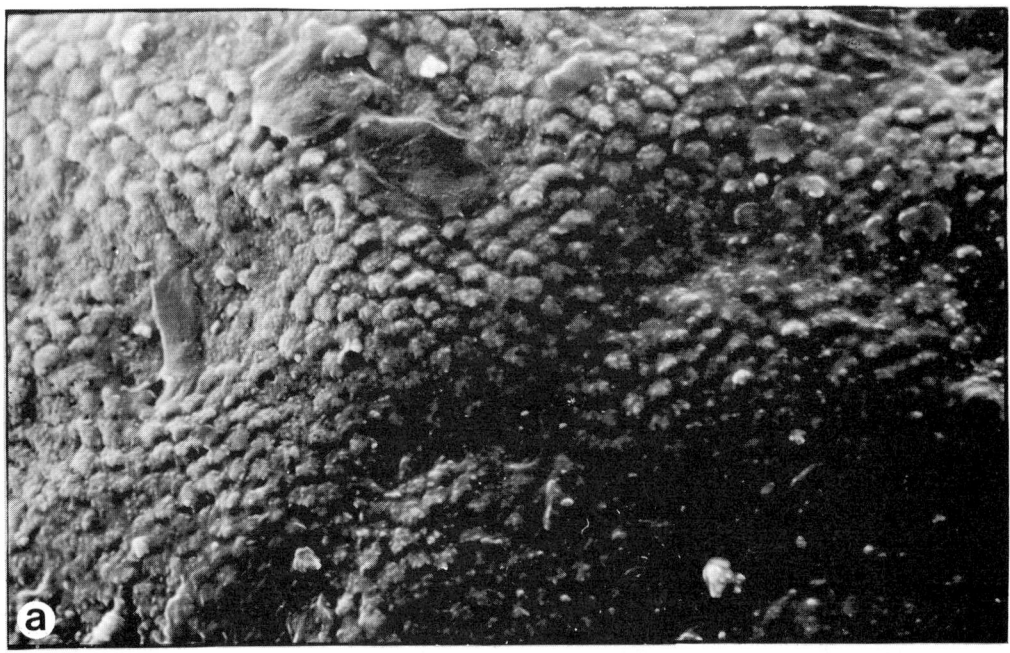

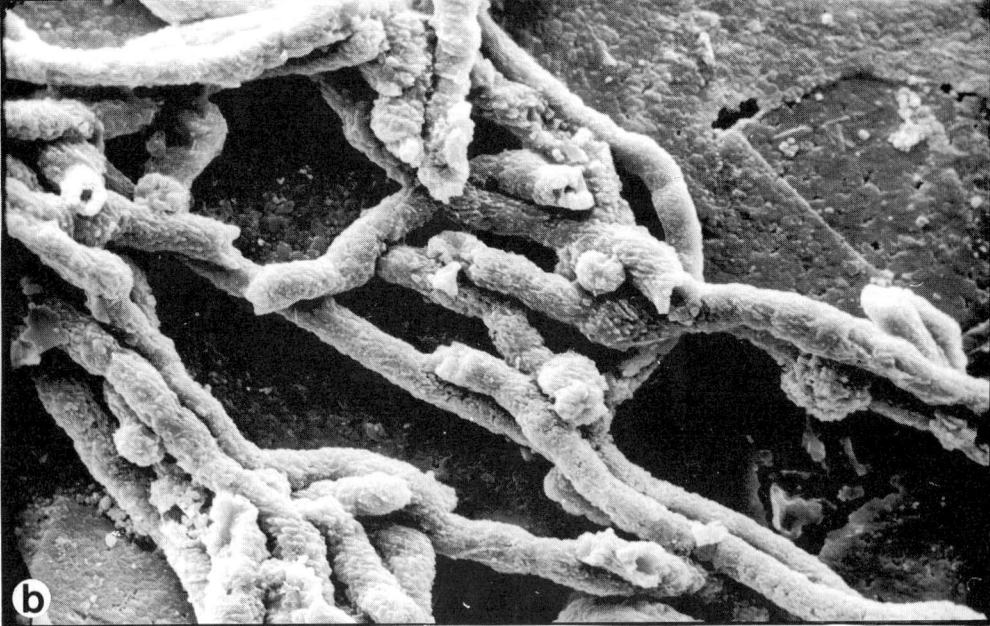

Fig. 4. Examples of phosphatized carbonate skeletal ultrastructure. (**a**) External view of a phosphatic steinkern of a monoplacophoran allied to *Xiandongoconus*, molding finer scale polygonal imprints of structures resembling prismatic carbonate, and broader scale lobate structures of the inner shell surface ($\times$ 600), Soltanieh Formation, Elburz Mountains, Iran. (**b**) External view of a phosphatic steinkern of a tube of *Anabarites*, molding filamentous endolithic borings which clearly show the longitudinal fibrous ultrastructure of the inner calcareous shell layer ($\times$ 1000), Soltanieh Formation, Elburz Mountains, Iran.

pendages (e.g. Müller 1979), vertebrate collagen (e.g. Martill 1988) and plant seeds (e.g. Allison 1988) but the phenomenon has barely been acknowledged in Precambrian–Cambrian boundary problematica. The following criteria may assist in recognition of such lagerstätten.

(i) Fossils in which remains of similar morphology are otherwise preserved as organic matter, as soft-bodied impressions, or are restricted to special conditions of preservation.
(ii) Remains showing irregular folds, tears, crumpling, wrinkle marks, compression or other soft tissue deformations (Fig. 3c).
(iii) Remains showing signs of post-mortem disaggregation or decay (e.g. in areas of thinner cuticle) prior to phosphate mineralization.
(iv) Remains in which the phosphatic skeleton varies widely in thickness between specimens of similar size, owing to variable degrees of mineralization.

Examples from the Precambrian–Cambrian boundary include faecal pellets, mucilaginous sheaths of ?cyanophytes such as *Spirellus* and *Cambricodium*, vesicles of ?tasmanitid prasinophytes such as *Olivooides* and acritarchs or sponges (*Archaeooides*), compressed annulated worm tubes (e.g. *Rushtonia* sp. of Nowlan et al. 1985), and the compressed and crumpled cuticles of *Arthrochites* (= *Barbitositheca*) and *Hexangulaconularia* (Fig. 3c) that are questionably related to the conulariids. These unusual forms have skeletal distributions largely restricted to phosphatic beds.

It appears from this brief survey that phosphatic early skeletal remains are extremely heterogenous, with the bulk of remains belonging in the second category (phosphatized carbonate skeletons). Much remains to be done on early biomineralization and skeletal taphonomy.

Phosphatic taphofacies

Once these various categories of phosphatic skeletons have been distinguished, differences in the preservational style of certain facies become apparent. First, there is the distinction between facies which contain or lack taxa with primary phosphate skeletons (Brasier 1980). Such taxa are widespread, for example, in condensed, glauconitic and phosphatic deposits of ocean-facing aspect, such as England, the Baltic, (see Ilyin & Heinsalu 1990) Newfoundland, China and Siberia, though they need not be abundant within phosphatic deposits *per se*. Such elements (e.g. phosphatic brachiopods, protoconodonts) were initially lacking from the carbonate banks and inner detrital belts of North America and the Morocco–southern European margins of Gondwana, either because of migrational barriers, water chemistry or higher depositional rates.

Within the phosphogenic provinces themselves, it is possible to distinguish three broad kinds of phosphatic taphofacies. Residual phosphatic dominant (RPD) taphofacies are typical of nearshore and littoral clastic deposits and contain accumulations of phosphatic inarticulate brachiopod shells, hyolithelminth tubes, tomotiid sclerites and problematica. A scarcity of calcareous shells may be attributed to post-mortem destruction and diagenesis in units near the base of transgressive cycles (Brasier 1979; Brasier & Hewitt 1979). The *Mobergella* beds of Baltica and '*Obolus*' *groomi* facies at the base of the Comley Sandstone and Home Farm Member in England are clear examples of this type. Early Nemakit–Daldynian assemblages of low diversity with *Hyolithellus* (Iran) or *Protohertzina* (southern Kazakhstan) may also belong here.

Phosphatized carbonate dominant (PCD) taphofacies are typical of nearshore and littoral carbonate deposits and contain accumulations of phosphatised carbonate shells (e.g. hyoliths, *Anabarites, Coleoloides*, monoplacophorans, paragastropods, chancellorids) as well as primary phosphate shells (e.g. brachiopods, *Hyolithellus*). Examples include the Xiaowaitoushan, Zhongyicun and Dahai Members of Meishucun (see Luo et al. 1982, 1984), the fossiliferous upper Yudomian and lower Tommotian strata of Siberia (e.g. Rozanov & Sokolov 1984; Khomentovsky 1986), most of the Soltanieh Formation of Iran (Hamdi et al. 1989), Bed 1 of the Hyolithes Limestone at Nuneaton (Brasier 1984, 1986) and many other deposits.

Phosphatized organic dominant (POD) taphofacies are known from nearshore to off-shore carbonates and pelletal phosphorites. They typically contain phosphatic faecal pellets, phosphatized cyanophytes and acritarchs (e.g. *Spirellus, Cambricodium, Olivooides*), ?conulariids (e.g. *Hexangulaconularia*) and well-preserved primary phosphate skeletons (e.g. protoconodonts). Such taphofacies are found locally within the boundary phosphorites at Maidiping and Maldeota (e.g. Luo et al. 1982, 1984; Xing et al. 1984; Brasier & Singh 1987) and occur sporadically elsewhere in the Lower to Middle Cambrian (see Nowlan et al. 1985; Peel 1988; Soudry & Southgate 1989). The POD taphofacies contains few primarily carbonate elements but POD elements may range into, or be admixed with, PCD taphophacies.

Those conditions which led to POD tapho-

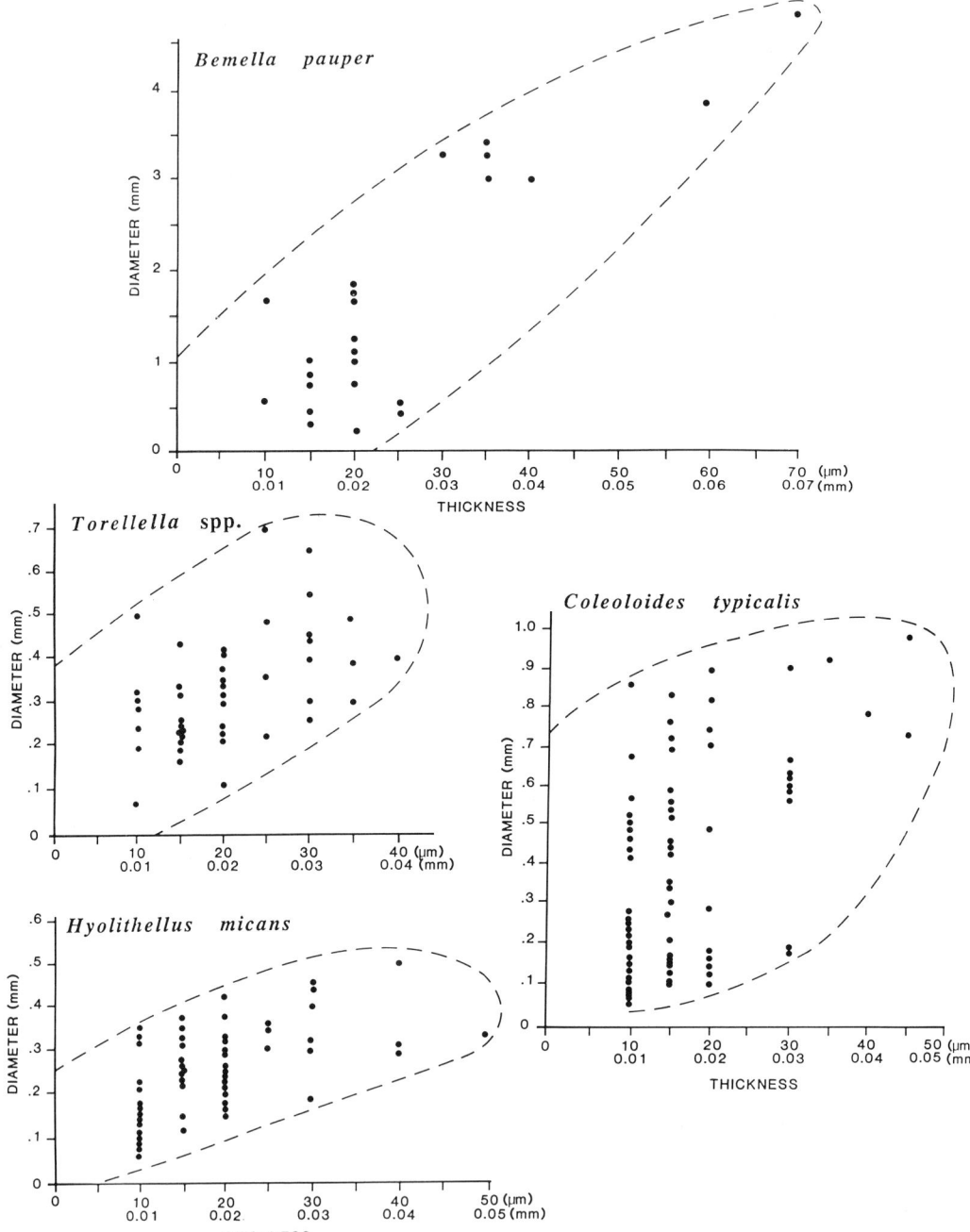

Fig. 5. Variations in shell thickness of calcareous, cap-shaped mollusc *Bemella pauper*, phosphatic tubes of *Torellella* sp. and *Hyolithellus micans*, and calcareous tubes of *Coleoloides typicalis* from the Hyolithes Limestone, English Midlands. Measurements determined from random thin sections are plotted against the maximum observable shell diameter (*Bemella pauper*) or tube diameter.

facies, and to variable qualities of preservation within PCD taphofacies, require much further research, but bacterial destruction of organic matter in the presence of calcium-enriched and magnesium-depleted micro-environments are implicated (Lucas & Prévôt 1985; Prévôt & Lucas 1986; Soudry & Southgate 1989).

The taphonomic window

Conditions were not good for the preservation of carbonate shells in the latest Precambrian to earliest Cambrian, for several reasons. First, it seems that the earliest shells were extremely thin walled, originating perhaps from minute granules and spicules set in soft tissues, or as weakly mineralized cuticles (e.g. Lowenstam & Margulis 1980; Conway Morris 1987). For example, phosphatic tubes of *Hyolithellus* and *Rugatotheca* and protoconodonts of *Protohertzina* from the *Anabarites trisulcatus – Protohertzina anabarica* interval in Iran have walls in the range of 10–30 µm thick. Associated phosphatized carbonate shells of early *Anabarites trisulcatus* appears to have been in the region of 10 µm thick, and one tube wall of *Tiksitheca licis* is merely some 5 µm thick. A more systematic study of skeletal thickness in the Hyolithes Limestone of the English Midlands (c. mid-Tommotian–lower Atdabanian) confirms this pattern, with *Hyolithellus*, *Torrellella*, *Bemella* and *Coleoloides* all falling in the 10–50 µm range, proportional to shell size (Fig. 5). Such thin carbonate shells were relatively vulnerable to neomorphism of the shell or matrix, to dolomitic recrystallization or to dissolution (cf. Bannér & Wood 1964).

A second taphonomic filter concerns the nature of the sediments themselves. Carbonates were widespread across the shallow platforms of Siberia, China, India, Iran, Kazakhstan (USSR) and North America in the late Precambrian, but dolomites (e.g. Tucker 1982) or carbonates of aragonitic origin (Sandberg 1983; Tucker 1987) appear to have been much more common than calcitic carbonates. Neomorphic and dolomitic destruction of shells may therefore have taken place, leaving little trace of very small or thin carbonate elements. The likely existence of such elements is illustrated by calcified sponge spicules from the Doushantuo Formation carbonates in Hubei, and calcareous tubes of *Cloudina* from Nama Group limestones in Namibia (both presumed to be older than Nemakit–Daldynian; Xing *et al.* 1985, Germs 1983). Calcareous tubes survived in partially dolomitized carbonates near the boundary in California, Nevada and Mexico (e.g. Signor *et al.* 1987) but these were recrystallized, thick (c. 100 µm) walled shells, locally enhanced by ankeritic rinds.

Clastic sediments close to the Precambrian–Cambrian boundary (e.g. the Chapel Island Formation of Newfoundland, Rovno Formation of Eastern Europe, Vampire Formation of northwestern Canada) contain relatively few carbonate remains. Such sediments might be expected to have a low preservation potential for early carbonate shells (Brasier 1979, fig. 5).

These adverse factors were diminished in settings where authigenic mineralization (especially of phosphate, glauconite and pyrite) was taking place in the sediment. As discussed above, the Iranian–Chinese margins of Gondawana, Kazakhstan and Siberia, were regions where environments favourable to phosphogenesis and to skeletal benthos lay in close proximity during the Precambrian–Cambrian boundary interval: oxygen depleted, phosphorus-enriched water masses of the outer shelf and slope; and well oxygenated coastal embayments and carbonate lagoons (e.g. Li 1986). A similar setting is known from the Ordian phosphogenic episode of Australia (Southgate 1988) where upwelling of nutrient enriched waters along palaeobathymetric highs resulted in increased organic productivity in the adjacent embayments and phosphogenesis in topographic traps.

Tubular carbonate shells (e.g. *Anabarites*, *Cambrotubulus*) appear to have provided microenvironments especially suitable to the activities of phosphogenic bacteria, as indicated by the appearance of phosphatic steinkerns and replicas in early assemblages. Indeed, a relationship may exist between shell morphology, size and authigenic mineralization in early skeletal fossils. In the Hyolithes Limestone, for example, very narrow carbonate tubes of *Coleoloides* are commonly found infilled with phosphate (Bed 1), calcite spar (Beds 2–12) or iron oxide and glauconite (Beds 12, 13); the narrow phosphatic tubes of *Hyolithellus* and *Torellella* are infilled with phosphate (Bed 1) or iron oxide/glauconite (Beds 11, 12); the narrow phosphatic protoconodont elements of *Hertzina* are infilled with phosphate (Bed 1); the enclosed tubular, carbonate sclerites of *Halkieria*, *Chancelloria* and hollow sponge spicules are infilled with glauconite (Bed 10). Larger hyoliths and molluscs have not been found so preserved. Müller & Walossek (1985) also found a relationship between phosphatized cuticle, small size (0.2 to 2.0 mm) and enclosed spaces in Upper Cambrian arthropods from POD taphofacies and the majority of phosphatized small shelly fossils from the Precambrian–Cambrian boundary interval are of about this size (e.g. Rozanov et al. 1969; Matthews & Missarzhevsky 1975; Luo et al. 1982, 1984; Runnegar 1985; Yu Wen 1987). Comparison can further be made with the precipitation of phosphate and glauconite in microfossils of more Recent date (e.g. Bjerkli & Ostrno-Saeter 1973; Kennedy & Garrison 1975; Prévôt & Lucas 1986; Carbone et al. 1987).

It follows that some skeletons larger than $c.$ 2 mm diameter may have been less amenable to phosphogenic preservation, lending yet a further bias to the fossil record. Larger hyolith remains seldom appear in small shelly fossil residues, though apical steinkerns (e.g. *Paragloborilus*), narrow tubes (e.g. *Circotheca*) or microconchs (e.g. *Microcornus*) are widely known.

Conclusion

Several factors may have cast a veil over the early skeletal fossil record: the solid skeletons of many early invertebrate fossils were thin ($c.$ 5 to 40 μm) or weakly mineralized, while facies unsuitable for the preservation of smaller carbonate skeletons were widespread. Tubular or confined shell spaces, however, provided sites suitable for the deposition of authigenic phosphate and glauconite minerals, and the fossil record of the first skeletal assemblages (the *Anabarites trisulcatus–Protohertzina anabarica* Assemblage) locally began as phosphatized carbonate skeletons in phosphatized carbonate-dominant (PCD) taphofacies. The climax of phosphatic preservation was reached during the Zhongyicun phosphogenic event, with the phosphatization of organic remains. It should even be considered whether the phosphogenic events themselves were related to the first widespread appearance of small shell spaces and abundant faecal/dead organic substrates, providing suitable traps for the accumulation of grainstone phosphorites for the first time in the stratigraphic record.

Part of the difficulty that palaeontologists have encountered in correlating early skeletal assemblages from phosphorites (e.g. Zone II of China; see review in Brasier 1989a) may relate to an inadequate appreciation of such taphonomic factors.

References

AHARON, P., SCHIDLOWSKI, M. & SINGH, I. B. 1987. Chronostratigraphic markers in the end-Precambrian isotope record of the Lesser Himalaya. *Nature*, **327**, 699–702.

ALLISON, P. 1988. Taphonomy of the Eocene London Clay biota. *Palaeontology*, **327**, 1079–1100.

AZMI, R. & PANCHOLI, V. P. 1983. Early Cambrian (Tommotian) conodonts and other shelly microfauna from the Upper Krol of Mussoorie Syncline, Garwhal Lesser Himalaya with remarks on the Precambrian–Cambrian boundary. *Himalayan Geology*, **11**, 360–372.

BANERJEE, D. M. 1986. Proterozoic and Cambrian phosphorites – regional review: Indian subcontinent. *In*: COOK, P. J. & SHERGOLD, J. H. (eds) *Proterozoic and Cambrian phosphorites*. Cambridge University Press, Cambridge, 70–90.

BANNER, F. T. & WOOD, G. V. 1964: Recrystallization in microfossiliferous limestones. *Geological Journal*, **4**, 21–34.

BJERKLI, K. and OSTRNO–SAETER, J. S. 1973. Formation of glauconite in foraminiferal shells on the continental shelf off Norway. *Marine Geology*, **14**, 169–178.

BRASIER, M. D. 1979. The Cambrian radiation event. *In*: HOUSE M. R. (ed.) *The origin of major invertebrate groups*. Systematics Association Special Volume, **12**, Academic Press, London, 103–159.

—— 1980. The Lower Cambrian transgression and glauconite-phosphate facies in western Europe. *Journal of the Geological Society London*, **137**, 695–703.

—— 1981. Sea-level changes, facies changes and the late Precambrian–early Cambrian evolutionary explosion. *Precambrian Research*, **17**, 105–123.

—— 1984. Microfossils and small shelly fossils from the Lower Cambrian *Hyolithes* Limestone at Nuneaton, English Midlands. *Geological Magazine*, **121**, 229–253.

—— 1986. The succession of small shelly fossils (especially conoidal microfossils) from English Precambrian–Cambrian boundary beds. *Geological Magazine*, **123**, 327–356.

—— 1987. 'Inner Tethyan' Precambrian–Cambrian boundary sequences from China, India, Pakistan and Iran. *Abstracts of the International Symposium on Terminal Precambrian and Cambrian Geology*, Yichang, 1–2.

—— 1989a. China and the Palaeotethyan Belt (India, Pakistan, Iran, Kazakhstan, and Mongolia). *In*: COWIE, J. W. & BRASIER, M. D. (eds) *The Precambrian–Cambrian boundary*. Clarendon Press, Oxford, 40–74.

—— 1989b. Towards a biostratigraphy of the earliest skeletal biotas. *In*: COWIE, J. W. & BRASIER, M. D. (eds) *The Precambrian–Cambrian boundary*. Clarendon Press, Oxford, 117–165.

—— 1989c. On mass extinction and faunal turnover near the end of the Precambrian. *In*: DONOVAN, S. (ed.) *Mass extinctions: processes and evidence*. Belhaven Press, London, 73–88.

—— 1990. Ocean-atmosphere chemistry and evolution across the Precambrian–Cambrian boundary. *In*: LIPPS, J. H. & SIGNOR, P. W. (eds) *Origins and early evolutionary history of the Metazoa*. Plenum Publishing Corporation, New

York. (In Press).
—— & COWIE, J. W. 1989. Concluding remarks. *In*: COWIE, J. W. & BRASIER, M. D. (eds) *The Precambrian–Cambrian boundary*. Clarendon Press, Oxford, 205–209.
—— & HEWITT, R. A. 1979. Environmental setting of fossiliferous rocks from the uppermost Proterozoic-Lower Cambrian of central England. *Palaeogeography, Palaeoclimatology, Palaeoecology*, **27**, 35–57.
—— & MAGARITZ, M. 1989. Towards an integrated carbon isotope – small shelly fossil stratigraphy for the Precambrian–Cambrian boundary. *Abstracts 28th International Geological Congress, Washington D.C.*
—— & SINGH, P. 1987. Microfossils and Precambrian–Cambrian boundary stratigraphy at Maldeota, Lesser Himalaya. *Geological Magazine*, **124**, 323–345.
——, MAGARITZ, M., CORHEILD, R., LUO HUILIN, WU XICHE, OUYANG LIN, JIANG ZHIWEN, HEMDI, B., HE TINGGUI & FRASER, A.G. 1990. The Carbon- and oxygen-isotope record of the Precambrian-Cambrian boundary internal in China and Iran and their correlation Geological Magazine, (in press).
CARBONE, S., GRASSO, M., LENTINI, F. & PEDLEY, H. M. 1987. The distribution and palaeoenvironment of early Miocene phosphorites of southeast Sicily and their relationships with the Maltese phosphorites. *Palaeogeography, Palaeoclimatology, Palaeoecology*, **58**, 35–53.
CHOUDHURI, R. 1990. Two decades of phosphorite investigations in India. *In*: NOTHOLT, A. J. G. & JARVIS, I. (eds) *Phosphorite Research and Development*. Geological Society, London, Special Publication, **52**, 305–311.
CONWAY MORRIS, S. C. 1987. The search for the Precambrian–Cambrian boundary. *American Scientist*, **75**, 157–167.
CONWAY MORRIS, S. C. & CHEN MENG E. 1990. Anabaritids from the Lower Cambrian of South China. *Geological Magazine*. (In press).
COOK, P. J. & McELHINNY, M. W. 1979. A reevaluation of the spatial and temporal distribution of sedimentary phosphate deposits in the light of plate tectonics. *Economic Geology*, **74**, 315–330.
—— & SHERGOLD, J. H. 1984. Phosphorus, phosphorites and skeletal evolution at the Precambrian–Cambrian boundary, *Nature*, **308**, 231–236.
—— & —— 1986a. Proterozoic and Cambrian phosphorites – nature and origins. *In*: COOK, P. J. & SHERGOLD, J. H. (eds) *Proterozoic and Cambrian phosphorites*. Cambridge University Press, Cambridge, 369–386.
—— & —— (eds) 1986b. *Proterozoic and Cambrian phosphorites*. Cambridge University Press, Cambridge.
COWIE, J. W. 1985. Continuing work on the Precambrian–Cambrian boundary. *Episodes*, **8**, 93–97.
—— 1989. Introduction. *In*: COWIE, J. W. & BRASIER, M. D. (eds) *The Precambrian–Cambrian boundary*. Clarendon Press, Oxford, 3–6.
COWIE, J. W. & BRASIER (eds) 1989. *The Precambrian–Cambrian boundary*. Clarendon Press, Oxford.
DONNELLY, T. H., SHERGOLD, SOUTHGATE, P. N. & BARNES, C. J. 1990. Events leading to global phosphogenesis around the Precambrian/Cambrian boundary. *In*: NOTHOLT, A. J. G. & JARVIS, I. (eds) *Phosphorite research and development*. Geological Society, London, Special Publication. **52**, 273–287.
GERMS, G. J. B. 1983. Implications of a sedimentary facies and depositional environmental analysis of the Nama Group in southwest Africa/Namibia. *Special Publication of the Geological Society of South Africa*, **11**, 89–114.
HAMDI, B., BRASIER, M. D. & JIANG ZHIWEN 1989. Earliest skeletal fossils from Precambrian–Cambrian boundary strata, Elburz Mountains, Iran. *Geological Magazine*, **126**, 283–289.
HASAN, M. T. 1986. Proterozoic and Cambrian phosphorites — deposits: Hazara, Pakistan. *In*: COOK, P. J. & SHERGOLD, J. H. (eds) *Proterozoic and Cambrian phosphorites*. Cambridge University Press, Cambridge, 190–201.
HOLSER, W. T. 1977. Catastrophic chemical events in the history of the ocean. *Nature*, **267**, 403–408.
HSÜ K. J., OBERHANSLI, H., GAO, J. Y., SUN SHU, CHEN, HAIHONG & KRAHENBUHL, U. 1985. 'Strangelove ocean' before the Cambrian explosion. *Nature*, **316**, 809–811.
HUTCHINSON, G. E. 1961. The biologist poses some problems. *In*: SEARS, M. (ed.) *Oceanography*. Publication of the American Association for the Advancement of Science. **67**, 85–94.
ILYIN, A. V. & HEINSALU, H. N. 1990. Early Ordovician shelly phosphorites of the Baltic Basin. *In* NOTHOLT, A. J. G. & JARVIS, I. (eds) *Phosphorite Research and Development*. Geological Society, London, Special Publication, **52**, 253–259.
JIANG ZHIWEN 1987. The nonisochroneity of phosphate deposition in earliest Cambrian of southern Asia. *Acta Geologica Sinica*, **3**, 201–210. (In Chinese).
——, BRASIER, M. D. & HAMDI, B. 1988. Correlation of the Meishucunian Stage in South Asia. *Acta Geologica Sinia*, **3**, 191–199. (In Chinese).
KAZMIERCZAK, J., ITTEKOT, V. & DEGENS, E. T. 1985. Biocalcification through time: environmental challenge and cellular response. *Palaeontologische Zeitschrift*, **59**, 15–33.
KENNEDY, W. J. & GARRISON, R. E. 1975. Morphology and genesis of nodular phosphates in the Cenomanian Glauconitic Marl of south-east England. *Lethaia*, **8**, 339–360.
KNOLL, A. H., HAYES, J. M., KAUFMAN, A. J., SWETT, K. & LAMBERT, I. 1986. Secular variation in carbon isotope ratios from Upper Proterozoic successions of Svalbard and East Greenland. *Nature*, **321**, 832–838.
KHOMENTOVSKY, V. V. 1986. The Vendian System of Siberia and a standard stratigraphic scale. *Geological Magazine*, **123**, 333–348.
LAMBERT, I. B., WALTER, M. R., ZANG WENLONG, LU

SONGNIAN & MA GUOGAN. 1987. Palaeoenvironment and carbon isotope stratigraphy of the Yangtze Platform. *Nature*, **325**, 140–142.

LI YUEYAN 1986. Proterozoic and Cambrian phosphorites — regional review: China. *In*: COOK, P. J. & SHERGOLD, J. H. (eds) *Proterozoic and Cambrian phosphorites*. Cambridge University Press, Cambridge, 42–62.

LOWENSTAM, H. A. & MARGULIS, L. 1980. Evolutionary prerequisites for early Phanerozoic calcareous skeletons. *BioSystems*, **12**, 27–41.

LUO HUILIN, JIANG ZHIWEN, WU XICHE, SONG XUELIANG & OUYANG LIN. 1982. *The Sinian–Cambrian boundary in Eastern China*. People's Publishing House, Yunnan, China.

——, ——, ——, —— & —— 1984. *Sinian–Cambrian boundary stratotype section at Meishucun, Jinning, Yunnan, China*. People's Publishing House, Yunnan, China.

MAGARITZ, M., HOLSER, W. T. & KIRSCHVINK, J. L. 1986. Carbon–isotope events across the Precambrian–Cambrian boundary on the Siberian Platform. *Nature*, **320**, 258–259.

MARTILL, D. M. 1988. The preservation of fishes in concretions from the Santona Formation (Cretaceous) of Brazil. *Palaeontology*, **31**, 1–18.

MATTHEWS, S. C. & MISSARZHEVSKY, V. V. 1975. Small shelly fossils of late Precambrian and early Cambrian age: a review of recent work. *Journal of the Geological Society, London*, **131**, 289–304.

MISSARZHEVSKY, V. V. 1982. Subdivision and correlation of the Precambrian–Cambrian boundry beds using some groups of the oldest skeletal organisms. *Byulleten Moskovskogo Obschestva Ispytatelei Prirody, Otdelenie Geologii*, **57**, 52–67. (In Russian).

—— 1983. Stratigraphy of oldest Phanerozoic deposits of Anabar Massif. *Soviet Geology*, **9**, 62–73. (In Russian).

MÜLLER, K. J. 1979. Phosphatocopine ostracodes with preserved appendages from the Upper Cambrian of Sweden. *Lethaia*, **12**, 1–27.

MÜLLER, K. J. & WALOSSEK, D. 1985. A remarkable arthropod fauna from the Upper Cambrian Orsten of Sweden. *Transactions of the Royal Society of Edinburgh*, **76**, 161–172.

NARBONNE, G. M., MYROW, P. M., LANDING, E. & ANDERSON, M. 1987. A candidate stratotype for the Precambrian–Cambrian boundary, Fortune Head, Burin Peninsula, southeastern Newfoundland. *Canadian Journal of Earth Sciences*, **24**, 1277–1293.

NOTHOLT, A. J. G. & BRASIER, M. D. 1986. Regional review: Europe. *In*: COOK, P. J. & SHERGOLD, J. H. (eds) *Proterozoic and Cambrian phosphorites*. Cambridge University Press, Cambridge, 9–19.

NOWLAN, G. S., NARBONNE, G. M. & FRITZ, W. H. 1985. Small shelly fossils and trace fossils near the Precambrian–Cambrian boundary in the Yukon Territory, Canada. *Lethaia*, **18**, 233–256.

PEEL, J. 1988. *Spirellus* and related helically coiled microfossils (cyanobacteria) from the Lower Cambrian of North Greenland. *Rapport Gronlands Geologiske Undersogelse*, **137**, 5–32.

QIAN YI & BENGTSON, S. 1989. Palaeontology and biostratigraphy of the early Cambrian Meishucunian stage of Yunnan Province, south China. *Fossils and Strata*, **24**.

RHODES, F. H. T. & BLOXAM, T. W. 1969. Phosphatic organisms in the Paleozoic and their evolutionary significance. *Proceedings of the North American Paleontological Convention*, 1475–1513.

RIDING, R. 1985. Calcareous algae near the Precambrian/Cambrian boundary. *NERC News Journal*, **3**, 11–12.

—— & VORONOVA, L. 1984. Assemblages of calcareous algae near the Precambrian/Cambrian boundary in Siberia and Mongolia. *Geological Magazine*, **121**, 205–210.

ROZANOV, A. YU. 1984. The Precambrian–Cambrian boundary in Siberia. *Episodes*, **7**, 20–24.

ROZANOV, A. YU & SOKOLOV, B. S. 1984. *Lower Cambrian stage subdivision. Stratigraphy*. Akademiya Nauk SSSR: Izdatelstovo 'Nauka', Moscow. (In Russian).

ROZANOV, A. YU., MIZZARZHEVSKY, V. V., VOLKOVA, N. A., VORONOVA, L. C., KRYLOV, I. N., KELLER, B. M., KOROLYUK, I. K., LENDZION, K., MICHNIAK, R., PYCHOVA, N. G. & SIDOROV, A. D. 1969. *The Tommotian Stage and the Cambrian lower boundary problem*. Akademiya Nauk SSSR, Trudy Geologicheskiy Institut, **206**, Izdatelystovo 'Nauka', Moscow, (In Russian; English translation, US Department of the Interior, 1981).

RUNNEGAR, B. 1985. Shell microstructures of Cambrian molluscs replicated by phosphate. *Alcheringa*, **9**, 245–257.

SANDBERG, P. A. 1983. An oscillating trend in Phanerozoic non-skeletal carbonate mineralogy. *Nature*, **305**, 19–22.

SHERGOLD, J. H. & BRASIER, M. D. 1986. Biochronology of Proterozoic and Cambrian phosphorites. *In*: COOK, P. J. & SHERGOLD, J. H. (eds) *Proterozoic and Cambrian phosphorites*. Cambridge University Press, Cambridge, 295–326.

SIGNOR, P. W., MOUNT, J. M. & ONKEN, B. R. 1987. A pretrilobite shelly fauna from the White-Inyo region of eastern California and western Nevada. *Journal of Paleontology*, **61**, 425–438.

SOKOLOV, B. S. & FEDONKIN, M. A. 1984. The Vendian as the terminal system of the Cambrian. *Episodes*, **7**, 12–19.

—— & —— 1986. Global biological events in the late Precambrian. *In*: WALLISER, O. H. (ed.) *Global bio–events*. Lecture notes in Earth Sciences, **8**, Springer-Verlag, Berlin, 105–108.

SOUDRY, D. & SOUTHGATE, P. N. 1989. Ultrastructure of a Middle Cambrian non-pelletal phosphorite and its early transformation into phosphate vadoids: Georgina Basin, Australia. *Journal of Sedimentary Petrology*, **59**, 53–64.

SOUTHGATE, P. N. 1988. A model for the development of phosphatic and calcareous lithofacies in the Middle Cambrian Thorntonia Limestone, northeast Georgina Basin, Australia. *Australian Journal of Earth Sciences*, **35**, 111–130.

STANLEY, S. M. 1976. Ideas on the timing of metazoan

diversification. *Paleobiology*, **2**, 209–219.

THORPE, R. S., BECKINSALE, R. D., PATCHETT, P. J., PIPER, J. D. A., DAVIES, G. R. & EVANS, J. A. 1984. Crustal growth and late Precambrian–early Palaeozoic plate tectonic evolution of England and Wales. *Journal of the Geological Society, London*, **141**, 521–536.

TUCKER, M. E. 1982. Precambrian dolomites: petrographic and isotopic evidence that they differ from Phanerozoic dolomites. *Geology*, **10**, 7–12.

—— 1986. Carbon isotope excursions in Precambrian/Cambrian boundary beds, Morocco, *Nature*, **319**, 48–49.

—— 1987. Changes in carbonate mineralogy across the Precambrian–Cambrian boundary. *Abstracts of the International Symposium on the Terminal Precambrian and Cambrian Geology, Yichang*, 78–79.

TYSON, R. V. 1987. The genesis and palynofacies characteristics of marine petroleum source rocks. *In*: BROOKS, J. & FLEET, A. J. (eds) *Marine petroleum source rocks*. Geological Society, London, Special Publication, **26**, 47–67.

XING YUSHENG, DING QUIXIU, LUO HUILIN, HE TINGGUI, & WANG YANGENG 1984. The Sinian–Cambrian boundary of China. *Bulletin of the Institute of Geology, Chinese Academy of Geological Sciences*, 10 (for 1983). (In Chinese).

XING YUSHENG, DUAN CHENGHUA, LIANG YUZUO, CAO RENGUAN *et al.* 1985. *Late Precambrian palaeontology of China*. People's Republic of China, Ministry of Geology and Mineral Resources, Geological Memoirs, Series 2, No. 2.

XU DAOYI, ZHANG QINWEN, SUN YIYING, & YAN ZHENG 1985. Three main mass extinctions – significant indicators of major natural divisions of geological history in the Phanerozoic. *Modern Geology*, **9**, 1–11.

YU WEN 1987. Yangtze micromolluscan fauna in Yangtze region of China with notes on Precambrian-Cambrian boundary. *In*: ZHANG WENTANG (ed.) *Stratigraphy and Palaeontology of Systemic boundaries in China. Precambrian–Cambrian boundary* (1). Nanjing University Publishing House, 19–344.

Two decades of phosphorite investigations in India

R. CHOUDHURI

Rajasthan State Mines and Minerals Ltd, Udaipur, Rajasthan, India

Abstract: Phosphate rock has been known for many years in some of the sedimentary basins of India. Several discoveries in the 1960s and 1970s have proved of economic importance although the country continues to remain dependent upon imports for most of its phosphate requirements. Scope for further finds of phosphate rock have probably not been exhausted and a much more co-ordinated exploration programme is recommended. Proterozoic basins continue to offer the best exploration potential and currently account for the bulk of the known domestic phosphate rock resources. Nevertheless, geologically younger basins should not be discounted in terms of their phosphate potential.
Geological ages for the main phosphorite sequences in India are not well established and there is no precise stratigraphic correlation with Proterozoic-Cambrian phosphorite provinces in other parts of the world. Since the discovery of phosphorite beds of Palaeozoic age near Dehra Dun, Uttar Pradesh, and Jaisalmer, in southeastern Rajasthan, considerable progress has been made in both the exploration for and exploitation of phosphate deposits in India. One notable, promising development in recent years has been that associated with research, in India, on the beneficiation of low-grade and highly dolomitic phosphate rock.

Occurrences of phosphate nodules in some sedimentary basins of the south Peninsular and north extra-Peninsular regions of India have been known for well over a century. For example, in 1884, phosphatic nodules and shales were reported by the Geological Survey of India from near Mussoorie in the northern state of Uttar Pradesh (King 1884). Many other discoveries have been made in India since that time (Table 1) and the geology of the main deposits and occurrences has been well documented (e.g. Banerjee 1971, 1986; Banerjee *et al.* 1980; Choudhuri 1989; Geological Survery of India 1968, 1981 *a*, *b*, 1984; Khan *et al.* 1989; Pant 1980; Choudhuri & Roy 1986; Sant 1980; Sant & Pant 1979; Shanker 1989; Chauhan & Sisoclia 1989), although in relation to the great size of the sub-continent there are relatively limited resources of phosphate rock of either sedimentary or igneous origin. Thus, the country's proven resources continue to be insufficient to meet the growing domestic demand for phosphate rock and India remains heavily dependent on imports not only of rock, but also of phosphate in the form of phosphoric acid and high-analysis fertilizers such as diammonium phosphate.

The initial discoveries during the 1960s and 1970s, several of which proved of considerable commercial importance, raised hopes that the country's phosphate rock resource base might be augmented substantially, particularly as the additional geological information then available gave support to the view that further discoveries were very likely. Unfortunately, such expectations have not yet been realized, with Indian phosphate rock resources currently amounting to only some 190 Mt (Table 1). Nevertheless, phosphate investigations carried out in India during the 1978–1988 period, notably under the aegis of Project 156, have been characterized by intense collaborative effort between the Geological Survey of India, Universities and other Governmental and private agencies in India, mainly through the National Working Group, and with many participants in Project 156 from countries overseas. As a result, much progress was made in understanding the geology of Indian phosphorites in terms of lithofacial variations, petrography and geochemistry, their sedimentary environment and regional tectonic setting. Useful comparisons between ancient and modern phosphorites have been made and detailed studies of the palaeogeographical aspects of phosphogenesis carried out. Discoveries such as those of Udaipur in Rajasthan and Mussoorie in Uttar Pradesh were based on the recognition of favourable lithological associations. The depositional models that were subsequently developed for the phosphorite deposits of these two areas have been found to be applicable to other phosphogenic basins. In spite of significant progress, large tracts of India remain unexplored in detail in terms of their phosphate potential and occurrences will undoubtedly continue to be found as systematic phosphate exploration proceeds.

Table 1. *Phosphate resources of India*

State/District	Deposit	Grade %P$_2$O$_5$	Reserves Mt	Characteristics
Rajasthan				
Udaipur	Jhamarkotra	+30	16.8	High grade, associated with quartz (SiO$_2$, 8%–12%)
	Jhamarkotra	25–30	5.0	High grade, associated with quartz (SiO$_2$, 12%–20%)
	Jhamarkotra	10–20	54.0	High dolomitic (MgO, 8%–15%; SiO$_2$, 3%–5%)
	Maton	+30	1.0	High grade, associated with quartz (SiO$_2$, 10%–15%)
	Maton	20–30	8.1	Siliceous phosphorite (SiO$_2$, 15%–30%)
	Kanpur	+30	0.8	High grade, associated with quartz (SiO$_2$, 10%–15%)
	Kanpur	20–30	1.0	Associated with quartz (SiO$_2$, 15%–30%), partly dolomitic (MgO, 2%–5%)
	Kanpur	11.6	6.2	Dolomitic phosphorite
	Kanpur	10–30	3.0	Dolomitic and siliceous
	Smaller deposits	6–9	1.0	Apatite in carbonatite
	Newania*			
Jaisalmer	Birmania	10–15	4.5	Calcitic phosphorite
Jaipur	Achraul	15–20	1.0	Siliceous, ferruginous phosphorite
Banswara	Sallopat	20–25	0.05	Dolomitic phosphorite
Madhya Pradesh				
Sagar, Chattarpur	Sagar-Chattarpur	–30	1.0	Siliceous phosphorite
	Hirapur-Bassia	25–30	8.0	Siliceous-ferruginous phosphorite
	Mardeora-kachar	13–35	4.1	R$_2$O$_3$, 5%–13%
	Bassia	10–30	4.9	Siliceous
Jhabua	Kelkua	10–30	3.2	Siliceous, dolomitic phosphorite
	Khatamba	10–15	6.8	Dolomitic phosphorite
Uttar Pradesh				
Dehra Dun	Mussoorie	+30	1.0	Siliceous, carbonate-rich phosphorite
	Mussoorie	25–30	3.9	Siliceous, carbonate-rich phosphorite
	Mussoorie	10–30	26.3	Siliceous, carbonate-rich phosphorite
Lalitpur	Pisneri	10–30	23.9	Siliceous
Andhra Pradesh				
Visakhapatnam	Kasipatnam*	15–35	3.0	Associated with magnetite
West Bengal				
Purulia	Beldih*	15–30	1.5	Associated with magnetite
Tamilnadu				
Dharmapuri	Dharmapuri*	4–11	0.5	Apatite in carbonatite

* Igneous deposit

Proterozoic

Several significant phosphorite discoveries of Late Proterozoic age have been made in northern India since systematic exploration was begun in 1966 by the Geological Survey of India (1968) and other organizations (Banerjee 1986; Banerjee & Saigal 1987; Choudhuri 1989; Khan et al. 1989; Pant 1980; Pant et al. 1979; Saigal & Banerjee 1987). The best known of these are situated in the Udaipur area of southeastern Rajasthan, as well as the deposits investigated by the Geological Survey of India in the Lalitpur and Bijawar basins of Uttar Pradesh and Madhya Pradesh, and near Jhabua in Madhya Pradesh (Fig. 1). Proterozoic phosphorites have been recorded from many other parts of India, but the occurrences are generally too small to be of commercial interest.

Udaipur

By far the most important Proterozoic phosphorites in India are those forming the Aravalli phosphogenic province, notably in the Maton Formation or its stratigraphical equivalents. The Maton Formation is best developed near Udaipur, in southeastern Rajasthan, where phosphorites were first discovered by the Geological Survey of India in 1967 (Geological Survey of India 1968; Choudhuri & Roy 1986). The phosphorites, which have been studied in detail by Banerjee (1971), Banerjee et al. (1980) and Ehauhan (1979) are essentially columnar, laminated, brecciated and fragmented stromatolites, the intercolumnar portions of which being normally occupied by recrystallized dolomite. The stromatolites are characterized by alternate convex laminae of dark grey phosphorite and grey dolomitic limestone, all in a sheath of phosphorite. Total production of phosphate rock from the Udaipur area is currently around 380 000 t per annum, accounting for about 20% of India's total phosphate rock requirements.

At Jhamarkotra, SSE of Udaipur and the largest deposit of high-grade rock in India, the phosphorite bed has an average thickness of about 15 m and can be traced for a strike length of 16 km, with a northward dip of 30–60°. The phosphorites around Udaipur are chemically similar (Banerjee et al. 1980), being characterized also by their crystallinity, grain sizes ranging from 10–50 μm. Resources are estimated at around 76 Mt with more than 10% P_2O_5, of which some 17 million tonnes is of high grade, 30% P_2O_5 or more (Choudhuri 1989, see also Table 1). The much smaller openpit mine at Maton came into operation in 1972. The ore, which contains 20–28% and averages 25% P_2O_5, is beneficiated to obtain a flotation concentrate with 32–33% P_2O_5 and 8–12% SiO_2. The Maton deposits are estimated to total about 9 million tonnes over a strike length of 3.5 km and to a depth of 200 m. The phosphorite horizon averages 4–7 m in thickness, with about 3 Mt averaging 25% P_2O_5 being available to openpit mining to a depth of 100 m (Choudhuri 1989). Reserves of high-grade and siliceous phosphorite amount to some 9 Mt (Table 1).

The Jhamarkotra deposit has been divided areally into seven blocks, A–G, from west to east, of which block D is situated in the central part of the outcrop. Openpit mining began in 1969 in block D, taking advantage of the availability of high-grade ore suitable for fertilizer manufacture, the mine being capable of producing 1 million tonnes of phosphate ore annually. The high-grade ore has a relatively high silica content (5–10% SiO_2) and low magnesia (about 0.5% MgO). In contrast, low-grade Jhamarkotra rock generally contains less than 20% P_2O_5, with 2–5% SiO_2 and 7–14% MgO. To test this rock for eventual commercial use, a semi-industrial pilot plant came into operation in 1979. The most favoured beneficiation route is the froth flotation of carbonate and phosphate which has yielded a product with about 35% P_2O_5 and a MgO content of 1.2–1.6%.

It is notable also that mining operations near Udaipur are confined to leached or weathered zones. These are highly variable in depth and result from pronounced and selective leaching of the more soluble carbonate fraction by meteoric waters circulating in structurally disturbed zones. Within these zones phosphatic stromatolitic columns have consequently been reduced to fragmented and powdery residual masses of 'high-grade' ore. In the weathered zone, the higher-grade rock generally contains higher silica and lower MgO, whereas the unweathered rock is of lower grade, with low silica and very high MgO (Choudhuri & Balasubramian 1980). Thin encrustations and vein fillings of secondary phosphorite also occur locally in pockets underlying the main phosphorite bed (Choudhuri 1981).

Hirapur and Lalitpur

Radioactive phosphorite was discovered by the Indian Atomic Energy Commission in central India in 1976, in Madhya Pradesh, within the Proterozoic Bijawar Group (Banerjee et al. 1982; Pant et al. 1989). The group overlies the Archaean Bundelkhand and consists of quartzites, shales and dolomites occurring in two isolated basins about 25 km apart: the Hirapur Basin in Madhya Pradesh, where the Bijawar Group occurs over a strike length of about 80 km, and the Lalitpur Basin in the southern part of neighbouring Uttar Pradesh, where phosphorite was discovered in 1977 during mapping by the Geological Survey of India. In the Hirapur Basin, the Sagar deposits have been worked since 1978 at the rate of about 22 000 t per annum. There are at least four distinct phosphorite horizons, each about 5 m in thickness and containing 15 to 35% P_2O_5 (Pant et al. 1989). The phosphorite deposits of the Lalitpur Basin occur as lenticular and detached bodies ranging from a few metres to about 4 km in length and up to 125 m in width. As in the Hirapur Basin, there are four phosphorite horizons, the most important being the uppermost, generally brecciated, horizon. The deposits contain 5–25%, averaging about 20% P_2O_5 (Pant et al. 1979). However, in weathered and leached zones as much as 37% P_2O_5 may be present.

Jhabua

Phosphorite occurring in a geological environment similar to that around Udaipur was first discovered in the Jhabua District of north-western Madhya Pradesh in 1973 (Banerjee & Basu 1979; Khan 1979; Munshi & Khan 1973). The economic potential of the Jhabua deposits, which lie about 200 km south of Udaipur, has been investigated in detail by the Geological Survey of India (Khan et al. 1976; Khan 1979; Khan et al. 1989). The area forms the southern extension of the main belt of Aravalli rocks in Rajasthan, the rocks consisting mainly of phyllite, quartzite, and dolomitic limestone. Stromatolitic phosphorite is associated with the dolomitic limestone as separate and elongated bodies up to 1500 m long. The phosphorite unit averages about 15 m. There are two distinct beds, the upper forming the main high-grade deposit of the Jhabua District, averaging around 28% P_2O_5. Production of phosphate rock began in 1976, but to date there has been only limited output.

Palaeozoic

Marine, fossiliferous Palaeozoic rocks are well developed in the extra-Peninsular region, mainly in the Kashmir, Uttar Pradesh and Himachal Pradesh regions. Palaeozoic rocks occur also in western Rajasthan in the Jaisalmer Basin, an undeformed structure situated at the north-western margin of the Indian Craton (Chaube 1980). Mesozoic and Tertiary sediments of the basin thicken to the west into the Indus basin of Pakistan. The most extensive phosphate deposits are near Mussoorie, in Uttar Pradesh, about 20 km northwest of Dehra Dun (Fig. 1).

Mussoorie Syncline

By far the most important Palaeozoic phosphate-bearing formation in India is the Lower Tal sequence in the western Kumaun Himalaya of Uttar Pradesh (Shanker 1971, 1989). In the Dehra Dun and Tehri districts, near Mussoorie, beds of phosphorite have been mapped over a distance of 120 km in the Mussoorie Syncline, a doubly-plunging structure much dislocated by cross-faulting. Drilling, trenching and development of exploratory adits by the Geological Survey of India and other agencies have led to the delineation of 11 blocks which together have been estimated to contain about 45 million tonnes of rock ranging from

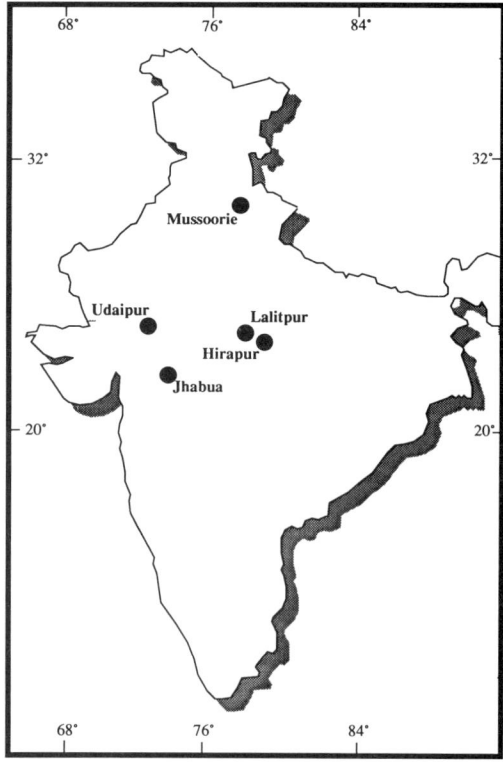

Fig. 1. Location of major phosphorite deposits in India.

16–18% P_2O_5 (Geological Survey of India 1981). Deposits in the Maldeota and Durmala blocks near Dehra Dun account for 10.64 million tonnes averaging 19% P_2O_5 and 9.14 million tonnes with 19.9% P_2O_5, respectively. Proved and probable reserves in the two blocks are 4.98 million tonnes and 3.69 million tonnes. The Maldeota and Durmala deposits were brought into commercial production in 1970 and 1976, respectively. In recent years, production has exceeded 100 000 t per annum.

Distinctive lithology and easy recognition of contacts permit a division of the Tal rocks into the Lower Tal and Upper Tal formations. The Lower Tal, which is about 200 m thick, is further divisible into four members, depending on the dominance of chert, shale, sand/silt or carbonate in the sediments (Geological Survey of India 1981). It contains the main phosphate unit which is 10 m thick. Fresh bulk samples from the phosphate unit contain averages ranging from about 8% to 19% P_2O_5 (by weight), but rock from the weathered zones, such as those mined at Maldeota and Durmala, may analyse up to 33% P_2O_5, depending upon the degree of alteration.

Beneficiation tests have shown that the unweathered Mussoorie phosphorite, containing as mined 17.8–19.5% P_2O_5, can be beneficiated by calcination to give a product with 31% P_2O_5 (Shanker 1989). Similarly, a concentrate with 28–29% P_2O_5 has been obtained by flotation, but the content of 3% Fe_2O_3 is too high for its use in fertilizer manufacture. A major difficulty has been the liberation of the phosphate grains from the predominantly carbonate matrix, with pyrites and organic matter hindering successful flotation. Currently, Mussoorie rock is finely ground and sold with 18–22% P_2O_5 for direct application to the soil as fertilizer.

Jaisalmer Basin

The Mesozoic sequence in the Jaisalmer Basin begins with the continental Lathi Formation, which rests on Palaeozoic sandstones and is overlain by limestones, shales and sandstones of Jurassic and Cretaceous age, and a thin but relatively continuous shallow-water, marine sequence of Palaeocene and Eocene limestones and shales. Quaternary clastic rocks overlie Middle Eocene rocks.

Among the earliest phosphate discoveries in this basin are those in the Upper Palaeozoic Birmania Formation, an unfossiliferous formation exposed over an area of only 4 km^2 (Deshmukh 1979) and comprising a folded sequence of limestone with sandstone, quartzitic sandstone, variegated shale, and dark shale (Sant 1980), believed to have been deposited during a Palaeozoic marine transgression over western Rajasthan. The Birmania rests, probably unconformably, on the Randa Formation and is, in turn, overlain by the Lathi Formation of Triassic–Jurassic age. Overlying the quartzitic sandstone is a phosphorite series 0.56–9 m thick and averaging about 13% P_2O_5 that consists of thinly banded phosphatic calcareous shaly sandstones laterally grading into thinly banded phosphatic limestone.

Discussion

Phosphate exploration and research carried out in India over the last two decades has provided a wealth of data on the country's phosphate resources in terms of mode of occurrence, facies variations, structure, petrography, geochemistry, and palaeogeography. Most of the occurrences belong to the Late Proterozoic, the three main areas being those of near Udaipur, Jhabua and Hirapur–Lalitpur. There, the phosphorite sections all overlie Archaean basement rocks and comprise essentially phosphorite-carbonate sequences with basal quartzite units. These basins account for more than 80% of known domestic resources but others of Proterozoic age remain to be fully assessed for their phosphate potential, as do some of the geologically younger sedimentary basins. To this end, a systematic programme of exploration and research with the collaboration of such national agencies as the Geological Survey of India and the Oil and Natural Gas Corporation, is highly recommended. For example, the full potential of the Cuddapah Basin in Andhra Pradesh, southern India, as a phosphogenic province has not yet been thoroughly evaluated. Within the basin, the occurrence of a phosphatic quartzite horizon within the Cumbum Formation of the Cuddapah Supergroup near Chelima, together with the possible correlation of the Cuddapah Supergroup with phosphate-bearing, Upper Proterozoic–Cambrian rocks elsewhere in India, suggests that further discoveries over the 200 km extent of the basin seem likely. Similarly, in the Kutch Basin in Gujarat of northwestern India, ferruginous sandstones and limestones of the Bhuj Formation are reported to contain thin phosphatic concretions and nodules (Sant 1980).

Unfortunately, in spite of the considerable amount of geological research on Proterozoic–Lower Palaeozoic deposits in India, a major as yet unresolved problem centres on the absence of precise stratigraphical correlation, particularly with those deposits forming part of the vast Proterozoic–Cambrian Phosphogenic Province that embraces also the USSR, China, the Mongolian People's Republic, and Australia. The Udaipur phosphorites lie in the basal part of the Aravalli geological event and have conventionally been considered to be of Early Proterozoic age. However, study of the stromatolite morphology has variously indicated a middle–upper Proterozoic or even post-late Proterozoic (Vendian) age (Choudhuri & Roy 1986; Banerjee 1986).

Similarly, the age of the phosphate-bearing Lower Tal continues to be a controversial topic. Variously regarded as of Proterozoic, Permian, or Jurassic–Cretaceous in age, the most recent investigations by the Geological Survey of India have yielded microfaunas similar to the Tommotian and Meishucun stages of the Russian and Chinese platforms respectively, indicative of a lowermost Cambrian age (Shrivasta & Batt, pers. comm. 1983, in Shanker 1989; Brasier 1990). A Lower Palaeozoic age is supported by well-documented descriptions of Cambro-Ordovician conodonts from the phosphorite unit of the Lower Tal (Azmi et al. 1981).

In economic terms, one of the most notable developments in phosphorite research has been the successful beneficiation of the low grade dolomitic phosphate ores. Proterozoic phosphorites are generally of the unreactive type and the dense and interlocking nature of the constituent minerals, including apatite, dolomite and quartz, usually renders such rocks difficult to beneficiate by physical methods. Recent agronomical investigation indicates the phosphorites are not too unreactive to be suitable for direct application as a fertilizer. However, considerable progress has been made in the development of a commercial beneficiation process capable of providing good-quality phosphate concentrates.

The author is extremely grateful, in his capacity as Convener of the Indian National Working Group of IGCP Project 156, to his colleagues for their valuable help and suggestions, and to successive Chairmen of both the Indian National Committee for IGCP and the National Working Group. The assistance of P. J. Cook and J. H. Shergold, both of the Australian Bureau of Mineral Resources, Geology and Geophysics, Canberra, S. R. Riggs, East Carolina University, Greenville, A. Notholt, Mineral Resources Consultant, London and W. C. Burnett, Department of Oceanography, Florida State University, Tallahassee, is also gratefully acknowledged. The encouragement and support of the Rajasthan State Mines & Minerals Limited, Udaipur, is much appreciated.

References

AZMI, R. J., JOSHI, M. A. & JUIAL, K. P. 1981. Discovery of the Cambrian-Ordovician conodonts from the Mussoorie Tal Phosphorite: its significance in the correlation of the Lesser Himalaya. *In*: SINHA, A. K. (ed.) *Contemporary Geo-Scientific Researches in the Himalaya*, Vol. 1, Singh Publication, Dehra Dun, 245–250.

BANERJEE, D. M. 1971. Precambrian stromatolitic phosphorites of Udaipur, Rajasthan, India. *Geological Society of America Bulletin*, **82**, 2319–2330.

—— 1986. Proterozoic and Cambrian phosphorites: Regional review: Indian Subcontinent. *In*: COOK, P. J. & J. H. SHERGOLD, J. H. (eds) *Phosphate Deposits of the World, Vol. 1, Proterozoic and Cambrian Phosphorites*, Cambridge University Press, Cambridge, 70–90.

—— & BASU, P. C. 1979. Geology and structure of Precambrian Jhabua phosphorite deposit, Madhya Pradesh. *Indian Mineralogist*, **20**, 32–42.

—— & SAIGAL, N. 1987. Proterozoic phosphorites of India: updated information. Part II: Geochemistry. *In*: *Purana Basins of Peninsular India*. Memoir of the Geological Society of India, **6**, 487–510.

——, BASU, P. C. & SRIVASTAVA, N. 1980. Petrology, mineralogy, geochemistry, and origin of the Precambrian Aravallian phosphorite deposits of Udaipur and Jhabua, India. *Economic Geology*, **75**, 1181–1199.

——, KHAN, M. W. Y., SRIVASTAVA, N. & SAIGAL, G. C. 1982. Precambrian phosphorites in the Bijawar rocks of Hirapur–Bassia areas, Sagar District, Madhya Pradesh, India. *Mineralium Deposita*, **17**, 349–362.

BRASIER, M. D. 1990. Phosphogenic events and skeletal preservation across the Precambrian/Cambrian boundary interval. *In*: NOTHOLT, A. J. G. & JARVIS, I. (eds) *Phosphorite Research and Development*. Geological Society, London, Special Publication, **52**, 289–303.

CHAUBE, A. N. 1980. Geologic framework of the oil and gas fields of the Indian Platform. *In*: SHELDON, R. P. & BURNETT, W. C. (eds) *Fertilizer Mineral Potential in Asia and the Pacific*, East-West Resource Systems Institute, Honolulu, 13–28.

CHAUHAN, D. S. 1979. Phosphate-bearing stromatolites of the Precambrian Aravalli phosphorite deposits of the Udaipur region, their environmental significance and genesis of phosphorite. *Precambrian Research*, **8**, 95–126.

—— & Sisodia, M. S. 1989. Phosphorites of Rajasthan. *In*: Banerjee, D.M. (ed.) *Phosphorites in India*. Memoir of the Geological Society of India, **9**, 9–13.

CHOUDHURI, R. 1979. The characteristics of the Jhamarkotra phosphorite deposit in Udaipur District, Rajasthan, India. *In*: COOK, P. J. & SHERGOLD, J. H. (eds) *Proterozoic–Cambrian Phosphorites*, Canberra Publishing & Printing

Co. Pty. Ltd., Canberra, 40–42.

—— 1981. Botryoidal apatite (staffelite) in the Jhamarkotra area of Udaipur district, Rajasthan (abstract). *In*: *Symposium on three decades of developments in petrology, mineralogy and geochemistry in India*, May 1981, Geological Survey of India, Jaipur, 52.

—— 1989. Proterozoic phosphorites around Udaipur, Rajasthan, India. *In*: NOTHOLT, A. J. G., SHELDON, R. P. & DAVIDSON, D. F. (eds) *Phosphate Deposits of the World, Vol. 2: Phosphate Rock Resources*, Cambridge University Press, Cambridge, 461–466.

—— & BALASUBRAMIAN, 1980. Grade distribution pattern and its bearing on the formation of high and low grade in the Jhamarkotra phosphorite deposit, Udaipur, Rajasthan. *Indian Journal of Earth Sciences*, **7**, 89–93.

—— & ROY, A. B. 1986. The stromatolite bearing Precambrian phosphorite deposits of Jhamarkotra, Rajasthan, India. *In*: COOK, P. J. & SHERGOLD, J. H. (eds) *Phosphate deposits of the World, Vol. 1, Proterozoic and Cambrian phosphorites*, Cambridge University Press, Cambridge, 202–219.

DESHMUKH, G. P. 1979. Palaeozoic phosphorites, Rajasthan, India. *In*: COOK, P. J. & SHERGOLD, J. H. (eds) *Proterozoic-Cambrian Phosphorites*, Canberra Publishing & Printing Co. Pty. Ltd., Canberra, 36–39.

GEOLOGICAL SURVEY OF INDIA. 1968. In search of phosphorite deposits in India. *In*: Proceedings of the Seminar on Sources of Raw Materials for the Fertilizer Industry in Asia and the Far East. *Mineral Resources Development Series*, United Nations, New York, **32**, 255–263.

—— 1981a. *Excursion Guide Book on Dehradun-Mussoorie Area.* (Fourth International Field Workshop and Seminar on Phosphorite, India, 25 November–6 December, 1981.)

—— 1981b. *Guidebook for Excursion. Aravalli Phosphorites around Udaipur, Rajasthan, India.* (Fourth International Field Workshop and Seminar on Phosphorite, India, 25 November–6 December, 1981.)

—— 1984. *Phosphorite.* Special Publication of the Geological Survey of India, 17. (Fourth International Field Workshop and Seminar on Phosphorite, India, 25 November–6 December, 1981.)

KHAN, H. H. 1979. An appraisal of Precambrian phosphorites of JHABUA, M. P., India. *In*: COOK, P. J. & SHERGOLD, J. H. (eds) *Proterozoic-Cambrian Phosphorites*, pp. 42–43. Canberra Publishing & Printing Co. Pty. Ltd., Canberra.

——, GHOSH, D. B., SONI, M. K., SONAKIA, A. & ZAFAR, M. 1989. Phosphorite deposits of the Jhabua District, Madhya Pradesh, India. *In*: NOTHOLT, A. J. G., SHELDON, R. P. & DAVIDSON, D. F. (eds) *Phosphate Deposits of the World, Vol. 2: Phosphate Rock Resources*, Cambridge University Press, Cambridge, 467–472.

——, ZAFAR, M. & GHOSH, D. B. 1976. Geochemical characteristics as applied to the genesis of Precambrian phosphorite, Jhabua District, M. P. *Special Publication of the Geological Survey of India*, **3**, 175–183.

KING, W. 1884. Phosphatic beds, Musuri. *Records of the Geological Survey of India*, **17**, 198–199.

MUNSHI, R. L., KHAN, H. H. & GHOSH, D. B. 1974. Algal structure and phosphorite in the Aravalli rocks of Jhabua, M. P. *Current Science (India)*, **43**, 446–447.

PANT, A. 1980. Resource status of rock phosphate deposits in India and areas of future potential. *In*: SHELDON, R. P. & BURNETT, W. C. (eds) *Fertilizer Mineral Potential in Asia and the Pacific*, East-West Resource Systems Institute, Honolulu, 331–357.

——, DAYAL, B., JAIN, S. C. and CHAKRAVARTY, T. K. 1979. Status of phosphorite investigations in Uttar Pradesh, India, and approach for future work. *In*: COOK, P. J. & SHERGOLD, J. H. (eds) *Proterozoic-Cambrian Phosphorites*, Canberra Publishing & Printing Co. Pty. Ltd., Canberra, 35–36.

——, KHAN, H. H. & SONAKIA, A. 1989. Phosphorite resources in the Bijawar Group of central India. *In*: NOTHOLT, A. J. G., SHELDON, R. P. & DAVIDSON, D. F. (eds) *Phosphate Deposits of the World, Vol. 2: Phosphate Rock Resources*, Cambridge University Press, Cambridge, 473–477.

SAIGAL, N. & BANERJEE, D. M. 1987. Proterozoic phosphorites of India: updated information. Part I: Petrography. *In*: *Purana Basins of Peninsular India*, Memoir of the Geological Society of India, **6**, 471–486.

SANT, V. N. 1979. Precambrian phosphorites of Rajasthan, India: a case history. *In*: COOK, P. J. & SHERGOLD, J. H. (eds) Proterozoic-Cambrian Phosphorites, Canberra Publishing & Printing Co. Pty. Ltd., Canberra, 39–40.

—— 1980. Geology of the Indian Platform and its phosphate occurrences. *In*: SHELDON, R. P. & BURNETT, W. C. (eds) *Fertilizer Mineral Potential in Asia and the Pacific*, East-West Resource Systems Institute, Honolulu, 29–47.

—— & PANT, A. 1980. A note on the known resources of phosphate and potash in India. *In*: LEE, A. I. N. (ed.) *Fertilizer Mineral Occurrences in the Asia-Pacific Region*, East-West Resource Systems Institute, Honolulu, 20–33.

SHANKER, R. 1971. Stratigraphy and sedimentation of the Tal Formation, Mussoorie Syncline, U. P. *Journal of the Palaeontological Society of India*, **16**, 1–15.

—— 1989. The Mussoorie Phosphate Basin. *In*: NOTHOLT, A. J. G., SHELDON, R. P. & DAVIDSON, D. F. (eds) *Phosphate Deposits of the World, Vol. 2: Phosphate Rock Resources*, Cambridge University Press, Cambridge, 443–448.

The influence of magnesium ions during the formation of stromatolitic phosphorites of Udaipur, Rajasthan, India

M. S. SISODIA & D. S. CHAUHAN

Department of Geology, University of Jodhpur, Jodhpur 342001, India

Abstract: The stromatolites of the Precambrian Aravalli Supergroup outcropping around Udaipur City are composed of carbonate (dominantly dolomite) and/or phosphate (carbonate-fluorapatite). The phosphate present in these stromatolites occurs in varying proportions, from traces levels to a major constituent. Compositional variation is attributed to the differences in the original magnesium contents stromatolites which reflects their different of the depositional environments, Intertidal stromatolites had primary magnesium calcite mineralogies, while those deposited in subtidal areas were composed of aragonite and/or lower magnesium calcite. The Mg/Ca ratio in the stromatolites appears to have been controlled by microorganisms which, it is argued, also played an important role in the process of phosphate fixation. It is inferred that the high Mg content of intertidal stromatolites prevented extensive phosphatisation in this facies.

Phosphorites around Udaipur, Rajasthan, India (Fig. 1) are unusual in being strictly confined to stromatolites. They were deposited in marine epicontinental subtidal to supratidal environments and now occur as a series of isolated deposits around the city (Chauhan 1979; Banerjee *et al.* 1980; Roy & Paliwal 1981; Sisodia 1986). The stromatolites are composed of carbonate and phosphate with both components in quite variable proportions, such that they can be classified as stromatolites with dominantly phosphate or dominantly carbonate compositions, or as stromatolites having alternate phosphate and carbonate laminae. The present investigation involved separation and geochemical analysis of these different categories of stromatolite. In this paper we present the results of these analyses and discuss the distribution of major elements in the deposits.

Geological setting

The Udaipur phosphorites (see Choudhuri 1990) belong to the Precambrian Aravalli Supergroup which rests with a profound unconformity on a gneissic basement known as the 'Banded Gneissic Complex' (Heron 1953). Controversy prevails regarding the age of Aravalli Supergroup. Crawford (1970) and Sarkar (1972), on the basis of geochronological work gave a depositional age of 2500–2000 Ma, assigning the unit to the Lower Proterozoic, while Banerjee (1971), using biostratigraphic evidence from the stromatolites in the phosphorite deposits, inferred a mid to late Riphean age (1600–1900 Ma). More recently, Raaben (1981) reinterpreted the stromatolite as being of early Proterozoic age, a conclusion which is in agreement with the earlier geochronological studies.

The rocks of the Aravalli Supergroup include volcanics, arkosic polymictic conglomerates, arkose and orthoquartzites which form a basal sequence passing up into dolomitic limestones, stromatolitic phosphorites (Fig. 2) carbonaceous pyrite phyllites, phyllitic slates, greywackes and lithic arenites (Fig. 1). The sequence was deposited in shelf and epicontinental sea environments and displays four depositional stages: an early volcanic phase, a peneplanation phase, a main clastic phase and a late clastic phase (Roy & Paliwal, 1981).

Petrography

Petrographic study of the stromatolites using light and scanning electron microscopy (SEM) indicates that they consist predominantly of very small unidentifiable particles less than 1 μm in diameter (Fig. 3). However, some larger pores are filled by structureless sparite of 10–20 μm size. Primary biofabrics have been almost completely destroyed and the most pronounced feature is now a series of contorted alternations of dark phosphate, (collophane) and light carbonate, forming either laminae or irregular masses. X-ray diffraction (XRD) analysis shows that the microcrystalline matrix consists of a variety of minerals including carbonate-fluorapatite, calcite, dolomite, quartz and rare mica (Frost 1988, pers. comm.). Individual minerals cannot be distinguished using SEM (Fig. 3).

The intercolumnar material is an assorted

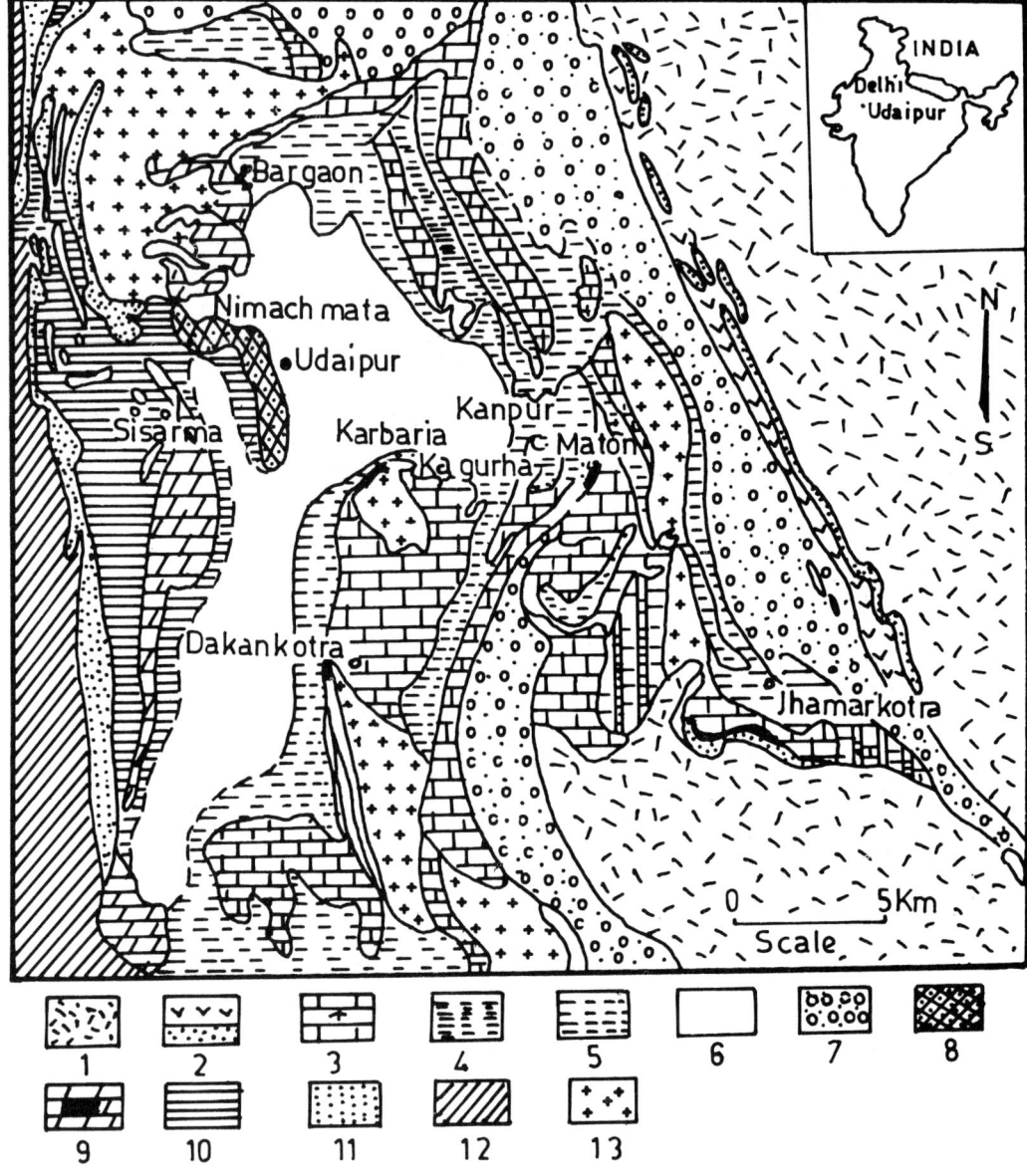

Fig. 1. Geological map of Udaipur showing stromatolitic phosphorite occurrences. 1, banded gneissic complex; 2, basal orthoquartzite−volcanic rocks; 3, lower dolomitic limestone with phosphorite; 4, carbonaceous phyllite; 5, lower phyllite slate; 6, greywacke phyllite and lithic arenite; 7, conglomerate arkose orthoquartzite; 8, orthoquartzite silty arenite; 9, upper dolomitic limestone with phosphorite; 10, upper phyllite and slates; 11, upper orthoquartzite; 12, phyllite with bands of quartzite; 13, post-Aravalli granites.

mixture of clasts derived from the stromatolites which is partially or completely cemented by sparry calcite of 20−100 μm size. This sparry calcite cement may exhibit a conspicuous radial fibrous fabric. Some small euhedral crystals of dolomite up to 100 μm size occur in the cement. The petrography of the deposits is presented in more detail elsewhere (Chauhan, 1979; Chauhan & Sisodia 1984, in press; Sisodia 1986).

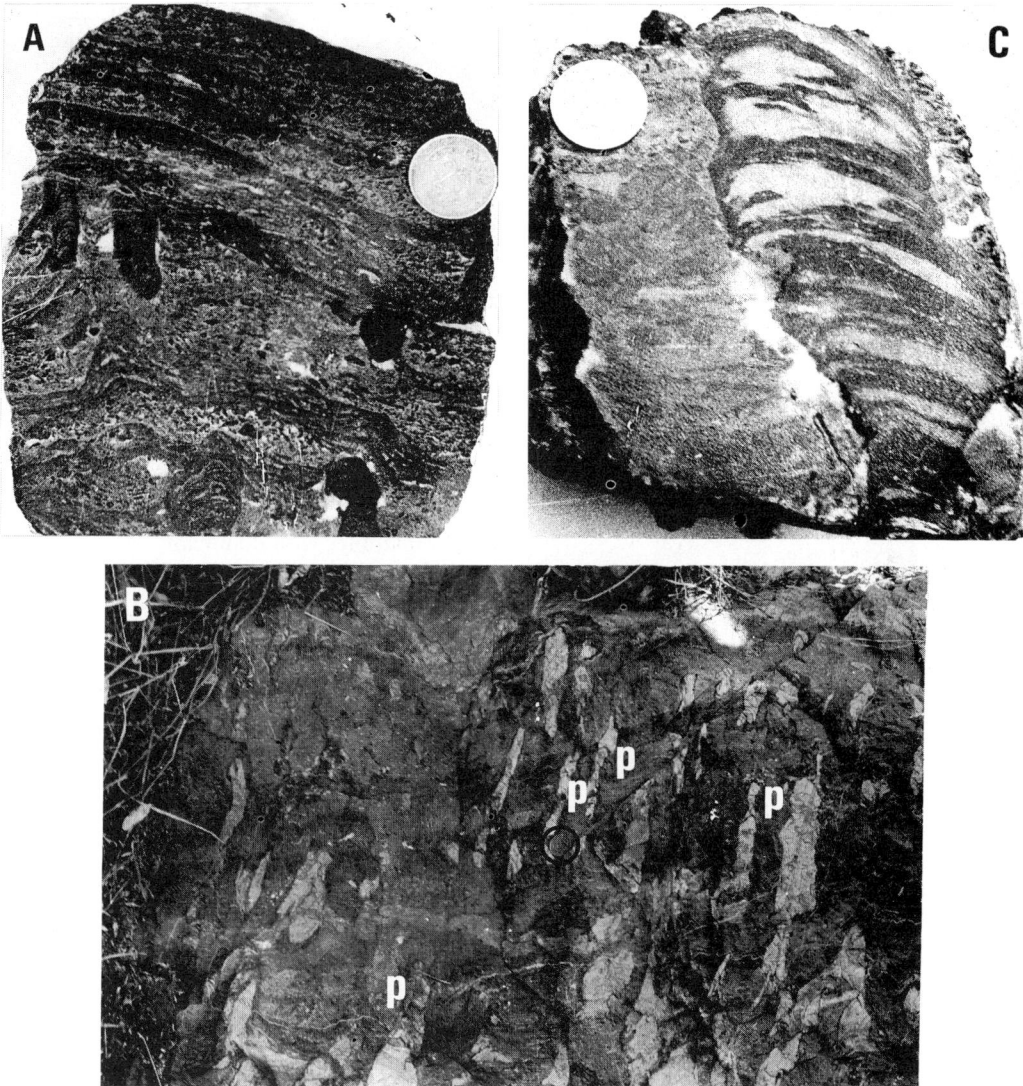

Fig. 2. (A) Stromatolite columns containing phosphate as the dominant constituent of the framework, Dakankotra. Coin is 2 cm diameter. (B) Stromatolite columns having phosphate on the periphery and also rarely as thin laminae (p) within the main body of carbonate, Nimachmata. Coin (circled) is 2 cm diameter. (C) Stromatolite column composed of alternate layers of phosphate (black) and carbonate (white), Dakankotra. Coin is 2 cm diameter.

Sampling and analytical techniques

Four different lithologies were selected, for chemical analysis (Fig. 2).
(1) Stromatolite columns composed dominantly of phosphate.
(2) Stromatolite columns composed dominantly of carbonate.
(3) Carbonate laminae of stromatolites having alternate phosphate and carbonate laminae.
(4) Dolomitic limestone host rocks.

Samples were obtained from freshly dug pits and trenches to avoid weathered samples. The stromatolites with dominantly phosphate and dominantly carbonate compositions were carefully separated from the host rocks. Longitudinal sections of stromatolites composed of alternating laminae of carbonate and phosphate were chosen and samples of the carbonate laminae were obtained using a portable drill.

Samples were analysed by wet chemical methods

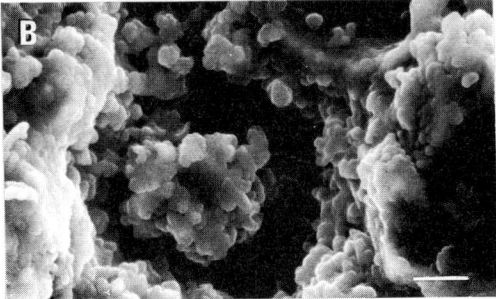

Fig. 3. Scanning electron microscope photomicrographs of the phosphate-bearing stromatolites showing small, less than 1 μm sized crystals of apatite, calcite and dolomite. Minerals are morphologically indistinguishable. Scale bars, (a) 10 μm (b) 1 μm.

using colorimetry, spectrophotometry, atomic absorption spectrometry and titrimetry. The oxide values were determined using calibrations based on artificial standard solutions made from Analar-grade reagents. Results were compared with multiple determinations of phosphate rock reference material carried out independantly at Rajasthan State Mines and Geology Laboratory, Udaipur and the Department of Geology, Rajasthan University, Udaipur. P_2O_5 was determined using a Beckmen Model DU2 Spectrophotometer at the Department of Geology, Udaipur. Alkali metals (Na and K) were determined using an EEL flame-photometer. Silica, calcium, magnesium and trace metals were determined using a Perkin Elmer Model 603 Atomic Absorption Spectrophotometer at the Central Arid Zone Research Institute, Jodhpur. Each analysis reported is an average of 3 to 10 determinations.

Results

Geochemical results are presented in Table 1. Stromatolites with dominantly phosphate composition contain 16–28% P_2O_5 and 40–48% CaO (see also Banerjee *et al*. 1984), no systematic relationship was observed between these two elements. These data demonstrate that even the most intensely phosphatized stromatolite columns have not been transformed fully into apatite since CaO is present in excess of that occurring in sedimentary carbonate-fluorapatite (cf. Deer *et al*. 1979; p. 506). Apparently 10–19% CaO must be present in the form of carbonate minerals, predominantly calcite. The MgO content of these stromatolites varies from 2–10%, and when plotted against P_2O_5, it shows a crude negative correlation with that element. The carbonate laminae of stromatolites with alternate carbonate–phosphate laminae show high MgO content of 17 to 19%, indicating the presence of stoichiometric dolomite as the dominant carbonate phase. The columns with dominant carbonate composition also show high, (17–20%) MgO contents. Figure 4 shows the distribution of major oxides in stromatolites with dominantly phosphate (a) and dominantly carbonate (b) compositions. The phosphatic stromatolites display high CaO and P_2O_5 contents and contain relatively little MgO. In contrast MgO is very high, CaO is low, and P_2O_5 negligible in the carbonate stromatolites.

The Na_2O and K_2O contents are 0.1–0.3% and 0.1–0.4% respectively in both the phosphate and carbonate dominant stromatolites. They show no definite relationship to variations in P_2O_5 content. The MnO content of phosphate-bearing stromatolite and the host rocks varies from 0.01 to 0.09%. There is not much difference in the content of manganese of stromatolites and host rocks. Manganese also shows no correlation with P_2O_5.

Discussion

It is considered by many workers that the formation of the carbonate component in stromatolites is controlled predominantly by microbial communities living in the algal mats. This is due to the biogenic precipitation of calcite both intercellularly and intracellularly (Monty 1965, 1967, 1976), termed in situ precipitation by Monty (1967). Depositional environment, however, also plays a vital role in the morphology and composition of stromatolites.

If different phosphorite deposits of Udaipur are compared, it is found, that the stromatolites at Jhamarkotra, Matoon, Kanpur and Bargaon are predominantly composed of phosphate, whereas, stromatolites at Dakankotra, Nimachmata and Sisarma are dominantly of carbonate composition and contain little phosphate. The stromatolites in the former area are mainly

Table 1. *Chemical composition of phosphate-bearing stromatolites*

P_2O_5	CaO	MgO	SiO_2	Fe_2O_3	MnO	Na_2O	K_2O	LOI
Phosphate-dominant stromatolites								
21.7	48.3	6.63	3.40	0.28	0.02	0.16	0.18	nd
23.6	42.0	5.37	3.18	0.98	0.03	0.18	0.40	nd
23.5	42.7	3.26	1.15	0.50	0.02	0.10	0.25	nd
23.0	44.5	5.10	9.80	0.28	0.01	0.34	0.10	nd
22.0	42.8	8.18	5.64	0.21	0.02	0.23	0.10	nd
27.7	45.9	2.25	2.42	0.83	nd	nd	nd	10.14
23.2	43.0	8.82	3.99	0.45	nd	nd	nd	19.40
16.0	40.0	9.84	5.49	1.15	nd	nd	nd	28.50
24.3	46.3	7.65	4.00	0.21	0.02	0.18	0.25	nd
Carbonate-dominant stromatolites								
4.60	35.0	17.9	2.60	0.36	0.02	0.16	0.25	nd
1.06	26.6	17.9	9.16	0.40	0.02	0.12	0.23	nd
2.53	32.5	18.6	9.40	1.10	0.20	0.24	0.20	nd
3.50	26.5	18.6	4.40	0.52	0.02	0.14	0.19	nd
1.16	31.6	19.8	8.20	1.65	0.09	0.18	0.40	nd
0.37	28.9	19.8	4.00	0.47	0.03	0.14	0.20	nd
0.50	30.2	19.6	3.69	0.26	nd	nd	nd	45.60
0.50	29.2	18.8	6.92	0.20	nd	nd	nd	42.68
0.68	27.9	17.6	13.5	nd	nd	nd	nd	40.72
Carbonate laminae of stromatolites having alternate phosphate/carbonate laminae								
nd	35.4	18.3	nd	nd	nd	nd	nd	nd
nd	33.9	17.4	nd	nd	nd	nd	nd	nd
nd	32.3	17.8	nd	nd	nd	nd	nd	nd
nd	33.2	18.5	nd	nd	nd	nd	nd	nd
nd	25.6	18.6	nd	nd	nd	nd	nd	nd
Dolostone host rocks								
nd	32.8	20.0	7.20	0.45	0.03	0.14	0.25	nd
nd	32.6	20.4	0.20	1.24	0.05	0.12	0.20	nd
nd	30.7	20.6	2.40	1.24	0.09	0.18	0.35	nd
nd	32.2	18.9	3.40	nd	nd	nd	nd	nd
nd	34.2	19.2	5.20	nd	nd	nd	nd	nd

LOI, loss on ignition; nd, not determined.

columnar structures showing uniform shape, size and maximum growth density per unit area. The inter-columnar material is usually free from stromatolitic and siliciclastic material. There is complete absence of erosional structures and features indicative of subaerial exposure, such as desiccation cracks or fenestral fabrics, suggesting that these stromatolites were deposited in deep to moderately deep subtidal conditions with only moderate wave action (Chauhan 1979; Chauhan & Sisodia 1984). The high population density of stromatolites, their richly phosphatic composition, and an association with carbonate sediments reflect that the depositional environment was favourable for both the prolific growth of the stromatolites and their phosphatization (Chauhan 1979).

In contrast, at Dakankotra and Nimachmata stromatolites show considerable variation in their morphology and phosphate content. The intercolumnar material is full of stromatolite intraclasts. These intraformational conglomerates exhibit desiccation cracks and saucer structures, and the adjacent orthoquartzites and silty arenites display flaser and lenticular bedding, and climbing ripple cross-lamination. All of these features indicate deposition in an intertidal environment with the growth of the algal mat probably continuing into the low supratidal zone. It has been argued (Chauhan 1979; Chauhan & Sisodia in press; Sisodia 1986) that hypersaline conditions prevailed, due to restricted circulation of water on the frequently exposed tidal flats.

Monty (1965) has demonstrated that the composition of microbial communities in intertidal and supratidal environments, where hypersaline conditions prevail, are different

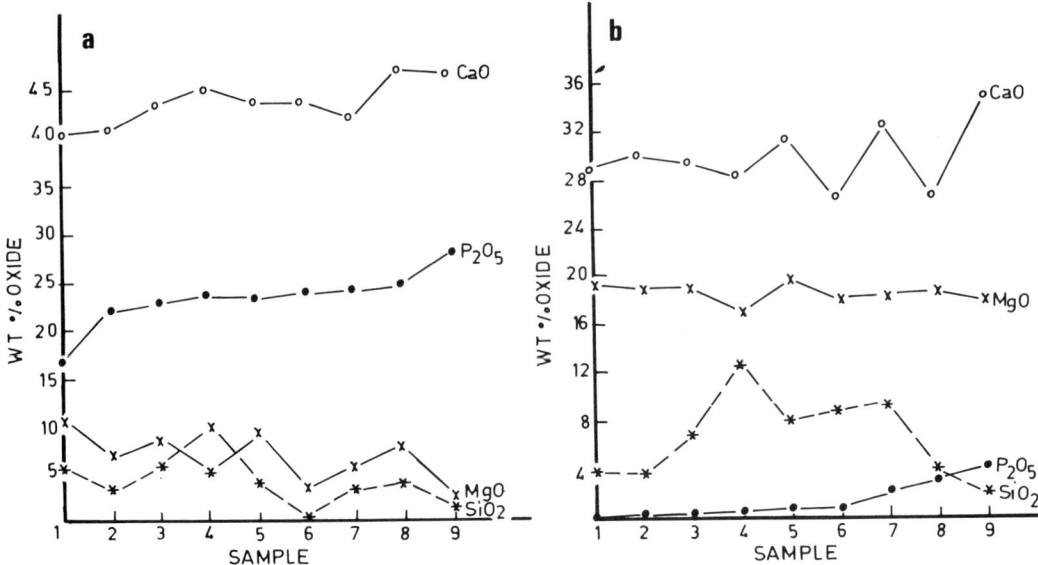

Fig. 4. Plots showing the major oxides compositions of (**a**) phosphate dominant and (**b**) carbonate dominant stromatolites.

from those thriving in subtidal conditions. The microorganisms in hypersaline environments may concentrate very large amounts of magnesium in different laminae of the stromatolites, precipitating high-magnesium calcite with Mg/Ca ratios as high as 4.5:1 (Gebelein & Hoffman 1973). In contrast, in subtidal environments they precipitate calcite and aragonite. Similarly, on Andros Island in the Bahamas, under hypersaline conditions the cyanobacterial communities have been observed to precipitate high-magnesium calcite containing up to 16 mole percent $MgCO_3$ (Monty & Hardie, 1976). It is likely that similar processes occurred in the Dakankotra and Nimachmata areas during the Proterozoic, and the microbial communities precipitated high-magnesium calcite within the intertidal stromatolites. It could be argued that the high magnesium content of these stromatolites may have aided their early dolomitization but this does not appear to have taken place in the Udaipur deposits. Gebelein & Hoffman (1973) deduced that, in general, dolomite in stromatolites cannot be penecontemporaneous because its formation is a long term process. These authors argued that the calcareous algal sheath is very stable and does not dissolve until long after decomposition of the organic matter and lithification of the stromatolite. Consequently, magnesium is only available for dolomite formation during the later history of the sediment, and is not involved in early diagenetic processes. The absence of microcrystalline dolomite or grumelous structure in the Udaipur stromatolites supports these conclusions, and hence it can be suggested that no penecontemporaneous dolomitization has occurred in them. However, petrographic evidence (cf. Rickets 1983) indicates that penecontemporaneous dolomitization may have occurred in the associated dolostone host rocks.

Marten & Harris (1970) and Lucas & Prevot (1981, 1984) have proven experimentally the inhibitory role of magnesium ions in the formation of apatite. The presence of abundant magnesium ions in the stromatolites, therefore, requires further consideration since it can reasonably be assumed that they might have inhibited the process of phosphatization. Bacterial mediation in the transformation of organic compounds into inorganic phosphorus is now very well documented (O'Brien et al. 1981; Riggs 1982; Soudry & Champetier 1983; Prévôt & Lucas 1986; Lucas et al. 1990). However, the controversy still exists as whether the apatite is formed by direct precipitation or by replacement of calcium carbonate. Many workers have concluded that most phosphorite deposits have formed due to accumulation of carbonate fluorapatite formed by the *post mortem* alteration of phosphorus-rich bacterial cells (O'Brien, et al.

1981; Riggs 1982; Prévôt & Lucas 1986; Lamboy 1990; Lewy 1990). On this basis it can be suggested that microorganisms were responsible for the phosphatized stromatolites of the Jhamarkotra and other deposits, and that mineralization took place preferentially in subtidal areas, in part because here stromatolites only originally contained low-Mg calcite, which was easily transformed into apatite. In contrast, the presence of high quantity of Mg ions in the intertidal stromatolites at Dakankotra and Nimachmata prevented their transformation into apatite. Following this argument, it can be suggested that the high magnesium content observed in the carbonate laminae of stromatolites with alternating carbonate–phosphate laminae, also reflects an originally high magnesium calcite mineralogy which prevented the complete transformation of the original carbonate structures into apatite.

Conclusions

Variations in the Mg:Ca ratio of carbonate phases in stromatolites around Udaipur reflect differences in the primary mineralogy of the sediments. The relative amount of these minerals was controlled by the microbial communities involved in the growth of stromatolites. Intertidal stromatolites deposited in hypersaline conditions were characterized by high-Mg calcite, while stromatolites forming in subtidal environments developed calcite/aragonite dominated mineralogies. The Mg/Ca ratio of the sediment played a crucial role in the process of phosphorite formation at Udaipur, since the high Mg/Ca ratio of the intertidal stromatolites acted as an inhibiting factor in the formation of apatite in these deposits. Subtidal stromatolites, however, were extensively replaced as a result of their more favourable primary mineralogy, and consequently phosphorites in the Udaipur region belong predominantly to this facies.

The geochemical studies were carried out at the Geology Department, Rajasthan University, Udaipur and the Central Arid Zone Research Institute, Jodphur. The authors sincerely thank R. K. Shrivastava, M. K. Pandya and R. P. Dhir for this work. The authors are indebted to M. J. Frost for X-ray diffraction studies and J. Lucas for SEM studies. The University Grants Commission, India provided a grant to M.S.S. for this work.

References

BANERJEE, D. M. 1971. Precambrian stromatolitic phosphorites of Udaipur, Rajasthan, India. *Geological Society of America Bulletin*, **82**, 2319–2330.
—— BASU, P. C. & SRIVASTAVA, N. 1980. Petrology, mineralogy, geochemistry and origin of the Pre-cambrian Aravalli phosphorite deposits of Udaipur and Jhabua, India. *Economic Geology*, **75**, 1181–1191.
—— SAIGAL, G. C., SRIVASTAVA, N. & KHAN, M. W. Y. 1984. Element variation pattern in the Precambrian phosphorites and country rocks of Udaipur, Rajasthan. *Geological Survey of India, Special Publication*, **17**, 63–78.
CHAUHAN, D. S. 1979. Phosphorite bearing stromatolites of the Precambrian Aravalli phosphorite deposits of the Udaipur region, their environmental significance and genesis of phosphorite *Precambrian Research*, **8**, 95–126.
—— & SISODIA, M. S. 1984. Nature of Udaipur phosphorite and its genetic implications. *Geological Survey of India, Special Publication*, **17**, 79–104.
—— & —— Evolution of shallowing upward stromatolite sequences in the vicinity of Udaipur, Rajasthan, India. (In press).
CHOUDHURI, R. 1990. Two decades of phosphorite investigations in India. *In*: NOTHOLT, A. J. G. & JARVIS, I. (eds) *Phosphorite Research and Development*. Geological Society, London, Special Publication, **52**, 305–311.
CRAWFORD, A. R. 1970. The Precambrian geochronology of Rajasthan and Bundelkhand, northern India. *Canadian Journal of Science*, **7**, 91–110.
DEER, W. A., HOWIE, R. A. & ZUSSMAN, J. 1979. *An introduction to the rock forming minerals*. Longman, London.
GEBELEIN, G. D. & HOFFMAN, P. 1973. Algal origin of dolomitic laminations in stromatolitic limestones. *Journal of Sedimentary Petrology*, **43**, 603–613.
HERON, A. M. 1953. *The Geology of Central Rajputana*. Geological Survey of India, Memoir 79.
LAMBOY, M. 1990. Microbial mediation in phosphatogenesis: new data from the Cretaceous phosphatic chalks, Northern France. *In*: NOTHOLT, A. & JARVIS, I. (eds) *Phosphorite Research and Development*. Geological Society, London, Special Publication, **52**, 157–167.
LEWY, Z. 1990. Pebbly phosphate and granular phosphorite (Late Cretaceous, southern Israel) and their bearing on phosphatization processes. *In*: NOTHOLT, A. & JARVIS, I. (eds) *Phosphorite Research and Development*. Geological Society, London, Special Publication, **52**, 169–178.
LUCAS, J. & PREVOT, L. 1981. Synthese L'apatite a partir de matière organique phosphorée (ARN) et de calcite par voie bacterienne. *Compte Rendu de l'Academie des Sciences, Paris*, **292**, 1203–1208.
—— & —— 1984. Apatite synthesis by bacterial activity from phosphate organic matter and several calcium carbonates in natural fresh water and seawater. *Chemical Geology*, **42**, 101–118.
——, EL FALEH, E. M. & PREVOT, L. 1990. Exper-

imental study of the substitution of Ca by Sr and Ba in synthetic apatites. *In*: NOTHOLT, A. & JARVIS, I. (eds) *Phosphorite Research and Development*. Geological Society, London, Special Publication, **52**, 33–47.

MARTENS, C. S. & HARRIS, R. C. 1970. Inhibition of apatite precipitation in the marine environment by magnesium ions. *Geochimica et Cosmochimica Acta*, **34**, 621–625.

MONTY, C. L. V. 1965. Recent algal stromatolites in the wind-ward lagoon, Andros Island, Bahamas. *Annales de la Societe Geologique de Belgique*, **88**, 269–276.

—— 1967. Distribution and structure of recent stromatolitic algal mats. Eastern Andros Island, Bahamas. *Annales de Societe Geologique de Belgique*, **90**, 55–102.

—— 1976. The origin and development of cryptalgal fabrics. *In*: WALTER, M. R. (ed.) *Stromatolites*. Elsevier, Amsterdam.

—— & HARDIE, L. A. 1976. The Geological significance of the freshwater blue-green algal calcareous marsh. *In*: Walter M. R. (ed.) *Stromatolites*. Elsevier, Amsterdam.

O'BRIEN, G. W., HARRIS, J. R., MILNES, A. R. & VEEH, H. H. 1981. Bacterial origin of East Australian Continental margin phosphorites. *Nature*, **294**, 442–444.

PREVOT, L. & LUCAS, J. 1986. Microstructure of apatite-replacing carbonate in synthesized and natural samples. *Journal of Sedimentary Petrology*, **56**, 153–159.

RAABEN, M. E. 1981. Riphean stromatolites versus Lower Proterozoic. *Izverstiya Akademie Nauk, SSSR, Seriya Geologicheskaya*, **6**, 51–64.

RICKETTS, B. D. 1983. The evolution of a middle Precambrian dolostone sequence — a spectrum of dolomitization regimes. *Journal of Sedimentary Petrology*, **53**, 565–586.

RIGGS, S. R. 1982. Phosphatic bacteria in the Neogene phosphorites of Atlantic coastal plain — continental shelf system. *Geological Society of America, Abstracts with Program*, **14**, 77.

ROY, A. B. & PALIWAL, B. S. 1981. Evolution of Lower Proterozoic epicontinental deposits. Stromatolite-bearing Aravalli rocks of Udaipur, Rajasthan, India. *Precambrian Research*, **14**, 49–74.

SARKAR, S. N. 1972. Present status of Precambrian geochronology of peninsular India. *24th International Geological Congress*, Section 1, 260–272.

SISODIA, M. S. 1986. *Studies of Precambrian Aravalli stromatolitic phosphorites of Bargaon, Nimachmata and Dakankotra of Udaipur region*. PhD Thesis, University of Jodhpur, Jodhpur.

SOUDRY, D. & CHAMPETIER, Y. 1983. Microbial processes in the Negev Phosphorites (Southern Israel), *Sedimentology*, **30**, 411–423.

Index

acritarchs 297
aeration *see* oxygen status
Agulhas Bank 61, 80, 108
Al Hasa 193, 195
Alexandria-Wonarah High 262–3, 265–6, 269
algae and phosphogenesis 175, 176, 197
  see also micro-organisms
Algeria, phosphates of 2
allochthonous phosphates 240
Alpine Helvetic Unit *see* Helvetic Shelf
America (North) *see* Canada; USA
Amman Formation 193
anoxic events and phosphogenesis 9, 292, 299
apatite family
  sedimentary
    composition 23
    controls on formation 318–19
    skeletal 294
    X-ray analysis 23
  synthetic
    geochemistry 33–4
    synthesis experiments 34–5
      barium content 35–9
      strontium content 41–2
Aravalli Supergroup 307
  chemical analysis 315–16
  geological setting 313
  micro-organism effects 316–18
  petrography 313–14
Arthur Creek Formation 265, 266, 267, 270
Asia, phosphates of 291
Atdabanian Stage 292
Austral–Asiatic Phosphogenic Province 5–6
Australia, phosphates of
  continental *see* Daly Basin; Duchess Basin; Georgina Basin; Wiso Basin
  offshore Neogene/Quaternary compared 61, 62, 76–82
    chemistry/geochemistry 66–76
      isotopic composition 98–100
      oxidation state 93–6
      pore water 91
      sulphate reactions 96–8
    controls on formation 82–4
      bacterial effects 98
    diagenetic model 100–8
    mineralogy 65
    nodule characteristics 64–5
    sedimentology 63–4, 89–90
Austia 238–39, 151–53
authigenic minerals and phosphogenesis 218–19, 233

bacteria and phosphogenesis
  history of interpretation 11–12
  regional studies
    Australia
      methods 90–1
      results 98, 112
      synthesis and model proposed 107–8

France 160, 161, 164
Israel 177
Jordan 197
Siberia 225
see also micro-organisms
Baltic Coast phosphates
  climatic interpretation 258
  history of exploitation 253–4
  palaeogeography 256–8
  petrography 256
  stratigraphy and structure 254–6
barium in phosphates 34, 41–2
bathymetry and phosphogenesis 177
Baturin Cycles and phosphogenesis 240, 245, 247, 248
bed amalgamation and phosphogenesis 219
Beetle Creek Formation 265, 267
Belgium, phosphates of 2
berthierine 223
Bijawar Group 307–8
biodetrital phosphates 179, 180–4
biological effect on phosphogenesis *see* algae; bacteria; micro-organisms
biomineralization and phosphogenesis 292
biosparite and phosphogenesis 210
bioturbation and burrowing in phosphogenesis 174, 225, 240
Birmania Formation 309
bitumen and phosphogenesis 176
Bone Valley Formation 5
  depositional history 144–5
  mineralogy 149–50
  stratigraphy 141–3
Border Water Hole Formation 265, 267
brachiopods, phosphatized 294, 297
Britain, phosphates of 1–2 *see also* North Sea
Brunette Sub-Basin 262–3, 266
Burke River Embayment 262–3, 265–6
burrow fills and phosphogenesis 174, 225, 240

California, Southern Borderland phosphates of 80
Cambrian phosphates
  Austral–Asiatic Province 5–6
  Georgina Basin 4, 49
    exploration history 261–2
    isotopic analysis 274, 276
    methods of phosphogenesis 267–9
    organic matter analysis 56–7
    stratigraphy and sedimentology 263–7
    tectonic setting 262–3
  see also Proterozoic-Cambrian boundary
Camooweal Dolomite 266, 267
Canada, phosphates of 2, 291, 292
carbon, organic *see* organic carbon
carbon isotope studies and phosphogenesis
  global values for Proterozoic–Cambrian boundary 273, 279–82
  South Africa 120–3
carbonates and phosphogenesis
  dissolution reactions 161, 175–6

321

replacement reactions 161, 175–6
substitution reactions 179, 181–4, 187–8
Carolina Phosphogenic Province
  comparison with Togo deposits 51, 152–3
  depositional history 143–4
  mineralogy 147–9
  stratigraphy 139–41
Castle Hayne Formation 141
catastrophic burial and phosphogenesis 246–7
Chabalowe Formation 265, 267, 270
chalks, phosphatic 157, 160–1
chamosite and phosphogenesis 223, 231
Chapel Island Formation 299
Chatham Rise 80
China, phosphates of 274, 276, 291, 292
chlorapatite 23
chlorite and phosphogenesis 223, 227, 231
clay minerals and phosphogenesis 226–30
climatic effects and phosphogenesis
  Baltic Coast deposits 258
  global oceanic effects 279, 280, 284–5
coated grain phosphates 197, 225, 245
cocci bodies and phosphogenesis 225
colloidal transport and phosphogenesis 219
collophane 313
concretions
  controls on development 129
  distribution factors 129–33
  see also nodules
condensation processes and phosphogenesis 219, 237
  Helvetic Shelf studies
    depositional trends 238–40
    diagenesis 245–7
    genetic mechanisms 240–2
    geological setting 238
    redox reactions 247–9
    sedimentology 243–5
coprolites 1, 161
cortices of phosphate 160
crandallite 44
Cretaceous phosphates see Egypt; France; Helvetic Shelf; Israel; Jordan; Siberia
Croatan Formation 139
crusts of phosphate
  occurrence 157, 174, 245
  petrography 158–60
current systems and phosphogenesis 62–3, 89, 245, 247
cyanobacteria and phosphogenesis 197
cyanophytes and phosphogenesis 297

Dakhla Formation 207, 208
Daly Basin 262–3, 266
Danois Bank 61, 80
desorption and phosphogenesis 219
Devonian rocks and search for phosphates 13
differential thermal analysis 5, 50, 52
Doushantuo Formation 299
Drusberg Formation 238, 239
DTA 5, 50, 52
Duchess Basin 5
Duwi Formation 205, 207, 210–13

Egypt, phosphates of
  Formations
    Duwi 210–13
    Phosphorite 214–15
    Sibâîya 213–14
  sedimentology 209–10
  stratigraphy 205–8
  systems tract analysis 215–20
Eocene phosphates
  Israel 180
  Togo 56, 143
  Tunisia 49, 56
  USA 141
erosion rates and phosphogenesis 279
Esh-Shidiya 193, 195
Estonia, phosphates of
  climatic effects 258
  history of exploitation 253–4
  palaeogeography 256–8
  petrography 256
  stratigraphy and structure 254–6
eubacteria and phosphogenesis 197
eustasy and phosphogenesis see sea level change
evaporites and phosphogenesis 232
evolution and phosphogenesis 9, 284

facies analysis 10, 242
faecal pellets 161, 225, 297
fertilizer uses for phosphates 1–2
fish mortality and phosphogenesis 3
florencite 44
Florida Phosphogenic Province
  comparison with Togo deposits 152–3
  depositional history 144–5
  mineralogy 149–50
  stratigraphy 139, 141–3
fluorapatite 23
fluvial transport and phosphogenesis 219
France, phosphates of 2, 157
francolite 23, 24
  occurrence
    Jordan 200
    North Sea 129
    South Africa 120–3
  substitutions 24–7, 145–7
  x-ray analysis 27–9

Gafsa Basin 49–54
Garschella Formation 238, 239
gastroliths 174
genesis of phosphates
  history of research 1–5
  summary of mechanisms 193
geochemistry and phosphogenesis 12
  Holocene of Australia
    nodule studies 72–6
    sediment studies 66–72
  Proterozoic–Cambrian boundary isotope studies
    carbon 279–82
    strontium 274–9
Georgina Basin 4, 49
  exploitation history 261–2
  genetic methods 267–9
  isotopic analysis 274, 276
  organic matter analysis 56–7

stratigraphy and sedimentation 263−7
tectonic setting 262−3
Ghareb Formation 169−70, 180
nodule shape and origin 171−4
stratigraphy 170−1
gibbsite and phosphogenesis 231−2
glaciations and phosphogenesis 279, 284
glauconite and phosphogenesis
Australia 63, 64
Egypt 216−17
Helvetic Shelf 240, 245
Siberia 223
globular phosphate 200
goethite (iron oxyhydroxide)
Australia 61−2, 65, 72, 82
Helvetic Shelf 247
goyazite 44
grainstone phosphorite 11
granular phosphate
France 157, 160
Israel 174−5
Jordan 197, 200
Gum Ridge Formation 265, 266, 267

halite and phosphorites 199−200
hardgrounds 10−11, 157, 160
mode of formation 83
Hawthorn Formation 5, 141−3
depositional history 144−5
mineralogy 149−50
Helvetic Shelf
depositional trends 238−40
diagenesis 245−7
genetic mechanisms 240−2
geological setting 238
redox reactions 247−9
sedimentology 243−5
Himalaya phosphates 291
Hirapur Basin 307−8
Holocene phosphates 9−10
see also Quaternary
Hooker Creek Formation 266, 270
humic content of phosphates 49−50
methods of analysis 50
results 50−5, 57
hydroxyapatite 23
Hyolith tubes 292, 294, 297
Hyolithes Limestone 298, 299

IGCP156, history of 5−8
India, phosphates of 4
composition 51
distribution of resources 306
Mesozoic 309
Palaeozoic 308−9
Proterozoic 307−8
chemical analysis 315−16
geological setting 313
micro-organism effects 316−18
petrography 313−14
history of exploitation 305
skeletal studies 290, 291, 295
infra-red analysis
clay minerals 227, 230

humic content 50, 53
intraclast phosphate 197, 210
Iran, phosphates of 291, 296
Iraq, phosphates of 205
iron minerals and phosphogenesis
ocean anoxic events, role of 283
pore water modelling 100−4
see also chamosite; glauconite; goethite; laterite; siderite
isotope analyses of phosphorites 12
see also carbon; lead; oxygen; strontium; sulphur; thorium; uranium
Israel, Cretaceous−Eocene phosphates of 4, 169−70
carbonate content 175−6, 181−4, 185−90
granule origin 174−5
nodule origin 171−4
organic matter content 49, 56, 176
petrography 179−80
stratigraphy 170−1

Jaisalmer Basin 309
Jhabua 308
Jhamarkotra 307
Jordan, phosphates of 51
distribution 193−6
genesis 197−202
petrography 197

kaolinite and phosphogenesis 231−2
Kingisepp 254, 255

Lalitpur Basin 307−8
laminated phosphorites 197
laterite and phosphogenesis 219, 231−3
leaching and phosphogenesis 200
lead isotopes in phosphates
methods of measurement 91
results 98−100, 112−12
synthesis model 104
London Clay Formation 131, 132

Maardu 253, 255
magnesium in phosphogenesis 33, 318−19
manganese oxyhydroxides and phosphogenesis 247
mat model for phosphogenesis 199
Maton Formation 307
Meishucunian Stage 292
Menuha Formation 179, 180, 185
Mesozoic phosphates 309
see also Cretaceous
microbes see micro-organisms
micronodule formation 175
micro-organisms and phosphogenesis
France 158−61, 163, 164
Helvetic Shelf 245
India 316−18
Israel 174, 175, 177, 185
Jordan 197−202
Siberia 225
microstromatolites 160
mineralogy and mineral associations 11, 65
Miocene phosphates 9
Carolina Province 49, 56
Florida Province 141−3

Mishash Formation 49, 169–70, 180
  carbonate content 185
  nodule origin 174
  organic matter content 54
molluscs, phosphatized 292, 294, 298
Monastery Creek Phosphorite 49
Montejinni Limestone 265, 266, 267, 270
Mor Formation 179, 180, 185
Morocco, phosphates of 2, 51, 61, 80
moulds of phosphate 185
mudstone phosphorite 11
Mussoorie Syncline 108, 308–9

Namibia phosphates of 61, 80
Nemakit–Daldynian Stage 290, 291, 292
Neogene phosphates
  see Australia offshore; USA
  also Miocene; Pliocene
New Zealand phosphates 176
Newfoundland phosphates 292
nodule formation studies
  Australia
    controlling factors 82–4, 108
    distribution 64–5
    geochemistry 66–76, 93–5
    mineralogy 65
    Neogene and Quaternary compared 76–82
  Israel
    composition 185
    origin 172–4
    shape 171–2
  Siberia
    composition 224–6
    clay mineral associations 226–30
    conditions of accumulation 231–3
    lithology 223–4
  see also concretions
North Sea phosphates
  basin development 127–8
  composition 129
  controlling factors 129–33
  periods of formation 125–7
Nubi Formation 207

OAE see ocean anoxic events
Oban Sub-Basin 262–3, 265–6
Obolusconglomerate 255
ocean anoxic events and phosphogenesis 273, 279–82, 283, 284–5
oceanic circulation and phosphogenesis 132, 134
Oligocene phosphates 176
ooids and phosphogenesis 161
Ordovician phosphates 254–8
organic carbon and phosphogenesis
  analytical methods 50
  analytical results 50–5
  concentrations measured
    Australia 66, 68–9, 71, 96, 107
    Egypt 210, 219
    Jordan 200–1
  flux rate effects 106–7
  skeletal remains 295–7
  upwelling theories and recent phosphate formation 88–9

sediments and chemistry 89–91
  bacterial analyses 90–1, 98, 107–8
  diagenesis 100–4
  lead isotope analyses 91, 98–100, 104
  oxygen levels 93–6
  sulphate levels 96–8
ostracods, phosphatic 294
oxygen isotope analysis 120–3
oxygen status and phosphogenesis
  Australia 93–6, 110
  Egypt 219
  Helvetic Shelf 240–1
  Israel 176–7

packstone phosphorite 11
Pakistan, phosphates of 274, 276, 290
palaeoceanography and phosphogenesis 8–9, 291–2
palaeoclimate and phosphogenesis 8–9, 133
palaeoenvironment analysis 119–20
palaeogeography and phosphogenesis 10, 241, 256–8
Palaeozoic phosphates 308–9
  see also Cambrian; Ordovician; Devonian; Permian
pebbly phosphate 170–2
pelmicrite 210
peloidal phosphorite
  Egypt 210
  Israel 179–80, 185
  Jordan 197
Permian phosphates 49, 57–8
Peru-Chile phosphate area 61, 80
petrography and phosphate classification 11, 158–61
Phosphate Formation 205, 207, 213–14
phosphatized carbonate dominated (PCD) taphofacies 297
phosphatized organic dominant (POD) taphofacies 297
phosphogenesis
  history of research 1–5
  summary of mechanisms 193
Phosphoria Formation 3, 4, 9, 11, 49, 80
Phosphoria Model 3
Phosphorite Formation 214–15
physicochemical effects see redox reactions
plate tectonics and phosphogenesis 9, 284
Pleistocene phosphates 139, 141
Pliocene phosphates
  Britain 129
  South Africa 120
  USA 139, 141
pore water effects on phosphogenesis
  methods of measurement 90
  modelling of behaviour 100–4
  results 91, 110–11
porosity effects on phosphogenesis 161
Precambrian see Proterozoic
pristine phosphates 240, 243–5
Proterozoic (Precambrian) phosphates 13, 307–8
  boundary with Cambrian
    environmental changes 289–91
    isotope geochemistry
      carbon 279–82
      strontium 274–9
      sulphur 282–3

ocean anoxic events 273-4
  palaeoceanography 291-2
  skeletal record 292-7
  taphofacies 297-300
protoconodonts, phosphatized 292, 294
pseudoconodonts 294
pump and shuttle effect in phosphogenesis 247
Pungo River Formation 49, 141
  depositional history 143-4
  humic analyses 56
  mineralogy 148-9
  sea level effects 133
pyrite and phosphogenesis 284
pyrolysis measurements 69, 212

Qiongzhusi Formation 290, 291, 295
Quaternary phosphates *see* Australia offshore
Quseir Formation 207, 208, 213

Red Crag Formation 129
redox sensitivity and phosphogenesis 82, 220, 247, 284-5
residual phosphatic dominant (RPD) taphofacies 297
reworking and phosphogenesis 219
Rovno Formation 299
Ruseifa 51, 193, 195, 197
Russia, phosphates of 2

Saudi Arabia 205
Sayyarim Formation 185, 18
scanning electron microscopy
  microbial structures 158-9, 162, 199, 200, 201, 315
  pebbly phosphates 173
  peloidal phosphates 186
Schrattenkalk Formation 238, 239
sclerites, phosphatized 292, 294, 297
sea level change and phosphogenesis 9, 82-3, 201
  Australian evidence 268
  Egyptian evidence 215-20
  Helvetic Shelf evidence 245
  North Sea evidence 130-2, 133
Seewen Formation 238, 239
sequence stratigraphy model for phosphogenesis 215-20
Sibâîya Formation 205, 207, 213-14
Siberia, phosphates of
  Asian correlations 290, 291
  clay mineralogy 226-30
  composition 224-6
  environment of formation 271-3
  lithology 223-4
siderite and phosphogenesis 218-19
siphogonuchitids, phosphatized 294
skeletal phosphates
  Cambrian-Proterozoic boundary 292-4
    primary 294
    secondary 294-7
  Egypt 210
  Jordan 197
  Siberia 225
skins of phosphate 158
South Africa, phosphates of 61, 80, 120
Spain 61, 80
storm conditions and phosphogenesis 219

stromatolites and phosphogenesis
  France 158-60
  Helvetic Shelf 245
  India 313-18
strontium in phosphorite
  apatites
    natural 43-4
    synthetic 33-4
      concentration effects 35-6
      precipitation effects 36-40
      time effects 35
  ocean water geochemistry 274-9
sulphates and phosphogenesis 90, 96-8, 111
sulphur isotopes in phosphates 282-3
Switzerland 238-9
systems tract analysis and phosphogenesis 215-20

Tal phosphorite 290, 295, 308-9
taphofacies 297-300
temperature gradients and phosphogenesis 279, 280, 284-5
Tertiary phosphates
  North Sea 125-7
  composition 129
  Togo 143-5
  controls on formation 129-33
  see also Eocene; Miocene; Neogene; Oligocene; Pliocene
Tethyan phosphogenic province 6, 13
  depositional trends 238-40
  diagenesis 245-7
  geological setting 238
  marginal effects 217-18
  phosphogenetic mechanisms 240-2
  redox currents 247-9
  sediment classification 243-5
thermogravimetric analysis (TGA) 50, 52
thorium isotope analysis 9-10
Thorntoni Limestone 265, 267
tidal effect on phosphogenesis 219
tillite and phosphorites 11
Tindall Limestones 265, 266, 267
Togo, phosphates of
  comparison with Carolina/Florida Phosphogenic Province 152-3
  depositional history 145
  distribution 143
  humic analysis 50, 51, 516
  mineralogy 150-2
Tommotian Stage 290, 291, 292
tommotiids, phosphatic 294
Top Springs Limestone 265, 267
total organic carbon *see* organic carbon
transgressions *see* sea level change
trilobites, phosphatized 292
Tunisia, phosphates of 2, 49, 56

Udaipur phosphates
  chemical analysis 315-16
  composition 307
  geological setting 313
  micro-organism effects 316-18
  petrography 313-14
Undilla Sub-basin 262-3, 265-6, 269

Upwelling effects on phosphogenesis 3, 4
   Cretaceous evidence 217
   Proterozoic–Cambrian evidence 292, 299
   Recent evidence 63, 125
     sediments and chemistry 89–91
      bacterial analyses 90–1, 98, 107–8
      diagenesis 100–4
      lead isotope analyses 91, 98–100, 104
      oxygen levels 93–6
      sulphate levels 96–8
uranium isotopes in phosphates 120–1
USA *see* California; Carolina; Florida; Phosphoria Formation; Pungo River Formation Western Mountain Field
USSR *see* Estonia; Russia; Siberia

Vampire Formation 299
Varswater Formation 120
veneers of phosphate 157, 158–60
vivianite 218–19
volcanism and phosphogenesis 4, 132

weathering regimes and phosphogenesis 219, 231–2
Western Mountain Field phosphates 49, 55
winnowing and phosphogenesis 219, 240
Wiso Basin 262–3, 266, 268
worm tubes, phosphatized 297

XRD and phosphate compositions
   India 313–14
   Siberia 226–30
   USA 23, 50, 52–3
XRF 225

Yenisei Bay phosphates
   clay mineralogy 226–30
   composition 224–6
   conditions of accumulation 231–3
   lithology 223–4
Yorktown Formation 133, 139–41
   depositional history 143–4
   mineralogy 148–9

Zhongyicun phosphates 290